# NanoBioEngineering

# NanoBioEngineering

Editor-in-Chief
**Bhupinder Singh**

Editors
**Jagat R. Kanwar**
**Sanjay Garg**

CRC Press
Taylor & Francis Group
Boca Raton London New York

CRC Press is an imprint of the
Taylor & Francis Group, an **informa** business

CRC Press
Taylor & Francis Group
6000 Broken Sound Parkway NW, Suite 300
Boca Raton, FL 33487-2742



Printed on acid-free paper

International Standard Book Number-13: 978-0-8153-4835-1 (Hardback)
978-1-351-13890-1 (eBook)



**Library of Congress Cataloging-in-Publication Data**

Names: Bhoop, Bhupinder Singh, editor.
Title: Nanobioengineering / [edited by] Bhupinder Singh.
Description: Boca Raton : Taylor & Francis, 2019. | Includes bibliographical references and index.
Identifiers: LCCN 2018028395| ISBN 9780815348351 (hardback : alk. paper) | ISBN 9781351138901 (ebook)
Subjects: LCSH: Nanomedicine. | Nanotechnology. | Biomedical engineering.
Classification: LCC R857.N34 N32566 2019 | DDC 610.28--dc23
LC record available at https://lccn.loc.gov/2018028395

**Visit the Taylor & Francis Web site at**
**http://www.taylorandfrancis.com**

**and the CRC Press Web site at**
**http://www.crcpress.com**

# Contents

# Foreword

Nanobioengineering is a rational fusion of the disciplines of nanotechnology and bioengineering. It is an emerging frontier of twenty-first century science and technology that is unleashing novel opportunities for building the desired expertise to develop innovative materials, devices and processes, and for ameliorating the existing technologies for improved sustainability and health.

This new field is anticipated to significantly impact the biomedical research in exploiting unique properties (i.e., physical, chemical and biological) of materials and systems in their nanometric scale, and to direct their application to biological targets at the tissue and cellular levels to derive optimal clinical applications. As a multidisciplinary field, nanobioengineering is creating newer opportunities in a gamut of domains of sciences and technology, encompassing physics, chemistry, biology, drug development, applied sciences, engineering and biomedical technology.

In this context, I am quite fascinated and satiated that the long-felt void of a repository on this discipline of nanobioengineering is being brought about through this multi-volume book series on nanobiomedicine. Providing a handy overview of all the vital aspects of nanobioengineering, this volume will provide the desired impetus for developing next-generation diagnostics and therapeutics. In doing so, it contributes to the path to improve the quality of human life. The pioneering efforts to collate and bring research advances in this field to the forefront are indeed praiseworthy. The articles, contributed by scientific leaders across the globe, capture the essence of different vistas of this realm, while underpinning the latest headway in the applications of nanobioengineering in medicine.

This integrated treatise is slated to be a classic for the current and future generations of readership hailing from diverse disciplines. Not merely can it serve as a reference book to stimulate the interest of academic researchers, industrial scientists and entrepreneurs, as well as to foster win-win collaboration amongst them, but as a textbook for education and training purposes as well. With immense pleasure, I sincerely wish the editors, the authors and the publisher tremendous success in reaching out to the potential readers.

**Ajaz S. Hussain**
*President*
*National Institute for Pharmaceutical Technology & Education, USA*

# About the Series

Implying miniaturization, the term "nano" has become a household term. Nanotechnology encompasses the study and application of functional nanomaterials in diverse fields even at their molecular levels. Virtually, nothing remains to be "nanonized" at the moment, whatever the domain may be. Traversing a voyage from the origin of this idea by Richard Feynman in 1959, nanotechnology has now entered into our day-to-day living, society, and culture, shaping our future in a categorical manner. On one hand, nanotechnology is being explored to develop newer nanostructures, and on the other hand, the horizons of their promising applications are being widened manifold. Not merely an "evolution," this has lately been turning out to be a "revolution" across different scientific disciplines and industrial sectors across the globe.

Notwithstanding the promising benefits of the cutting-edge technology, its applications in healthcare are considered as paramount. This perspective assumes immense significance, as the whole world is experiencing unprecedented biomedical challenges and changes, particularly in context with "3-**D**" succession of **D**iseases, **D**iagnosis, and **D**rugs. In line with the futuristic predictions made by Robert Freitas in 1999 in his book *Nanomedicine*, today this is ostensibly the fastest emerging offshoot of nanotechnology. Further integration of nanomedicine with the biological sciences, such as biotechnology, biochemistry, pharmaceutical sciences, biomedical engineering, bioimaging, nanomedical robotics, and biomedical technology, has spurred the genesis of a whole contemporary field of "nanobiomedicine." Employing various nanostructures and nanodevices, the science of nanobiomedicine primarily endeavors to improve upon the patient safety, efficacy, economy, and compliance of varied therapeutic, preventive, and diagnostic strategies. Not merely dealing with the nuances of interactions of nanostructures with biomolecular receptors, the modern discipline of nanobiomedicine also embraces the methods of design, engineering, and technological transfer of nanoformulations of drugs and biomacromolecules, as well as the fabrication of advanced biomedical appliances. Drawing apt inputs from almost all the frontiers of science and technology, nanobiomedicine today is heralding a new era in medical care with lots of hopes and promises for the suffering society. Nanobiomedicine, has been making rapid strides recently, eventually providing significant impetus toward scientific enthusiasm, research efforts, industrial anticipation, and regulatory attention.

Of late, several nanomedicinal products have been introduced into commercial circulation, with a few thousand more in the pipeline and yet many more in the early stages of developmental research. These have been documented to exhibit stellar benefits not only of enhanced surface-area-per-unit volume, but also of improved aqueous solubility, target specificity, controlled-release potential, biocompatibility, stealth characteristics, precise control of particle size, and ability for drug combination therapy. Such nanoscaled products principally encompass different types of nanoparticles, nanocrystals, nanoemulsions, nanoconjugates, liposomes, dendrimers, nanogels, nanocosmeceuticals, nanocapsules, nanophytomedicines, lipid nanospheres, and nanocomposites. A huge proportion of these innovative nanotech-enabled products have demonstrated enormous patient benefits and plausible solutions to the erstwhile insurmountable medical problems. Alarmingly high incidences of inadequate efficacy, safety, and patient compliance reported with the current crop of drugs, have been calling for developing such newer and therapeutically superior nanomedicinal formulations to address the present patient needs and market demands. Driven by the technological updates, increased gubernatorial support, recent notifications of newer regulatory guidance for drug products containing nanomaterials and enhanced preference for targeted therapeutics, the world has now been witnessing a rapid surge in such nanomedicines. Nanopharmaceuticals, therefore, are an outsized industry today, with its market crossing a few hundreds of billions of US dollars, over 500 companies and over 80 products worldwide. In the forthcoming decades too, these nanomedicines are anticipated to massively impact almost all the terrains of patient care, personalized medicine, clinical and veterinary practice, and above all, the pharmaceutical market. The ardent need in the current hour of shifting paradigms, accordingly, is to acquaint the professionals on advances in this vibrant science of nanobiomedicine, growing at a phenomenally striking pace. A recent literature survey of the leading journals revealed a phenomenally high rate of publications in this multi- and interdisciplinary domain of nanobiomedicine. For a realm advancing as fast as this one, it is quite challenging to obtain a complete overview of the exciting developments, as most updated information on such nano-bio interfaces lies scattered in diverse journals. Owing to nonavailability of the standard treatise(s) on the subject, therefore, we have undertaken a grand task of bringing about a book series covering a gamut of subject topics spanning nanobiomedicine. Accordingly, the current book series, "**Emerging Trends in NanoBioMedicine**," an anthology of four volumes, aspires to dexterously present the pertinent facts and figures on an array of novel subject topics. Each individual volume of the compendium is an assortment of choicest reviews and research articles, contributed by illustrious scientists and experts selectively drawn from across the world. The primary motivation behind this Herculean effort has been to bring forth high quality insight into latest research in the field as a multi-volume digest. It comprises four volumes:

Volume 1: **NanoBioMaterials**
Volume 2: **NanoBioEngineering**
Volume 3: **NanoNutraceuticals**
Volume 4: **NanoAgroceuticals and NanoPhytochemicals**

**NanoBioMaterials**, the first volume of the series, maps out an exhaustive overview of the contemporary research accomplishments of variegated nanomaterials exhibiting the profound impact in improving treatment efficiency of various diseased states. This volume tacitly introduces and demonstrates the close association of biology with material science, with intent for effective and safe use of nanomaterials for different biological applications. Being a promising, though burgeoning, realm of science, the treatise accentuates diverse applications of nanomaterials in the fields of drug delivery, bioimaging, tissue engineering, and many more. It primarily focuses on innovations based on various nanobiomaterials, such as polymeric, metallic, lipidic, peptidal, carbon-based, and prodrugconjugate-based, leading finally to produce next-generation biomedicinal products.

Highlighting challenges for fundamental research and opportunities for ground-breaking advances in nano-enabled bioengineering, the second volume of this series successfully unfolds the topic of **NanoBioEngineering**. Being a rational fusion of biology, engineering and nanotechnology, it presents breakthrough techniques and tools amid the existing armamentarium to utilize novel phenomena like microfluidics and nanotooling and nanomaterials such as aptamers, nanocargos, nanofibers, nanocrystals, hyperbranched dendrimers, surface-engineered and superparamagnetic nanoparticles for meeting the desired healthcare needs. The integrated volume, herein, provides a lucid account on far-reaching topics from diverse disciplines of life sciences and technology, including physics, chemistry, biology, engineering, biomedical technology, and drug development. The enormous potential of nutraceuticals in fostering human healthcare, the burning issues confronting their poor bioavailability and usage of nanotechnology principles for resolving these issues, have been adroitly deliberated upon in the third volume of the book series, **NanoNutraceuticals**. Provision of the requisite knowledge on a gamut of modern hi-tech approaches to improve their activity, especially using nanotechnology through a single textual source enshrined, this book volume can be instrumental in ameliorating the subdued popularity of nutraceuticals and lack of serious consideration by health regulatory agencies. Thus, the contents of this book volume have the capability to metamorphose the biological performance of these functional foods through their skilful delivery to the specific target using apt nanocarriers for safer and more efficacious management of diseased states such as cancer and neurodegenerative diseases and for benefitting the human healthcare *per se*.

Volume four, **NanoAgroceuticals & NanoPhytochemicals**, is a testament to the sophisticated and unconventional extrapolation of nanotech applications in the field of agroceuticals and phytochemicals. Albeit still in its infancy, the subject matter encompassed in the book volume holds immense potential to bring forth tangible benefits to the farmers, food industry, patients, and consumers. This book presents a cogent account on various nano-enabled delivery technologies of phytochemicals in agriculture to improve the storage shelf-life of crops, reduce the amount of sprayed chemicals through smart delivery of phytochemicals, sustain delivery of fertilizers, decrease the loss of nutrient(s) in fertilizers, and augment the percent harvesting yield through optimized water and nutrient management. The role of nanotechnology in the development of novel and innovative products out of agriculture products and herbal supplements, and their subsequent production, processing, storage, and packaging, has also been explicitly highlighted in this book volume.

All these volumes put together highlight some of the most remarkable advances that have taken place recently in the vast field of nanobiomedicine and reflect new challenges and issues that confront their further progress. The series seeks to offer the readers an apt tête-à-tête with fresh and original perspectives, epitomizing a vast multiplicity of experience, wisdom, and vision, which is quite implausible to be made available collectively anywhere else.

Verily, in this context, I feel quite satiated and fascinated that the long-felt void of a repository on the much-sought-after discipline of nanobiomedicine is being filled up by spearheading this current multi-volume book series. Although the field of nanobiomedicine has become too vast today, meticulous endeavor has been made to customize the contents of the book series and steer the novitiates as well as experts on the recent and upcoming trends. Not only does this treatise embrace holistic applications of a plethora of promising nanoconstructs, especially in the fields of medical therapeutics and diagnostics, biotechnology, pharmacogenomics, human gene biology, clinical pharmacology, and so on, but it also touches upon their regulatory, safety, and industrial scale-up issues.

At this juncture, I would like to take this opportunity to express my heart-felt gratitude to several people whose pooled efforts brought forth this dream of a mega-series into reality. The list includes all the authors, editors, the publishing house team, my research group, and research funding agencies, besides others.

First and foremost, I put across my heart-felt gratitude to the eminent and erudite authors for dexterously working on the chapters by bringing together the most comprehensive and updated information on wide-ranging subject(s). Not merely did these contributors meet with the stringent timelines, they also abided by a battery of editorial suggestions sportingly. They garnished and flavoured these chapters with apt perspectives, emanating from their own experiential wisdom. Earnest efforts have been made by them to encompass a diversity of graphic illustrations, presented immaculately as expository graphs, photographs, methodology flow charts, hierarchical tabulations, bar charts, and so on. Each chapter has been duly referenced to current literature, also citing vital texts for further reading.

As the designated editor-in-chief of the book series, I fall short of apposite words to express my deep appreciation to the distinguished editors of the four volumes for pouring in their invaluable and pertinent inputs from time to time. The sagacious, prompt, and critical feedback provided by them, as well as by the worthy reviewers of the individual chapters, all of whom spent their priceless time, helped a great deal in ameliorating the quality standards of the manuscripts.

I put into record the highly painstaking, persistent, and meticulous efforts of the management and staff members of the publisher, *viz.*, M/s CRC Press (Taylor & Francis Group), USA, in bringing about this mega-compilation. Without their immense support and direction, it could not have been possible to bring forth this colossal series in a timely manner.

Our well-knit team of exceptionally diligent, efficient, and perseverant researchers and post-docs deserve special mention, as they were verily the spine of this massive and momentous project. Naming a few, while leaving many others, may demean their stellar contribution, for which I would always remain beholden.

I also seize this opportunity in acknowledging the University Grants Commission (UGC), New Delhi, India, for awarding this prestigious National UGC Centre of Excellence in "Applications of Nanomaterials, Nanoparticles & Nanocomposites" with focus area as "Biomedical Sciences" to the Panjab University, Chandigarh, India. In fact, the drive to take up this marathon task for writing the explicit treatise was initiated after my assuming the charge of this Centre of Excellence, dedicated to explore and establish various nanobiomedicinal precepts. Thanks are also due for the tangible financial support provided by several other governmental agencies and pharma industrial houses to enable us to bring out tangible outcomes on diverse innovative nanostructured systems and in characterizing them. Several of such systematic works on nanopatterned structures, primarily employing Quality by Design (QbD)-based paradigms, have fetched exalted global recognitions by the American Association of Pharmaceutical Sciences (USA), Select Biosciences (UK), StateEase Inc. (USA), Evonik AG (Germany), Minitab Ltd. (UK), CPhI-Asia, EBA-Oxford (UK), APTI (India), and many more.

With this, I invite and incite you to start this amusing expedition to the salubrious and picturesque world of nanobiomedicine, with a missionary message, "Think Tiny and Act Big." I earnestly wish, hope, and pray that this four-volume book series on **Emerging Trends in NanoBioMedicine** will serve as a classic repertoire of essential knowledge and know-how in the domain to address the unmet needs of the scientists and practitioners from current and future generations.

**Bhupinder Singh**

# Preface

Nanodelivery of drugs is an area of growing interest. Some key factors such as small size and high surface area, along with drug loading capability, make these ideal drug delivery systems. These can easily be modified to cater to diverse therapeutic and diagnostic applications. Nanoparticles bridge the gap between biomedical and physical sciences since they are applicable in a wide range of medical fields, such as nanobiotechnology, drug delivery, tissue microengineering and microfluidics.

Nanoparticles can be organic or inorganic in origin, mostly depending upon their applications. Organic nanoparticles are biocompatible and biodegradable in nature and promote the bioavailability of the therapeutic drug payload. Inorganic nanoparticles, on the other hand, can be employed for medical imaging and subsequent theranostic applications. Therefore, such nanoparticles not only conform to targeted drug delivery, but also provide a nanomedical tracking system from systemic to cellular levels. For efficient drug delivery, site-specific nanoparticles can be loaded with various hydrophilic or hydrophobic drugs and functionalised with one among numerous targeting ligands such as antibodies, peptides or aptamers. Moreover, the size range of these particles lying in nanometeric scale facilitates their ability to cross the blood-brain barrier, which is otherwise difficult for most drugs to cross. In line with this, nanoparticles are also capable of penetrating tissues and, finally, being taken up by cells.

This present book volume on NanoBioEngineering intends to provide an insight into diverse biotechnological and bioengineering applications of various types of nanomaterials. A rare assortment of invaluable book chapters, it lucidly highlights the versatile therapeutic and diagnostic potential of nanosized particles. The prominence of each of the book chapters is accentuated, as follows:

Chapter 1 discusses the properties of physical, chemical and surface characteristics of nanosized materials and their significant potential in nanomedicine. It lays emphasis on their utilisation for drug engineering, imaging, diagnostics, sensing and regenerative applications. Chapter 2 focuses on the inorganic magnetic nanoparticles and the essential characteristics required to make them biocompatible theranostic agents. Moreover, it discusses the surface coating of these particles to prevent agglomeration and facilitate surface charge modifications, efficient drug loading derived for effective therapeutics.

Chapter 3 explicitly delineates the salient characteristics and characterisation aspects of effective nanobiopolymers (i.e., dendrimers) as smart drug delivery vehicles with unique multifunctional and hyperbranched structural features, amenable to need-based customization. The overview chapter also highlights the stellar applications of dendrimers (especially PAMAM ones) in the targeted delivery of not only the small molecules, but of genes, polynucleotides and other biomacromolecules as well. The main focus of Chapter 4 is on using nanocargos as a novel strategy to combat multi-drug resistance, the major hurdle in current cancer therapy. It mentions various mechanisms involved in tumor multi-drug resistance and indicates the potential of nano-based drug-loaded delivery agents in facilitating effective targeting and therapeutics against resistant tumors.

Chapter 5 discusses the efficacy of aptamers as a novel targeted drug delivery tool. It reviews the strategy of conjugation of such targeting ligands with organic or inorganic nanoparticles to overcome current therapeutic drawbacks such as non-specific cytotoxicity and inability to cross the blood-brain barrier. Chapter 6 summarises a multitude of engineering strategies that have been employed for nanostructures in order to inculcate enhanced compatibility for numerous biopharmaceutical applications. Additionally, it mentions the various approaches such as ligand-based, metal-based, bioengineered or stimuli-sensitive strategies that promote binding of targeting ligands to achieve site-specific delivery.

Chapter 7 discusses an organic and biodegradable poly(lactic acid)-based polymer for bioengineering applications, providing information regarding the synthesis mechanism, conjugation techniques and stimuli-specific achievable modifications in this polymer. Chapter 8 emphasises the improvements of titanium implants through nanoporous surfaces (incorporated with tantalum) in order to improve the antibacterial performance.

Chapter 9 focuses on the challenges associated with drugs that have poor water solubility and the remedial capability of nanocrystals to overcome this drawback through increasing the dissolution of drugs. It discusses the key benefits and limitations of such nanocrystals for biomedical applications. Chapter 10 reviews nanofibers and their therapeutic significance since they are considered as smart and strong nanosystems. It explains the processes and parameters involved in the synthesis and characterisation of nanofibers, along with their useful efficacy as efficient drug delivery agents.

Chapter 11 emphasises the applicability of nano-tooling in the domain of clinical dentistry. It highlights the remarkable wear resistance properties of the composite materials in comparison to the basic filament. Chapter 12 discusses the role of nanotechnology in providing a face identification system. It unravels the recent strategies for the synthesis of a successful face recognition system under real-time situations.

Chapter 13 discusses the innovative strategies used for the brain-specific delivery of drugs through the oral route. It provides an insight into the brain delivery, associated limitations and the ability of nanotechnology to overcome these limitations through small size, stability and low toxicity. Chapter 14 reviews the toxicity issues associated with nanomaterial exposure. Further, it unearths the screening models based on the cutting-edge technology of microfluidics to monitor and examine nanotoxicity. Chapter 15 summarises the safety issues in context to chronic nanomaterial exposure. It outlines the long-term toxicology concerns that need to be addressed during nanodrug formulations.

Collectively, this book successfully provides a complete repertoire of knowledge regarding the efficacy of engineered, surface-modified or ligand-conjugated nanomaterials for drug delivery in variegated fields of biomedical science, such as oncology, clinical dentistry and brain delivery. Further, this book exquisitely highlights the current nanotherapeutic strategies, while throwing some light on the perils of nanotoxicology as well.

**Bhupinder Singh**
**Jagat R. Kanwar**
**Sanjay Garg**

# Editors

**Professor Bhupinder Singh Bhoop, MPharm, PhD, DSt, FPAS**, is globally acclaimed for his extensive research work on diverse nanostructured drug-delivery systems, developed primarily using Quality-by-Design (QbD) paradigms. His work encompasses variegated nanoengineered systems, naive as well as functionalized, such as SLNs, NLCs, SNELs, MWCNTs, polymeric nanoparticles, lipid-polymer hybrid nanoparticles, mixed nanomicelles, nanocapsules, nanosponges, nanobilosomes, liposomes and other vesicular systems, for a gamut of disease states, particularly cancers and cardiovascular, immunodeficiency and neurodegenerative disorders.

A prolific writer, he has over 280 original journal publications, 16 books, one directory, 62 book chapters, five patents and three technology transfers to his credit. He is on editorial and review boards of scores of journals. A widely travelled scientist, he has delivered more than 275 invited talks across the globe and has duly trained tens of hundreds of scientists from the pharmaceutical industry on nanomedicines and QbD. He has completed 16 research projects and guided 31 PhD (previously and currently), 3 postdoctoral and 60 postgraduate scholars. Currently, he is serving as founder and coordinator of UGC Centre for Excellence in Nano Biomedical Applications, and coordinator, UGC Centre of Advanced Studies (Pharmaceutical Sciences), both at Panjab University, Chandigarh, India. He has been Chairman, University Institute of Pharmaceutical Sciences, and Dean, Faculty of Pharm Sciences, as well as a Fellow of Panjab University. His work has earned Professor Singh numerous awards and accolades at global and national levels.

**Professor Jagat R. Kanwar** is the Group Leader of Nanomedicine-Laboratory of Immunology and Molecular Biomedical Research at Deakin University, Australia. He has more than 210 publications in peer-reviewed journals and 20 patents and has guided more than 25 PhD students. His work focuses on discovering novel and safe targeted nanomedicines for cancers, autoimmune disorders and inflammatory diseases. Besides, he has been working to investigate the underlying mechanisms involved in regeneration, apoptosis, autophagy and inflammation by targeting the production of cytokines, chemokines, oxygen radicals and matrix metalloproteinase. He has lately ventured into microfluidics, cutting-edge technologies like lab-on-a-chip techniques for cancer cells, stem cell capture, disease-specific biomarkers and exosomes.

**Professor Sanjay Garg, MPharm, PhD, MMgt,** is Director, Centre for Pharmaceutical Innovation and Development, University of South Australia, Adelaide, and Co-Director of China Australia Health Science Research Centre. His research work is focused on novel nanocarriers against cancer, superbugs and bacterial infections, medical devices, functional foods, diagnostics and cosmetics. He has to his credit over 170 peer-reviewed papers, 35 patents, 25 book chapters, four books and tech transfers for six products, as well as supervision of 188 research scholars including 21 PhDs. He has been bestowed with several awards including Australasian Pharmaceutical Science Association Medal, Butland Award for Excellence in Research Supervision, Fellowships of Indian Pharmaceutical Association and Academy of Translational Medicine and Adjunct appointment with International Medical University, Malaysia.

# Contributors

**Afrah Jalil Abd**
Nanomedicine-Laboratory of Immunology and Molecular Biomedical Research (NLIMBR)
School of Medicine (SoM)
Centre for Molecular and Medical Research (C-MMR)
Deakin University
Victoria, Australia

**Naznin Akhtar**
Nanomedicine-Laboratory of Immunology and Molecular Biomedical Research (NLIMBR)
School of Medicine (SoM)
Centre for Molecular and Medical Research (C-MMR)
Deakin University
Victoria, Australia

**Mays Alhamami**
Department of Chemical Engineering
Ryerson University
Toronto, Ontario, Canada

**Jalal Azadmanjiri**
School of Engineering
Faculty of Science, Engineering and Technology
Swinburne University of Technology
Victoria, Australia

**Raman Bahal**
Department of Pharmaceutical Sciences
University of Connecticut
Storrs, Connecticut

**Chelladurai Karthikeyan Balavigneswaran**
Polymer Engineering Laboratory
School of Biomedical Engineering
Indian Institute of Technology (BHU)
Varanasi, India

and

Department of Biotechnology
Sri Shakthi Institue of Engineering and Technology
Tamil Nadu, India

**Christopher C. Berndt**
School of Engineering
Faculty of Science, Engineering and Technology
Swinburne University of Technology
Victoria, Australia

and

Department of Materials Science and Engineering
Stony Brook University
Stony Brook, New York

**Saurav Bhandari**
Department of Quality Assurance
ISF College of Pharmacy
Moga, India

**Gurjit Kaur Bhatti**
UGC-Centre of Excellence in Applications of Nanomaterials, Nanoparticles and Nanocomposites (Biomedical Sciences)
Panjab University
Chandigarh, India

**Kamaljit Singh Boparai**
Department of Mechanical Engineering
Giani Zail Singh Campus College of Engineering and Technology
Maharaja Ranjit Singh Punjab Technical University
Bathinda, India

**Joana R. Campos**
Department of Pharmaceutical Technology
Faculty of Pharmacy
University of Coimbra (FFUC)
Coimbra, Portugal

**Suman Chaudhary**
Nanomedicine-Laboratory of Immunology and Molecular Biomedical Research (NLIMBR)
School of Medicine (SoM)
Centre for Molecular and Medical Research (C-MMR)
Deakin University
Victoria, Australia

**Indu Chhabra**
Department of Computer Science
Panjab University
Chandigarh, India

**Yaser Dahman**
Department of Chemical Engineering
Ryerson University
Toronto, Ontario, Canada

**Ana Rita Fernandes**
Department of Pharmaceutical Technology
Faculty of Pharmacy
University of Coimbra (FFUC)
Coimbra, Portugal

**Maria L. Garcia**
Department of Pharmacy, Pharmaceutical Technology and Physical Chemistry
Faculty of Pharmacy
University of Barcelona
Barcelona, Spain

**Avinash Gothwal**
Department of Pharmacy
School of Chemical Sciences and Pharmacy
Central University of Rajasthan
Ajmer, India

**Amit Kumar Goyal**
Institute of Research & Development
Gujarat Forensic Sciences University
Gandhinagar, India

**Umesh Gupta**
Department of Pharmacy
School of Chemical Sciences and Pharmacy
Central University of Rajasthan
Ajmer, India

**Rodney J. Y. Ho**
Department of Pharmaceutics
and
Department of Bioengineering
University of Washington
Seattle, Washington

**Rupinder Kaur Kanwar**
Nanomedicine-Laboratory of Immunology and Molecular Biomedical Research (NLIMBR)
School of Medicine (SoM)
Centre for Molecular and Medical Research (C-MMR)
Deakin University
Victoria, Australia

**Jagat R. Kanwar**
Nanomedicine-Laboratory of Immunology and Molecular Biomedical Research (NLIMBR)
School of Medicine (SoM)
Centre for Molecular and Medical Research (C-MMR)
Deakin University
Victoria, Australia

**Ajay Kapoor**
School of Engineering
Faculty of Science, Engineering and Technology
Swinburne University of Technology
Victoria, Australia

**Ranjot Kaur**
University Institute of Pharmaceutical Sciences
Panjab University
Chandigarh, India

and

University of Central Lancashire (UCLAN)
Preston, United Kingdom

**Ripandeep Kaur**
UGC-Centre of Excellence in Applications of Nanomaterials, Nanoparticles and Nanocomposites (Biomedical Sciences)
Panjab University
Chandigarh, India

**Lavanya Khanna**
Department of Physics
Panjab University
Chandigarh, India

**Rajneet Kaur Khurana**
University Institute of Pharmaceutical Sciences
UGC Centre of Advanced Studies
Panjab University
Chandigarh, India

**Pramod Kumar**
Department of Pharmaceutics
School of Pharmaceutical Education and Research
Jamia Hamdard (Deemed to be University)
Delhi, India

and

Department of Pharmacy
School of Chemical Sciences and Pharmacy
Central University of Rajasthan
Ajmer, India

**Balak Das Kurmi**
SLT Institute of Pharmaceutical Sciences
Guru Ghasidas Vishwavidyalaya
Bilaspur, India

**Lena Leopold**
Department of Pharmaceutics
University of Washington
Seattle, Washington

**Shikha Lohan**
UGC-Centre of Excellence in Applications of Nanomaterials, Nanoparticles and Nanocomposites (Biomedical Sciences)
Panjab University
Chandigarh, India

**Joanna MacDonald**
Nanomedicine-Laboratory of Immunology and Molecular Biomedical Research (NLIMBR)
School of Medicine (SoM)
Centre for Molecular and Medical Research (C-MMR)
Deakin University
Victoria, Australia

**Sanjeev Kumar Mahto**
Tissue Engineering and Biomicrofluidics Laboratory
School of Biomedical Engineering
Indian Institute of Technology (Banaras Hindu University)
Varanasi, India

**Ruchi Malik**
Department of Pharmacy
School of Chemical Sciences and Pharmacy
Central University of Rajasthan
Ajmer, India

**Lisa McConnachie**
Department of Pharmaceutics
University of Washington
Seattle, Washington

**Nira Misra**
Polymer Engineering Laboratory
School of Biomedical Engineering
Indian Institute of Technology (BHU)
Varanasi, India

**Jayanth Surya Narayanan Shankara Narayanan**
Nanomedicine-Laboratory of Immunology and Molecular Biomedical Research (NLIMBR)
School of Medicine (SoM)
Centre for Molecular and Medical Research (C-MMR)
Deakin University
Victoria, Australia

**Ehsan Jafari Nasr**
Department of Chemical Engineering
Ryerson University
Toronto, Ontario, Canada

**Rishi Paliwal**
Siddhi Vinayaka Institute of Technology and Sciences
Bilaspur, India

**Shivani Rai Paliwal**
SLT Institute of Pharmaceutical Sciences
Guru Ghasidas Vishwavidyalaya
Bilaspur, India

**Suruchi Poddar**
Tissue Engineering and Biomicrofluidics Laboratory
School of Biomedical Engineering
Indian Institute of Technology (Banaras Hindu University)
Varanasi, India

**Drashti Prekh**
Department of Chemical Engineering
Ryerson University
Toronto, Ontario, Canada

**Kalyan Ramesh**
Polymer Engineering Laboratory
School of Biomedical Engineering
Indian Institute of Technology (BHU)
Varanasi, India

**Goutam Rath**
Department of Pharmaceutics
ISF College of Pharmacy
Moga, India

**Kaisar Raza**
Department of Pharmacy
School of Chemical Sciences and Pharmacy
Central University of Rajasthan
Ajmer, India

**Ana Claudia Santos**
Department of Pharmaceutical Technology
Faculty of Pharmacy
University of Coimbra (FFUC)
Coimbra, Portugal

**Ashok Kumar Sharma**
Department of Pharmacy
School of Chemical Sciences and Pharmacy
Central University of Rajasthan
Ajmer, India

**Teenu Sharma**
University Institute of Pharmaceutical Sciences
UGC Centre of Advanced Studies
Panjab University
Chandigarh, India

**Sarah Shigdar**
Nanomedicine-Laboratory of Immunology and Molecular Biomedical Research (NLIMBR)
School of Medicine (SoM)
Centre for Molecular and Medical Research (C-MMR)
Deakin University
Victoria, Australia

**Bhupinder Singh**
University Institute of Pharmaceutical Sciences
and
UGC-Centre of Excellence in Applications of Nanomaterials, Nanoparticles and Nanocomposites (Biomedical Sciences)
Panjab University
Chandigarh, India

**Charan Singh**
University Institute of Pharmaceutical Sciences
Panjab University
Chandigarh, India

**Geetika Singh**
Department of Computer Science
Panjab University
Chandigarh, India

**Rupinder Singh**
Department of Production Engineering
Guru Nanak Dev Engineering College
Ludhiana, India

**Eliana B. Souto**
Department of Pharmaceutical Technology
and
REQUIMTE/LAQV
Group of Pharmaceutical Technology
Faculty of Pharmacy
University of Coimbra (FFUC)
Coimbra, Portugal

**Vijay K. Srivastava**
Department of Mechanical Engineering
Indian Institute of Technology (Banaras Hindu University)
Varanasi, India

**Pawan Tekchandani**
SLT Institute of Pharmaceutical Sciences
Guru Ghasidas Vishwavidyalaya
Bilaspur, India

**Surya Kant Tripathi**
Department of Physics
Panjab University
Chandigarh, India

**James Wang**
School of Engineering
Faculty of Science, Engineering and Technology
Swinburne University of Technology
Victoria, Australia

**Wai Hong Wong**
School of Engineering
Faculty of Science, Engineering and Technology
Swinburne University of Technology
Victoria, Australia

**Chaoyang Zhao**
Department of Pharmaceutics
University of Washington
Seattle, Washington

and

Xiangyang Hospital
Hubei University of Medicine
Hubei, People's Republic of China

# 1 Nanotechnology and its Applications in Biomedical Engineering

*Yaser Dahman,* Ehsan Jafari Nasr, Drashti Prekh, Mays Alhamami, and Ranjot Kaur*

## CONTENTS



* Corresponding author.

## 1.1 INTRODUCTION

Nanotechnology involves studying or working with matters at the nanometric scale. It includes active components of the biological molecules and structures inside living cells in the size range ranging from one to a few hundred nanometers. Nanomedicine is one of the widest disciplines of nanotechnology, with the focus of studies on therapeutics and diagnostics. It involves the usage of engineered medical devices and nanostructures in the medical domain to manage, supervise, and enhance the human biological system, particularly when diseased. The following bullets describe the main applications of nanomedicine (Boisseau and Loubaton, 2011).

- Medico-diagnostic applications, i.e., usage of various *in vitro* tools, assist in early-stage diagnostics and in monitoring the development of the treatment process. The synthetic particles allow for recognition and "signaling" of the presence, activity or concentration of biomolecules. On the other hand, *in vivo* diagnosis usually gets accomplished through a contrast agent for imaging purposes. It creates quite delicate and dependable sensing agents, which are able to fulfil the requirements and control the therapy.
- Strong development in drug delivery (i.e., nanopharmaceuticals) using an assemblage of nanoparticles is aimed at targeted drug delivery to enhance drug distribution to the diseased tissue only by avoiding damage to healthy tissues, where the targeted molecules are located.
- In regenerative medicine, nanotechnology promotes the formation of biocompatible materials, which assist in the maturation of cells and can be utilized in cell therapy. It replaces malfunctioning tissues that are old, diseased, harmed, or congenitally defective with new living functional tissues.

As shown in Figure 1.1, after 2001, a notable gap between the number of patents and publications indicates that companies are more interested in investing in patents compared with research. Figure 1.2 illustrates that drug delivery is the most productive research base area.

In 2016, the global market size of nanomedicine is estimated to reach USD 138.8 billion. Therapeutics (drugs, drug delivery systems, and devices) account for the major share of it, and the market is projected to rise meteorically by 2025 (www.grandviewresearch.com). Moreover, the presence of drug delivery projects in clinical phase trials is anticipated to result in commercialisation and eventually in revenue generation in the coming decades. Table 1.1 represents some of the anticancer molecules that have reached Phase III clinical trials (Meel *et al.*, 2017).

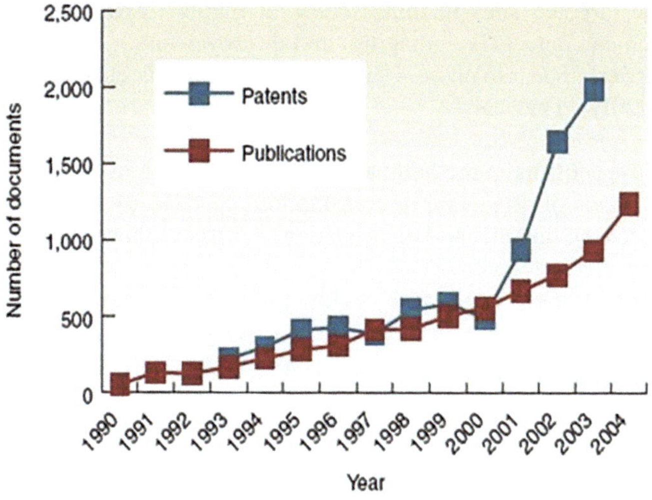


**FIGURE 1.1** Paper publication. (Reproduced from Wagner V *et al. Nat Biotechnol.* 2006; 24(10): 1211–1217. With permission.)

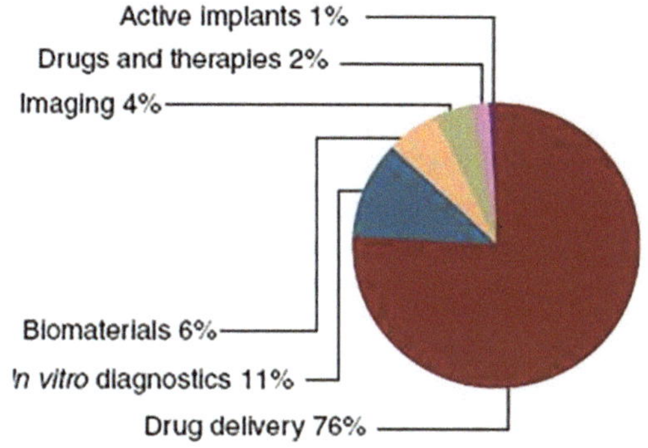


**FIGURE 1.2** Interest in the area has grown exponentially. (Reproduced from Wagner V *et al. Nat Biotechnol.* 2006; 24(10): 1211–1217. With permission.)

**TABLE 1.1**
**Some of the Anticancer Drugs in Phase III Clinical Trials**

| Drug(s) | Formulation | Company |
|---|---|---|
| Ado-trastuzumab emtansine | DM1 coupled to + pertuzumab | Roche (Genentech) |
| Doxorubicin | Liposomal formulation | Celsion, Inc. |
| Inotuzumab ozogamicin, CMC-544 | Calicheamicin coupled to CD22-targeted antibody | Pfizer, Inc. |
| CPX-351 | Cytarabine and daunorubicin liposomes | Jazz Pharmaceuticals |
| MM-398 | Liposomes containing irinotecan + 5-FU/LEU | Merrimack Pharmaceuticals, Inc. |

### 1.1.1 Nanocarriers

Traditional chemotherapy treatment utilizes drugs recognized to be functional in killing the cancer cells. Nevertheless, these cytotoxic drugs destroy both well cells and tumor cells, leading to unfavorably harmful side effects such as nausea, neuropathy, hair loss, fatigue, and compromised immune function. Similarly, nanocarriers can be utilized to surround the anticancer drugs and deliver these to the targeted sites in tumor cells for regulated release. Dual-mode theranostic nanoparticles, which can detect (i.e., imaging) and destroy a tumor (i.e., therapy), have also been developed. Nanocarriers appear to possess various advantages over chemotherapy (Velmurugan and Selvamuthukumar, 2012). They can

- Avoid drug dose splitting in the body along the desired path to the intentional target
- Enhance absorption of drugs into tumors and into the cancerous cells
- Assist in controlling the time and dispensation of drugs to the tissues, leading to a simpler monitoring process for oncologists
- Hinder the interactions of drugs with healthy cells to prevent any side effects

### 1.1.2 Passive Targeting

Paul Ehrlich proposed the theory of targeted drug delivery by introducing a "Magic Bullet" theory, composed of two parts—one part responsible for identifying and uniting the target, while the other is responsible for therapeutic effect in the intentional target. Recognizing the target or localizing using biological mechanisms is often termed the "enhanced permeability and retention (EPR)" effect. According to this property, certain sizes of molecules (typically liposomes, nanoparticles, and macromolecular drugs) are more likely to accumulate in tumors than in normal tissues (Bazak *et al.*, 2014). The reason for this phenomenon is said to be the stimulated production of blood vessels in order to allow the tumor cells to grow quickly. Because of their size and surface features, the particular nanoparticles have the ability to leak out of the blood vessel walls into tissues. In addition, Figure 1.3 shows tumors with perforated blood vessels and defective lymphatic drainage, resulting in the nanoparticles' accumulation inside the tumors. This method allows higher concentrations of the cytotoxic drug where needed and, consequently, prevents the damage to any normal tissues and considerably minimizes the unfavorable side effects (Provenzale and Silva, 2009; Bazak *et al.*, 2014).

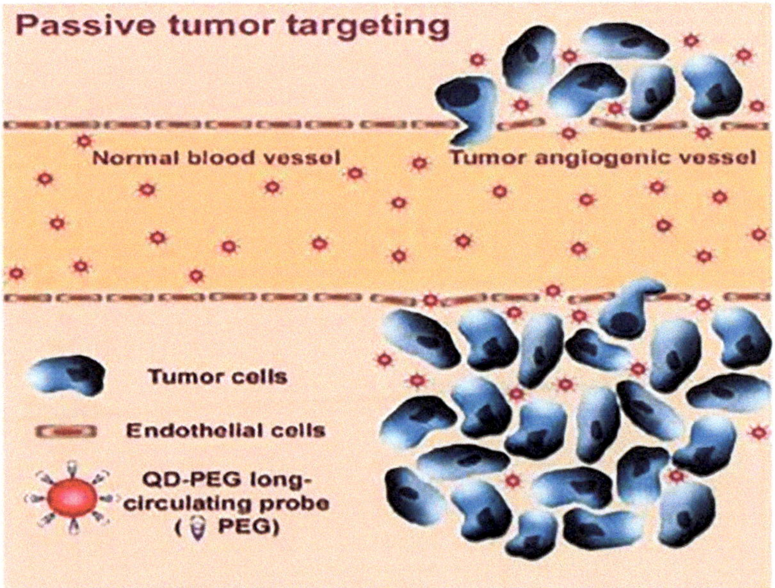


**FIGURE 1.3** Passive tumor targeting. (Reproduced from Provenzale JM, Silva GA. *AJNR Am J Neuroradiol.* 2009; 30(7): 1293–1301. With permission.)

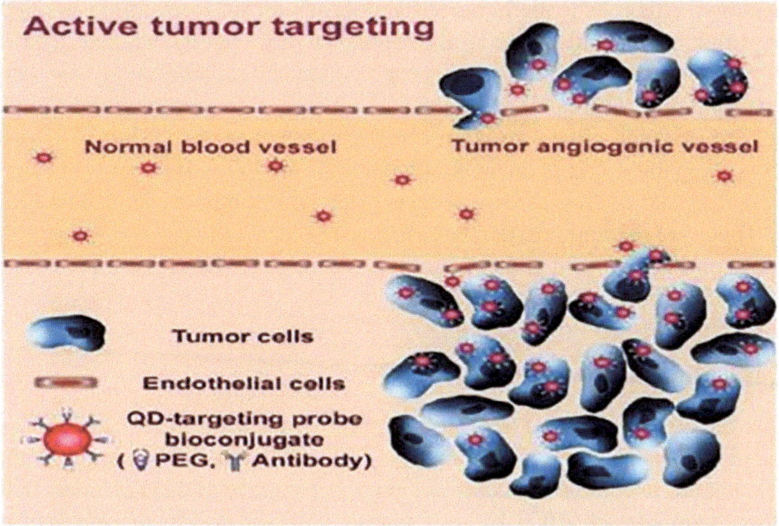


**FIGURE 1.4** Active tumor targeting. (Reproduced from Provenzale JM, Silva GA. *AJNR Am J Neuroradiol.* 2009; 30(7): 1293–1301. With permission.)

### 1.1.3 Active Targeting

There are existing nanoparticles that can deliver the desired drugs to the cancerous cells, followed by the cancerous molecules located on the cell surface. This is accomplished by designing an aimed nanoparticle that holds a molecule to selectively bind to the intentional cancerous receptors. Peripherally conjugated specific targeting moieties are utilized in active targeting for promoted transportation of nanoparticles. The targeting moieties are different ligands, inclusive of antibodies. Passive targeting alters the physicochemical features of the drug carrier complex. Then, it escapes the body's defense system and accumulates in the target tissue (Provenzale and Silva, 2009; Sagar *et al.*, 2012).

The fact that active targeting has superiority over passive targeting is due to the ability of the active nanoparticles to be particularly selective in orienting towards the intentional tumors/pathogenic tissues, and these nanoparticles have the ability to endure in the site for a considerable period of time, leading to higher accumulation in the desired sites. In addition, active targeting is utilized to deliver drugs into the cancerous cells by encouraging the cell to absorb the nanocarrier. A rational integration of both active and passive targeting can be utilized to further decrease the interaction of the transported drug with the healthy tissue. Nanotechnology-enabled active and passive targeting can also improve the performance of a chemotherapeutic agent, thus causing greater tumor depletion and, consequently, reduction in the drug doses (Sagar *et al.*, 2012; Bazak *et al.*, 2015) (Figure 1.4).

## 1.2 MEDICAL DIAGNOSTICS

Researchers are searching for techniques to physically damage the cancerous cells from inside as an alternative method to traditional chemotherapy methods. This technology includes the utilization of nanoshells in order to thermally damage the internal structure of the tumors. Nanoshells can be constructed to absorb light of various frequencies, thus producing heat (hyperthermia). When the cancerous cells grab the nanoshells (through active targeting), a near-infrared light is applied to be absorbed by the nanoshells, generating extreme heat within the tumor, leading eventually to only damage to the tumor cells without hurting any nearby cells. Similarly, targeted magnetic nanoparticles are in development and will both be recognizable through magnetic resonance imaging (MRI) and be capable of damaging cells by hyperthermia (Muller and Radtke, 2001; Velmurugan and Selvamuthukumar, 2012).

Nanotechnology in diagnostic medicine has a great advantage from a time standpoint: a period of five days is required for a fungal or bacterial infection to show results, whereas a new NMR (nuclear magnetic resonance) system with nanotechnology enhancement technology can detect the disease earlier with the right dose of treatment, as the surface area is so great that the signal gets amplified tremendously. For instance, Taxotere is an effective anticancer drug, but if it travels throughout the whole body, the patient can invariably come across problems like neutrophilia. By modifying the drug and adding nanoparticles, it can target only the tumor without damaging neighboring healthy cells. There are difficult challenges with some nanoparticles, e.g., getting the nanoparticle to target the right part of cells and understanding issues related to toxicity and environmental impact (Velmurugan and Selvamuthukumar, 2012).

### 1.2.1 Imaging and Cancer Detection

Antibodies that recognize particular receptors present in the cancerous cells can be tethered onto nanoparticles, which can produce a high contrast signal on MRI or CT (computed tomography) scans. The antibodies that are inside the nanoparticles bind particularly to the cancerous cells, thus making them visible to the scan (Mestel, 2017). Devices have the ability to occupy thousands of small molecules; devices that are smaller than 50 nm can easily enter most cells, while the ones smaller than 20 nm can transit out of blood vessels. This section summarizes some of the nanoparticles used in diagnostic imaging.

#### 1.2.1.1 Gold (Au) Nanoparticles

These are ideal for a broad range of research applications due to their resistance to surface oxidation. These particles are now being used in research to detect colorectal cancer, especially flat lesions that are difficult to be spotted during colonoscopy. Raman spectroscopy can also be used to detect the gold nanoparticles with around 20 nm diameter, where the surface can be adsorbed with ascorbate ion (Podsiadlo *et al.*, 2008; Li and Chen, 2015; Yang *et al.*, 2017).

#### 1.2.1.2 Photo-Acoustic Imaging Nanoparticles

Research is being conducted on these particles to enhance a new method of tumor detection. Once the light penetrates the body, the heat will result in the particles creating a sound that is not dangerous to the tissue, while this sound can be detected using an external source. Carbon nanotubes and gold nanoparticles can be aptly altered to produce sounds (Yang *et al.*, 2009).

#### 1.2.1.3 Superparamagnetic Iron Oxide Nanoparticles (SPIONs)

In recent years, numerous uses of superparamagnetic iron oxide nanoparticles (SPIONs), which are iron oxide nanoparticles (magnetite, $Fe_3O_4$; maghemite, $Fe_2O_3$) coated with dextran, have arisen in the biotechnology industry. Magnetization has the ability to arbitrarily overturn direction because of the temperature. The typical time taken for magnetization to flip is referred to as the Néel relaxation time (Hofmann *et al.*, 2009).

Magnetization is based on the Néel time; if the time exceeds that limit—an average of zero—it is said to be in the super-paramagnetic state. During this state, an external magnetic field is able to magnetize the nanoparticle. These particles are essentially utilized to aim lymph nodes for detecting cancer with MRI. Only the normal cells in the lymph node, but not the cancer cells, can grab such particles. Lymph nodes that extract the nanoparticles will alter their dye to black, and the cancer cells will remain bright. Nanoscaled systems can be made to heat up, which would lead to their use as hyperthermia agents. Different types of hyperthermia treatments include (Hofmann *et al.*, 2009)

- Local hyperthermia: Application of heat to tumor is accomplished using different techniques. Various categories of energy may be utilized, such as microwave, radiofrequency, and ultrasound. Using passive targeting, nanoparticles can pass through the leaky vessels in the tumors.

- Regional hyperthermia: Large areas of tissue, such as a body cavity, organ, or limb, are heated using different approaches, such as external applicators or regional perfusion.
- Whole body hyperthermia: This technique is usually utilized to treat metastatic cancer, which is distributed everywhere in the body.

By transforming and transferring drug to the intended site(s), a stimulus delicate shell is filled with the drug as enteric coating to preserve the distribution of drug. A growth in the temperature is observed immediately after the targeted site is attained.

#### 1.2.1.4 Nanoshells

Drs. Naomi Halas and Jennifer West at Rice University first developed nanoshells in 1994 (Haque *et al.*, 2010). These have a silica core and a metallic exterior layer. These nanoshells can be safely injected into the animal body, and because of their nano size, these tend to accumulate in cancer lesion sites, the physical selectivity happening via EPR. The nanoshells can be further modified to carry molecular conjugates to the antigens that are expressed on the cancer cells themselves or in the tumor microenvironment. This second degree of specificity preferentially links the nanoshells to the tumor and not to the neighboring healthy cells. Larger nanoshells (with 120–100 nm radius and 35–30 nm shell thickness) are used for imaging diagnostic. Smaller nanoshells (60 nm radius and 10 nm shell thickness) are used for photothermal therapy application (Muller and Radtke, 2001; Haque *et al.*, 2010; Velmurugan and Selvamuthukumar, 2012).

#### 1.2.1.5 Gold Nanoshelled Microcapsules (GNS-MCs)

Gold nanoshells are formed by a combination of electrostatic gold nanoparticle depositions onto microcapsules and the surface seeding method. This nano/microcomposite is capable of functioning as a theranostic agent for both contrast-enhanced ultrasonic imaging (i.e., diagnostic) and photo-hyperthermia (i.e., therapeutic) and, thus, carries enormous potential for photo-thermal therapy in cancer treatment. Ultrasound imaging is the most useful diagnostic imaging technique compared to others due to various properties such as real-time results, low cost, highly safe, and ease of incorporating into portable devices (Wang *et al.*, 2014).

Microcapsules composed of polylactic acid (PLA) have excellent biocompatibility and biodegradability and display great ultrasound contrast-enhancing abilities. Besides, due to their stable nature, they are able to hold either hydrophilic or hydrophobic species. They are surface-charged and have functional groups on the surface; hence, they can easily be modified to introduce further practical features. Fabrication processes of gold-nanoshelled microcapsules involve the use of a water-in-oil-in-water (W/O/W) double-emulsion method, where polymeric microcapsules are generated from PLA and polyvinyl alcohol (PVA) materials (www.sciencenewsline.com), as under

a. Electrostatic adsorption of (–25 mV) polyamine hydrochloride (PAH) onto microcapsules.
b. Deposition of gold nanoparticles onto PAH-coated microcapsules.
c. Formation of gold nanoshells by surface-seeding method.
d. Gold nanoparticles grow large enough to aggregate; nanoshells are formed around the microcapsule surface.

The nanoshells begin building around the microcapsule surface, when the nanoparticles are mature enough to aggregate. After lyophilisation, the encapsulated water in the inner aqueous phase of the microcapsules is sublimated to leave a small hollow space that can provide a basis for the ultrasound-responsive properties (Ke *et al.*, 2011) (Figure 1.5).

### 1.2.2 Contrast Agents

Contrast agents play an imperative role in molecular diagnostic imaging. Depending upon the suitability of imaging techniques, a wide range of contrast agents is available, as summarized in

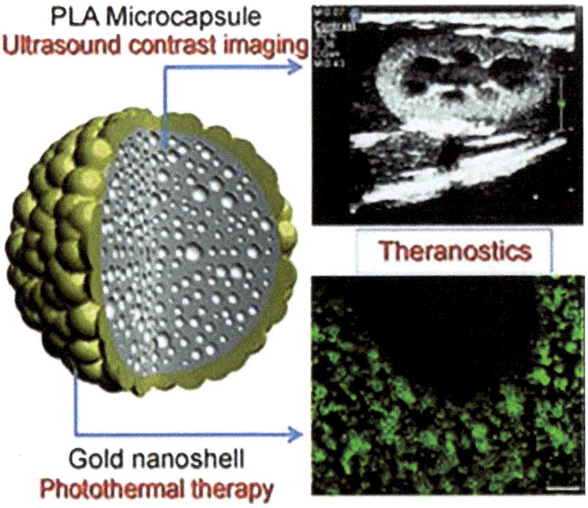


**FIGURE 1.5** Fabrication process microcapsulates. (Reproduced from Ke H *et al.*, *ACIEE Angewandte Chemie*, 2011; 50(17): 3017–3021. With permission.)

Table 1.2. Ideally, they should be able to significantly enhance the imaging contrast and breakdown the imaging depth limit. Contrast agents can also act as drug carriers that enhance the delivery of the drug to the target site and have the potential to minimize the adverse effects (Wu *et al.*, 2014).

## 1.2.3 Diagnostic Applications

### 1.2.3.1 Breathalyser Diagnosis of Lung Cancer

Gold nanoparticles are coated with an organic monolayer of thiol (R–SH) for application in chemiresistive sensors. These groups of sensors have elevated responsiveness and a reversible response to volatile organic compounds (VOCs), due to which these are widely pursued. Alteration in the Au nanoparticle film's electrical impedance is a consequence of sorption of organic vapor molecules onto the thin organic film when the thin-film sensor is open to the vapors, resulting in the film swell, which leads to an extended distance between the nanoparticles (Garg *et al.*, 2010).

The mechanism is based on conductivity ($\sigma$) of the nanoparticle film at a temperature of $T$, as described by Equation 1.1:

$$\sigma = \exp(-\beta s)\exp\left(-\frac{E_c}{k_B T}\right) \tag{1.1}$$

where $\beta$ is a decay constant related to the probability of electron tunnelling from one gold particle to another, $s$ is the interparticle distance, $E_c$ is the Activation energy, and $k_B$ is the Boltzmann constant.

Activation energy is given by Equation 1.2:

$$E_c = \frac{1}{4\pi\varepsilon_0\varepsilon_r}\frac{e^2}{r} \tag{1.2}$$

where $e$ is the fundamental charge, $\varepsilon_0$ is the vacuum permittivity, $\varepsilon_r$ is the dielectric constant of the film, and $r$ is the radius of the Au nanoparticle.

From the above equations, it is proved that the nanoparticle dimensions and surface coating significantly affect the sensing function. In addition, it is significant to account for the nanoparticles' distribution and purity through the film.

**TABLE 1.2**
**Contrast Agents Used in Imaging**

| Type | Contrast Name | Examples | Advantages | Reference |
|---|---|---|---|---|
| **MRI** | Gadolinium nanostructures | Gadoxetate disodium; Gadodiamide; Gadoversetamide; Gadoteridol; Gadobutrol; Gadoterate meglumine | No limitation in depth; high spatial resolution; safe; good soft tissue contrast; provides both anatomical and functional information | Wu *et al.* (2014), Caschera *et al.* (2016), Lahooti *et al.* (2016) |
| | Iron oxide nanoparticles | Ferumoxsil; Ferrixan; Ferumoxide | Non-toxic and is eventually metabolized by the body into hemoglobin | Lahooti *et al.* (2016), Wei *et al.* (2017), Shen *et al.* (2017) |
| **CT** | Iodinated nanostructures, gold nanoparticles, alkaline earth-based nanoparticles, bismuth sulphide (Bi2S3) nanoparticles | Diatrizoate; Amidotrizoate; Iothalamate; Ioxaglate; Iohexol | Best clinical spatial resolution; widely available; cost effective; highly efficient | Caschera *et al.* (2016), Lahooti *et al.* (2016) |
| **PET/SPECT** | Radiopharmaceuticals: Mostly emit positrons (PET) and gamma rays (SPECT) | Fludeoxyglucose Ioflupane I; C-11 L-methionine | No limitation in depth; high sensitivity and quantifiability; metabolic imaging PET: Better spatial resolution than SPECT SPECT: Potential to detect multiple probes simultaneously in contrast to PET | Mitchell *et al.* (2013), Lahooti *et al.* (2016) |
| **Fluorescence/ Optical Imaging** | Fluorescent silica particles, quantum dots | Combination of heavy metals (CdSe, CdTe, InP, InAs, PbSe, PbTe, etc.) | High sensitivity; vascular and intravascular signals; high temporal sensitivity | Tavares *et al.* (2011), Lahooti *et al.* (2016), Radhakrishnan *et al.* (2016) |

Comparison between synthesis of monothiol-coated Au nanoparticles and trithiol-capped Au nanoparticles shows that the long-term stability of the nanosensor is a critical consideration for deployment in real-world applications. The lifetime of the sensor should ideally be at least five years. Formerly, we discovered that the monothiol–Au sensor began to deteriorate noticeably in less than a year. The trithiol-Au nanoparticle sensor was presumed to show more stability than the monothiol–Au nanoparticle sensor. Thiol molecules (R–SH) are prone to oxidation, with the degradation rate of the trithiolate group, being completely oxidized, is lower than monothiol Au nanoparticles. Equivalent to the monothiol–Au nanosensors, the trithiol–Au nanoparticle sensors display an exceptional response to VOC vapors. However, the stability of such trithiol–Au sensors is significantly improved compared with monothiol–Au sensors. Therefore, the trithiol–Au nanoparticles in sensor design possess

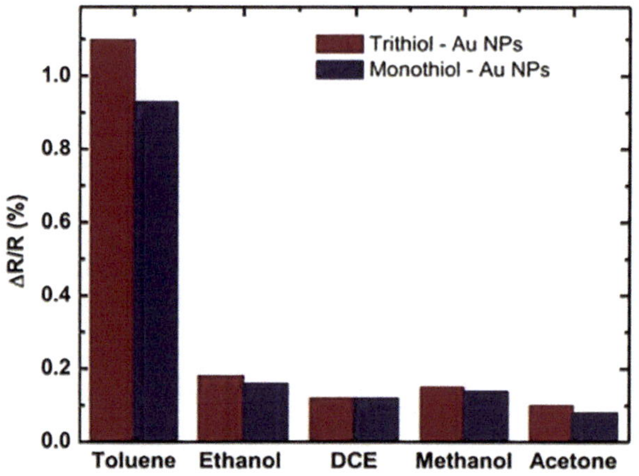


**FIGURE 1.6** Responses of trithiol- and monothiol-capped Au nanoparticle sensors to various VOCs. (Reproduced from Garg N *et al.*, *Nanotechnol.* 2010; 21: 1–6. With permission.)

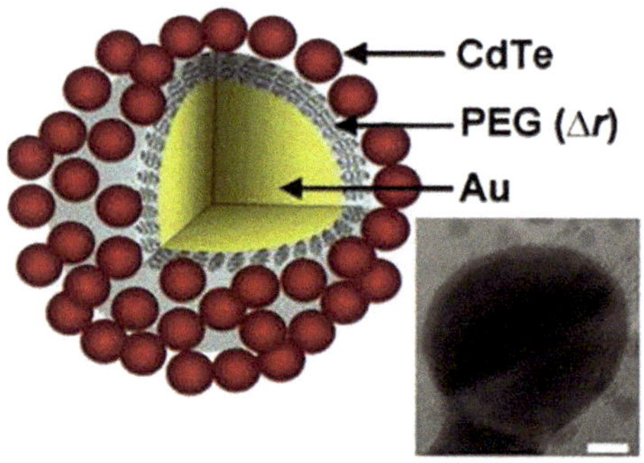


**FIGURE 1.7** A gold nanoparticle is functionalized with a PEG (polyethyleneglycol) coating. (Reproduced from Lee J, Kotov NA. *Nano Today.* 2007; 2(1): 48–51. With permission.)

promising potential in safety applications, especially for safety equipment used by employees in dangerous environments (Garg *et al.*, 2010) (Figure 1.6).

#### 1.2.3.2 Using Nano-Thermometers for Breast Cancer

Since cancer cells are known to have a higher temperature as compared to other cells, local temperature mapping can be used to determine the spread of tumor. Variation in the temperature and emission is based upon the intensity. Gold nanoparticles utilize a polyethyleneglycol (PEG) covering layer leading to a smaller coating of quantum dots (QDs). Because of the stretching/contraction of the PEG, the emission features of the nanoparticles are altered with temperature (Lee and Kotov, 2007) (Figure 1.7).

## 1.3 DRUG DELIVERY

### 1.3.1 Nanoparticles Employed for Drug Delivery

#### 1.3.1.1 Liposomes

Liposomes consist of one or more vesicular bilayers with an aqueous core surrounded by a hydrophobic lipid bilayer, created by the extrusion of phospholipids. The liposomal lipid bilayer is composed of biocompatible and biodegradable lipids, present in the biological membranes. Drugs

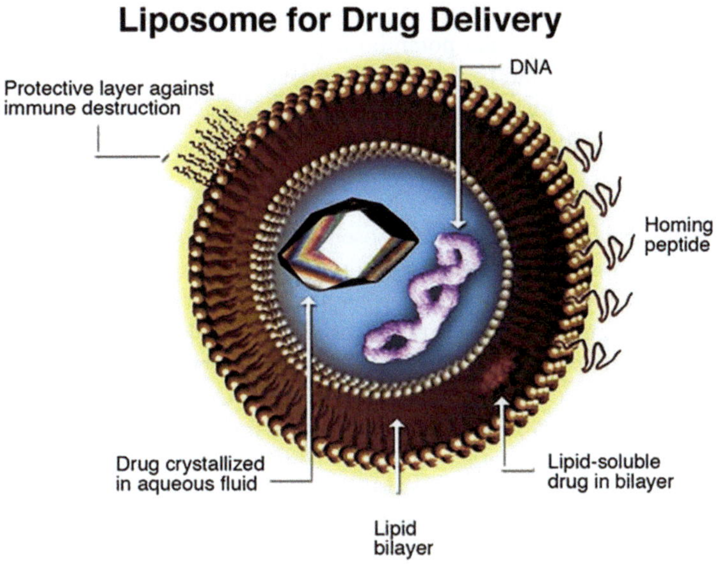


**FIGURE 1.8** Structure of liposome. (Reproduced from Debnath B *et al. PharmaTutor.* 2015: 3(8): 30–37. With permission.)

are usually unable to pass through the lipid bilayer, but hydrophobic molecules get adsorbed on the bilayer surfaces. Accordingly, liposomes can carry both hydrophobic and hydrophilic molecules. The lipid bilayer of liposomes fuses with other bilayers, such as the cell membrane, resulting in the release of its contents that make liposomes suitable for delivery applications of drugs and cosmetics (Debnath *et al.*, 2015) (Figure 1.8).

Liposomes are classified based on their size and the number of lamellae:

- Small unilamellar vesicles (SUV) contain one bilayer with a size <100 nm.
- Large unilamellar vesicles (LUV) contain one bilayer with a size >100 nm.
- Multilamellar vesicles (MLV) are made of many concentric lipid bilayers, with a size of several μm.

Liposomes are mainly used in brain drug delivery for the treatment of cerebral ischemia and brain tumors and in delivery of opioid peptides. Sphingomyelin, phosphatidylcholine, and glycerol phospholipids are examples of some common liposomal constituents. A common addition to the liposomal formulations is cholesterol, which is considered a significant part of the cell membrane due to its ability to lower the bilayer permeability and enhance the stability of the liposomes *in vivo.* These versatile properties of liposomes allow them to be used as powerful carriers for various drugs like antibacterials, antivirals, insulin, antineoplastics, and plasmid DNA.

#### 1.3.1.2 Solid Lipid Nanoparticles

Solid lipid nanoparticles (SLNs) are steady lipid-based structures of nanocarriers containing solid hydrophobic lipid cores in which drugs can be dissolved or dispersed. They are composed of biocompatible lipids such as triglycerides, fatty acids, or waxes. Their size generally varies from around 40 to 1000 nm, making these capable of crossing tight endothelial cells of the blood–brain barrier. SLNs offer a wide range of properties, such as small size, large surface area, high drug loading, and the interactions of phases at the interfaces and, hence, offer their immense potential to improve performance of pharmaceuticals, nutraceuticals, and other materials. The methods used to prepare SLNs incorporate spray drying, high shear mixing, ultra-sonication, and high-pressure homogenization (Velmurugan and Selvamuthukumar, 2012; Naseri *et al.*, 2015).

Some advantages of SLNs are the use of physiological lipids, avoidance of organic solvents in the preparation process, and the wide potential application spectrum (dermal, oral, intravenous). Moreover, SLNs exhibit better stability and easier upgradability to production scale compared with liposomes. SLNs set the basis for colloidal drug delivery systems, where their biodegradability and stability are preserved for a minimum one-year period (Velmurugan and Selvamuthukumar, 2012).

#### 1.3.1.3 Nanostructured Lipid Carriers

Nanostructured lipid carriers (NLCs) are composed of blended solid lipid particles with liquid-lipid particles creating a chain-like structure. The production of NLCs is based upon solidified emulsion (dispersed phase) technologies. After polymorphic transition during storage, the NLCs can present an insufficient loading capacity due to drug expulsion, particularly if the lipid matrix consists of similar molecules. The main objective to invent NLCs is to overwhelm the drawbacks affiliated with SLNs, as the second lifetime of lipid nanoparticles. The NLCs have a higher loading capability for hardworking compounds compared with SLNs by conceiving a less organized solid lipid matrix. For instance, when a fluid lipid blends with a solid lipid, a higher element drug stacking can be achieved. Therefore, an NLC has an expanded drug stacking capacity as compared to a SLN. Moreover, NLCs have lower water content in their elemental suspension and are less susceptible to unpredictable gelation (Velmurugan and Selvamuthukumar, 2012). Some other benefits of NLCs, in contrast to the other colloidal carrier schemes, include

- Controlled drug release characteristics,
- Increased chemical stability of the drugs,
- Protective milieu to drugs,
- Easy scalability,
- Improved bioavailability, and
- Enhanced drug loading capacity.

Drug release from lipid particles occurs mainly by diffusion and simultaneously by lipid particle degradation in the body. NLCs tend to be solid at normal body temperature, and they accommodate the drug due to their highly unordered lipid structures. To initiate the desired burst drug release, a trigger impulse is invariably applied to the matrix to convert it in a more ordered structure (Muller and Radtke, 2001). Major application areas of NLCs in pharmaceutics include drug delivery through topical, oral, and parenteral routes. In addition, NLCs find applications in cosmetics, food, and agricultural products too (Sagar *et al.*, 2012).

#### 1.3.1.4 Polymeric Nanoparticles

Polymeric nanoparticles contain a core polymer matrix, which allows the embedding of drugs inside. These are solid colloidal particles with a diameter ranging from 1 to 1000 nm. In recent years, some of the polymers have been designed mainly for medical applications and especially for attaining controlled release of bioactive agents (Sagar *et al.*, 2012; Natarajan, 2013). Polymeric nanoparticles are used therapeutically as adjuvants in vaccines and as drug carriers, wherein the active drug is dissolved, entrapped, encapsulated, adsorbed, or chemically attached (Natarajan, 2013). Some of the popular polymeric nanoparticles that are designed to degrade within the body are polylactides (PLA), polyglycolides (PGA), poly(lactide-co-glycolides) (PLGA), polyanhydrides, polycyanoacrylates, and polycaprolactone.

Controlled and sustained release of the drug is provided by polymeric nanoparticles during transportation, altering the organ distribution of the drug at the targeted site to achieve increased drug therapeutic efficacy and reduced side effects. These also provide lower toxicity, less frequent adverse drug reactions, and better drug utilisation. By the attachment of targeting ligands to the surface of the particles or through use of magnetic guidance, site-specific targeting can be achieved. Application of these particles has been documented through various routes of administration, including oral, nasal, parenteral, intra-ocular, etc. (Natarajan, 2013).

In terms of drug delivery, some research works describe enhanced drug delivery to the brain tissue, mediated by these particles. Antitubercular drugs such as rifampicin, isoniazid, pyrazinamide, and ethambutol were inserted into nanoparticles made up of PLGA for cerebral drug delivery in mice. High drug levels were maintained for 5–8 days in plasma and for 9 days in brain, which is much longer than that observed with free drugs. In mice infected with *Mycobacterium tuberculosis*, five consecutive doses of these nanoparticle formulations resulted in total removal of bacteria from the blood (Sagar *et al.*, 2012).

#### 1.3.1.5 Polymeric Micelles

Polymeric micelles, made up of amphiphilic copolymers containing spheroidal structures with a hydrophilic shell and a hydrophobic core, tend to exhibit high stability, which can further be improved by cross-linking between the shell and the core chains. The hydrophilic part of the polymeric micelles is significant due to its structure of brush-like formations, allowing the hydrophilic part to protect the hydrophobic part from the biological invasion (Shah *et al.*, 2014) (Figure 1.9).

Another feature of polymeric micelles includes the possibility to be responsive to external stimuli such as pH, light, temperature, and ultrasound, while triggering controllable release of captured drugs. One of the most utilized polymers is Pluronic®, block-copolymer based on ethylene oxide and/or propylene oxide. Compared to other drug carriers, polymeric micelles show remarkable potential due to their large solubilization power, higher loading capacity and higher stability in the blood stream, therapeutic potential, and longevity.

Immediate engineering of core and shell have resulted in the development of multifunctional polymeric micelles performing several functions through one nano-formulation. These systems, therefore, provide an infinite control over spatial and temporal drug delivery. Researchers are utilizing different methods to improve the design of the drug-loaded micelles in the tumors, the vital strategy being ligand-mediated smart drug delivery systems (Shah *et al.*, 2014; Herzberger *et al.*, 2016). The most useful application of this type of nanoparticle in drug delivery includes delivery to brain cells.

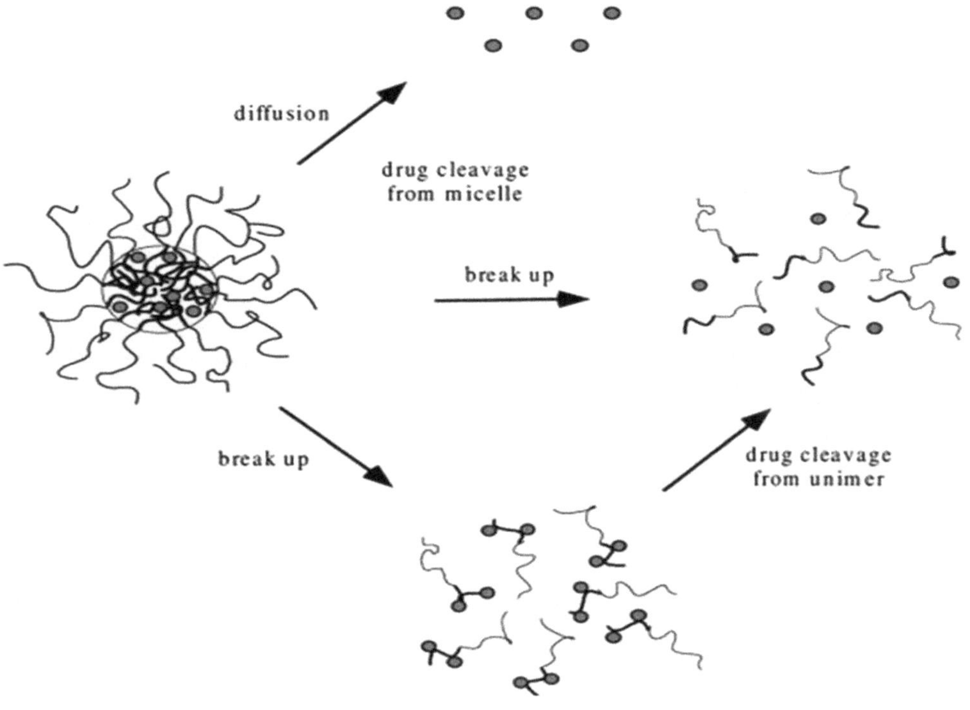


**FIGURE 1.9** Mechanisms of drug release from polymeric micelles. (Reproduced from Shah A *et al. RSC.* 2014; 4: 17028–17038. With permission.)

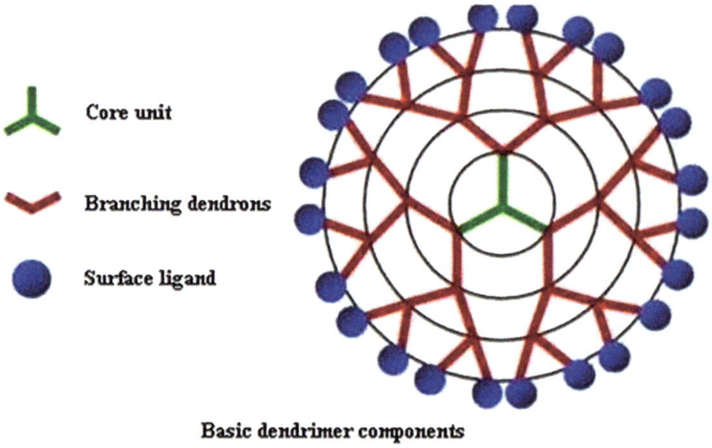


**FIGURE 1.10** Basic structure of dendrimer. (Reproduced from Sayani B *et al. IRJP.* 2011; 2(9): 25–32. With permission.)

### 1.3.1.6 Dendrimers

Dendrimers are unimolecular, monodispersed, and micellar nanostructures that are around 20 nm in size. These are well-defined, regularly branched, and symmetrical structures with high density of functional end groups at their periphery. A dendrimer structure consists of three distinct architectural regions, a focal moiety or a core, layers of branched repeat units emerging from the core, and functional end groups on the outer layer of repeat units (Sagar *et al.*, 2012). These are known to be robust, covalently-fixed, three-dimensional structures possessing both a solvent-filled interior core (nanoscale container) as well as a homogenous, mathematically defined, and exterior surface functionality (Frechet and Grayson, 2001) (Figure 1.10).

In terms of drug delivery, the applications of dendrimers are most commonly through parenteral administration, either straight into the tumor tissue or intravenously via systemic delivery. Dendrimers, used in drug delivery studies, typically incorporate one or more of the polymers such as (Sayani *et al.*, 2011) polyamidoamine (PAMAM), melamine, poly L-glutamic acid (PG), polyethyleneimine (PEI), polypropyleneimine (PPI), polyethylene glycol (PEG), and/or chitin.

The functions of dendrimers include diagnostic imaging, drug delivery, gene transfection, and therapeutic treatment of cancer by (i) passive targeting—nanodimension mediated via the EPR effect involving primary tumor vascularization or organ-specific targeting and (ii) active targeting—receptor-mediated cell-specific targeting involving receptor-specific targeting groups (Hofmann *et al.*, 2009).

### 1.3.1.7 Superparamagnetic Nanoparticles

Superparamagnetic molecules are attracted to a magnetic field but do not retain any residual magnetism after removal of the field. Magnetic drug targeting using superparamagnetic iron oxide particles (SPIONs) as carrier systems is used in accessible body parts targeting magnetic fields to show high gradients (Hofmann *et al.*, 2009). For selective magnetic bioseparation purposes, iron oxide nanoparticles, with their diameter ranging between 5 and 10 nm, are utilized. SPIONs are inorganically based particles with an iron oxide core coated by either inorganic materials such as silica or gold; organic materials such as phospholipids, fatty acids, polysaccharides, and peptides; and other surfactants and polymers (Hofmann *et al.*, 2009) (Figure 1.11).

Because of the superparamagnetic nanoparticle characteristics, such nanoparticles can be seen on MRI, which is a major advantageous feature of interest, and thus can be guided to a specific site using a magnetic field. Furthermore, by heating the magnetic field, the incorporated drugs can be released to the targeted site. These distinctive properties make these an ideal choice for a variety of applications, ranging widely from various separation techniques and contrast-enhancing agents for

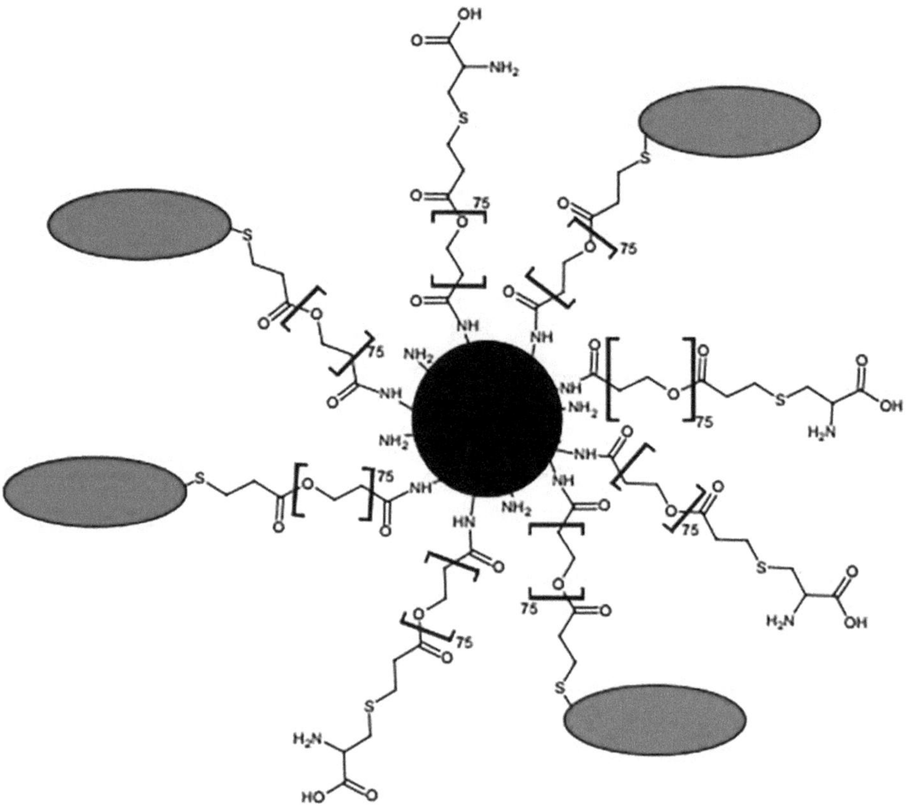

**FIGURE 1.11** Structure of a typical SPION particle coated with a polymer. (Reproduced from Hofmann M *et al.*, 2009. Superparamagnetic nanoparticles for biomedical applications. In: MC Tan, eds., *Nanostructured Materials for Biomedical Applications*, Transworld Research Network, India, pp. 483–496. With permission.)

MRI to drug delivery systems, magnetic hyperthermia, and magnetically assisted transfection of cells (Sagar *et al.*, 2012; Lee *et al.*, 2015).

Beads are micron polymer particles filled with SPIONs. For analytical intention, beads can be purposed with molecules that provide limited adsorption of proteins or other biomolecules and subsequent separation in a magnetic field gradient (Sagar *et al.*, 2012). However, certain applications like imaging of single cells or tumors, delivery of drugs or genes, local heating and separation of peptides, or signaling molecules or organelles from a single living cell or from a living (human) body are yet to be researched upon.

### 1.3.1.8 Quantum Dots

The quantum dots are semiconducting nanocrystals and core-shell nanocrystals containing an interface between different semiconductor materials. The size of quantum dots ranges from 2–10 nm. After polymer encapsulation, it normally increases to 5–20 nm in diameter. In the typical structure of the quantum dots, the core acts as the structural scaffold and the imaging contrast agent, and the small molecule hydrophobic drugs are embedded between the inorganic core and the amphiphilic polymer coating layer (Zhao and Zhu, 2016) (Figure 1.12). This fully integrated nanostructure may behave like a magic bullet and play a vital role in identifying and connecting to diseased cells to provide therapeutic treatment. Some of the recent literature instances of quantum dots in drug delivery are given in Table 1.3.

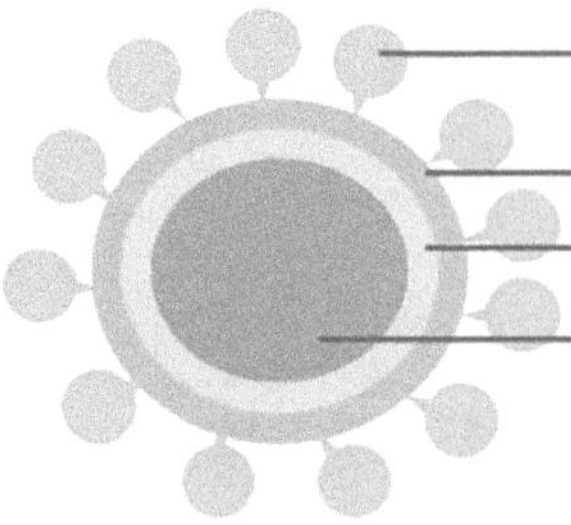


**FIGURE 1.12** Structure of quantum dot.

**TABLE 1.3**
**Recent Literature Reports on Quantum Dots in Biomedicine**

| Quantum Dots | Drug Molecule | Target | Reference |
|---|---|---|---|
| Biotin-linked graphene quantum dots | Doxorubicin | Lung Cancer | Iannazzo *et al.* (2017) |
| Blue fluorescent nitrogen-doped graphene quantum dots | Methotrexate | Breast cancer | Khodadadei *et al.* (2017) |
| Carbon quantum dot tailored calcium alginate hydrogel | Vancomycin | Osteomyelitis | Sarkar *et al.* (2017) |
| Gadolinium-doped quantum dots nanocomposites | Doxorubicin | Cancer | Chen *et al.* (2017) |
| Inulin tethered silver-graphene quantum dots | 5-Fluorouracil | Pancreatic Cancer | Joshi *et al.* (2017) |
| CdSe type quantum dots and carbon dots | – | TB biomarkers detection | Bhattacharyy *et al.* (2017) |
| Graphene quantum dots-chitosan hybrid xerogels | Sodium salicylate | Cancer, wound healing, arthritis | Lv *et al.* (2016) |
| Carbon dot modified mesoporous silica nanocarriers | Doxorubicin | Cancer | Jiao *et al.* (2016) |
| CdTe quantum dots loaded nanogels | Methotrexate | Cancer | Li *et al.* (2015) |

## 1.3.2 Drug Delivery across the Blood–Brain Barrier (BBB)

The main function of the BBB is to prevent noxious agents from entering the brain. There are various methods of drug delivery across this barrier, but some of these have resulted in damaging the structure of the barrier by forcibly opening it in order to allow the drugs to pass through. Therefore, an ideal method to deliver the drug across this barrier is considered to be controlled and damage-free, employing apt nanocarriers (Jain, 2012; Wei *et al.*, 2014).

### 1.3.2.1 Nanoparticles Used to Cross the BBB

Using nanoparticles to deliver drugs to the brain has far more advantages than any other mode of drug delivery. The main advantage is that nanoparticles compensate for the BBB-limiting characteristics of the therapeutic drug molecules (Jain, 2012). Some of the factors that affect the efficiency of drug delivery in the brain include type of polymer or surfactant used and sizes of the nanoparticles and the drug molecules. The important mechanisms by virtue of which nanoparticles are involved in the drug delivery to the brain are as follows (Jain, 2012):

- Tight junctions between endothelial cells are opened by the nanoparticles, enabling the drug to enter the BBB.

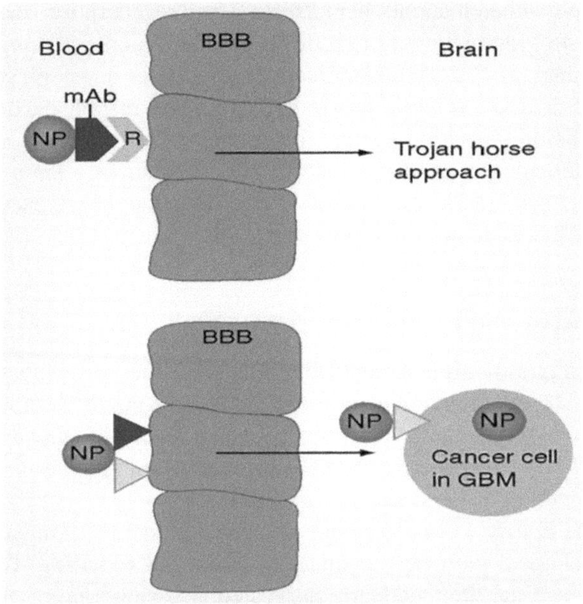


**FIGURE 1.13** Concept of targeted drug delivery to glioblastoma multiforme across the blood–brain barrier. (Reproduced from Jain KK. *Nanomedicine.* 2012; 7(8): 1225–1233. With permission.)

- Nanoparticles are transcytosed through the endothelial cell layer.
- They are endocytosed by endothelial cells and thus release the drug inside the cell.
- Their coating agents, such as polysorbates, inhibit the transmembrane efflux systems.

#### 1.3.2.2 Nanoparticle-Based Targeted Delivery of Chemotherapy across the BBB

G-technology has been applied to anticancer drugs to improve their targeted delivery to brain tumor cells. In G-technology, PEGylated liposomes coated with glutathione—an endogenous tripeptide transporter expressed on the BBB—are used to facilitate the delivery of drugs to the brain (Jain, 2012). G-technology has been applied for the delivery of methylprednisolone to treat several diseases with a neuro-inflammatory component. It is applicable for targeted chemotherapy of brain cancer.

The ideal anticancer drug delivery system for glioblastoma multiforme (GBM) should consist of an extra component to target the tumor cells across the BBB, in addition to crossing the BBB. An intravenously administered polymeric nanobioconjugate containing a biodegradable as well as non-immunogenic agent, polymalic acid, has been used for the delivery of a monoclonal antibody (mAb) targeting the tumor cells as well as for antisense oligonucleotides to block the synthesis of laminin-411, a protein involved in angiogenesis (Jain, 2012).

Figure 1.13 illustrates the concept of targeted drug delivery to GBM across the BBB by combining a monoclonal antibody (mAb) using the Trojan horse approach. The main idea of the whole process targeting the brain tumor using a nanostructured drug delivery system is to focus on the treatment of brain tumors. Among various nanoparticles, therefore, biocompatible polymer nanoparticles are considered to be the safest and most versatile vehicles for drug delivery to the brain. There are still some limitations to this, and further research on the topic is currently in progress (Jain, 2012).

## 1.4 REGENERATIVE MEDICINE

Tissue engineering focuses mainly on the repair of damaged and worn-out tissues and organs in the body. Regenerative medicine allows physicians to grow tissues and organs in the laboratory and safely implant these in the body when it cannot heal itself. Tissue regeneration occurs using a combination of living cells, which provide biological functionality, and materials acting as scaffolds to support cell proliferation (Khang *et al.*, 2008). Nanotechnology allows developing biocompatible material that enhances tissue regeneration while minimizing immune responses, decreasing inflammation, increasing the initial tissue formation needed to prolong implant lifetime, and inhibiting infection. Tissue biomaterial interaction at the nanoscale level creates surface features applicable to various medical fields, such as regenerative strategies for bone, vascular, heart, cartilage, bladder, and brain tissue (Khang *et al.*, 2008; Mao and Mooney, 2015).

### 1.4.1 Wound Healing and Regenerative Strategies

Regenerative medicine mainly aims to replace damaged tissues with functional tissues, as well to accelerate healing. Tissue regeneration can be applied by cell-based therapy, use of biomaterials, or a combination of the two (biomaterials seeded with cells). Wound healing and scar formation affects various tissues in the body. For instance, people suffering from diabetes are more likely to get oral infections and periodontal diseases (Nauta *et al.*, 2011).

Stem cells have the ability to differentiate and replace lost or damaged tissues, as well as to influence multiple biological pathways at once by paracrine signaling. Hence, these are used to promote regenerative healing. Stem cells are harvested at various stages of differentiation and can be implanted either directly or grown. These are ideal candidates for promoting tissue repair. Thus, regenerative medicine can be approached through the following cell-based therapy sources (Nauta *et al.*, 2011; Duscher *et al.*, 2016).

#### 1.4.1.1 Embryonic Stem Cells (ESCs)

Embryonic stem cells are one of the best sources for cell-based therapy due to their theoretically unlimited supply and pluripotent potential. Embryonic stem cells have allowed researchers to study various disorders from discarded *in vitro* fertilized embryos. Researchers have also attempted to make artificial skin from ESCs directed along the epidermal (outermost layer of skin) lineage pathway *in vitro*. Despite several of their advantages, ESCs are difficult to control, and chromosomal stability is difficult to maintain when ESCs are maintained in culture over multiple passages. Additionally, ethical controversy poses barriers to embryonic stem cell research, as it involves destroying the embryo while extracting the cells. This process is frowned upon and has prompted objection from political, religious, and social groups (Nauta *et al.*, 2011; Mehta, 2014).

#### 1.4.1.2 Induced Pluripotent Stem Cells (iPS Cells)

Induced pluripotent stem cells are favorable due to their potential clinical applications. Somatic cells could be isolated from a patient, reprogrammed, and then used in the same patient, theoretically evading the immune system. It can be used to screen new drug therapies and study the differentiation of both normal and diseased cells. The controversy linked to the ESC research can be dismissed using somatic cells. However, iPS cell technology is limited, and its reprogramming efficiency is exceedingly low, estimated to range between 0.01% and 10%. In addition, the reprogramming process has traditionally involved the use of modified viruses, which may be unsafe to use in therapeutics (Nauta *et al.*, 2011).

#### 1.4.1.3 Adult Mesenchymal Stem Cells (MSCs)

Adult mesenchymal stem cells (MSCs) have a potential cell population to develop cell-based therapies. The cells are harvested from adult tissues of many types, such as bone marrow-derived mesenchymal stem cells (BM-MSCs), which have attracted the most attention. MSCs tend to be more lineage specific and are more restrictive than ESCs and iPS cells (Mehta, 2014).

### 1.4.2 Ophthalmic Strategies

The eye is a complicated organ that is connected to the brain, where it senses the light from the neighboring environment and transfers the information to the brain. It consists of various functional components or tissues, with each comprising different varieties of cells responsible for executing a particular job (Willoughby *et al.*, 2010). The light entrance window is known as the cornea and is found in the front of the eye. The lens focuses the entering light, like a camera lens. However, the most complicated part of the eye is the retina, which is composed of several types of cells, each with its determined function. The electrical signal is the specific function of the photoreceptor cells, which produce the signal as these detect the light. The connection between the eye and the brain is via a biological wire, known as the optic nerve, and it is in-charge of the electrical impulse delivery from the retina to the brain. The brain is responsible for receiving these signals and translating them to provide the images of our surroundings. The optic nerve is closely associated with the retina. Retinal pigment epithelium is a sheet of cells that sit beneath the retina, supporting it, and executes a number of important roles, including the processing of nutrients. Eye diseases tend to occur if any of these parts are harmed, injured, or impaired, resulting in malfunctioning of the eye. Stem cell technology is used as a treatment for vision loss. These cells can serve as a substitute for the impaired cells in the eye, providing healthy cells, thus unleashing a new life to the patient. Different varieties of stem cells are available that can be utilized as cures for various diseases, such as repairing the cornea, replacing retinal pigment epithelial cells, and replacing retinal cells (www.eurostemcell.org; Zarbin *et al.*, 2012).

#### 1.4.2.1 Repairing the Cornea: Limbal Stem Cells

The cornea is the window of the eye that is constantly damaged by blinking and exposure to the outside world. In order to repair the damaged cornea and make a new corneal cell, a small number of stem cells at the edge of the cornea, known as limbal stem cells, are used. The cornea can no longer be repaired once the stem cells are lost due to an injury or disease, and this affects the ability of light to enter the eye, resulting in loss of vision. Transplantation of limbal stem cells from a healthy donor eye is not an ideal approach, as the success rates are lower and donor cells are usually only effective in the short to medium term.

New discoveries show improvement in techniques for growing limbal stem cells in the laboratory and preferable transplantation methods. However, there are still limitations in the number of new cells that can be obtained from a sample taken from the eye. Currently, studies are conducted to explore the feasibility of utilizing various sources like ESCs or iPS cells to produce new limbal stem cells in the laboratory. Stem-cell-based therapy would ease the complex surgery for the donor and provide an endless source of large quantities of limbal stem cells for the patient. This approach has been proved to work in clinical trials and is anticipated to sustain in the future too (Yamato and Okano, 2004) (Figure 1.14).

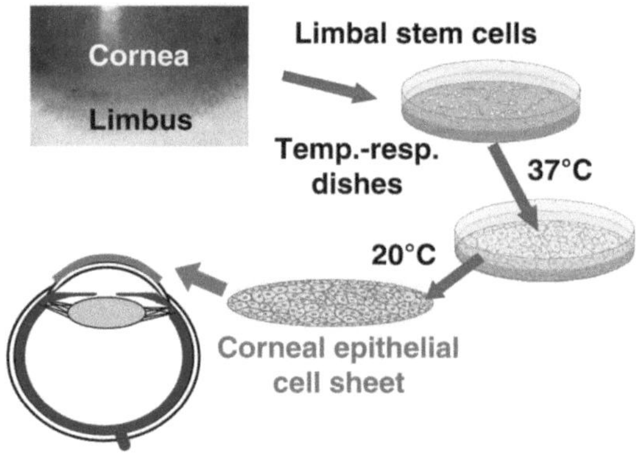


**FIGURE 1.14** Repairing the cornea. (Reproduced from Yamato M, Okano T. *Mater Today*. 2004; 7(5): 42–47. With permission.)

#### 1.4.2.2 Replacing Retinal Pigment Epithelial Cells

The retina is the component of the eye responsible for detecting light. Retinal pigment epithelial (RPE) cells have a role to look after the adjacent retina. The retina can be damaged in a variety of diseases such as age-related macular degeneration (AMD), retinitis pigmentosa, and Leber's congenital neurosis. To treat such diseases, damaged RPE cells can be replaced with transplanted healthy cells. However, it is not possible to take healthy RPE cells from donors, thus RPE cells are produced from both embryonic stem cells and iPS cells in the lab. The approach is being tested currently and similar trials will be conducted on patients. It is important to note that RPE is not responsible for "seeing," but is actually responsible for supporting the "seeing" retina. Therefore, replacing a damaged RPE will only be helpful to patients with a functioning retina (i.e., during early stages of the disease) (www.eurostemcell.org).

#### 1.4.2.3 Replacing Retinal Cells

Various disorders occur in which particular specialized cells in the circuit either stop working properly or die off. Stem cell technology provides a source of replacement of retinal cells using embryonic and iPS stem cells. In the retina, a type of cell known as Muller cell acts as a stem cell in some species, such as zebra fish. It has been suggested that the Muller cell is also able to act as a stem cell in humans.

This provides an additional source of retinal cells for the repair of the retina. This approach provides new treatment for people who suffer from retinal diseases like retinitis pigmentosa and glaucoma. Currently, there are no patient clinical trials planned to use this type of approach. There is enormous potential in stem cell technology to produce new biological components that can be used to repair and improve the lives of people who suffer from visual disorders. Research studies are currently being undertaken to develop new therapies to treat and/or prevent a loss of vision (Ouyang *et al.*, 2016).

### 1.4.3 Endodontics

Tooth regenerative therapy is a useful study model for organ replacement therapies. Tooth loss causes fundamental problems in terms of oral functions. Bioengineered teeth restore oral and physiological tooth functions by the replacement of an entire and fully functioning organ achieved though the transplantation of bioengineered organ germ. Several major areas of research, which have found an application in the development of regenerative endodontics are summarized in Table 1.4.

Bioengineered tooth research focuses primarily on producing immature teeth that can grow into adult teeth. This is accomplished by isolating adult human gum tissue (i.e., epithelial cells) samples from patients and growing them in a culture in the laboratory. The gum tissue cells are then combined with embryonic tooth mesenchymal cells (i.e., stem cells) that can induce tooth formation. This gum-tooth combination creates teeth with surrounding gum tissue, which can then be transplanted into the gums (Segura-Egea *et al.*, 2017).

### 1.4.4 Human Bladder

Tissue engineering focuses on restoring the functions of damaged organs and tissues by using apt aspects of biomedical engineering, cell biology, and material science to develop substitutes. This involves the injection of functional cells into a non-functional site to stimulate regeneration and the use of biocompatible materials to produce new tissues and organs. These biomaterials are called scaffolds and can either be natural or synthetic. They boost the body's ability to regenerate new tissues by assisting in the determination of the orientation and direction of new tissue growth (Tabata, 2009; Atala, 2011).

**TABLE 1.4**
**Techniques in the Development of Regenerative Endodontics**

| Technique | Approach | Advantage | Disadvantage |
|---|---|---|---|
| Stem cell therapy | Autologous or allogeneic stem cells are delivered to teeth via injectable matrix | • Quick<br>• Easy to delivery<br>• Least painful<br>• Cells are easy to harvest | • Low cell survival<br>• Cells do not produce new functioning pulp<br>• High risk of Complications |
| Root canal revascularization via blood clotting | Open up tooth apex to 1 mm to allow bleeding into root canals | • Lowest risk of immune rejection<br>• Lowest risk of pathogen transmission | • Potential risk of necrosis if tissue becomes re-infected |
| Pulp implantation | Pulp tissue is grown in the laboratory in sheets and implanted surgically | • Sheet of cells are easy to grow<br>• More stable than an injection of dissociated cells | • Sheets lack vascularity so only small constructs are possible<br>• Must be engineered to fit root canal precisely |
| Scaffold implantation | Pulp cells are seeded onto a 3-D scaffold made of polymers and surgically implanted | • Structure supports cell organization<br>• Some materials may promote vascularization | • Low cell survival after implantation<br>• Must be engineered to fit root canal precisely |
| Injectable scaffold delivery | Polymerizable hydrogels, alone or containing cell suspension are delivered by injection | • Easy delivery<br>• May promote regeneration by providing substitute for extracellular matrix | • Limited control over tissue formation<br>• Low cell survival |
| 3-D cell printing | Inket-like device dispenses layers of cells in a hydrogel which is surgically implanted | • Multiple cell types can be precisely positioned | • Must be engineered to fit root canal precisely<br>• Early stage research has yet to prove functional *in-vivo* |
| Gene therapy | Mineralizing genes are transfected into the vital pulp cells of necrotic and symptomatic teeth | • May avoid cleaning and shaping root canals<br>• May avoid the need to implant stem cells | • Most cells in a necrotic tooth are already dead<br>• Difficult to control<br>• Risk of health hazards |

In bladder reconstruction, the cell transplantation procedure relies on the capability to utilize donor tissue effectively and to supply the convenient conditions for enduring survival, differentiation, and growth. For this purpose, different varieties of cell sources have been researched, but the most preferable source is the native cells, since they can be used without rejection (Atala, 2011).

Human urothelial and muscle cells can be expanded *in vitro*, seeded onto polymer scaffolds, and allowed to attach and form sheets of cells. The cell-polymer scaffold can then be implanted *in vivo*. Histologic analysis indicates that viable cells are able to self-assemble back into their respective tissue types and retain their native phenotype(s) (Atala, 2011).

Bladder-wall cells and smooth muscle cells, which can regenerate, are cultured *in vitro* after obtaining these from a biopsy of the bladder. Cells are then seeded onto a bladder-shaped scaffold, where they are allowed to reproduce in the laboratory for about seven weeks. After tens of thousands of cells are allowed to reproduce, approximately 1.5 billion cells are obtained. After the regrown cells are surgically inserted into the organ, they keep maturing *in vivo* (Hodges and Atala, 2008).

An alternative cell source for bladder regenerative is the stem cells. Bone marrow stromal cells, for instance, can develop a phenotype similar to bladder smooth muscle cells *in vitro* and are seeded on Small intestinal submucosa (SIS) and implanted in a model. The role of stem cells in regenerative tissue engineering remains a fertile area for ongoing investigations (Hodges and Atala, 2008).

### 1.4.5 Nanomedicine for Treating Spinal Cord Injury

The spinal cord is very complex to treat acutely and is difficult to regenerate. It is, therefore, not an easy task to treat a spinal cord injury. Nanomaterials can be used to provide effective treatments and provide platforms to simulate regrowth of the damaged tissues. Regenerative therapy focuses on regaining neural circuitry and functionality in the damaged tissue(s). The main goal of neural regeneration assures that the environment is supplied for axon growth. In order to bridge the physiological gap in tissue, caused by lesion formation, or to prevent a lesion from forming, a scaffold can be incorporated into the damaged portion of the spinal cord (Tyler *et al.*, 2013).

Scaffolds give constructional support for the harmed spinal cord and also physical surface for regeneration, guiding and supporting cell growth from migration or transplantation. Several different nanomaterial approaches to scaffolding have been explored such as nanofiber scaffolds, self-assembled peptide systems, and nanofiber conduits. For all regenerative techniques, environmental cues are very important to ensure that cells differentiate and develop in physiologically relevant ways. Because of this, it is mostly favorable that the mechanical and chemical features of the regenerative scaffolds are nearly equivalent to those of the original tissue. A summary of the regenerative methods is provided in the next sections (Tyler *et al.*, 2013; Chen and Liu, 2016).

#### 1.4.5.1 Electrospun Nanofiber Scaffolds

Nanofiber orientation can affect the regeneration process within a scaffold. These scaffolds can be prepared by electrospinning. Scaffolds with aligned fibers not only discourage the differentiation of embryonic stem cells into astrocytes, prevalent in glial scarring, but also promote outgrowth of neurites parallel to the direction of fiber alignment. Random orientation of fibers in the fabric impede the forward movement of the neurites and, subsequently, regeneration is not very successful (Tyler *et al.*, 2013).

#### 1.4.5.2 Conduits

Conduits are tubes that facilitate communication between the proximal and distal ends of the nerve gap, offering a physical guidance for regrowth. Conduits have been successful in peripheral nerve regeneration and these nanofibrous can be formed via electrospinning. Even though collagen and nanofibrous conduits in peripheral nerve regeneration have been known to be successful, nanoscale conduits have been less effective in repairing spinal cord injury (SCI). Further improvement on these nanofibrous conduit systems can be made through surface functionalization, and perhaps in the future, the designs will be more successful. Although these electrospun scaffolds and conduits have the orientation benefits of aligned fibers and directional guidance, a limitation of these systems is that they can only be invasively implanted into the subject. Disrupting a spared tissue might cause further damage that can lead to infection or other surgical complication (Tyler *et al.*, 2013).

#### 1.4.5.3 Self-Assembling Systems

Self-assembling peptide systems are synthetic amino acid-based system that transition from a solution to a gel within seconds under *in vivo* pH and ion concentration conditions, thus resulting in the formation of a nanofibrous mesh. The gels are generally biocompatible and noncytotoxic, although in some cases, pH must be buffered prior to implantation. The gels can also safely encapsulate cells or drugs for combination therapies (Chen and Liu, 2016).

### 1.4.6 Bone Regeneration

The main cellular unit of embryologic bone formation is the mesenchymal stem cell (MSC). This can be manipulated to regenerate bone combined with the following technique: osteogenesis—the transfer of cells; osteoinduction—the induction of cells to become bone; osteoconduction—providing a scaffold for bone forming cells; or osteopromotion—the promotion of bone healing and regeneration by encouraging the biologic or mechanical environment of the healing or regenerating tissues (Kraus and Kirker-Head, 2006).

#### 1.4.6.1 Osteogenesis

Osteogenic is a graft that supplies and supports bone forming cells. Bone formation requires the cellular machinery to fabricate its structural components. An instance of osteogenic material is bone marrow. Bone marrow aspirates contain mesenchymal cells, already committed to osteogenic or chondrogenic lineage, and some biologically active proteins that stimulate bone regeneration in a similar manner to a naturally occurring fracture clot, many from platelet degranulation (Kraus and Kirker-Head, 2006; Henkel *et al.*, 2013).

#### 1.4.6.2 Osteoinduction

Osteoinductive materials are the materials which are able to induce bone formation when placed into a site where no bone formation will occur. The material does not work alone, but recruits MSC or progeny to infiltrate the material, then induces the multipotential cells to multiply and become cells that comprise the regenerating bony callus. An instance of an osteoinductive material is demineralized bone matrix (DBM). Processes that do not inactivate the bone's biologically active components are used to decalcify it, exposing the matrix along with a plethora of bone stimulatory cytokines trapped within the organic matrix. These are naturally occurring initiators and mediators of bone healing, which are released when a bone gets fractured (Kraus and Kirker-Head, 2006).

#### 1.4.6.3 Osteoconduction

Osteoconductive material is a material providing a scaffold for MSC and their progeny to migrate into and proliferate within. These are the physical materials of three-dimensional shapes, which offer appropriate framework and surface characteristics for adherence of MSCs, osteoblasts, osteocytes, chondroblasts, and chondrocytes. Furthermore, it offers interconnecting porosity for cellular proliferation and vascular ingrowth. Instances of naturally occurring osteoconductive materials include autogenous cancellous bone and some forms of DBM or deproteinized bone. Their use has been best demonstrated when there is a very active MSC population. However, their potential use should not be discounted, because any of these materials have been shown to be essential in MSC or cytokine delivery in bone regeneration (Kraus and Kirker-Head, 2006).

#### 1.4.6.4 Osteopromotion

Osteopromotion is a material or physical impetus that results in the enhancement of regenerating bone. Osteopromotion can function at various stages during healing and provide different stimulatory signals to bone regenerating tissues. Osteopromotion and osteogenesis/osteoconduction differ from each other as bone formation is enhanced without cells or a scaffold. However, an osteopromotive stimulus alone cannot induce bone formation. Osteopromotion can be achieved by introducing substances or materials that enhance bone regeneration or by physical or mechanical strategies that induce proliferation and differentiation of MSC and their progeny. An instance of an osteopromotive substance is platelet-rich plasma (PRP), which is produced by centrifugation of the patient's own blood, yielding a suspension of concentrated platelets. Not only needed in hemostasis, the platelets also play an essential role in healing, especially in bone healing (Kraus and Kirker-Head, 2006).

## 1.5 CONCLUSION

The main research areas of nanomedicine that are highlighted in this chapter include drug delivery using nanoparticles, medical diagnostics, imaging, therapeutic applications, and regenerative medicine. Nanoparticles can be functionalized with antibodies to target specific cells and can be used in diagnosis through the detection of biomarkers. Application of nanotechnology in medicine, such as to diagnose, cure, and regenerate organs from tissues, leads to a great deal of possibilities and future advancements against ongoing major diseases like cancer and brain disorders. The researches have proved that targeted nanoparticles offer a light of hope for the continued global fight against cancer, as ideal nanoparticles perform very well in detecting, diagnosing, and attaching to the tumorous cell, *vis-à-vis* the traditional chemotherapy and radiotherapy. Therefore, there is a clear possibility that nanotechnology will change the course of healthcare, dentistry, and human life, per se much more profoundly than several of the developments in the past. As is with other technological advancements, the immensely powerful tool of nanotechnology needs to be harnessed with caution, lest it may be misused to cause any harm to society, health, and environment.

## REFERENCES


Atala A. Tissue engineering of human bladder. *Br Med Bull.* 2011; 81–104.

Bazak R, Houri M, Achy SE, Hussein W, Refaat T. Passive targeting of nanoparticles to cancer: A comprehensive review of the literature. *Mol Clin Oncol.* 2014; 2(6): 904–908.

Bazak R, Houri M, El Achy S, Kamel S, Refaat T. Cancer active targeting by nanoparticles: A comprehensive review of literature. *J Cancer Res Clin Oncol.* 2015; 141(5): 769–784.

Bhattacharyy D, Sarswat PK, Michael L. Quantum dots and carbon dots based fluorescent sensors for TB biomarkers detection. *Vacuum.* 2017.

Boisseau P, Loubaton B. Nanoscience and nanotechnologies: Hopes and concerns. Nanomedicine, nanotechnology in medicine. *C R Physique.* 2011: 620–636.

Caschera L, Lazzara A, Piergallini L, Ricci D, Tuscano B, Vanzulli A. Contrast agents in diagnostic imaging: Present and future. *Pharmacol Res.* 2016; 110: 65–75.

Chen FM, Liu X. Advancing biomaterials of human origin for tissue engineering. *Prog Polym Sci.* 2016; 53: 86–168.

Chen ML, Pang SC, Chen XM, Zhang Y, Li L. Synthesis of permeable yolk-shell structured gadolinium-doped quantum dots as a potential nanoscale multimodal-visible delivery system. *Talanta.* 2017; 175: 280–288.

Debnath B, Uddin MJ, Maiti D. Nanoparticle as a targeting drug delivery system to blood-brain barrier: A review. *Pharmatutor.* 2015; 3(8): 30–37.

Duscher D, Barrera J, Wong VW, Maan ZN, Whittam AJ, Januszyk M, Gurtner GC. Stem cells in wound healing: The future of regenerative medicine? A mini-review. *Gerontology.* 2016; 62(2): 216–225.

Frechet JJM, Grayson SM. Convergent dendrons and dendrimers: From synthesis to applications. *J Am Chem Soc.* 2001; 101(12): 3819–3868.

Garg N, Mohanty A, Lazarus N, Schultz L, Rozzi TR, Santhanam S, Weiss L, Snyder JL, Fedder GK, Jin R. Robust gold nanoparticles stabilized by trithiol for application in chemiresistive sensors. *Nanotechnol.* 2010; 21: 1–6. doi:10.1088/0957-4484/21/40/405501

Haque N, Khalel RR, Parvez N, Yadav S, Hwisa N, Al-Sharif MS, Awen BZ, Molvi K. Nanotechnology in cancer therapy: A review. *J Chem Pharm Res.* 2010; 2(5): 161–168.

Henkel J, Woodruff MA, Epari DR, Steck R, Glatt V, Dickinson IC *et al.* Bone regeneration based on tissue engineering conceptions–A 21st century perspective. *Bone Res.* 2013 Sep; 1(3): 216–248.

Herzberger J, Niederer K, Pohlit H, Seiwert J, Worm M, Wurm FR *et al.* Polymerization of ethylene oxide, propylene oxide, and other alkylene oxides: Synthesis, novel polymer architectures, and bioconjugation. *Chem Rev.* 2016; 116: 2170–2243.

Hodges S, Atala A. Regenerative medicine and the neurogenic bladder. *Curr Bladder Dysfunct Rep.* 2008; 6(2): 67–74.

Hofmann M, Newberger T, Schopf B, Hofmann H, Rechenberg B. 2009. Superparamagnetic nanoparticles for biomedical applications. In: MC Tan, eds., *Nanostructured Materials for Biomedical Applications* (ISBN: 978-81-7895-397-7), Transworld Research Network, India, pp. 483–496.

Iannazzo D, Pistone A, Salamò M, Galvagno S, Romeo R, Giofré SV *et al.* Graphene quantum dots for cancer targeted drug delivery. *Int J Pharm.* 2017; 518: 185–192.

Jain KK. Nanobiotechnology-based strategies for crossing the blood–brain barrier. *Nanomedicine.* 2012; 7(8): 1225–1233.

Jiao J, Liu C, Li X, Liu J, Di D, Zhang Y *et al.* Fluorescent carbon dot modified mesoporous silica nanocarriers for redox-responsive controlled drug delivery and bioimaging. *J Colloid Interface Sci.* 2016; 483: 343–352.

Joshi PN, Agawane S, Athalye MC, Jadhav V, Sarkar D, Prakash R. Multifunctional inulin tethered silver-graphene quantum dots nanotheranostic module for pancreatic cancer therapy. *Mater Sci Eng C.* 2017; 78: 1203–1211.

Lahooti A, Sarkar S, Laurent S, Shanehsazzadeh S. Dual nano-sized contrast agents in PET/MRI: A systematic review. *Contrast Media Mol Imaging.* 2016; 11: 428–447.

Lee J, Kotov NA. Thermometer design at the nanoscale. *Nano Today.* 2007; 2(1): 48–51.

Lee N, Yoo D, Ling D, Cho MH, Hyeon T, Cheon. Iron oxide based nanoparticles for multimodal imaging and magnetoresponsive therapy. *Chem Rev.* 2015; 115(19): 10637–10689.

Li W, Chen X. Gold nanoparticles for photoacoustic imaging. *Nanomedicine (London, England)*. 2015; 10(2): 299–320.

Li Z, Xu W, Wang Y, Shah BR, Zhang C, Chen Y *et al.* Quantum dots loaded nanogels for low cytotoxicity, pH-sensitive fluorescence, cell imaging and drug delivery. *Carbohydr Polym*. 2015; 121: 477–485.

Lv O, Tao Y, Qin Y, Chen C, Pan Y, Deng L *et al.* Highly fluorescent and morphology-controllable graphene quantum dots-chitosan hybrid xerogels for *in vivo* imaging and pH-sensitive drug carrier. *Mater Sci Eng C*. 2016; 67: 478–485.

Kraus KH, Kirker-Head C. Mesenchymal stem cells and bone regeneration. *Vet Surg*. 2006; 35(3): 232–242.

Khang D, Carpenter J, Chun YW, Pareta R, Webster TJ. Nanotechnology for regenerative medicine. *Biomed Microdevices*. 2008; 12(4): 574–587.

Ke H, Wang J, Dai Z , Jin Y, Qu E, Xing H, Guo C, Yue X, Liu J. Gold-Nanoshelled Microcapsules: A Theranostic Agent for Ultrasound Contrast Imaging and Photothermal Therapy. *ACIEE Angewandte Chemie*, International Edition in English. 2011; 50(17): 3017–3021.

Khodadadei F, Safarian S, Ghanbari N. Methotrexate-loaded nitrogen-doped graphene quantum dots nanocarriers as an efficient anticancer drug delivery system. *Mat Sci Eng C*. 2017; 79: 280–285.

Mao AS, Mooney DJ. Regenerative medicine: Current therapies and future directions. *Proc Natl Acad Sci (U S A)*. 2015; 112(47): 14452–14459.

Meel RV, Lammers T, Hennink WE. Cancer nanomedicines: Oversold or underappreciated? *Expert Opin Drug Deliv*. 2017; 14(1): 1–5.

Mehta RH. Sourcing human embryos for embryonic stem cell lines: Problems & perspectives. *Indian J Med Res*. 2014; 140(Suppl 1): S106–S111.

Mestel R. Cancer: Imaging with antibodies. *Nature*. 2017; 543: 743–746.

Mitchell N, Kalber TL, Cooper MS, Sunassee K, Chalker SL, Shaw KP *et al.* Incorporation of paramagnetic, fluorescent and PET/SPECT contrast agents into liposomes for multimodal imaging. *Biomaterials*. 2013; 34(4): 1179–1192.

Muller RH, Radtke H. Novel concept of topical cyclosporine delivery with supersaturated SLN™ creams. *Saudi Pharm*. 2001; 28: 470–471.

Naseri N, Valizadeh H, Zakeri-Milani P. Solid lipid nanoparticles and nanostructured lipid carriers: Structure, preparation and application. *Adv Pharm Bull*. 2015; 5(3): 305–313.

Natarajan J. Polymeric nanoparticles for drug delivery and targeting: A comprehensive review. *Int J Pharm Res Allied Sci*. 2013; 1(4): 217–223.

Nauta A, Gurtner G, Longaker MT. Wound healing and regenerative strategies. *Oral Dis*. 2011; 17(6): 541–549.

Ouyang H, Goldberg JL, Chen S, Li W, Xu GT, Li W *et al.* Ocular stem cell research from basic science to clinical application: A report from Zhongshan Ophthalmic Center Ocular Stem Cell Symposium. *Int J Mol Sci*. 2016 Mar; 17(3): 415.

Podsiadlo P, Sinani VA, Bahng JH, Kam NW, Lee J, Kotov NV. Gold nanoparticles enhance the anti-leukemia action of a 6-mercaptopurine chemotherapeutic agent. *Langmuir*. 2008; 24(2): 568–574.

Provenzale JM, Silva GA. Uses of nanoparticles for central nervous system imaging and therapy. *AJNR Am J Neuroradiol*. 2009; 30(7): 1293–1301.

Radhakrishnan S, Samanta A, Sudeep PM, Maldonado KL, Mani SA, Acharya G *et al.* Metal-free dual modal contrast agents based on fluorographene quantum dots. *Part Part Syst Charact*. 2016; 34: 1–8.

Sarkar N, Sahoo G, Das R, Prusty G, Swain SK. Carbon quantum dot tailored calcium alginate hydrogel for pH responsive controlled delivery of vancomycin. *Eur J Pharm Sci*. 2017; 109: 359–371.

Sagar RM, Amol D, Patil S, Bhalgat CM. Nanoparticles: Emerging carriers for drug delivery. *Saudi Pharm J*. 2012; 19(3): 129–141.

Sayani B, Sujitha V, Kalyani P. Dendrimers and its application. *IRJP*. 2011; 2(9): 25–32.

Segura-Egea JJ, Gould K, Şen BH, Jonasson P, Cotti, Mazzoni A, Sunay H, Tjäderhane L, Dummer PMH. Antibiotics in Endodontics: A review. *Int Endontic*. 2017.

Shah A, Ahmad Z, Siddiq M, Kraatz H-B. Polymeric micelles as drug delivery vehicles. *RSC*. 2014; 4: 17028–17038.

Shen Z, Wu A, Chen X. Iron oxide nanoparticle based contrast agents for magnetic resonance imaging. *Mol Pharm*. 2017; 14(5): 1352–1364.

Tabata Y. Biomaterial technology for tissue engineering applications. *J R Soc Interface*. 2009; 6(Suppl 3): S311–S324.

Tavares AJ, Chong L, Petryayeva E, Algar WR, Krull UJ. Quantum dots as contrast agents for *in vivo* tumor imaging: Progress and issues. *Anal Bioanal Chem*. 2011; 399(7): 2331–42.

Tyler J, Xu X, Cheng J. Nanomedicine for treating spinal cord injury. *Nanoscale*. 2013; 5(19): 8821–8836.

Velmurugan R, Selvamuthukumar S. Nanostructured lipid carriers: A potential drug carrier for cancer chemotherapy. *Lipids Health Dis.* 2012; 11(159): 1–8.

Wang S, Dai Z, Ke H, Qu E, Qi X, Zhang K, Wang J. Contrast ultrasound-guided photothermal therapy using gold nanoshelled microcapsules in breast cancer. *Eur J Radiol.* 2014; 83(1): 117–122.

Wagner V, Dullaart A, Bock AK, Zweck A. The emerging nanomedicine landscape. *Nat Biotechnol.* 2006; 24(10): 1211–1217.

Wei H, Bruns OT, Kaul MG, Hansen EC, Barch M, Wisniowska A *et al.* Exceedingly small iron oxide nanoparticles as positive MRI contrast agents. *Proc Natl Acad Sci U S A.* 2017; 114(9): 2325–2330.

Wei X, Chen X, Ying M, Lu W. Brain tumor-targeted drug delivery strategies. *Acta Pharm Sin B.* 2014; 4(3): 193–201.

Willoughby CE, Ponzin D, Ferrari S, Lobo A, Landau K, Omidi Y. Anatomy and physiology of the human eye: effects of mucopolysaccharidoses disease on structure and function: A review. *Clin Exp Ophthalmol.* 2010; 38: 2–11.

Wu D, Huang L, Jiang MS, Jiang H. Contrast agents for photoacoustic and thermoacoustic imaging: A review. *Int J Mol Sci.* 2014; 15(12): 23616–236139.

www.eurostemcell.org/factsheet/eye-and-stem-cells-path-treating-blindness (Accessed on August 28, 2017).

www.grandviewresearch.com (Accessed on June 29, 2017).

www.sciencenewsline.com/summary/2014021409380009.html (Accessed on 12 February, 2017).

Yamato M, Okano T. Cell sheet engineering. *Mater Today.* 2004; 7(5): 42–47.

Yang G, Nanda J, Wang B, Chen G, Hallinan DT. Self-assembly of large gold nanoparticles for surface-enhanced Raman Spectroscopy. *ACS Appl Mater Interfaces.* 2017; 9(15): 13457–13470.

Yang X, Stein EW, Ashkenazi S, Wang LV. Nanoparticles for photoacoustic imaging. *Wiley Interdiscip Rev Nanomed Nanobiotechnol.* 2009; 1(4): 360–8.

Zhao MX, Zhu BJ. The research and applications of quantum dots as nano-carriers for targeted drug delivery and cancer therapy. *Nanoscale Res Lett.* 2016; 11: 207.

Zarbin M, Montemagno C, Leary JF, Ritch R. Regenerative nanomedicine and the treatment of degenerative retinal diseases. *WIRES Nanomed Nanobi.* 2012; 4(1): 113–137.

# 2 Biomedical Applications of Superparamagnetic Nanoparticles–An Insight

*Lavanya Khanna* and Surya Kant Tripathi*

## CONTENTS



## 2.1 INTRODUCTION

### 2.1.1 Nanotechnology in Biology

The fascination and widespread applications of nanotechnology upsurge from the fact that the properties (mechanical, thermal, electrical, chemical, optical and magnetic) of a material are significantly altered on reducing its size from bulk to nano dimension. These changes in properties occur due to size-dependent phenomena, such as high surface energy, discrete energy bands and quantum confinement (Gupta, 2006; Gao and Bing, 2009). Nanotechnology has developed as an interdisciplinary field, rationally integrating the fundamentals of physics, chemistry and biology. This can be explained by the following account.

Synthesis of nanomaterials is accomplished using chemical methods, where reaction parameters play a crucial role. Underlying physics of the final product, so obtained, is dependent on its physical parameters, such as size and shape. Finally, the product obtained finds applications in biology and biomedicine. Therefore, it brings all the streams of sciences together and requires extensive as well as intensive analyses.

Much of the fascination of nanotechnology in biology stems from their facts and factors (Gupta, 2006):

- Small size (as a human arm cannot be used as a tool for treating an ant)
- High surface/volume ratio
- Long shelf life (nanoparticle suspensions do not settle)
- Surface tailoralibility

* Corresponding author.

- Multifunctionality
- Cellular processes get selectively influenced at their natural scales

Nanoscaled structures and materials have been extensively explored in many biological applications, which encourage many new possibilities in several domains:

- Biological sensing techniques
- Molecular imaging
- Targeted gene/drug delivery
- Diagnosis and therapeutic treatment

MNPs have been expansively employed and studied for these biomedical applications. For targeted drug delivery applications, the basic idea is to encapsulate or attach any therapeutic drug to magnetic nanoparticles/nanocomposites, which by the effect of externally applied magnetic field, could be directed to the desired site, where the drug could be released. This is similar to the concept of the "magic bullet", as popularized by Dr. Paul Ehrlich, a German scientist and Noble Laureate in Physiology/Medicine in 1908 (Arruebo *et al.*, 2007). He won accolades for his findings in chemotherapy, haematology and immunology. He is the one who crafted the term "chemotherapy" and proposed that if a compound could be made such that it specifically aims the organism producing disease, then a toxin/drug attached to that agent for targeted delivery could also be delivered for destroying it. Ehrlich referred to it as a "magic bullet", an idea which has been further conceptualized in designing various strategies for targeted drug delivery. Initially, magnetic microparticles were used as contrast agents in radiation and anti-angiogenic therapy (Arruebo *et al.*, 2007; McBain *et al.*, 2008). But in the 1970s, more research was directed towards their drug delivery applications. Zimmermann and Pilwat (1976) delivered cytotoxic drugs using magnetic erythrocytes. Animal models with anticancer drug-loaded magnetic albumin microspheres were studied (Arruebo *et al.*, 2007). This work gained higher prevalence in the 1980s when more researchers delivered different drugs using micron-sized magnetic particles. The first report of MNPs used in animal models was by Lübbe *et al.* (1996), in which epirubicin conjugated MNPs were studied for the first Phase I clinical trial in patients in advanced stage cancers and ineffectively treated cancers. More than 50% of the MNPs are accumulated in the liver (Arruebo *et al.*, 2007).

## 2.2 MAGNETIC NANOPARTICLES

MNPs, as reflected in the name, correspond to the class of magnetic materials which are nanosized. Being inherently magnetic, these materials can be managed by an externally applied magnetic field (Reiss and Hutten, 2007). Magnetic properties are exceptionally responsive to size, composition and local atomic environment (Hosokawa *et al.*, 2007). MNPs are best suited for biomedical applications because of their several inherent characteristics, discussed in the next section.

### 2.2.1 Why MNPs?

#### 2.2.1.1 Essential Features

The following features are the essential features of MNPs, which have resulted in their extensive applications in biomedicine (Pankhurst *et al.*, 2003; Gupta and Gupta, 2005; Gupta, 2006; Arruebo *et al.*, 2007; McBain *et al.*, 2008).

1. *Size matching*: This is the foremost advantage and requisite for using nanoparticles (NPs) for biomedical applications. When one refers to a human body, it actually implies to the cellular level, which is in the nanometric size range. Hence, the simple rationale used here is that any tool required to fix anything must also be of the same size, i.e., if something needs to be

altered or treated at the cellular/tissue level, which is nanosized, it must also be in nanometers. This is where nanotechnology gains importance. Size of MNPs can be controlled by varying the synthesis parameters. Their size lies close to that of a gene (2 nm wide and 10–100 nm long), cell (10–100 μm), protein (5–50 nm) and virus (20–450 nm) (Pankhurst *et al.*, 2003). This size-matching verily enables MNPs to go near a biological entity or malignant site.

2. *Possibility of surface functionalization*: The MNP surface can be modified/functionalized with surface coatings (e.g., biodegradable inorganic and organic polymers, silica, etc.). Surface functionalizations further enable attachment of therapeutic drug/biological entity/florescent dye to the MNP complex. In other words, this MNP complex can be labeled and targeted for a specific application.
3. *Magnetic drive*: Being magnetic, MNPs can be guided magnetically to the target site, where they can deliver a drug (anticancer) or respond resonantly to a changing magnetic field as in hyperthermia, chemotherapy and radiotherapy, wherein an adequate degree of heat is required for destroying the tumor site effectively (Pankhurst *et al.*, 2003). This property gives them the power to be driven along when injected in the bloodstream. For the sites closer to the surface of the body, the presence of an externally applied magnetic field can be efficiently realized, but on going deeper inside the body, this falls off rapidly. Therefore, this factor needs to be carefully thought while planning the overall design of the MNP complex for the desired application.
4. *Nontoxicity, biocompatibility and biodegradability*: This is an obvious requirement since the MNP complex has to be introduced in the human body, where safety is a supreme necessity. MNPs considered for biomedical applications are generally iron-based compounds, which are inherently nontoxic in nature and are thus safe for *in vivo* applications. It should be well ascertained that MNPs are biocompatible, such that cell death occurs only because of the attached drug, but not due to the MNP complex. When something so complex is introduced inside the body, there is a possibility of its components to break and escape out. In that case, it is highly desired that the components be biodegradable, so that there is no harmful effect on the body. MNPs serve both of these purposes. Being nontoxic, they exhibit biocompatibility, and on metabolism, the iron ions accumulate safely in the body's iron storage, which is further integrated as haemoglobin by erythrocytes (Sun *et al.*, 2008).
5. *Injectability*: Aqueous and colloidal suspensions of MNP complexes can be prepared and injected in the human body intravenously (Sun *et al.*, 2008).

#### 2.2.1.2 Essential Requirements

In context of the these essential features, there are requirements for the practical usefulness of MNPs in biomedicine.

1. *Optimum size and shape*: Shape of the MNPs plays a central role, as it influences its toxicity and overall performance, for which even some models are reported having been studied (Akbarzadeh *et al.*, 2012). It was proposed that spherically shaped NPs displayed higher diffusion rates. This in turn resulted in increased concentration of NPs at the centre of the blood vessel and less interaction with endothelial cells. Endothelial cells are present at the interior surface of all blood vessels in the body, forming an interface between blood and the vessel wall. Since the spherically shaped NPs have restricted interaction with the endothelial cells, they have prolonged blood circulation time (Akbarzadeh *et al.*, 2012). It is highly essential that the MNP complex, introduced into the body, remains in the body. As soon as anything foreign is detected, it activates the body's defense system, i.e., reticuloendothelial system (RES). Therefore, the first challenge the MNP complex has to overcome inside the body is to escape the RES. One of the initial detection mechanisms of the RES is based on the size. NPs having a diameter >200 nm get removed by the spleen via mechanical filtration. Also, there are chances that larger particles clog the small capillaries (Zheng *et al.*, 2005; Akbarzadeh *et al.*, 2012). NPs with a size <5 nm

get rapidly eliminated by the kidneys. In context of intravenous injection, particles in the size range of 5–100 nm, are considered most advantageous, since they exhibit the most expanded blood circulation times. In addition, particles with a size in the 5–100 nm range display higher surface areas, lesser settling rates and improved diffusion (Zheng *et al.*, 2005). Apart from size, other factors that decide the fate of the NP's blood circulation time and internalization are the charge and surface chemistry of the MNP complex (Gupta and Curtis, 2004). MNPs are prone to agglomeration pertaining to high surface energy (high surface/volume) and magnetic interactions. On agglomeration, MNPs adsorb plasma proteins and are removed by the process called opsonization (Arruebo *et al.*, 2007). Also, particles with hydrophobic surfaces rapidly adsorb plasma proteins due to hydrophobic–hydrophobic interactions. Opsonized NPs are released out of the body by the RES. And, it has been reported that a hydrophobic surface is more cytotoxic than a hydrophilic group (McBain *et al.*, 2008). A hydrophilic surface resists the plasma protein coating process. In order to escape opsonization and the RES, the MNP surface needs to be coated. The overall size depends upon nature and geometric arrangement of the coating. It is also significant for determining biokinetics and biodistribution of MNPs inside the body. A "perfect" coating is also required for obtaining the preferred "zero premature release" of conjugated drug to avoid leaking through the matrix (Slowing *et al.*, 2008). Various other physiological conditions, such as pH, chemical environment and structurally unstable soft materials, affect the undesired and uncontrollable drug release (Slowing *et al.*, 2008).

All of these limitations can be reduced by functionalizing/coating the NP's surface using appropriate functional groups or stabilizers (Sun *et al.*, 2005). The coating should not only facilitate their selective binding to biological entities, but should also be nontoxic, hydrophilic, biocompatible, stable in aqueous suspensions and must not alter the superparamagnetic properties of the core material (Sun *et al.*, 2005). The surface coatings/stabilizers on the MNPs are required to

i. Provide stabilization of MNPs against aggregation (Yin *et al.*, 2005)
ii. Increase biocompatibility (Pankhurst *et al.*, 2003)
iii. Protect against corrosion (Pankhurst *et al.*, 2003)
iv. Escape the RES (Koo *et al.*, 2005)
v. Evade opsonization
vi. Enable binding of biological ligands to the NP's surface (Rana *et al.*, 2007)

There are four methods for functionalization of MNPs (Zhang and Olin, 2012):

1. Covalent Bonding
2. Electrostatic interaction
3. Adsorption
4. Hydrophobic/hydrophilic interactions

There has been an escalating inclination of researchers towards functionalizing/coating the surfaces of MNPs with inorganic and organic materials. Many surface coatings have been employed, comprising natural polymers such as carbohydrates and proteins; synthetic polymer coatings such as polyethylene glycol (PEG), dextran, chitosan, polyethyleneimine, polyvinyl alcohol, copolymers, etc.; and inorganic coatings such as silica and gold (Pankhurst *et al.*, 2003; Tartaj *et al.*, 2003; Gupta and Gupta, 2005; Arruebo *et al.*, 2007; McBain *et al.*, 2008; Sun *et al.*, 2008).

2. *Superparamagnetism*: The other characteristic required is that MNPs must exhibit superparamagnetism, which occurs due to finite-size effect, as is described here. Mainly, it appears in ferromagnetic/ferrimagnetic NPs, in which a long-range ordering phenomenon takes place pertaining to dipole interactions. Dipoles line up in parallel orientation. The size range of this parallel orientation is limited due to energetic reasons. These ranges are known

as magnetic domains. Within a grain, these domains are separated by "Bloch" walls. The direction of magnetization is changed by moving the Bloch walls. Retentivity/remanence is the magnetism retained on removal of the magnetic field, and coercivity refers to the magnetic field intensity for reducing to zero magnetization from saturation. Hysteresis determines the magnetic history of a material.

In bulk materials, magnetic domains are energetically controlled and separated by "Bloch" walls. In it, remanence and coercivity are largely independent of the particle size. As the particle size reduces, long range order diminishes in the material. A single magnetic domain exists, when the particle size becomes smaller than the critical diameter, $D_c$, as shown in Figure 2.1 (Mody *et al.*, 2013). Further reduction of size to the superparamagnetic diameter, $D_{sp}$, results in coercivity and remanence values rapidly approaching zero, i.e., the magnetization curve exhibits no hysteresis. In simple words, it is an S-shaped curve, depicting zero remanence and coercivity.

Basically, the condition of superparamagnetism is of typical thermal instability and magnetic anisotropy. The energy required for magnetizing a ferromagnetic/ferrimagnetic crystal depends on the magnetic field direction relative to the crystal orientation. This leads to "easy" and "hard" directions, corresponding to direction of easy magnetization and lower magnetization, respectively. In superparamagnetic materials, the vector of magnetization fluctuates between the easy magnetic directions, overcoming the hard directions. Corresponding to the easy axis, the magnetic moment usually has only two stable orientations, separated by an energy barrier of height $\Delta E = Kv$. If $Kv \gg kT$, then the moment cannot switch spontaneously, pertaining to permanent magnet. However, if $Kv \sim kT$ or less, then spontaneous switching can occur on the timescale of the experiment pertaining to superparamagnetic materials. Coupling of the electron spins as well as angular momentum of the electron orbital gives rise to magneto-crystalline anisotropy. This is further related to L-S coupling. As superparamagnetic behaviour is directly dependent on the magneto-crystalline anisotropy; therefore it is well connected to L-S coupling as well (Gubbala *et al.*, 2004; Misra *et al.*, 2004). It is further directly proportional to V, volume of particles ($E_A = KV \sin^2 \theta$). Here, $E_A$ represents the energy barrier required to block the flips of magnetic moments, K represents the magneto-crystalline anisotropy constant (associated with L-S coupling strength) and $\theta$ corresponds to the angle between direction of the easy axis and magnetization (Gubbala *et al.*, 2004; Zheng *et al.*, 2005).

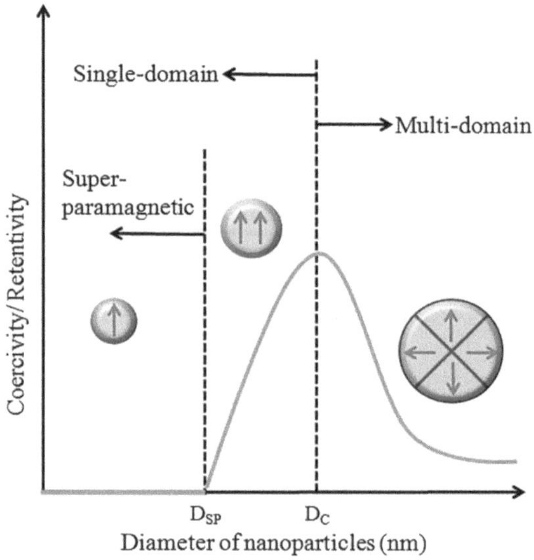


**FIGURE 2.1** Size-dependent variation of coercivity/retentivity. (Based on Mody VV *et al.*, *Eur J Nanomed.* 2013; 5(1): 11–21.)

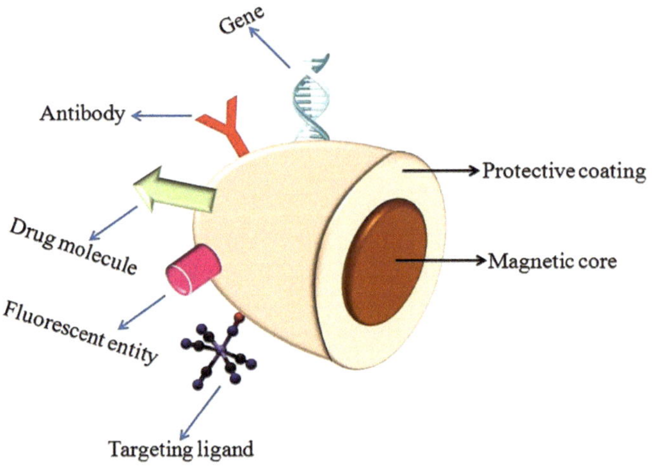


**FIGURE 2.2** Magnetic nanocomposite comprising of a magnetic core, protective coating and different surface functionalities. (Based on Bull E *et al.*, *Int J Nanomedicine.* 2014; 9(1): 1641–1643.)

When kept under an external magnetic field, a superparamagnet magnetizes in a similar manner as a paramagnet but possesses a much larger susceptibility than the former. When a magnetic field is applied, magnetic moments of superparamagnetic NPs align parallel to the applied field direction, resulting in a net magnetization. Generally, a reversible S-shaped curve is obtained. It is well known that above Curie temperature, ferromagnetic/ferrimagnetic material changes to paramagnetic. Unlike this phenomenon, superparamagnetism occurs below the Curie temperature of the material. In superparamagnetic NPs, there are two stable orientations with a small energy difference between them. Magnetization of each superparamagnetic domain swiftly jumps between these two, thus leading to the absence of retentivity. This energy difference is so little that thermal energy balances the total magnetization. Thus, it can be said that, in the case of superparamagnetic NPs, thermal fluctuations spontaneously demagnetize a magnetized state. Less squareness value, $M_R/M_S$ ratio, where $M_R$, $M_S$ are the remanent magnetization, saturation magnetization, respectively, corresponds to superparamagnetic behaviour (Zheng *et al.*, 2005). The squareness value pertaining to a noninteracting superparamagnetic assembly is 0.5 (Franco *et al.*, 2005). For interacting superparamagnetic particles, this value decreases due to demagnetizing effects of the dipolar interactions (Schulz *et al.*, 2009). In general, its value for exhibiting superparamagnetism is 0.1, implying that, on removing the external magnetic field, more than 90% magnetism is lost in the NPs (Schulz *et al.*, 2009). In simple words, it can be said that superparamagnetic NPs exhibit magnetism only when the magnetic field is on, and as soon as it is removed, they lose their magnetism. This means they are active only when the field is applied and such behaviour is ideal for biomedical applications. This whole concept of using superparamagnetic NPs is based on the following procedure. In an externally applied magnetic field, NPs get attracted and retained to a selected tumor site. The drug attached is released by some triggering chemical process according to the dosage concentration and time duration. These are then removed from the body (Koo *et al.*, 2005; Schulz *et al.*, 2009). This allows for effectively controlling the drug release externally by simply switching the magnetic field on or off.

In a nutshell, it can be said that a functional multicomponent magnetic nanocomposite is composed of a magnetic core material and protective coating, along with a surface functionality (according to the desired application), as depicted in Figure 2.2 (Bull *et al.*, 2014).

### 2.2.2 Active Targeting by Folic Acid Conjugation

MNPs target tumors mainly by two ways, i.e., passive and active strategies (Arruebo *et al.*, 2007; Kumari *et al.*, 2016). Passive targeting is due to extravasation of MNPs at the malignant site (tumor).

Here, the microvasculature is hyperpermeable and leaky, thus leading to the enhanced permeation and retention (EPR) effect. It utilizes the anatomical and physiological abnormalities of the tissue/cells (Kumari *et al.*, 2016). Active targeting is more specific, as it is done by identifying the specific receptor/proteins present at the surface of the target. Basically, overexpression of receptor-activated ligands such as folic acid, fatty acids, proteins and polypeptides on the target surface occurs (Arruebo *et al.*, 2007; Kumari *et al.*, 2016). This principally works on lock and key phenomenon, i.e., the MNP complex binds only to the specific target or releases the drug only when it binds to the desired target. Folic acid (FA) conjugation is carried out to enhance the targeting efficiency of the MNPs complex. The folate receptor (FR) is a glycosyl-phosphatidylinositol attached with high-affinity folate-binding protein, which is overexpressed in different varieties of tumors, while its expression does not appear in healthy tissues and organs (Zwicke *et al.*, 2012). Therefore, folate receptors have preferential sites, which distinguish tumor cells from normal cells and impede the nourishment to the rapidly dividing tumor cells. Overexpression of folate receptors is present in brain tumors, ovarian epithelial, lymphomas, neuroendocrine carcinomas, colorectal cancer, sarcomas and cancers of the lung, pancreas, breast, testicles, bladder, cervix, kidney and prostate (Mornet *et al.*, 2006; Mohapatra *et al.*, 2007; Zwicke *et al.*, 2012; Jiang *et al.*, 2014).

## 2.3 MNPs FORM COMPLEXES IN BIOMEDICAL APPLICATIONS

NPs of iron oxide and ferrites of Zn, Ni, Mn and Co have been extensively researched for biomedical applications. Mostly, PEG and silica are used as the coating material. Gupta and Curtis (2004) studied the influence of PEG-coated superparamagnetic iron oxide NPs (SPIONs) on human fibroblasts. They observed that PEG-coated SPIONs did not affect cell adhesion behaviour. Basically, the interaction of NPs with cells depends on their surface chemistry, and more importantly, it reflects on its hydrophobic/hydrophilic characteristics. On interaction with NPs, cells first bind, stick and then spread on the surfaces. Factors that affect cell adhesion encompass morphology, cyctoskeleton, organization and capability of proliferation and differentiation. In addition, interaction of surface proteins (integrins) with proteins governs cell adhesion. This is critical for determining cellular functions such as cell growth, migration, differentiation and survival. As observed, PEG-coated particles exhibit no considerable difference as compared to control on cell growth. This is attributed to low toxicity and high solubility of PEG. In the study by Cole *et al.* (2011a), starch-coated MNPs were cross-linked and aminated, as well as modified with PEG (5 and 20 kDa), using N-hydroxysuccinimide (NHS) chemistry. The final products obtained exhibited 61–98 fold longer plasma half-life (hours) in rats as compared to unmodified starch MNPs. After intravenous administration, higher plasma residence led to 100–150 greater tumor exposures to MNPs. The same group (Cole *et al.*, 2011b) developed a long-circulating, PEG-modified, cross-linked starch-MNP (PEG-MNP). This also had the same application as the earlier study, i.e., magnetic targeting, and a 9L-glioma rat model was used. This work explored the biodistribution patterns of PEG-MNPs in organs of exclusion such as lung, liver, spleen and kidney. It also showed that a long circulation lifetime of the NPs led to enhancement in magnetic brain tumor targeting. Singh *et al.* (2012) used the sol–gel method to coat a silica layer of thickness 1.5–129 nm on the surface of 12 nm-sized NPs. On reducing the silica-layer thickness, zeta potential moved towards negative values, but magnetic properties increased. Due to good dispersibility, it displayed superior colloidal stability as compared to uncoated MNPs. Also, silica-coated MNPs exhibited lesser cellular toxicity and decreased reactive oxygen species (ROS) generation *in vitro,* as compared to uncoated MNPs, thus enabling biomedical applications. Lee *et al.* (2008) synthesized uniformly sized silica-coated MNPs. A simple one-pot process was used in which reverse micelles acted as nanoreactors. Silica-coated and uncoated magnetite NPs exhibited 20 and 39.6 emu/g magnetic saturation values respectively. Their enzyme activity and stability revealed that the hybrid composites were magnetically separable, highly active, and stable under harsh shaking conditions for more than 15 days. Hwang *et al.* (2016) studied the effectiveness of drug delivery using a minimum amount of drug (MPA, mycophenolic acid) on silica-coated iron oxide NPs. Injumpa *et al.* (2017) studied size- and concentration-dependent cytotoxicity and inflammation responses of PEGylated silica-iron oxide nanocomposites and observed more toxicity at higher concentrations. Huang *et al.* (2017) studied doxorubicin-loaded PEG and polyethylenimine (PEI) co-coated SPIONs for drug delivery applications and magnetic resonance imaging (MRI). The synthesized nanocomposite effectively inhibited the tumor growth in mice. Stocke *et al.* (2017) demonstrated hyperthermia studies by a low, as well as high, dose of MNPs in the presence of an alternating magnetic field (AMF) using 3D-tumor tissue analogs (TTA) for secondary lung tumors. They observed that, in the presence of AMF, more cell damage occurred at the high concentration of MNPs (1 mg/mL). Liana *et al.* (2017) studied *E. coli* capturing efficiency of silica-coated $Fe_3O_4$ NPs functionalized with amine, carboxylic and methyl functionalization groups. They observed that incubation temperature and solution chemistry (presence of tryptone) plays a vital role in determining its capturing efficiency.

The characteristics of ferrites that make them suitable for biomedical applications include (Rana *et al.*, 2007; Tadic *et al.*, 2014)

i. High chemical stability
ii. Narrow size distribution
iii. High colloidal stability
iv. Tunable magnetic moment depending on size
v. Superparamagnetic properties
vi. Amenability for easy and simple covalent functionalization on their surface

Yang *et al.* (2010) synthesized $MnFe_2O_4$ (manganese ferrite) NPs and investigated their MRI contrast agent capabilities. The synthesized $MnFe_2O_4$ NPs were nontoxic at a concentration of 200 μg/mL. Sahoo *et al.* (2014) prepared silica-coated superparamagnetic $MnFe_2O_4$ NPs conjugated with fluorescent targeting moieties. The shell thickness was ~20–25 nm with overall size ~200–300 nm. These NPs exhibited specific adhesion to HeLa cancer cells as compared to normal cells. Their drug loading and release studies were investigated. A considerable amount of an anticancer drug, doxorubicin (DOX), was attached at the outer mesoporous silica shell (MSS). Release of DOX occurred at lysosomal pH. *In vitro* studies showed that DOX-loaded folate-targeted NPs attained exceptional efficiency for concurrently targeting, as well as for destroying, cancer cells. Mohapatra *et al.* (2011) synthesized multifunctional folic acid-conjugated superparamagnetic mesoporous $CoFe_2O_4$-based nanocarriers. The particle size obtained was 35–40 nm, and exceptional aqueous dispersion stability in physiological pH was observed. No reduction in hydrodynamic size or zeta potential occurred. HeLa cells were used to evaluate their cytotoxicity and internalization efficiency. Using simple organic coupling reactions, the fluorescent molecule rhodamine isothiocyanate (RITC) and anticancer drugs such as methotrexate and doxorubicin were effectively conjugated to the surface amine group, thus reflecting on their therapeutic and optical properties. Bano *et al.* (2016) studied paclitaxel (PTX) containing, folic acid (FA), bovine serum albumin (BSA), chitosan (CS)/carboxymethyl cellulose (CMC)-coated nanocomposites. Enhanced PTX release and effective tumor inhibition rate were observed in the presence of an external magnetic field. Tomitaka *et al.* (2009) studied the biocompatibility of $Fe_3O_4$ (20–30 nm), $ZnFe_2O_4$ (15–30 nm) and $NiFe_2O_4$ (20–30 nm) NPs using a cytotoxicity cell viability assay. At a concentration of 10 μg/mL, biocompatibility values of HeLa cells exposed to $Fe_3O_4$, $ZnFe_2O_4$ and $NiFe_2O_4$ NPs for 24 h and 48 h were more than 80%. However, at a higher concentration of 100 μg/mL, the cells exposed to $ZnFe_2O_4$ and $NiFe_2O_4$ NPs exhibited lower viability of 77.6% and 70.5%, respectively for 24 h and 59.4% and 52.4%, respectively, for 48 h. At higher concentrations, $ZnFe_2O_4$ and $NiFe_2O_4$ NPs showed lower cell viability, and at concentrations lower than 10 μg/mL, the three ferrites exhibited nontoxicity. Ahmad *et al.* (2015) reported about the feasibility studies of chitosan-coated nickel-ferrite ($NiFe_2O_4$) NPs to be used as $T_1$ and $T_2$ contrast agents in MRI. Obtained NPs were cylindrical in shape with an average length and width of 17 nm and 4.4 nm, respectively. The obtained values for $T_1$ and $T_2$ relaxivities were 0.858 ± 0.04 and 1.71 ± 0.03 $mM^{-1}s^{-1}$, respectively. About 25% signal enhancement in the $T_1$-weighted image and 71% signal loss in the $T_2$-weighted image were obtained in an animal study. Yang *et al.* (2013) successfully synthesized folic acid-functionalized magnetic $ZnFe_2O_4$ hollow microsphere core/MSS (MZHM-MSS-NHFA) composite with a higher drug storage capacity. The amount of drug loading of ibuprofen reached 28%. This system displayed enhanced and sustained drug-release ability along with targeting efficacy of the FA molecule. Sun *et al.* (2010) synthesized magnetic, florescent, folic acid-conjugated silica nanocomposites. The cytotoxicity was found to be greater than 90% at the concentration of 70 μg/mL. Rastogi *et al.* (2011) synthesized magnetic nanocomposites and coated them with PEG as well as folic acid. The magnetic saturation values of MNPs and FNPs (folic acid-conjugated composite MNPs) were 36 and 32 emu/g, respectively. Functionalization was studied by Fourier Transform Infrared (FT-IR) spectroscopy. The cytotoxicity results of polymer-coated NPs and polymer alone at 20 and 40 μg/mL were studied. Treatment with 20 and 40 μg/mL polymer NPs showed 93.41% and 82.35% cell viabilities, respectively, whereas with polymer alone, at the same concentrations, it was more than 90% and ~82%, respectively. Zhou *et al.* (2013) synthesized highly biocompatible and superparamagnetic folate-conjugated $Fe_3O_4$ NPs. The particle size was ~8 nm, hydrated diameter was 25.7 nm and $M_S$ value was 51 $Am^2/kg$. Additionally, these NPs exhibited good physiological stability, less cytotoxicity on human skin fibroblasts and insignificant effect on Wistar rats, even at high concentrations of body mass. The synthesized NPs successfully interceded into the human hepatoma Bel 7402 cells due to binding of the folic acid receptor. Zhang *et al.* (2002) synthesized PEG and folic acid modified superparamagnetic magnetite NPs. The synthesized nanocomposite internalized into macrophage and breast cancer cells. In BT20 cells, PEGylation resulted in enhanced uptake, though it was suppressed by RAW 264.7 cells. The presence of PEG and folic acid resisted the protein adsorption, which further enabled the avoidance of removal by macrophage cells. In addition, it also facilitated NP uptake to specific cancer cells. Noninvasive *in vivo* detection of vulnerable atherosclerotic plaque in cardiovascular disease based on $Fe_3O_4$ NPs and an osteopontin (OPN)-specific MRI/optical dual-modality probe was studied by Qiao *et al.*, 2017. These probes exhibited excellent targeting efficacy due to specific recognition towards foamy cells. A distinctive approach for MR molecular imaging of thrombosis by means of a dual mode strategy utilizing positive and negative contrast iron oxide NPs was reported by Ta *et al.* (2017). Andhariya *et al.* (2013) studied the drug release kinetics of doxorubicin-loaded folic acid-PEG functionalized $Fe_3O_4$ NPs. In the synthesized drug delivery system, 52% of drug loading was observed. Doxorubicin drug was released over a sustained period of 7 days, which is practically helpful and applicable when recurrent dosing is not feasible. Hwang *et al.* (2017) developed heparin-immobilized superparamagnetic iron oxide (HSPIO) NPs and studied their MRI sensitivity with anticoagulant activity for islet implantation for applications in diabetes therapy.

Ovarian cancer is one of the most lethal causes of gynaecologic malignancies. Liu *et al.* (2016) have reported an effective method for separating and detecting ovarian cancer cells from female whole blood. This was accomplished by folic acid-conjugated magnetic iron oxide NPs (IO–FA NPs). Owing to overexpressed folate receptor, these NPs firmly bind to the ovarian surface and, being magnetic, were separated from the blood under the influence of an applied magnetic field. Even with a lower quantity of spiked ovarian cancer SKOV3 cells, $5 \times 10^{-5}\%$, a high separation efficiency of 61.3% was obtained. Another achievement was that the synthesized NPs successfully detected half (5/10) of the metastatic ovarian cancer. This suggests the possibility of early detection of metastatic ovarian cancer cells, which would increase the survival rate for clinical applications. An early detection would lead to early treatment of the disease. This would save and render an improved healthy life to the affected patients. Richard *et al.* (2017) presented a novel and active targeting tool made up of $\gamma Fe_2O_3$ and aspartic acid-glycine-arginine (or cRGD peptides) for MRI of brain tumors. Hou *et al.* (2017) synthesized chitosan and indocyanine green dye (IR820)-coated $Fe_3O_4$ NPs (IR820-CS-$Fe_3O_4$), which exhibited considerable MRI ability along with huge potential for melanoma diagnosis as well as treatment. To improve the MRI sensitivity in lung cancer, Wang *et al.* (2017) successfully synthesized epidermal growth factor receptor monoclonal antibodies (anti-EGFR mAb) modified PEGylated $Fe_3O_4$ NPs. These synthesized NPs extensively improved the efficiency of magnetic resonance-guided focused ultrasound surgery (MRgFUS), thereby also reducing the side effects to nearby normal tissues. In addition, antibacterial studies have also been conducted with MNPs. Esmaeili and Ghobadianpour (2016) investigated antibacterial activity, drug loading and release properties of PEG- and chitosan-coated $MnFe_2O_4$ NPs. The nanocomposite exhibited significant antibacterial effect on gram-negative bacteria. Table 2.1 summarises the nanocomposites explored for various biomedical applications.

Some advantages of targeted drug delivery (Arruebo *et al.*, 2007) include

- Increasing efficacy of the drug simply by not changing anything in the drug itself, but changing only how it is administered into the cell.
- Due to high surface area, more amount of drug can be loaded into the NPs.
- Optimised drug therapy, i.e., right dose at the right site of action, for correct duration of time and not affecting other healthy sites.
- This decreases the undesired side effects on the body's healthy cells and kills only the targeted tumor cells.
- Targeting or localizing the drug exactly at the place where it needs to go.

## 2.4 SOME CHALLENGES

There are still many obstacles as well as challenges, which need to be addressed and controlled for successfully using magnetic NPs for biomedical applications and their use *in vivo* (Peng *et al.*, 2008; Sensenig *et al.*, 2012):

1. Chemistry of the NP–biomolecule interface, as magnetic carriers have to be administered systemically, for instance, with regards to the cellular barrier (e.g., blood–brain barrier (BBB), round window membrane (RWM) and endothelial luminal layer lining the walls of blood vessels.
2. Reduction of saturation magnetization of the core magnetic material during multistep chemical reaction for functionalization of stabilizers, biological entities and drug conjugation. The surface coatings which are indispensable, being magnetically dead, reduce the overall saturation magnetization of the composite as a whole and might also modify their magnetic property. Drug conjugation may alter their chemical properties. Therefore, all of these factors need to be carefully ratified.
3. The overall aim of drug targeting is not only to guide the therapeutic drug, but also to withhold the therapeutic agent at the tumor site, so effective tissue penetration of the drug is also a vital aspect.
4. Another factor is to govern the drug release kinetics; optimum drug dosage needs to be ascertained, as lower drug loading efficiency will hinder the whole process, and similarly, rapid release of encapsulated drugs from NPs in the blood might occur even before entering into the tumor site.
5. The targeting ability might be decreased due to the embedding part of the ligand.

Though encouraging results have been accomplished demonstrating the effectiveness of magnetic targeting, several challenges, uncertainties and apprehensions prevail for successful translation of this technology into clinical applications. Considerable development, ranging from magnetic core material to strategic targeting, is of utmost necessity to make this a viable clinical treatment modality. MNPs definitely have enormous possibility for clinical applications. Further extensive clinical studies are required to completely comprehend the medical competence of these MNP systems. This requires cooperative efforts of scientists from various disciplines and backgrounds.

**TABLE 2.1**
**Nanocomposites Used for Various Biomedical Applications**

| Magnetic Core Material | Coating Material | Application/Study Performed | Reference |
|---|---|---|---|
| Magnetite | PEG (Polyethylene Glycol), folic acid | Internalization studies in RAW 264.7 and human breast cancer (BT20) cells | Zhang *et al.* (2002) |
| Ni–Zn, Zn–Mn, Ni–Mn ferrites | – | Magnetic properties | Gubbala *et al.* (2004) |
| Iron oxide | PEG | Interaction studies with human fibroblasts | Gupta and Curtis (2004) |
| Nickel ferrite, zinc ferrite, Manganese ferrite | – | Magnetic characteristics | Misra *et al.* (2004) |
| $Fe_3O_4$ | Silica | Superparamagnetic study | Sun *et al.* (2005) |
| Nickel ferrite | Oleic acid | Cytotoxicity studies depending on size and coating | Yin *et al.* (2005) |
| Magnetite | Sodium dodecyl sulfate (SDS), oleic acid | Characterizations and magnetic study | Zheng *et al.* (2005) |
| Magnetite | Folic acid | Folate-conjugated superparamagnetic nanoparticles exhibited noncytotoxicity and receptor-mediated internalization cancer cells (HeLa, B16 melanoma F0) | Mohapatra *et al.* (2007) |
| Nickel ferrite | Polyvinyl alcohol, polyethylene oxide and polymethacrylic acid | Superparamagnetism and drug (doxorubicin) release kinetics | Rana *et al.* (2007) |
| Magnetite | Silica | Exhibited stability, magnetic separability, very active | Lee *et al.* (2008) |
| $Fe_3O_4$, $ZnFe_2O_4$ and $NiFe_2O_4$ | – | Biocompatibility studies | Tomitaka *et al.* (2009) |
| $MnFe_2O_4$ | Tetra-ethylene glycol | Magnetic resonance imaging (MRI) | Yang *et al.* (2010) |
| $Fe_3O_4$ | Silica, folic acid | Magnetic manipulation, fluorescence imaging, hollow structure and cell targeting | Sun *et al.* (2010) |
| Iron oxide | Starch, PEG | Improved plasma stability and enhanced tumor MNPs (Magnetic Nanoparticles) exposure | Cole *et al.* (2011a) |
| Iron oxide | Starch, PEG | Long circulation lifetime of the nanoparticles led to enhancement in magnetic brain tumor targeting | Cole *et al.* (2011b) |
| $CoFe_2O_4$ | Folic acid | Exhibited high cytotoxicity and apoptosis in HeLa cells | Mohapatra *et al.* (2011) |
| Iron oxide | NIPAAM (N-isopropylacrylamide), AA (acrylic acid) and PEGMA (polyethylene glycol methacrylate), folic acid | *In vivo* imaging and hyperthermia-based drug delivery | Rastogi *et al.* (2011) |

(*Continued*)

**TABLE 2.1 (*Continued*)**
**Nanocomposites Used for Various Biomedical Applications**

| Magnetic Core Material | Coating Material | Application/Study Performed | Reference |
|---|---|---|---|
| $Fe_3O_4$ | Silica | Lesser cellular toxicity and decreased reactive oxygen species (ROS) generation *in vitro* | Singh *et al.* (2012) |
| $ZnFe_2O_4$ | Silica, folic acid | Exhibited high drug (ibuprofen) storage and sustained drug release | Yang *et al.* (2013) |
| $Fe_3O_4$ | Folic acid | Tumor cell labelling, MRI target tracking and detection and magnetic hyperthermia | Zhou *et al.* (2013) |
| $Fe_3O_4$ | PEG, folic acid | Controlled and sustained release of doxorubicin | Andhariya *et al.* (2013) |
| $Fe_3O_4$ | Folic acid | MRI study | Jiang *et al.* (2014) |
| $MnFe_2O_4$ | Silica | MRI | Sahoo *et al.* (2014) |
| Maghemite | Silica | Magnetic properties | Tadic *et al.* (2014) |
| $NiFe_2O_4$ | Chitosan | $T_1$ and $T_2$ contrast agents in MRI | Ahmad *et al.* (2015) |
| $NiFe_2O_4$ | Paclitaxel, BSA (Bovine Serum Albumin) , CS/ CMC (chitosan/ carboxymethyl cellulose), folic acid | MRI and biocompatibility studies | Bano *et al.* (2016) |
| Iron oxide | Silica | Drug delivery studies with minimum amount of drug | Hwang *et al.* (2016) |
| Iron oxide | Folic acid | Early detection of ovarian cancer | Liu *et al.* (2016) |
| $MnFe_2O_4$ | PEG, chitosan | Antibacterial, drug loading and release studies | Esmaeili and Ghobadianpour (2016) |
| Iron oxide | PEG, PEI (Polyethylenimine), folic acid | MRI and drug delivery | Huang *et al.* (2017) |
| Iron oxide | PEG, Silica | Size-dependent cytotoxicity and inflammatory studies | Injumpa *et al.* (2017) |
| $Fe_3O_4$ | Silica with $-NH_2$, $-COOH$, $-CH_3$ surface functionalizations | *E.Coli (Escherichia coli)* capturing efficiency studies | Liana *et al.* (2017) |
| Iron oxide | – | Hyperthermia studies | Stocke *et al.* (2017) |
| $Fe_3O_4$ | Cy5.5-OPN (osteopontin)-DMSA (meso-2,3-dimercaptosuccinic acid)-MNPs(COD-MNPs) and Cy5.5-IgG-DMSA-MNPs (CID-MNPs) | Detection of vulnerable atherosclerotic plaque | Qiao *et al.* (2017) |
| $Fe_3O_4$ | PEG, anti-EGFR (epidermal growth factor receptor) monoclonal antibodies | MRI and MR-guided focused ultrasound (MRgFUS) ablation of lung cancer | Wang *et al.* (2017) |
| Iron oxide | Single-chain antibodies | Molecular imaging | Ta *et al.* (2017) |
| Iron oxide | Heparin | MRI sensitivity and anticoagulant activity | Hwang *et al.* (2017) |
| $Fe_3O_4$ | Chitosan, indocyanine green dye | Bio-imaging and cancer therapy | Hou *et al.* (2017) |
| $\gamma Fe_2O_3$ | cRGD (aspartic acid-glycine-arginine) peptides | MRI of brain tumors | Richard *et al.* (2017) |

## 2.5 CONCLUSIONS

This chapter endeavors to highlight some exciting research progress on the utilization of magnetic NPs/nanocomposites as promising tool for biomedical applications. The quintessential features and requisites for such applications have been discussed. The nanomaterials employed for biomedical applications must be nanosized, superparamagnetic and biocompatible. Surface coating is yet another indispensable requirement for their efficiency which not only reduces undesired agglomeration, but also facilitates bioconjugation and longer blood circulation time. In context to drug delivery applications, active targeting using folic acid makes the MNPs complex more specific and reduces the off-target effects. Further enhancements are probable for augmenting the catalogue of such materials for controlled release mechanisms, improved blood circulation properties, precise accumulation and so on. The research conducted so far is very promising and encouraging. Using better and sophisticated techniques can help in anticipating novel breakthroughs. Cell type specificity is another challenge that also needs to be addressed. The scientists look forward to experiencing many more exhilarating research activities in this growing field of nanobioengineering and nanostructured systems to deliver different types of guest molecules.

## ACKNOWLEDGMENT

Dr. Lavanya Khanna gratefully acknowledges University Grants Commission, Government of India, New Delhi, India, for awarding her Dr. D. S. Kothari Postdoctoral Fellowship to carry out this research work.

## REFERENCES

Ahmad T, Bae H, Iqbal Y, Rhee I, Hong S, Chang Y *et al.* Chitosan-coated nickel-ferrite nanoparticles as contrast agents in magnetic resonance imaging. *J Magn Magn Mater.* 2015; 381(1): 151–157.

Akbarzadeh A, Samiei M, Davaran S. Magnetic nanoparticles: Preparation, physical properties, and applications in biomedicine. *Nanoscale Res Lett.* 2012; 7: 144.

Andhariya N, Upadhyay R, Mehta R, Chudasama B. Folic acid conjugated magnetic drug delivery system for controlled release of doxorubicin. *J Nanopart Res.* 2013; 15: 1416.

Arruebo M, Pacheco RF, Ibarra MR, Santamaria J. Magnetic nanoparticles for drug delivery. *Nano Today.* 2007; 2: 22–32.

Bano S, Afzal M, Waraich MM, Alamgir K, Nazir S. Paclitaxel loaded magnetic nanocomposites with folate modified chitosan/carboxymethyl surface; a vehicle for imaging and targeted drug delivery. *Int J Pharm.* 2016; 513: 554–563.

Bull E, Madani SY, Sheth R, Seifalian A, Green M, Seifalian AM. Stem cell tracking using iron oxide nanoparticles. *Int J Nanomedicine.* 2014; 9(1): 1641–1643.

Cole AJ, David AE, Wang J, Galbán CJ, Hill HL, Yang VC. Polyethylene glycol modified, cross-linked starch-coated iron oxide nanoparticles for enhanced magnetic tumor targeting. *Biomaterials.* 2011a; 32: 2183–2193.

Cole AJ, David AE, Wang J, Galbán CJ, Yang VC. Magnetic brain tumor targeting and biodistribution of long-circulating PEG-modified, cross-linked starch-coated iron oxide nanoparticles. *Biomaterials.* 2011b; 32: 6291–6301.

Esmaeili A, Ghobadianpour S. Vancomycin loaded superparamagnetic $MnFe_2O_4$ nanoparticles coated with PEGylated chitosan to enhance antibacterial activity. *Int J Pharm.* 2016; 501: 326–330.

Franco V, Conde CF, Conde A. Relationship between coercivity and magnetic moment of superparamagnetic particles with dipolar interaction. *Phys Rev B.* 2005; 72: 174424 1–10.

Gao J, Bing X. Applications of nanomaterials inside cells. *Nano Today* 2009; 4: 37–51.

Gubbala S, Nathani H, Koizol K, Misra RDK. Magnetic properties of nanocrystalline Ni-Zn, Zn-Mn, and Ni-Mn ferrites synthesized by reverse micelle technique. *Physica B* 2004; 348: 317–328.

Gupta AK, Curtis ASG. Surface modified superparamagnetic nanoparticles for drug delivery: Interaction studies with human fibroblasts in culture. *J Mater Sci: Mater Med.* 2004; 15: 493–496.

Gupta AK, Gupta M. Synthesis and surface engineering of iron oxide nanoparticles for biomedical applications. *Biomaterials.* 2005; 26: 3995–4021.

Gupta RB. 2006. Fundamentals of drug nanoparticles. In: RB Gupta and UB Kompella, eds., *Nanoparticle Technology for Drug Delivery* (ISBN: 10:1-57444-857-9), Taylor & Francis group LLC, NY, pp. 1–19.

Hosokawa M, Nogi K, Naito M, Yokoyama T. 2007. Basic properties and measuring methods of nanoparticles. In: *Nanoparticle Technology Handbook* (ISBN-978-0-444-56336-1), Elsevier B.V., UK, pp. 5–47.

Hou X, Zhou H, Wang L, Tang J, Chen C, Jiang G *et al.* Multifunctional near-infrared dye-magnetic nanoparticles for bioimaging and cancer therapy. *Cancer Lett.* 2017; 390: 168–175.

Huang Y, Mao K, Zhang B, Zhao Y. Superparamagnetic iron oxide nanoparticles conjugated with folic acid for dual target-specific drug delivery and MRI in cancer theranostics. *Mater Sci Eng C.* 2017; 70: 763–771.

Hwang J, Lee E, Kim J, Seo Y, Lee KW, Hong JW *et al.* Effective delivery of immunosuppressive drug molecules by silica coated iron oxide nanoparticles. *Colloids Surf B: Biointerfaces.* 2016; 142: 290–296.

Hwang YH, Kim MJ, Lee DY. MRI-sensitive contrast agent with anticoagulant activity for surface camouflage of transplanted pancreatic islets. *Biomaterials.* 2017; 138: 121–130.

Injumpa W, Ritprajak P, Insin N. Size-dependent cytotoxicity and inflammatory responses of PEGylated silica-iron oxide nanocomposite size series. *J Magn Magn Mater.* 2017; 427: 60–66.

Jiang QL, Zheng SW, Hong RY, Deng SM, Guo L, Hu RL *et al.* Folic acid-conjugated $Fe_3O_4$ magnetic nanoparticles for hyperthermia and MRI *in vitro* and *in vivo*. *Appl Surf Sci.* 2014; 307: 224–233.

Koo OM, Rubinstein I, Onyuksel H. Role of nanotechnology in targeted drug delivery and imaging: A concise review. *Nanomed Nanotech.* 2005; 1: 193–212.

Kumari A, Sinla R, Guliani A, Walia S, Acharya A, Yadav SK. 2016. Nanoscale materials in targeted drug delivery. In: SK Yadav, ed. *Nanoscale Materials in Targeted Drug Delivery, Theragnosis and Tissue Regeneration* (ISBN-978-981-10-0818-4), Springer Science+Business media, Singapore, pp. 1–15.

Lee J, Lee Y, Youn Jk, Na HB, Yu T, Kim H *et al.* Simple Synthesis of functionalized superparamagnetic Magnetite/Silica Core/Shell nanoparticles and their application as magnetically separable high-performance biocatalysts. *Small.* 2008; 4: 143–152.

Liana AE, Marquis CP, Gunawan C, Gooding JJ, Amal R. T4 bacteriophage conjugated magnetic particles for E. coli capturing: Influence of bacteriophage loading, temperature and tryptone. *Colloids Surf B: Biointerfaces.* 2017; 151: 47–57.

Liu W, Nie L, Li F, Aguilar ZP, Xu H, Xiong Y *et al.* Folic acid conjugated magnetic iron oxide nanoparticles for nondestructive separation and detection of ovarian cancer cells from whole blood. *Biomater Sci.* 2016; 4: 159–166.

Lübbe AS, Bergemann C, Huhnt W, Fricke T, Riess H, Brock JW *et al.* Predinical experiences with magnetic drug targeting: Tolerance and efficacy. *Cancer Res.* 1996; 56: 4694–4701.

McBain SC, Yiu HHP, Dobson J. Magnetic nanoparticles for gene and drug delivery. *Int J Nanomed.* 2008; 3(2): 169–180.

Misra RDK, Gubbala S, Kale A, Egelhoff Jr. WF. A comparison of the magnetic characteristics of nanocrystalline nickel, zinc, and manganese ferrites synthesized by reverse micelle technique. *Mater Sci Eng B.* 2004; 111: 164–174.

Mody VV, Singh A, Wesley B. Basics of magnetic nanoparticles for their application in the field of magnetic fluid hyperthermia. *Eur J Nanomed.* 2013; 5(1): 11–21.

Mohapatra S, Mallick SK, Maiti TK, Ghosh SK, Pramanik P. Synthesis of highly stable folic acid conjugated magnetite nanoparticles for targeting cancer cells. *Nanotech.* 2007; 18: 385102.

Mohapatra S, Rout SR, Maiti S, Maiti TK, Panda AB. Monodisperse mesoporous cobalt ferrite nanoparticles: Synthesis and application in targeted delivery of antitumor drugs. *J Mater Chem.* 2011; 21: 9185–9193.

Mornet S, Vasseur S, Grasset F, Veverka P, Goglio G, Demourgues A *et al.* Magnetic nanoparticle design for medical applications. *Prog Solid State Ch.* 2006; 34: 237–247.

Pankhurst QA, Connolly J, Jones SK, Dobson J. Applications of magnetic nanoparticles in biomedicine. *J Phys D: Appl Phys.* 2003; 36: R167–R181.

Peng XH, Qian X, Mao H, Wang AY, Chen Z, Nie S *et al.* Targeted magnetic iron oxide nanoparticles for tumor imaging and therapy. *Int J Nanomed.* 2008; 3(3): 311–321.

Qiao H, Wang Y, Zhang R, Gao Q, Liang X, Gao L *et al.* MRI/optical dual-modality imaging of vulnerable atherosclerotic plaque with an osteopontin-targeted probe based on $Fe_3O_4$ nanoparticles. *Biomater.* 2017; 112: 336–345.

Rana S, Gallo A, Srivastva RS, Misra RDK. On the suitability of nanocrystalline ferrites as a magnetic carrier for drug delivery: Functionalization, conjugation and drug release kinetics. *Acta Biomater.* 2007; 3: 233–242.

Rastogi R, Gulati N, Kotnala RK, Sharma U, Jayasundar R, Koul V. Evaluation of folate conjugated pegylated thermosensitive magnetic nanocomposites for tumor imaging and therapy. *Colloids Surf B: Biointerfaces.* 2011; 82: 160–167.

Reiss G, Hutten A. 2007. Magnetic nanoparticles. In: KD Sattler, ed. *Handbook of Nanophysics: Nanoparticles and Quantum Dots* (ISBN: 13:978-1-4200-7545-8), CRC Press, Taylor & Francis group LLC, US, pp. 2(1–13).

Richard S, Boucher M, Lalatonne Y, Mériaux S, Motte L. Iron oxide nanoparticle surface decorated with cRGD peptides for magnetic resonance imaging of brain tumors. *Biochim Biophys Acta*. 2017; 1861: 1515–1520.

Sahoo B, Devi KSP, Dutta S, Maiti TK, Pramanik P, Dhara D. Biocompatible mesoporous silica-coated superparamagnetic manganese ferrite nanoparticles for targeted drug delivery and MR imaging applications. *J Colloid Interface Sci*. 2014; 431: 31–41.

Schulz DL, Sailer RA, Caruso AN. Superparamagnetic transition metal iron oxygen nanoparticles. US 20090194733 A1; 2009.

Sensenig R, Sapir Y, MacDonald C, Cohen S, Polyak B. Magnetic nanoparticle-based approaches to locally target therapy and enhance tissue regeneration *in vivo*. *Nanomed Lond*. 2012; 7(9): 1425–1442.

Singh RK, Kim TH, Patel KD, Knowles JC, Kim HW. Biocompatible magnetite nanoparticles with varying silica-coating layer for use in biomedicine: Physicochemical and magnetic properties, and cellular compatibility. *J Biomed Mater Res Part A*. 2012; 100A: 1734–1742.

Slowing II, Vivero-Escoto JL, Wu CW, Lin VSY. Mesoporous silica nanoparticles as controlled release drug delivery and gene transfection carriers. *Adv Drug Deliv Rev*. 2008; 60: 1278–1288.

Stocke NA, Sethi P, Jyoti A, Chan R, Arnold SM, Hilt JZ, Upreti M. Toxicity evaluation of magnetic hyperthermia induced by remote actuation of magnetic nanoparticles in 3D micrometastatic tumor tissue analogs for triple negative breast cancer. *Biomaterials*. 2017; 120: 115–125.

Sun C, Lee JSH, Zhang M. Magnetic nanoparticles in MR imaging and drug delivery. *Adv Drug Deliv Rev*. 2008; 60: 1252–1265.

Sun L, Zang Y, Sun M, Wang H, Zhu X, Xu S. Synthesis of magnetic and fluorescent multifunctional hollow silica nanocomposites for live cell imaging. *J Colloid Interface Sci*. 2010; 350: 90–98.

Sun Y, Duan L, Guo Z, DuanMu Y, Ma M, Xu L *et al*. An improved way to prepare superparamagnetic magnetite-silica core-shell nanoparticles for possible biological application. *J Magn Magn Mater*. 2005; 285: 65–70.

Ta HT, Li Z, Hagemeyer CE, Cowin G, Zhang S, Palasubramaniam J *et al*. Molecular imaging of activated platelets via antibody-targeted ultrasmall iron oxide nanoparticles displaying unique dual MRI contrast. *Biomater*. 2017; 134: 31–42.

Tadic M, Kralj S, Jagodic M, Hanzel D, Makovec D. Magnetic properties of novel superparamagnetic iron oxide nanoclusters and their peculiarity under annealing treatment. *Appl Surf Sci*. 2014; 322: 255–264.

Tartaj P, Morales MDP, Verdaguer SV, Carreno TG, Serna CJ. The preparation of magnetic nanoparticles for applications in biomedicine. *J Phys D: Appl Phys*. 2003; 36: R182–R197.

Tomitaka A, Hirukawa A, Yamada T, Morishita S, Takemura Y. Biocompatibility of various ferrite nanoparticles evaluated by *in vitro* cytotoxicity assays using HeLa cells. *J Magn Magn Mater*. 2009; 321: 1482–1484.

Wang Z, Qiao R, Tang N, Lu Z, Wang H, Zhang Z *et al*. Active targeting theranostic iron oxide nanoparticles for MRI and magnetic resonance-guided focused ultrasound ablation of lung cancer. *Biomaterials*. 2017; 127: 25–35.

Yang D, Wei K, Liu Q, Yang Y, Guo X, Rong H *et al*. Folic acid-functionalized magnetic $ZnFe_2O_4$ hollow microsphere core/mesoporous silica shell composite particles: Synthesis and application in drug release. *Mater Sci Eng C*. 2013; 33: 2879–2884.

Yang H, Zhang C, Shi X, Hu H, Du X, Fang Y *et al*. Water-soluble superparamagnetic manganese ferrite nanoparticles for magnetic resonance imaging. *Biomaterials*. 2010; 31: 3667–3673.

Yin H, Too HP, Chow GM. The effects of particle size and surface coating on the cytotoxicity of nickel ferrite. *Biomater*. 2005; 26: 5818–5826.

Zhang R, Olin H. 2012. Magnetic nanoparticles in biomedical application. In: D Barh, K Blum and MA Madigan, eds., *OMICS-Biomedical Perspectives and Applications* (ISBN-13:978-1-4398-5009-1), CRC Press, Taylor & Francis Group LLC, NW, pp. 119–140.

Zhang Y, Kohler N, Zhang M. Surface modification of superparamagnetic magnetite nanoparticles and their intracellular uptake. *Biomaterials*. 2002; 23: 1553–1561.

Zheng W, Gao F, Gu H. Magnetic polymer nanospheres with high and uniform magnetite content. *J Magn Magn Mater*. 2005; 288: 403–410.

Zhou CJ, Wang SH, Zhou Y, Rong PF, Chen ZZ, Liu JY *et al*. Folate-conjugated $Fe_3O_4$ nanoparticles for *in vivo* tumor labelling. *Trans Nonferrous Met Soc China*. 2013; 23: 2079–2084.

Zimmermann U, Pilwat G. Organ specific application of drugs by means of cellular capsule systems (author's transl)]. *Zeitschrift fur Naturforschung. Section C, Biosci*. 1976; 31(11–12): 732–736.

Zwicke GL, Mansoori GA, Jeffery CJ. Utilizing the folate receptor for active targeting of cancer nanotherapeutics. *Nano Rev*. 2012; 3: 18496 1–11.

# 3 Dendrimers as Promising Nanocarriers in Drug Delivery

*Avinash Gothwal, Ashok Kumar Sharma, Gurjit Kaur Bhatti, and Umesh Gupta**

## CONTENTS



* Corresponding author.

## 3.1 INTRODUCTION

Dendrimers are hyper-branched, unimicellar, monodispersed, globular, versatile and synthetic macromolecules having high molecular weight (Tomalia *et al.*, 1985; Kesharwani and Iyer, 2015). Their physicochemical properties resemble those of biomacromolecules such as proteins and oligomeric macromolecules with globular shapes (Boas and Heegaard, 2004). The structure of a dendrimer can be discriminated in a core, branches and consequent radially developed terminal functionality (Esfand and Tomalia, 2001; Tomalia, 2005). Size of the dendrimers generally depends upon the number of generations; as the generation size increases, the molecular weight of the dendrimers also increases accordingly (Congur *et al.*, 2014). Surface functional groups offer tailor-made functionality and controllable size, which attracts researchers the most. Introduced for the first time by Vogtle and his group, the term 'dendrimer' basically originated from the Greek word *dendron*, meaning 'tree' or 'branch', and *meros* meaning 'part'. Earlier, dendrimers were known as 'arborols' or 'cascade polymers' too. Despite their complex and large molecular architecture, the dendrimer molecules offer low polydispersity (Gupta and Nayak, 2015).

Dendrimers demonstrate immense potential in drug delivery (Tomalia *et al.*, 2007; Medina and El-Sayed, 2009; Wang *et al.*, 2010; Buczkowski *et al.*, 2014), mainly as carriers of anticancer drugs, oligonucleotides (i.e., non-viral vectors of genetic materials) and contrast media for medical imaging. It is quite well known that a huge volume of revenue has to be expended in improving the bioavailability of several poorly bioavailable active pharmaceutical ingredients (APIs), and such expenditure is steeply growing with time (Svenson, 2009). A lot of liquid money is wasted or is unrecoverable because of the undesired physical properties such as aqueous solubility, due to which about 40% of newly synthesized APIs are rejected by the pharmaceutical industries owing primarily to their low aqueous solubility, leading invariably to poor bioavailability. It has also been noticed that about 17% of launched APIs demonstrate suboptimal bioperformance due to their unfriendly aqueous behavior. Similarly, gene delivery also faces some challenges such as stability, targeted delivery of the genetic material, etc. Dendrimers can be used to surmount such aforementioned hurdles. Verily, dendrimers have proved their ability in the drug delivery of various diseases such as cardiovascular disease, cutaneous, brain tumors and genetic disorders (Wang *et al.*, 2001; Katare *et al.*, 2015). The positive charge on the dendrimer's surface can be used to deliver the negatively charged genetic materials through complex formation. PAMAM dendrimers have been used as one of the gold standards in delivering drugs, DNA or RNA across the membrane due to the cationic surface.

PAMAM dendrimers are one of the most exhaustively studied and established dendrimers as drug delivery carriers and for solubilization. PAMAM dendrimers were first synthesized by a divergent method developed by Tomalia *et al.* (1985). Micellar solubilization, ionic interactions, hydrophobic interactions and hydrogen bonding are the main mechanisms responsible for dendrimer-mediated solubility enhancement. Solubilization efficiency of dendrimers can be easily adjusted by tailoring core, branching units and surface functionalities with hydrophilic moieties (Kesharwani *et al.*, 2014; Devarakonda *et al.*, 2005a,b). It was documented that solubilization efficacy of polypropylene oxide-core PAMAM dendrimers depends on their generation and concentration. Several research studies on solubility have pointed out that dendrimers are effective solubility enhancers for nonsteroidal anti-inflammatory drugs (NSAIDs) and other drugs (Koc and Senel, 2013). Dendrimers also emerge as outstanding carriers in respect to bioavailability enhancement of difficult chemical entities. It has been shown that PAMAM dendrimer macromolecules of integral generations bind thread which resembles interactions with histones (Percec *et al.*, 1998; Dutta *et al.*, 2006), unbalance biological membranes (Roberts *et al.*, 1996; Malik *et al.*, 2000; Padilla De Jesús *et al.*, 2002; Jevprasesphant *et al.*, 2003a; Morgan *et al.*, 2003; Chen *et al.*, 2004) and affect the conformation of polypeptides and their ability to aggregate.

The present chapter highlights the properties and applications of dendrimers in solubility enhancement of bioactives, particularly their chemical properties in facilitating an improvement of solubility of the hydrophobe(s), besides correlating the nature of binding between hydrophobic drug and dendrimer carriers.

## 3.2 STRUCTURE, SYNTHESIS AND PROPERTIES OF DENDRIMERS

### 3.2.1 Structure

Despite the fact that dendrimers have a large molecular size, they are structurally well-defined and monodisperse *vis-à-vis* the traditional polymers. Due to the dendritic branching, the structure of the dendrimer results in a semi-globular to globular shape and provides higher functionalities on the surface that depend on the branching molecules such as amines and hydroxyls, as well as terminal functionalities. The globular structure of the dendrimer leads to residing in a less hydrodynamic volume than the equivalent linear polymers (Boas and Heegaard, 2004). Branching of the dendrimer resembles many arms radiating from the core molecule, as depicted in the typical dendrimer structure (Figure 3.1).

Components of the dendrimer:

#### 3.2.1.1 Multi-Functional Initiator Core

The core is basically an initiator and has at least two ends with similar or modifiable functionalities. In case of a PAMAM dendrimer, ethylene-diamine or diaminobutane is mostly used as an initiator.

#### 3.2.1.2 Inner Emanating Branches/Generation

This consists of repeating branched units or branches. The generation number is basically the number of focal points while moving towards the surface from the core. If the focal number of focal point is five, it will be a fifth generation of the dendrimer.

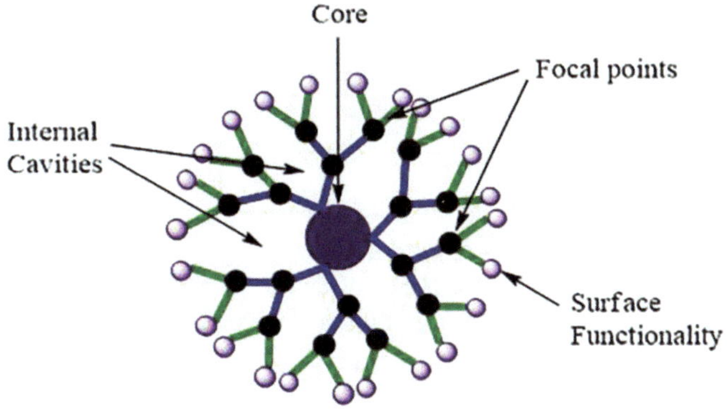


**FIGURE 3.1** A typical structure of a dendrimer.

#### 3.2.1.3 Outer Shell

In dendrimers, pincers before reaching the dendrimer surface and the outer shell consist of a changeable number of the same. Polypropyleneimine (PPI) and PAMAM dendrimers have half the number of pincers to the number of surface groups due to the bifurcation of each focal point (Svenson and Tomalia, 2005).

#### 3.2.1.4 Exterior or Terminal Surface Groups

Exterior or terminal surface groups are attached to the outermost surface. This exterior provides a platform to conjugate the bioactives. Dendrimers acquire a globular shape as the number of generations increases, with lower generation dendrimers being more asymmetric than the higher generation ones. The acquired shape and nanometric size (1–20 nm) play a key role in most of its unmatched properties and applications in solubilization and targeted delivery of drugs.

### 3.2.2 Synthesis

The first reports on dendrimer research were published about three decades ago (Tomalia *et al.*, 1985), which mainly explored their synthesis and their chemical and physical properties. As indicated in Figure 3.2, there are several methods of dendrimer synthesis:

#### 3.2.2.1 Divergent Method

Divergent synthesis was proposed by Tomalia *et al.* (1984). They cross-linked oligonucleotides, which led to stabilization of the self-assembled dendrimer. The divergent approach has two steps, the first being the activation of functional groups on the surface and second being an addition of monomer branching units. Later, this approach was also proposed for the synthesis of biomolecules diagnostics (Boas *et al.*, 2006). The approach is successful for large-scale production of dendrimers, with the first synthesized dendrimers using this technique being the PAMAMs.

#### 3.2.2.2 Convergent Method

Convergent synthesis proceeds centripetally, meaning inwards from the surface of the dendrimer and conjugating in the end to a core to give a whole new dendrimer. The convergent approach is mostly used for the synthesis of symmetric dendrimers in which different parts are coupled together to create assorted morphologies (Boas *et al.*, 2006). The advantages of convergent growth are that only two coexisting reactions are needed for any higher generation adding a step, and the consequent dendrimers are easier to purify. Herein, this is

- Possible to exactly locate the functional groups at the edge or surface of the dendrimer.
- Easy to synthesize asymmetric dendrimers, but difficulties remain in the development of higher generation dendrimers due to steric hindrances taking place during construction of the dendrons and the core molecule.

Despite these advantages, the convergent method is a low-yielding approach during the synthesis of large structures.

#### 3.2.2.3 Lego Chemistry

Lego chemistry is mainly used for the synthesis of phosphorus-based dendrimers using highly functionalized cores and branched monomers (Svenson and Tomalia, 2005). Many alterations to this method have been developed. These types of dendrimers need just one step for every higher generation, use the least volume of solvent and allow easy purification.

#### 3.2.2.4 Click Chemistry

The term "click chemistry" was given by Sharpless in 2001 to describe the kinds of reactions which are high-yielding, easy to carry out, have broad aspects, are stereospecific and produce the least

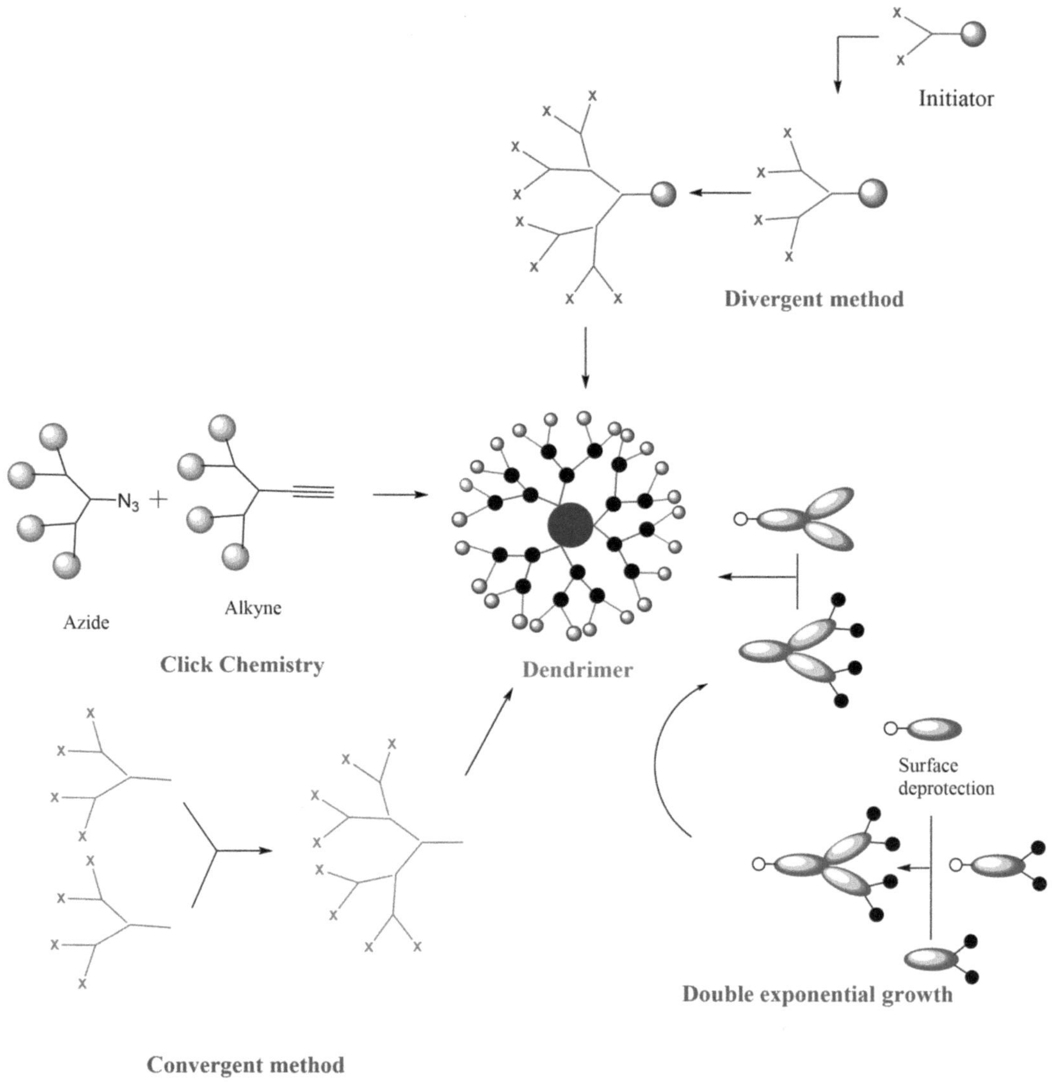


**FIGURE 3.2** Different synthetic approaches for dendrimer synthesis.

amount of byproducts, to be removable without any chromatography. This approach was introduced for pharmaceutical materials and industries which were capable of creating libraries of compounds for screening in research. Researchers synthesized dendrimers with different surface functionalities by Cu (I) mediated catalysis of 1, 2, 3-triazoles from azides and alkynes with purity and high yield (Wu *et al.*, 2004).

### 3.2.2.5 Hypercores and Branched Monomers Approach

This approach is based on the pre-assembly of oligomeric species and to accelerate the rate of dendrimer synthesis. This method has fewer steps of synthesis with higher yield. In this method of dendrimer synthesis, oligomeric species are conjugated together (Frechet and Tomalia, 2001).

### 3.2.2.6 Double Exponential Approach

This approach is a mixed approach and resembles a rapid-growth approach for linear polymers. It allows the construction of monomers for both divergent and convergent growth from a single starting material. An orthogonally protected trimer can be achieved by reacting two products; repetition of the process can be used to grow again for the higher generation of the dendrimer (Kesharwani *et al.*, 2014).

### 3.2.3 Properties of Dendrimers

Dendritic macromolecules (in particular, dendrimers) present outstanding properties, which determine their distinctive features, functionalities and biomedical applications as a marvelous nanocarrier. As mentioned earlier, dendrimers possess several unrivaled properties apart from those features of dendrimers such as structural perfection, compactness, nanometric size and a highly multivalent surface, which determines the chemical properties of this macromolecule (e.g., solubility, miscibility and reactivity). In solution form, they typically show lower viscosity and different rheological properties, unlike linear polymers. The presence of internal cavities, the capability of encapsulation and conjugation with the guest molecules and hybridization with linear polymers can lead to novel and remarkable properties, which make dendrimers very efficient and effective nanocarriers (Klajnert and Bryszewska, 2001).

## 3.3 DENDRIMER TOXICITY AND BIOSAFETY PROFILE

Beyond doubt, the dendrimers have caused remarkable developments in the field of drug delivery. Owing to its nanometric size, that is, 1–100 nm, these may interact efficiently, especially with the cell organelles, plasma membranes and proteins such as enzymes (Lee *et al.*, 2008; Cancino *et al.*, 2013). In contrast, the nanoparticles (NPs) are relatively nonselective. Irrespective of the wide pharmaceutical and biomedical applications of dendrimers, toxicity is the *Janus face* of the dendrimers, restricting their use in medical applications (Malik *et al.*, 2000). Dendrimers, including PAMAM, PPI and poly-L-lysine (PLL), exert significant *in vitro* cytotoxicity because of surface cationic groups (Agashe *et al.*, 2006). Past research on dendrimers reflects the concentration and generation-dependent toxicity due to surface amino functional groups (Chen *et al.*, 2004). This section will delineate the toxicity associated with the dendrimers during their drug-delivery applications.

### 3.3.1 Cytotoxicity

A plethora of literature is available supporting the enhanced efficacy of the bioactives when conjugated or encapsulated in the dendrimer. But the dendrimer itself has a cytotoxicity profile which limits the application in biological systems. According to Duncan and Izzo (2005), some general trends are clearly associated with dendrimer toxicity, as abounding literature reports are available for other polycations and dendrimers that surface $-NH_2$ terminus and usually generation-dependent cytotoxicity. Using the study conducted against Caco-2 cell lines, Jevprasesphant *et al.* (2003a) reported that PAMAM dendrimers are cytotoxic in nature. Similarly, Agashe *et al.* (2006) also investigated cytotoxicity of G5.0 PPI dendrimer and amino acid-decorated and carbohydrate-protected PPI dendrimers and concluded that the cell viability decreases as the concentration and incubation time increase. Later, Stasko *et al.* (2007) also compared the cytotoxicity and membrane disruption of polyethyleneglycol (PEG)-decorated or without decorated and peripheral neutral acetamide-conjugated PPI dendrimers against the cultured human umbilical vein endothelial cells (HUVEC) and found that the naive PPI dendrimer showed severe time-dependent cytotoxicity. Roberts *et al.* (1996) reported reduction of the cell viability as a consequence of the cytotoxicity of the cationic PAMAM dendrimer on the V79 Chinese hamster lung fibroblasts when the dendrimer was incubated with the cell lines for 24 hours. Later, similar results were experienced by Malik *et al.* (2000), who explored cytotoxicity of PAMAM, PLL and polyethyleneimine (PEI) dendrimers against different cells and compared with dextran. Duncan and Izzo (2005) counteracted the results of Kissel and colleagues (Fischer *et al.*, 1999; Fischer *et al.*, 2003) regarding their reported cytotoxicity of a series of polycations in L929 mouse fibroblasts and also readjustment back to physiological pH to minimize the effect of polymers in medium osmolarity and pH. Duncan and Izzo (2005) finally concluded that such manipulations are irrelevant to the *in vivo* situation, where it is not possible to eliminate pH and osmotic effects.

Chemistry of the core of the dendrimer plays a vital role in the dendrimer toxicity, though the nature of the surface influences it more. Chen *et al.* (2004) reported that surface modification of the dendrimer alters its cytotoxicity. Similarly, Lee *et al.* (2003) documented less cytotoxic effect of quaternized PAMAM–OH derivatives than PAMAM–$NH_2$, due to masking of the internal cationic charges by peripheral hydroxyl groups. There are several more reports which support that the surface modification of the dendrimer minimizes the cytotoxicity of the dendrimer. But it is not universally true that surface modification reduces the cytotoxicity. As evidence, Choi *et al.* (2004) reported G4.0 PAMAM dendrimer decorated with lysine or arginine led to reduction of toxicity than naive PAMAM. The authors also concluded that the core is more accessible in lower generations due to open structure, while in higher generations it becomes more strictly hindered. Hong *et al.* (2004) reported the mechanism of cationic PAMAM-induced membrane damage using dendrimers of different generations. They also concluded that G5.0 PAMAM dendrimer does not induce cytotoxicity within the size range size up to 500 nM.

Some cytotoxic reports of dendrimers have also been documented in the last decade. Mukherjee *et al.* (2010) documented generation-dependent cytotoxicity against mammalian cells in terms of reactive oxygen species (ROS), apoptosis and DNA damage. The authors suggested higher generations cause higher DNA damage. As the dose increased, the number of apoptotic cells decreased, while the number of necrotic cells increased. Cancino *et al.* (2013) evaluated the toxicity of single-walled carbon nanotubes (SWCNT), PAMAM dendrimers and PAMAM–SWCNT complexes in a mouse myoblast cell line (C2C12). The authors found that nanomaterial damaged the DNA and was significantly toxic towards C2C12 cells. It was concluded that the toxicity of nanomaterial was strongly correlated to their surface charge. Surface charge of dendrimers, therefore, plays an important role in *in vivo* exploration of dendrimers with regards to safety. Oral toxicity of higher generation PAMAM dendrimers was also documented by Thiagarajan *et al.* (2013b). They evaluated acute oral toxicity and physicochemical disposition of PAMAM G6.5 dendrimers via *in vivo* oral translocation in CD-1 mice. On the basis of the results, the authors decided that dendrimers have high potential to permeate the gut epithelial barrier. Later, Thiagarajan *et al.* (2013a) investigated the *in vivo* toxicity of PAMAM dendrimers in CD-1 mice, the effect of charge and size on the permeability via the epithelial barrier and acute toxicity on oral administration. Interestingly, Lazniewska *et al.* (2013) demonstrated cytotoxicity of cationic phosphorus dendrimers in murine neural cell lines. The authors reported that concentration above 1 μM led to decreased cell viability. The scientists investigated the cationic as well as anionic dendrimers for toxicity and found that the maximum tolerated dose was found to be 10-fold higher than for cationic dendrimers. Recent research has focused on the development of biocompatible dendrimers, which will further strengthen the era of dendrimer-mediated drug delivery.

### 3.3.2 Hemolytic Toxicity

Hemolysis is a simple method for polymer-membrane interaction investigation. It is widely used, as it tells about the quantity of hemoglobin (Hb) released from the RBC (red blood corpuscles) (Duncan *et al.*, 1991) and it also indicates the hemolytic toxicity. Cationic dendrimers lead to hemotoxicity due to free terminal groups on the surface which interact with erythrocytes (Bhadra *et al.*, 2003; Asthana *et al.*, 2005). Malik *et al.* (2000) reported that PPI and PAMAM dendrimers of 3.0 generations do not cause hemolysis below concentrations of 1 mg/mL. Asthana *et al.* (2005) also evaluated hemolytic toxicity of the dendrimers, and the results were found to be quite close to those reported by Bhadra *et al.* (2003) with hemolysis up to 18%. Meanwhile, Bhadra *et al.* (2005) also documented quantitative hemolysis of G4.0 and G5.0 PPI dendrimers to be 35.7% and 49.2%, respectively. Agashe *et al.* (2006) also reported the hemolytic toxicity of G5.0 PPI dendrimers in 1 mg/mL concentration as 34.2 ± 0.2%, 51.6 ± 0.3%

and 86.2 ± 0.6% following incubation of 1, 2 and 4 h, respectively. Agrawal *et al.* (2007) investigated hemolytic toxicity of PLL dendrimers *in vitro,* which was observed to depend on the generation of the dendrimer, with higher generation leading to higher toxicity. It must also be considered that Fischer *et al.* (2003) failed to find any significant hemolysis of G3.0 PAMAM, even up to a concentration of 10 mg/mL and incubation of 1 h, observing <7% hemolysis. After incubation with dendrimers, RBCs turned rounded in appearance and aggregated may ostensibly be due to cross-linking of dendrimers. A higher concentration of dendrimers (i.e., 1 mg/mL) accelerated this 'clumping' behavior and membrane damage was evidenced. Malik *et al.* (2000) also concluded that dendrimer administration causes damage to RBC membranes. Klajnert *et al.* (2010) demonstrated that though hemolysis by PAMAM dendrimers depends on generation as well as time of incubation, human serum albumin (HSA) decreases the Hb release due to a higher affinity of the dendrimer towards HSA.

### 3.3.3 Hematological Toxicity

With dendrimers known to cause hemolysis and affect hematological parameters, Agashe *et al.* (2006) observed the effect of G5.0 PPI dendrimers on hematological parameters. A noteworthy reduction in erythrocyte count, increase in leucocyte count, lowering of Hb content and mean corpuscular hemoglobin (MCH) value and an extensive difference between haematocrit (HCT) values of the control and PPI dendrimers were observed. Similarly, Bhadra *et al.* (2005), during *in vivo* studies of primaquine phosphate-loaded dendrimers in albino rats, demonstrated a severe decrease in RBC count, while the white blood corpuscle (WBC) count increased. Thereafter, Agrawal *et al.* (2007) also experienced the similar results, that is, WBCs count was significantly increased and RBCs count was decreased with G4.0 PLL dendrimers. Based on these documented reports, one can decipher the need for some parallel strategy to make these incredible nanocarriers as more biocompatible.

### 3.3.4 Immunogenicity

There are some reports which indicate that dendrimers show no or only weak immunogenicity (Roberts *et al.*, 1996; Rajananthanan *et al.*, 1999; Agashe *et al.*, 2006). Roberts *et al.* (1996) investigated the immunogenicity induced by PAMAM dendrimers and concluded that dendrimers do not show immunogenicity up to the dose range of 0.1–0.0001 μM. Rajananthanan *et al.* (1999) also studied the immune-potential properties of a novel G5.0 dendrimeric polymer-based formulation and found the formulation to be nontoxic to mice. Later, Agashe *et al.* (2006) also revealed that G5.0 PPI dendrimers do not produce any detectable immune response in Balb/C mice. This indicates that dendrimers do not provoke any immunogenicity *in vivo*, which is a positive sign to use the caliber of dendrimer in drug delivery. Similarly, Majoros *et al.* (2006) also suggested that the widely used PAMAM dendrimer exhibits minimal immunogenicity.

### 3.3.5 *In Vivo* Toxicity

As *in vivo* toxicity concerns are critical to mark the safety of any developed formulation or drug, the researchers have investigated systematic studies on *in vivo* toxicity of dendrimers. Precisely, Roberts *et al.* (1996) documented *in vivo* toxicity of dendrimers in Swiss–Webster mice and observed that only G7.0 PAMAM dendrimers showed latent biological complications. On the basis of the results, they concluded that dendrimers are safe and did not demonstrate any harmful properties that would avert their use in biological applications. Later, Malik *et al.* (2000) studied the *in vivo* toxicological profile of $^{125}$I-labelled PAMAM dendrimers and observed that the dendrimer was rapidly eliminated from the circulation, and only 0.1%–1.0% of the dose was recovered after 1 h of administration. Similarly, an exhaustive study on *in vivo* behavior of melamine dendrimers was demonstrated by Neerman *et al.* (2004). During acute toxicity studies, they observed 100% mortality in 6–12 h after administration of 2, 5, 10, 40 and 160 mg/kg doses of melamine dendrimer. At the 40 mg/kg dose, hepatic toxicity was observed, while no toxicity was observed at 2.5 and 10 mg/kg. Additional studies were reported. For instance, Bourne *et al.* (2000) documented that 10 mg/kg PAMAM concentrations (up to G5.0) does not lead to any toxic effect, and no change in the safety profile was observed even after modifications on the surface of the dendrimer. Furthermore, Rajananthanan *et al.* (1999) also did not observe any toxicity of unmodified amino-terminated PAMAM dendrimers, together with ovalbumin in mice, though these mixtures showed adjuvant activity. Later, De Jesus *et al.* (2002) and Ihre *et al.* (2002) investigated the toxicity profile of hydroxyl- or methoxy-terminated dendrimers based on a polyester scaffold that was nontoxic. On the basis of these reports, it can be delineated that the dendrimers need to be thoroughly investigated on their *in vivo* disposition behavior. Chauhan *et al.* (2009) investigated preclinical and behavioral changes of G4.0 PAMAM dendrimers with different surface functionality, that is, amine and hydroxyl, in Swiss albino mice and observed no significant changes in feeding or body and organ weight in various dendrimer-fed groups but observed decreased glucose levels and toxicity related to kidney and liver in high doses of $NH_2$ dendrimer. Similarly, Ziemba *et al.* (2011) demonstrated *in vivo* toxicity of poly (propyleneimine) dendrimer and observed behavioral changes, decreased food and water intake and lowered body weight gain when PPI G4.0 was administered additionally. When the surface of the dendrimer was decorated with maltotriose, the toxicity was amazingly diminished.

## 3.4 DENDRIMER-DRUG INTERACTIONS

Cationic charges, surface terminal functionalities and void spaces of dendrimer create many possibilities of the interaction with the biological active substances. Drugs or bioactives, therefore, can be chemically conjugated to the terminal groups of a dendrimer or encapsulated in the void cavities via weak interactions, such as hydrogen bonding, hydrophobic interaction or electrostatic interactions, between guest molecules and dendrimers (Hughes, 2005). As the dendrimer generation increases, the drug loading also increases correspondingly, with the higher generation having more space to encapsulate the drug through host–guest interaction. Many dendrimers such as PPIs, PAMAMs, PLLs, polypeptides, polyesters, polyethers and dendrimers based on PEG or carbohydrates have been investigated and reported for delivery of anticancer drugs (Shi *et al.*, 2015).

The aforementioned dendrimers have been frequently explored as drug carriers due to possession of various unique properties themselves. In addition, cationic charges of PPI, PAMAM and PLL dendrimers enable the delivery of nucleic acids. Host–guest interaction between drug–dendrimer plays an important role in drug release from the dendrimer (Howell and Fan, 2010). Different host–guest interaction mechanisms, for example, physical encapsulation, electrostatic interaction and covalent conjugation, are discussed in the next sections, while Figure 3.3 portrays different approaches for drug loading.

### 3.4.1 PHYSICAL ENCAPSULATION

Certain ideal properties of dendrimers with terminal functionalities, that is, $-NH_2$ groups and $-COOH$ groups in large numbers on the periphery of dendrimers, have latent use in solubility enhancement of poorly water-soluble drugs through electrostatic interactions. A dendrimer has a spherical shape and unfilled internal void space, thus making it suitable for direct entrapment of guest molecules into the macromolecule interior. Internal cavities of dendrimers create hydrophobic drag due to their own hydrophobic nature and interact with less water-soluble drugs through hydrophobic interactions (Liu and Frechet, 1999). Nitrogen or oxygen atoms present in the void cavities of dendrimers can forge hydrogen bonds within the interacting guest molecule (Kesharwani *et al.*, 2011). As it is very clearly defined, the dendrimers are charged macromolecules; hence, there is the possibility of electrostatic interaction with the oppositely charged guest molecules (Medina and El-Sayed, 2009). It is also reported that a hydrophilic drug can also be encapsulated in the

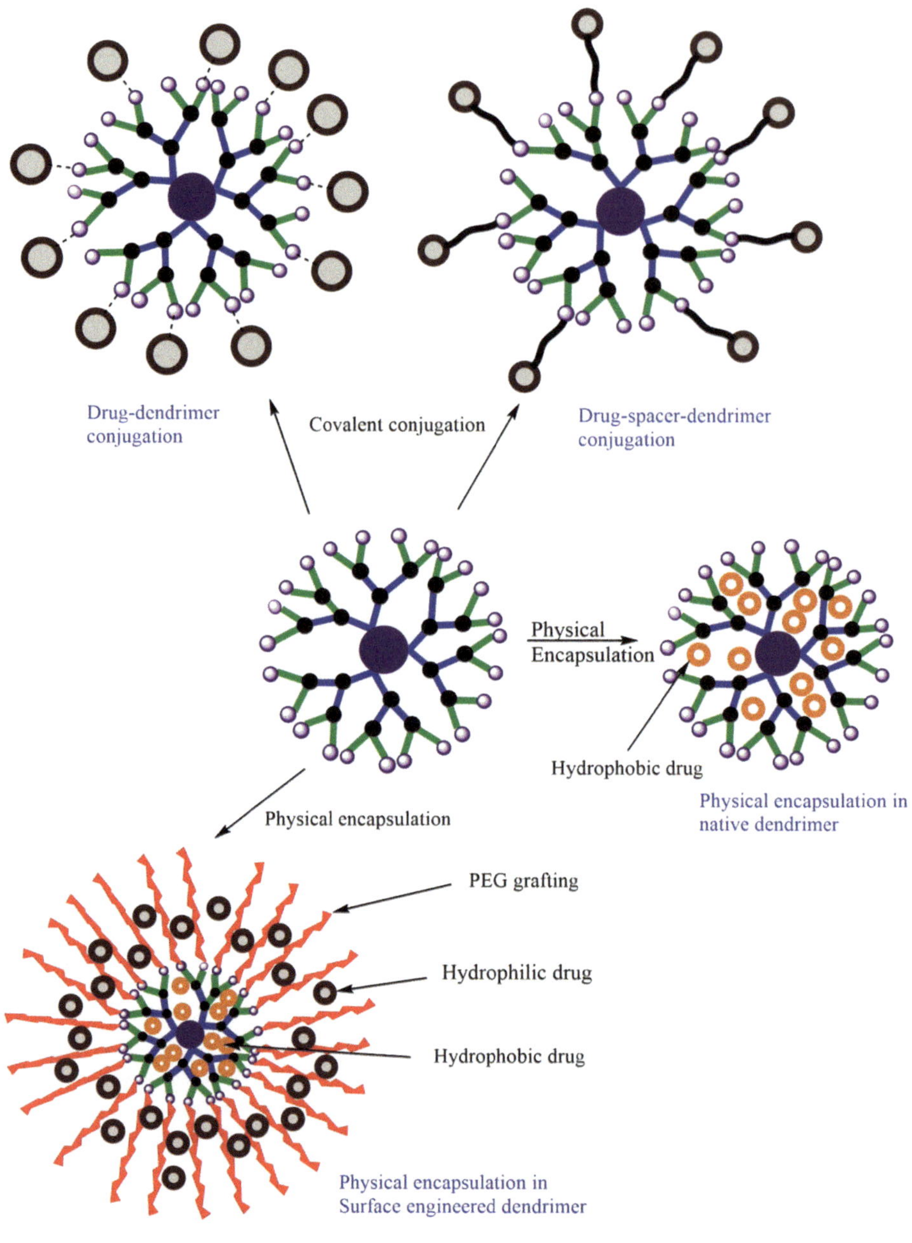


**FIGURE 3.3** Different approaches for drug loading in the dendrimer.

dendrimer. There are some advantages of physical encapsulation over chemical conjugation, that is, rapid preparation and inertness with respect to drug pharmacological activity. Like every coin has two faces, it also has disadvantages too, such as truncated stability in terms of payload, untimely release, low-drug loading capacity and batch-to-batch variation in amount of solubilized drug (Jain and Jain, 2015). An abundance of literature is available related to drugs with –COOH functional groups complexed with dendrimers, the driving force being electrostatic interactions. Like an aforementioned drug, various ionizable drugs also form complexes with the multifunctional surface groups. On the surface of PAMAM dendrimers, both the primary amine and tertiary amine groups are there within the architecture, having different pKa values of 10.7 and 6.5, respectively. Hence, dendrimers acquire ionizable groups in the architecture, which offer latent sites for the interaction with drugs. Several NSAIDs have been shown to form stable complexes with the dendrimers through electrostatic interactions, including piroxicam, ibuprofen and indomethacin (Madaan *et al.*, 2014).

### 3.4.2 Covalent Interactions

A large number of the terminal functionalities on the surface of dendrimers provide them the possibility of covalent conjugation with several drugs having suitable functional groups of the drug molecules (Patri *et al.*, 2005). Release from the covalently attached drug with the dendrimer takes place through chemical or enzymatic cleavage of labile bonds. For instance, if a drug is attached with an ester bond, then the release will depend on an esterase enzyme (Scutaru *et al.*, 2010). Interestingly, the drugs can be chemically conjugated to dendrimers via some spacers. This prodrug approach stabilizes the drug much better than the naïve drugs, with the drug release kinetics also getting improved. Many researchers have successfully conjugated drugs with dendrimers, including penicillin V, venlafaxine, 5-aminosalicylic acid, naproxen, propranolol and several anticancer drugs, and these conjugates have significantly improved solubility and drug release kinetics *vis-à-vis* the plain drug, for example, an epirubicin-PRG dendrimer conjugate demonstrated enhanced blood residual time, improved efficacy and higher stability against chemical degradation. The authors suggested that this can be an effective approach in the development of prodrugs of large molecules (Madaan *et al.*, 2014). Drawbacks of physical drug encapsulation can be counteracted by covalent conjugation of a drug with the dendrimer, as it can surmount untimely drug release from the conjugate. As drug release from the conjugate depends upon the environment, it can be changed by altering different aspects of the environment such as pH, temperature or concentration of a specific enzyme (Shukla *et al.*, 2008).

## 3.5 DENDRIMERS AS SOLUBILITY ENHANCERS

Low aqueous solubility of drug molecules limits their therapeutic applications in biological systems, and it is the foremost constriction in the development of a safer and more effective and stable drug formulation. Even after employing several novel methods, solubility problems remain unresolved per se. Dendrimers have proved themselves as promising candidates for solubilization of diverse drugs, including anticancer, antimalarials, antivirals, anti-tuberculars, antimicrobials, NSAIDs, steroids, anti-hypertensives, dexamethasone and bioactive flavonoids (Gupta *et al.*, 2007; Avila-salas *et al.*, 2017; Diaz *et al.*, 2017; Pentek *et al.*, 2017). PAMAM dendrimers are unimolecular micelles due to their hydrophilic exteriors and interiors. Dendrimer-based drug delivery enhances the aqueous solubility of drugs and bioactives, thereby improving the overall bioavailability of drugs (Tomalia *et al.*, 1986).

### 3.5.1 Factors Affecting Solubility

Solubility of pharmacologically active substances can be affected by a myriad of factors like generation size, three-dimensional polymeric architecture, pH, internal branching units, surface functionality and temperature. In this section, we will focus on the major factors affecting dendrimer-mediated solubilization.

#### 3.5.1.1 Effect of Generation Size

A wealth of literature is available on the effect of generation of dendrimers on solubilization, and it can be concluded from most reports that solubility of poorly soluble bioactives get enhanced as the generation rises. There is propensity to select dendrimers as a solubilizer enhancer up to G3.0 due to lower immunogenic and biocompatible characteristics than higher generations (Gupta *et al.*, 2006).

Devarakonda *et al.* (2005a,b) demonstrated enhanced nifedipine solubility with PAMAM dendrimers. The scientists demonstrated solubilization of the hydrophobic drugs with both amine- and ester-terminated dendrimers and found aqueous solubility was enhanced several-fold with higher generations of both types of dendrimers. They proposed that the possible reason of enhanced solubility was hydrogen bonding between guest and host molecules, and the hydrophobic environment of dendritic cavities may also have contributed in solubility enhancement. Similarly, Chauhan *et al.* (2003) also demonstrated solubility enhancement of indomethacin with G4-$NH_2$, G4-OH and G4.5 ester-terminated PAMAM dendrimers and found that maximum solubility could be achieved by the G4.0 $NH_2$-terminated dendrimer, followed by OH-terminated and the minimum solubility enhancement was achieved by the G4.5 ester-terminated dendrimer. Later, Namazi and Adeli (2005) synthesized PEG dendrimers as a solubilizer enhancer for different hydrophobes. They concluded that the higher the dendrimer generations, the higher is the solubility of hydrophobes accomplished, while the size of guest molecules also influences the solubility. Solubility of larger molecules was less, while smaller molecules exhibited higher solubility. Kaanumalle *et al.* (2004) reported that pyrene solubility in water was increased several fold with poly(alkyl aryl ether) dendrimers. Pistolis and Malliaris (2002) documented PPI dendrimers for pyrene solubilization. They concluded that, as the generation of the dendrimer increased, solubility of hydrophobes also linearly increased, suggesting the possible mechanism for solubility enhancement through exciplex formation. Yiyun *et al.* (2005) successfully enhanced the solubility of the different NSAIDs using PAMAM dendrimers and also studied the effect of different generations (2, 3 and 4) and concentrations of the dendrimer. The scientists observed that the solubilization was proportional to the dendrimer concentration, and higher generations tended to solubilize relatively more number of drugs. Several more reports support the dendrimer-mediated solubilization of hydrophobes, including NSAIDs, anticancer drugs and beta-blockers (Milhem *et al.*, 2000; Asthana *et al.*, 2005; Devarakonda *et al.*, 2005a,b). The literature reveals that PAMAM dendrimers are the most investigated dendrimers in solubility enhancement. Recently, Gurbuz *et al.* (2016) synthesized new-generation PAMAM dendrimers and reported a several-fold increase in solubility of poorly soluble sulfamethoxazole (SMZ). Similarly, Ertürk *et al.* (2016) reported several-fold enhanced aqueous solubility of candesartan cilexetil (CC) with dendrimers. All of the aforementioned reports support the hypothesis of increased solubility with increase in the dendrimer generation number. As one moves towards lower generation numbers, dendrimer structure becomes more open and globular. Generation number may attribute the conformational changes in the solubility enhancement.

#### 3.5.1.2 Effect of Polymeric Architecture

Polymeric construction of the dendrimers affects solubilization and controlled release of a drug. Ooya *et al.* (2003) studied the effect of polymeric architecture on solubilization and basically concluded that density of ethylene glycol chains of synthesized polyglycerol dendrimers affects the solubility of paclitaxel.

#### 3.5.1.3 Effect of pH

A major mechanism involved in dendrimer-mediated solubilization was an electrostatic interaction between the hydrophobic guest and amines of the dendrimers. Protonation and deprotonation states of amines of dendrimers also play a vital role in the dendrimer-mediated solubility. Generally, these protonation states are influenced by pH. Devarakonda *et al.* (2004) reported that nifedipine solubility depends upon the protonation state of the amine-terminated dendrimer concentration. Similarly, Asthana *et al.* (2005) reported that solubility of the weakly acidic drug flurbiprofen depends on the pH milieu. Milhem *et al.* (2000) also studied the effect of pH on the solubility enhancement of weakly acidic ibuprofen with G4.0 PAMAM and concluded that that ibuprofen gets ionized at highly basic pH, leading ultimately to electrostatic interactions with terminal amine groups of dendrimers, while this interaction could not be possible at pH 2 where ibuprofen remains unionized. Later, Yiyun *et al.* (2005) reported that solubility of nicotinic acid and ketoprofen with PAMAM dendrimers depends on pH. The authors suggested that the electrostatic interaction between dendrimers and the guest molecules, as well as hydrogen bonding, were the possible mechanisms for the increased solubility. Devarakonda *et al.* (2005a,b) demonstrated solubility enhancement of niclosamide using

PAMAM dendrimers and cyclodextrins (HPβ-CD, β-CD).The authors concluded that the electrostatic interaction between primary amines of dendrimers and the niclosamide molecule was the driving force underlying the solubility enhancement.

#### 3.5.1.4 Effect of Temperature

There is not much literature available on the effect of temperature on drug solubility, though temperature plays a crucial role in the same. Milhem *et al.* (2000) have studied the effect of temperature on solubility of ibuprofen using G4.0 PAMAM dendrimers and performed the solubility enhancement studies of ibuprofen at different temperatures and observed that the solubility enhancement was not favored by increments in temperature, though no explanation was cited for the unusual trend.

## 3.6 SURFACE ENGINEERING

Dendrimers are documented to exhibit certain toxicity profiles, for example, hematological, wherein the cationic charge is considered mainly responsible for toxicity due to primary and tertiary amines in the dendrimer molecule. Masking cationic charge, however, can reverse the situation. This can be quite easy due to the accessibility of plenty of tailor-made functionalities on the surface of dendrimers, offering an opportunity for conjugation to any ligand, for example, PEG and folate. All such alterations can affect the toxicity issues of dendrimers. Twyman *et al.* (1999) tailored ester-terminated PAMAM dendrimers to hydroxyl-terminated dendrimers, and they observed there were two significant changes in the dendrimer properties, the first being the reduced inherent haemolytic toxicity and the second being the dramatic increase in the aqueous solubility of the dendrimer due to formation of three hydroxyl end-groups through one ester terminus. The authors demonstrated that these terminals influence the solubility of an acidic molecule. Hence, it can be said that the terminal modification in the dendrimer can overrule the toxicity and increase the solubility. There are an abundance of reports available on terminal modification, with some of these discussed in this section.

### 3.6.1 PEGylation

The literal meaning of PEGylation (Figure 3.4) is conjugation of PEG to the surface of nanocarriers like dendrimers. PEGylation of dendrimers has led to drastic changes in properties such as transfection

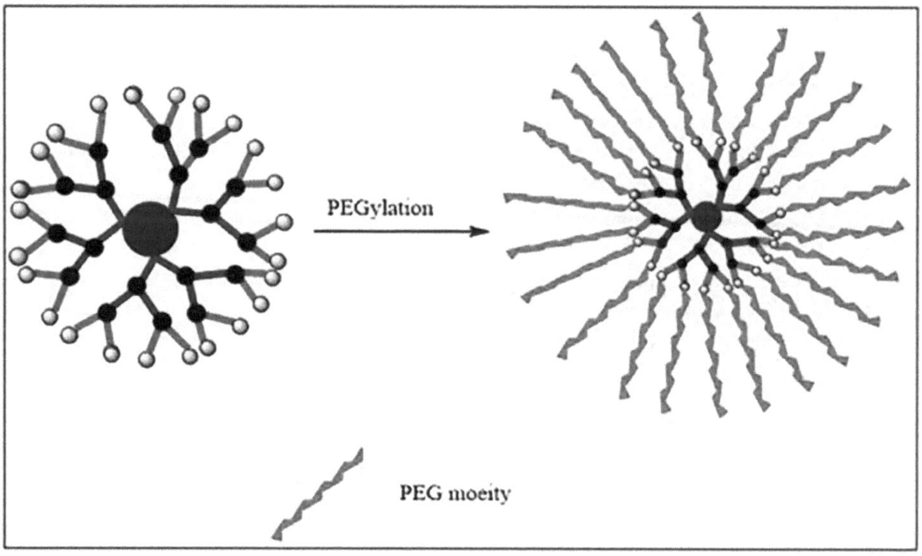


**FIGURE 3.4** Surface modification of a dendrimer using PEGylation.

capacity and solubilization efficacy of the dendrimers. For instance, Bhadra *et al.* (2003) synthesized PEGylated G4 PAMAM dendrimers and used them as solubilizers for 5-fluorouracil. Similarly, Kojima *et al.* (2000) studied the effect of an arm length of PEG on dendrimer-mediated solubility. PAMAM dendrimers were grafted with PEG monomethyl ether (MPEG), followed by loading of two anticancer drugs, methotrexate (MTX) and adriamycin (ADR). They concluded that ADR was complexed on the surface of MPEG chains, while MTX molecules show electrostatic interaction with dendrimers. Interestingly, Yang *et al.* (2004) grafted PEGs on the surface of G3 PAMAM and observed that arm-length PEG affects the aqueous solubility of pyrene. A longer PEG arm produces a dense network of PEG chains, leading to significant solubility enhancement. Meanwhile, as per Sideratou *et al.* (2001), use of PEGylated DAB-64-poly(propyleneimine) dendrimer successfully resulted in solubility enhancement of pyrene, through the plausible mechanism of pyrene entrapment in the dendrimer and in PEG chains. Recently, Shen *et al.* (2014) explored G5 and G7 PEGylated PAMAM dendrimers for gene delivery in C2C12 and HepG2 cells and found that the syndecan-4 was downregulated in both the cell lines, while caveolin-1 was upregulated intentionally. The authors concluded that the transfection efficiency was increased and suggested the strategy can be used to develop a better gene delivery system in transgenic therapy. During the same time period, Sun *et al.* (2014) reported that toxicity of PAMAM dendrimers can be reduced using PEGylation. The authors reported that cell viability was quite higher in PAMAM-PEG than native dendrimers in 239 T cells, and the transfection efficiency was also increased when a mixture of native PAMAM and PAMAM–PEG was used. Later, Liu *et al.* (2015) studied the biocompatibility of dual-functionalized (by PEG conjugation or thiolation) PAMAM G4.0 dendrimer. The authors found that thiolation of dendrimer alone did not reduce haemotoxicity or functions, whilst PEGylation of the dendrimer reduced the toxicity and interactions of the unmodified and thiolated dendrimer on various blood components and functions. Some recent reports have also been documented evidence that toxicity can be reduced by surface modification of the dendrimer. Nie *et al.* (2016) synthesized a nanoconjugate of PEG-functionalized hyperbranched PAMAM dendrimer, followed by conjugation with doxorubicin and acronyms it as PPCD. The authors used the nanoconjugate in effective delivery of doxorubicin in stomach cancer. On the basis of the results, the authors suggested that the synthesized nanosystem can be a better option in gastric chemotherapy.

### 3.6.2 Amphiphilic Dendrimers

The amphiphilic dendrimers were first introduced by Hawker *et al.* in 1993. Much later, Vutukuri *et al.* (2004) synthesized 'amphiphilic dendrimers'. The synthesized dendrimers were able to accommodate both hydrophobic and hydrophilic compounds. Recently, Chen *et al.* (2016) documented the conformation of homodendrimers and amphiphilic dendrimers in various solvents using classical density functional theory (DFT). The synthesized unimolecular dendrimer could be used as a unique candidate for encapsulation applications. Quite interestingly, Liu *et al.* (2015) modified PAMAM dendrimers decorated with arginine terminals, and the dendrimer proved its potential in delivering siRNAs in human prostate cancer PC-3 cells. Wei *et al.* (2015) established an amphiphilic dendrimer (AmDM) loaded with doxorubicin (DOX) with higher drug payload. The resulting AmDM/DOX nanomicelles were more potent to combat DOX resistance in breast cancer models along with significantly enhanced cellular uptake and lower efflux of the drug.

## 3.7 TARGETING AND THERAPEUTIC APPROACHES OF DENDRIMERS

### 3.7.1 Dendrimer-Based Targeting

Targeted delivery of bioactives is much needed in the treatment of lethal diseases like cancer and neurodegenerative ailments. As the anticancer agents are equally cytotoxic to the normal cells apart from the cancerous cells, the carrier which can deliver the agents to the targeted site are sought. As a part of targeted drug delivery, nanocarriers still face an acute challenge regarding their commercial exploitation in treatment, as there are not many FDA-approved nanocarriers for targeted delivery available to date (Perez-Herrero and Fernandez-Medarde, 2015).

An exhaustive literature survey unveils that a dendrimer can deliver the drug to the targeted site due to its nanometric size. The mechanism of the vascular permeability of the tumor was thoroughly studied by Matsumura and Maeda (1986), as the phenomenon of a leaky vasculature of tumor vessels' enhanced permeability and retention (EPR) (Matsumura and Maeda, 1986). The EPR effect is verily a phenomenon of solid tumors due to the exceptional anatomical and pathophysiological characteristics. Solid tumors show higher vascular density than normal tissues and organs; electron microscopy of tumor blood vessels showed that they are structurally different from normal blood vasculature (Iyer *et al.*, 2006; Greish, 2007). Tumor nodules show angiogenesis even at the small size of 0.2 mm (Yin *et al.*, 2014). This EPR phenomenon plays an important role in the targeted delivery of drugs to the site, subdivided as follows:

#### 3.7.1.1 Active Targeting

There are preclinical promising *in vitro* and *in vivo* reports available with active targeting using dendrimers as carriers (Hong *et al.*, 2007). Even antibodies were delivered using dendrimers as a carrier with improved efficacy over free antibodies (Wu *et al.*, 2004), peptide (RGD) dendrimer conjugates are also reported with enhanced tumor targeting (Zhang *et al.*, 2011), and folic acid-mediated tumor targeting with dendrimers showed better uptake than free drug, resulting in stronger tumor size reduction (Hong *et al.*, 2007). Despite the fact that dendrimers have excellent potency to deliver the bioactives to the targeted site, these have not been able to move forward towards clinical trials, ostensibly due to their associated toxicity.

#### 3.7.1.2 Passive Targeting

Many preclinical reports are documented as the evidence of the capability of the dendrimer to deliver the drug to the targeted site passively through the EPR effect (Kojima *et al.*, 2000). The EPR effect enables dendrimers to assemble and be retained by a tumor. Generally, the passive targeting approach is based on the EPR effect, whereas a ligand-anchored based is active targeting. Malik *et al.* (1999) produced an early example of *in vivo* passive targeting by the conjugation of cisplatin with sodium carboxyl terminated G3.5 PAMAM dendrimer.

### 3.7.2 Dendrimer-Based Drug Delivery

The size of dendrimers, coupled with three-dimensional structure and terminal functionalities make unique identification of dendrimers in the drug delivery era. Since the last few decades, dendrimers have emerged as carriers of choice for delivery of various bioactives (Luna *et al.*, 2014). Surface engineering of dendrimers is possible through conjugation of several biologically active substances, ligands or antibodies. Among various types of dendrimers, PAMAMs have been extensively examined for drug delivery, due to their favorable water solubility and compact architecture. Applications of dendrimers in drug delivery via a different route have been summarized in Table 3.1.

#### 3.7.2.1 Oral Drug Delivery

The oral route of administration of pharmacologically active substances is the preferred route because of its convenience to patients, cost-effectiveness of administration, dose flexibility and facilitation of sophisticated treatment regimens. Beyond popularity, there are some limitations with oral/peroral route, including low aqueous solubility and inadequate permeability across biological membranes (Kesharwani *et al.*, 2014). A few past researches related to dendrimers reveal promising results for its oral delivery implication in various types of diseases. Both the types of dendrimers, that is, plain and surface-engineered, have been found to increase the transepithelial permeability. For proving its oral efficiency, Jevprasesphant *et al.* (2003b) studied the permeation of both PAMAMs

**TABLE 3.1**
**Literature Instances on Diverse Application of Dendrimers by Different Routes**

| Route of Administration | Dendrimer Type(s) | Limitation(s) | Drug(s)/Bioactive(s) | Advantage(s) | Reference(s) |
|---|---|---|---|---|---|
| **Topical** | PAMAM | Poor water solubility & stability issues | Resveratrol | Improved solubility & stability of resveratrol in water | Pentek *et al.* (2017) |
| | PAMAM | Poor aqueous solubility | Sulfamethoxazole | Enhanced aqueous solubility | Gurbuz *et al.* (2016) |
| | Porphyrin glycodendrimer | Low targeting & reduced toxicity | Concanavalin A | Porphyrin glycodendrimers | Makky *et al.* (2011) |
| **Brain through intranasal route** | PAMAM | Limited aqueous solubility | Haloperidol | Improved delivery through intranasal route | Katare *et al.* (2015) |
| **Ocular** | PAMAM | Side effects | Dexamethasone | Higher penetration & cell viability | Yavuz *et al.* (2015) |
| | PAMAM-coated iron oxide NPs | Poor aqueous solubility | Doxorubicin & polyinosinic: polycytidylic acid | Improved efficacy | Khodadust *et al.* (2014) |
| | Surface-engineered PAMAM | Poor retention & patient inconvenience | Pilocarpine | Increased retention of bioactive in eye | Tolia and Choi (2008), Vandamme and Brobeck (2005) |
| | PAMAM | Dysphagia, bleeding & renal side effects | Ketoprofen, diflunisal | Improved transpermeation through skin | Cheng *et al.* (2007) |
| **Oral** | PAMAM | Poor absorption | Camptothecin | Improved absorption | Sadekar *et al.* (2013) |
| | PAMAM | Poor solubility & bioavailability | Puerarin | Enhanced solubility & bioavailability | Gu *et al.* (2013) |
| | PAMAM | Low aqueous solubility, bioavailability, low protein affinity | Risperidone | Antipsychotic agent | Prieto *et al.* (2011) |
| | PEGylated PAMAM | Low bioavailability | Silybin | Hepato-protective agent | Huang *et al.* (2011) |

*(Continued)*

**TABLE 3.1** (*Continued*)
**Literature Instances on Diverse Application of Dendrimers by Different Routes**

| Route of Administration | Dendrimer Type(s) | Limitation(s) | Drug(s)/Bioactive(s) | Advantage(s) | Reference(s) |
|---|---|---|---|---|---|
| **Pulmonary** | Cationic PAMAM | Extensive side effects on lungs or at mucocillary transport rate | Enoxaparin & low mol. weight heparin | Increased rate & extent of absorption | Bai *et al.* (2007) |
| | Surface-functionalized PAMAM | Poor permeability | Indomethacin | Higher blood concentrations in wistar rat | Chauhan *et al.* (2003) |
| **Transdermal** | PAMAM | Immunotoxiciy | Ketoprofen & Diflunisal | Anti-inflammatory | Cheng *et al.* (2006) |
| | PAMAM | Side effects | Tamsulosin HCl | Transdermal mechanism | Wang *et al.* (2003) |
| **Intravenous** | PAMAM | Limited aqueous solubility | Rifampicin | Anti-tuberculosis | Bellini *et al.* (2015) |
| | PAMAM | Limited aqueous solubility | Amphotericin | Antifungal agent | Jose and Charyulu (2015) |
| | EGF conjugated PAMAM | Nausea | Boron | BNCT | Yang *et al.* (2015) |
| | Antibody conjugated PAMAM | Gastric discomfort | Boron | BNCT | Wu *et al.* (2004) |
| **Intraperitoneal** | PAMAM | Limited aqueous solubility | Haloperidol | Antipsychotic agent | Katare *et al.* (2015) |
| | PAMAM | Increased intestinal discomfort | Glucosamine | Glcosamine Biological modulators | Vannucci *et al.* (2003) |

and surface-engineered PAMAMs using Caco-2 cell monolayers by measuring [14C] and TEER in both the directions. The authors concluded that plain and surface-modifieddendrimers with lauryl groups could proficiently traverse epithelial monolayers via paracellular and transcellular pathways. Najlah *et al.* (2007) studied the oral delivery from PAMAM dendrimers of naproxen prodrug, low aqueous solubility and poor transepithelial permeability of which are known to hamper its oral bioavailability. The authors also reported high transepithelial permeability of naproxen–dendrimer conjugates and their stability in 50% liver homogenates and in 80% human plasma.

#### 3.7.2.2 Transdermal Drug Delivery

There are a few fundamental merits underlying transdermal utility of dendrimers, such as the presence of hydrophobic groups, leading to poor aqueous solubility, ultimately restricting the entry in biological compartment and resulting in increased solubility and biocompatibility. A few researches have favored the potential utility of dendrimers in transdermal delivery of drugs with the added advantages of improved pharmacokinetic profiles (Nanjwade *et al.*, 2009). Various drugs like NSAIDs have been investigated for the transdermal efficacy of dendrimers. In one of the study, Chauhan *et al.* (2003) investigated the transdermal efficacy of indomethacin using G4.0 PAMAM, G4.5 PAMAM dendrimers. Researchers have also studied the transdermal potentiality of dendrimers with different kinds of drugs like ketoprofen and disulfinal and found significantly better results in terms of transdermal permeabilities (Cheng *et al.*, 2007).

#### 3.7.2.3 Ocular Drug Delivery

Treatment of ocular disorders raises many questions regarding types of formulation and significant residence time in the cul-de-sac of the eye, which majorly determines the bioavailability of an ocular drug. Thus, ocular delivery of drugs and bioactives necessitates the use of a novel formulation which can overcome the said hurdles. Additionally, ocular formulation should be nonirritating, biodegradable, isotonic and biocompatible, with favorable retention time in the eye (Tolia and Choi 2008; Nanjwade *et al.*, 2009).

#### 3.7.2.4 Pulmonary Drug Delivery

Dendrimers have been recognized as novel nanocarriers for pulmonary drug delivery of pharmacologically active compounds like heparin. For this study, a number of generations like G2.0, G2.5 and G3.0 of PAMAM dendrimers were assessed for their absorption at pulmonary site of enoxaparin by determining the anti-factor Xa activity and analyzing the deep vein thrombosis preclusion efficacy in a rodent model. These studies revealed that the cationic dendrimers were found to increase the relative bioavailability (40%) of enoxaparin without any serious effect on lung tissue and mucociliary transport rate. It was concluded that positively charged dendrimers can serve as promising vehicles for pulmonary delivery of various bioactives (Bai *et al.*, 2007; Nanjwade *et al.*, 2009).

#### 3.7.2.5 Parenteral Drug Delivery

Several reports document the exploration of the parenteral route for dendrimer-mediated drug delivery. He *et al.* (2011) conjugated doxorubicin using dendrimer as a carrier for targeted delivery to the brain. The authors synthesized a nanoconjugate PAMAM-PEG-WGA-Tf, and assay of transport across the blood–brain barrier revealed that PAMAM-PEGWGA-Tf delivered the drug efficiently to the targeted site. This is the only report among a plethora of literature available on the parenterally administered nanoconjugates in which dendrimers have been used as a carrier for parenteral administration.

Table 3.2 describes the different types of commercialized dendrimers and their products.

#### 3.7.2.6 Miscellaneous

Due to exclusive properties like nanometric size range, monodispersity, permeability across the membrane and container properties, dendrimers can assist as suitable host candidates for the guest

**TABLE 3.2**
**Some Commercially Available Dendrimers and Their Products**

**Commercialized Dendrimers**

| Brand Name | Dendrimer Types | Manufacturer | Reference(s) |
|---|---|---|---|
| Priostar® | PEHAM/PEA | Starpharma | Tomalia and Rookmaker (2009) |
| Polylysine | Poly-L-lysine | | Tolia and Choi (2008) |
| Astramol® | PPI | | Patton *et al.* (2006) |
| Starburst® | PAMAM | | Tomalia and Frechet (2002) |

**Dendrimer-Based Products**

| Brand Name | Dendrimer Types | Manufacturer | Application | Reference(s) |
|---|---|---|---|---|
| Superfect® | PAMAM | Qiagen | Transfection agent | Liu *et al.* (2012) |
| Priofect™ | PAMAM | Starpharma | Transfection agent | Liu *et al.* (2012) |
| Startus CS® | PAMAM | Dade Behring | Cardiac Assay Diagnostic | Mintzer and Grinstaff (2011) |
| Texotere | Not defined | Sanofi-Aventis | Anti-Cancer | Peeyush Kumar *et al.* (2010) |
| Vivagel® | Poly-L-lysine | Starpharma | In HIV & STDs | Patton *et al.* (2006) |
| Alert ticket™ | PAMAM | U.S. Army Lab | Anthrax-detecting agent | Spangler (2006) |

molecules, either in the interior cavities or on the periphery of the dendrimers. These amino-terminated dendrimers increase the transfection efficiency into the nucleus by the endocytosis process (Ferruti *et al.*, 2002). Dendrimers also have been found as promising candidates for diagnostic application due to their unique morphology and distinct characteristics (Wolinsky and Grinstaff, 2008). The role of dendrimers has also been investigated as an imaging agent, as X-ray and MRI contrast agents in radiotherapy and as molecular probes (Jang *et al.*, 2009).

## 3.8 CONCLUSION AND FUTURE PROSPECTS

Selection of nanocarriers for drug delivery is an important factor in the last few years. Among all the nanocarriers, dendrimers emerge as a key tool for drug delivery and discovery owing to their several unique properties such as highly branched three-dimensional architecture, tailor-made functionalities, uniformity of structure, multivalency and compatibility with different bioactives. Surface engineering of dendrimers facilitates its utility in targeted drug delivery. Apart from drug delivery, dendrimers have also been employed as solubility enhancers, gene transfecting agents, diagnostic aids and for other biomedical applications. Its controlled architecture and opportunities to work with this nanocarrier allures many more scientists from different arenas. Host–guest interaction of dendrimers allows the possible attachment of a foreign group on its surface and finally takes part in conjugation or encapsulation strategies. Both of the targeting approaches can use dendrimers, whether active targeting by attachment of suitable ligand on the surface of dendrimers or passive targeting through EPR effects. In a nutshell, dendrimers are versatile nanocarriers for a diversity of applications in the domains of drug delivery, gene delivery and biomedical science.

## REFERENCES

Agashe HB, Dutta TD, Garg M, Jain NK. Investigations on the toxicological profile of functionalized fifth-generation poly (propylene imine) dendrimer. *J Pharm Pharmacol.* 2006; 58: 1491–8.

Agrawal P, Gupta U, Jain NK. Glycoconjugated peptide dendrimers-based nanoparticulate system for the delivery of chloroquine phosphate. *Biomaterials.* 2007; 28: 3349–59.

Asthana A, Chauhan AS, Diwan P, Jain NK. (PAMAM) dendritic nanostructures for controlled site-specific delivery of acidic anti-inflammatory active ingredient. *AAPS Pharm Sci Tech.* 2005; 6: E536–42.

Avila-Salas F, Pereira A, Rojas M, Saavedra-Torres M, Montecinos R, Bonardd S *et al.* An experimental and theoretical comparative study of the entrapment and release of dexamethasone from micellar and vesicular aggregates of PAMAM-PCL dendrimers. *Eur Polym J.* 2017.

Bai S, Thomas C, Ahsan F. Dendrimers as a carrier for pulmonary delivery of enoxaparin, a low molecular weight heparin. *J Pharm Sci.* 2007; 96: 2090–6.

Bellini RG, Guimarães AP, Pacheco MA, Dias DM, Furtado VR, de Alencastro RB, Horta BA. Association of the anti-tuberculosis drug rifampicin with a PAMAM dendrimer. *J Mol Graph Model.* 2015; 60: 34–42.

Bhadra D, Bhadra S, Jain S, Jain NK. A PEGylated dendritic nanoparticulate carrier of fluorouracil. *Int J Pharm.* 2003; 257: 111–24.

Bhadra D, Yadav AK, Bhadra S, Jain NK. Glycodendrimeric nanoparticulate carriers of primaquine phosphate for liver targeting. *Int J Pharm.* 2005; 295: 221–33.

Boas U, Christensen JB, Heegaard PMH. Dendrimers: Design, synthesis and chemical properties. *J Mater Chem.* 2006; 16: 3785–98.

Boas U, Heegaard PM. Dendrimers in drug research. *Chem Soc Rev.* 2004; 33: 43–63.

Bourne N, Stanberry LR, Kern ER, Holan G, Matthews B, Bernstein DI. Dendrimers, a new class of candidate topical microbicides with activity against herpes simplex virus infection. *Antimicrob Agents Chemother.* 2000; 44: 2471–4.

Buczkowski A, Urbaniak P, Belica S, Sekowski S, Bryszewska M, Palecz B. Formation of complexes between PAMAM-NH2 G4 dendrimer and L-a-tryptophan and L-α-tyrosine in water. *Spectro Acta Part A: Mole Biomol Spect.* 2014; 128: 647–52.

Cancino J, Paino IMM, Micocci KC, Selistre-de-Araujo HS, Zucolotto V. *In vitro* nanotoxicity of single-walled carbon nanotube-dendrimer nanocomplexes against murine myoblast cells. *Toxicol Lett.* 2013; 219: 18–25.

Chauhan AS, Jain NK, Diwan PV. Preclinical and behavioural toxicity profile of PAMAM dendrimers in mice. *Proc R Soc A.* 2009; 1–16.

Chauhan AS, Sridevi S, Chalasani KB, Jain A, Jain SK, Jain NK. Dendrimer-mediated transdermal delivery: Enhanced bioavailability of indomethacin. *J Control Release.* 2003; 90: 335–43.

Chen C, Tang P, Qiu F, Shi AC. density functional study for homodendrimers and amphiphilic dendrimers. *J Phys Chem B.* 2016; 120: 5553–63.

Chen HT, Neerman MF, Parrish AR, Simanek E. Cytotoxicity, haemolysis, and acute *in vivo* toxicity of dendrimer based on melamine, candidate vehicles for drug delivery. *J Am Chem Soc.* 2004; 126: 10044–8.

Cheng Y, Man N, Xu T, Fu R, Wang X, Wang X *et al.* Transdermal delivery of nonsteroidal anti-inflammatory drugs mediated by polyamidoamine (PAMAM) dendrimers. *J Pharm Sci.* 2007; 96: 595–602.

Cheng YY, Man N, Xu TW. Transdermal delivery of nonsteroidal anti inflammatory drugs mediated by PAMAM dendrimers. *J Pharm Sci.* 2006; 96: 595–602.

Choi JS, Nam K, Park J, Kim JB, Lee JK, Park J *et al.* Enhanced transfection efficiency of PAMAM dendrimer by surface modification with l-arginine. *J Control Release.* 2004; 99: 445–56.

Congur G, Eksin E, Mese F, Erdem A. Succinamic acid functionalized PAMAM dendrimer modified pencil graphite electrodes for voltammetric and impedimetric DNA analysis. *Sens Actuators B.* 2014; 201: 59–64.

De Jesus OLP, Ihre HR, Gagne L, Fréchet JML, Szoka FC. Olyester dendritic system for drug delivery applications: *In vitro* and *in vivo* evaluations. *Bioconjug Chem.* 2002; 13: 453–61.

Devarakonda B, Hill RA, De Villiers MM. Effect of polyamidoamine (PAMAM) dendrimers on the *in vitro* release of water-insoluble nifedipine from aqueous gels. *Int J Pharm.* 2004; 284: 133–40.

Devarakonda B, Hill RA, Liebenberg W, Brits M, De Villiers MM. Comparison of aqueous solubilization of practically insoluble niclosamide by polyamidoamone (PAMAM) dendrimers and cyclodextrins. *Int J Pharm.* 2005a; 300: 193–209.

Devarakonda B, Li N, De Villiers MM. Effect of polyamidoamine (PAMAM) dendrimers on the *in vitro* release of water-insoluble nifedipine from aqueous gels. *Pharm Sci Tech.* 2005b; 6: E504–12.

Diaz C, Guzmán J, Jiménez VA, Alderete JB. Partially PEGylated PAMAM dendrimers as solubility enhancers of Silybin. *Pharm Dev Technol.* 2017; 19: 1–8.

Duncan R, Bhakoo M, Riley ML, Tuboku-Metzger A. Soluble polymeric drug carriers: Haematocompatibility. In: J.C. Gomez-Fernandez, D. Chapmanand, L. Packer, eds., *Progress in Membrane Biotechnology.* Basel: Birkhauser Verlag; 1991; pp. 121–9.

Duncan R, Izzo L. Dendrimer biocompatibility and toxicity. *Adv Drug Del Rev.* 2005; 57: 2215–37.

Dutta T, Aghase HB, Vijayarajkumar P, Joshi M, Jain NK. Dendrosome-based gene delivery. *J Exp Nanosci.* 2006; 1: 235–48.

Ertürk AS, Gurbuz MU, Tulu M. The effect of PANAM dendrimer concentration, generation size and surface functional group on the aqueous solubility of candesartan cilexetil. *Pharm Dev Technol.* 2016; 8: 1–11.

Esfand E, Tomalia DA. Poly(amidoamine) (PAMAM) dendrimers: From biomimicry to drug delivery and biomedical applications. *Drug Discov Today.* 2001; 6: 6427–36.

Ferruti P, Marchisio MA, Duncan R. Poly(amido-amine)s: Biomedical applications. *Macromol Rapid Commun.* 2002; 23: 332–55.

Fischer D, Bieber T, Li Y, Elsasser HP, Kissel T. A novel non-viral vector for DNA delivery based on low molecular weight, branched polyethylenimine: Effect of molecular weight on transfection and cytotoxicity. *Pharm Res.* 1999; 16: 1273–9.

Fischer D, Li YX, Ahlemeyer B, Krieglstein J, Kissel T. *In vitro* cytotoxicity testing of polycations: Influence of polymer structure on cell viability and hemolysis. *Biomaterials.* 2003; 24: 1121–31.

Frechet JMJ, Tomalia DA. *Dendrimers and Other Dendritic Polymers.* Chichester: Wiley Publishers; 2001; p. 688.

Greish K. Enhanced permeability and retention of macromolecular drugs in solid tumors: A royal gate for targeted anticancer nanomedicines. *J Drug Targeting.* 2007; 15: 457–64.

Gu L, Wu Z, Qi X, He H, Ma X, Chou X, Wen X, Zhang M, Jiao F. Polyamidomine dendrimers: An excellent drug carrier for improving the solubility and bioavailability of puerarin. *Pharm Dev Technol.* 2013; 18: 1051–7.

Gupta U, Agashe HB, Asthana A, Jain NK. Dendrimers: Novel polymeric nanoarchitectures for solubility enhancement. *Biomacromolecules.* 2006; 7: 649–58.

Gupta U, Agashe HB, Jain NK. Poly (propylene imine) dendrimer mediated solubility enhancement, effect of pH and functional groups of hydrophobes. *Int J Pharm Pharm Sci.* 2007; 10: 358–67.

Gupta V, Nayak S. Dendrimers: A review on synthetic approaches. *J App Pharm Sci.* 2015; 5: 117–22.

Gurbuz MU, Ertürk AS, Tulu M. Synthesis of surface modified TREN cored PAMAM dendrimers and their effects on the solubility of sulfamethoxazole (SMZ) as an analog antibiotic drug. *Pharm Dev Technol.* 2016; 8: 1–12.

Hawker CJ, Wooley KL, Frechet JMJ. Unimolecular micelles and globular amphiphiles: Dendritic macromolecules as novel recyclable solubilization agents. *J Chem Soc Perkin Trans.* 1993; 1: 1287–97.

He H, Li Y, Jia X-R, Du J, Ying X, Lu WL. PEGylated Poly(amidoamine) dendrimer-based dual-targeting carrier for treating brain tumors. *Biomaterials.* 2011; 32: 478–87.

Hong S, Leroueil PR, Majoros IJ, Orr BG, Baker Jr JR, Banaszak Holl MM. The binding avidity of a nanoparticle-based multivalent targeted drug delivery platform. *Chem Biol.* 2007; 14: 107–15.

Hong SP, Bielinska AU, Mecke A, Keszler B, Beals JL, Shi XY. Interaction of poly(amidoamine) dendrimers with supported lipid bilayers and cells: Hole formation and the relation to transport. *Bioconjug Chem.* 2004; 15: 774–82.

Howell BA, Fan D. Poly(amidoamine) dendrimer-supported organoplatinum antitumour agents. *Proc R Soc. A* 2010; 466: 1515–1526.

Huang X, Wu Z, Gao W, Chen Q, Yu B. Polyamidoamine dendrimers as potential drug carriers for enhanced aqueous solubility and oral bioavailability of silybin. *Drug Dev Ind Pharm.* 2011; 37(4): 419–27.

Hughes GA. Nanostructure-mediated drug delivery. *Nanomed Nanotechnol Biol Med.* 2005; 1: 22–30.

Ihre HR, Padilla De Jesús OL, Szoka FC, Fréchet JMJ. Polyester dendritic systems for drug delivery applications: Design, synthesis, and characterization. *Bioconjug Chem.* 2002; 13: 443–52.

Iyer AK, Khaled G, Fang J, Maeda H. Exploiting the enhanced permeability and retention effect for tumor targeting. *Drug Discov Today.* 2006; 11: 812–8.

Jain K, Jain NK. *Dendrimers as Nanobiopolymers in Cancer ChemoTherapy: Nanobiomedicine.* New Delhi: Studium Press; 2015; pp. 65–90.

Jang WD, Selim KMK, Lee CH, Kang IK. Bioinspired application of dendrimers: From biomimicry to biomedical applications. *Prog Polym Sci.* 2009; 34: 1–23.

Jevprasesphant R, Penny J, Attwood D, McKeown NB, D'Emanuele A. Engineering of dendrimer surface to enhance transepithelial transport and reduce cytotoxicity. *Pharma Res.* 2003a; 20: 1543–50.

Jevprasesphant R, Penny J, Jalal R, Attwood D, McKeown N, D'Emanuele A. The influence of surface modification on the cytotoxicity of PAMAM dendrimers. *Int J Pharm.* 2003b; 252: 263–6.

Jose J, Charyulu N. Solubility enhancement of an antifungal agent by association with dendrimers. *IJRPB.* 2015; 3: 2320–471.

Kaanumalle LS, Nithyanandhan J, Pattabiraman M, Jayaraman N, Ramamurthy VJ. Water-soluble dendrimers as photochemical reaction media: Chemical behavior of singlet and triplet radical pairs inside dendritic reaction cavities. *Am Chem Soc.* 2004; 126: 8999–9006.

Katare KY, Daya PR, Gray CS, Luckman RE, Bhandari J, Chauhan AS, Mishra RK. Brain targeting of a water insoluble antipsychotic drug haloperidol via the intranasal route using PANAM dendrimer. *Mol Pharm.* 2015; 12: 3380–8.

Kesharwani P, Iyer AK. Recent advances in dendrimer-based nanovectors for tumor-targeted drug and gene delivery. *Drug Discov Today*. 2015; 20: 536–47.

Kesharwani P, Jain K, Jain NK. Dendrimer as nanocarrier for drug delivery. *Prog Polym Sci*. 2014; 39: 268–307.

Kesharwani P, Tekade RK, Gajbhiye V, Jain K, Jain NK. Cancer targeting potential of some ligand-anchored poly(propylene imine) dendrimers: A comparison. *Nanomedicine*. 2011; 7: 295–304.

Khodadust R, Unsoy G, Gunduz U. Development of poly (I:C) modified doxorubicin loaded magnetic dendrimer nanoparticles for targeted combination therapy. *Biomed Pharmacother*. 2014; 68: 979–87.

Klajnert B, Bryszewska M. Dendrimers: Properties and applications. *Acta Biochimica Polonica*. 2001; 48: 199–208.

Klajnert B, Pikala S, Bryszewska M. Haemolytic activity of polyamidoamine dendrimers and the protective role of human serum albumin. *Proc R Soc A*. 2010; 466: 1527–34.

Koc FE, Senel M. Solubility Enhancement of non-steroidal anti-inflammatory drugs (NSAIDs) using polypolypropylene oxide core PAMAM dendrimers. *Int J Pharm*. 2013; 451: 18–22.

Kojima C, Kono K, Maruyama K, Takagishi T. Synthesis of polyamidoamine dendrimers having poly(ethylene glycol) grafts and their ability to encapsulate anticancer drugs. *Bioconjug Chem*. 2000; 11: 910–7.

Lazniewska J, Milowska K, Zablocka M, Mignani S, Caminade AM, Majoral JP, Bryszewska M, Gabryelak T. Mechanism of Cationic Phosphorus Dendrimer Toxicity against Murine Neural Cell Lines. *Mol Pharm*. 2013; 10: 3484–96.

Lee H, Lim YB, Choi JS, Lee Y, Kim TI, Kim HJ. Polyplexes assembled with internally quaternized PAMAM–OH dendrimer and plasmid DNA have a neutral surface and gene delivery potency. *Bioconjug Chem*. 2003; 14: 1214–21.

Lee SH, Choi SH, Kim SH, Park TG. Thermally sensitive cationic polymer nanocapsules for specific cytosolic delivery and efficient gene silencing of siRNA: Swelling induced physical disruption of endosome by cold shock. *J Control Release*. 2008; 125: 25–32.

Liu H, Wang H, Yang W, Cheng Y. Disulfide cross-linked low generation dendrimers with high gene transfection efficacy, low cytotoxicity, and low cost. *J Am Chem Soc*. 2012; 134: 17680–7.

Liu M, Frechet JM. Designing dendrimers for drug delivery. *Pharm Sci Technol Today*. 1999; 2: 393–401.

Liu Y, Pang Y, Toh MR, Chiu GN. Dual-functionalized poly(amidoamine) dendrimers with poly(ethylene glycol) conjugation and thiolation improved blood compatibility. *J Pharm Pharmacol*. 2015; 67: 1492–502.

Luna BN, Godinez LA, Rodriguez FJ, Rodriguez A, Zaldivar-Lelo de Larrea G, Sosa-Ferreyra CF *et al*. Applications of dendrimers in drug delivery agents, diagnosis, therapy, and detection. *J Nanomat*. 2014; 2014: 1–19.

Madaan K, Kumar S, Poonia N, Lather V, Pandita D. Dendrimers in drug delivery and targeting: Drug-dendrimer interactions and toxicity issues. *J Pharm Bioallied Sci*. 2014; 6: 139–50.

Majoros IJ, Myc A, Thomas T, Mehta CB, Baker JR. PAMAM dendrimer-based multifunctional conjugate for cancer therapy: Synthesis, characterization, and functionality. *Biomacromolecules*. 2006; 7: 572–9.

Makky A, Michel JP, Maillard P, Rosilio V. Biomimetic liposomes and planar supported bilayers for the assessment of glycodendrimeric porphyrins interaction with an immobilized lectin. *Biochim Biophys Acta*. 2011; 1808(3): 656–66.

Malik N, Evagorou EG, Duncan R. Dendrimer-platinate: A novel approach to cancer chemotherapy. *Anticancer Drugs*. 1999; 10: 767–76.

Malik N, Wiwattanapatapee R, Klopsch R, Lorenz K, Frey H, Weener JW *et al*. Dendrimers: Relationship between structure and biocompatibility *in vitro*, and preliminary studies on the biodistribution of I-125-labelled polyamidoamine dendrimers *in vivo*. *J Controlled Release*. 2000; 65: 133–48.

Matsumura Y, Maeda H. A new concept for macromolecular therapeutics in cancer chemotherapy: Mechanism of tumoritropic accumulation of proteins and the antitumor agent SMANCS. *Cancer Research*. 1986; 46: 6387–92.

Medina SH, El-Sayed ME. Dendrimers as carriers for delivery of chemotherapeutic agents. *Chem Rev*. 2009; 109: 3141–57.

Milhem O, Myles C, Mckeown N, Attwood D, Demanuele A. Polyamidoamine starburst dendrimers as solubility enhancers. *Int J Pharm*. 2000; 197: 239–41.

Mintzer MA, Grinstaff MW. Biomedical applications of dendrimers: A tutorial. *Chem Soc Rev*. 2011; 40: 173–90.

Morgan MT, Carnahan MA, Immoos CE, Ribeiro AA, Finkel-stein S, Lee SJ *et al*. Dendritic molecular capsules for hydrophobic compounds. *J Am Chem Soc*. 2003; 125: 15485–9.

Mukherjee SP, Lyng FM, Garcia A, Davoren M, Byrne HJ. Mechanistic studies of *in vitro* cytotoxicity of poly(amidoamine) dendrimers in mammalian cells. *Toxicol Appl Pharmacol*. 2010; 248: 259–68.

Najlah M, Freeman S, Attwood D, D'Emanuele A. *In vitro* evaluation of dendrimer prodrug for oral drug delivery. *Int J Pharm*. 2007; 336: 183–90.

Namazi H, Adeli M. Dendrimer of citric acid and poly(ethylene glycol) as the new drug-delivery agent. *Biomaterials*. 2005; 26: 1175–83.

Nanjwade BK, Bechra HM, Derkara GK, Manvi FV, Nanjwade VK. Dendrimers: Emerging polymers for drug-delivery systems. *Euro J Pharm Sci.* 2009; 38: 185–96.

Neerman MF, Zhang W, Parrish AR, Simanek EE. *In vitro* and *in vivo* evaluation of a melamine dendrimer as a vehicle for drug delivery. *Int J Pharm.* 2004; 281: 129–32.

Nie J, Wang Y, Wang W. *In vitro* and *in vivo* evaluation of stimuli-responsive vesicle from PEGylated hyperbranched PAMAM-doxorubicin conjugate for gastric cancer therapy. *Int J Pharm.* 2016; 509: 168–77.

Ooya T, Lee J, Park KJ. Effects of ethylene glycol-based graft, star-shaped, and dendritic polymers on solubilization and controlled release of paclitaxel. *J Controlled Rel.* 2003; 93: 121–7.

Padilla De Jesús O, Ihre H, Gagne L, Fréchet J, Szoka FJ. Polyesterdendritic systems for drug delivery applications: *In vitro* and *in vivo* evaluation. *Bioconjug Chem.* 2002; 13: 453–61.

Patri AK, Kukowskalatallo JF, Baker JR. Targeted drug delivery with dendrimers: Comparison of the release kinetics of covalently conjugated drug and non-covalent drug inclusion complex. *Adv Drug Deliv Rev.* 2005; 57: 2203–14.

Patton DL, Cosgrove Sweeney YT, McCarthy TD, Hillier SL. Preclinical safety and efficacy assessments of dendrimer-based (SPL7013) microbicide gel formulations in a nonhuman primate model. *Antimicrob Agents Chemother.* 2006; 50: 1696–700.

Peeyush Kumar, Meena KP, Pramod Kumar, Pranav Bajpayee. Dendrimer: A novel polymer for drug delivery. *JITPS.* 2010; 1(6): 252–69.

Pentek T, Newenhouse E, O'Brien B, Chauhan AS. Development of a topical resveratrol formulation for commercial applications using dendrimer nanotechnology. *Molecules.* 2017; 22: 1–16.

Percec V, Cho WD, Mosier PE, Ungar G, Yeardley DJP. Structural analysis of cylindrical and spherical supramolecular dendrimers quantifies the concept of monodendron shape control by generation number. *J Am Chem Soc.* 1998; 120: 11061–70.

Perez-Herrero P, Fernandez-Medarde A. Advanced targeted therapies in cancer: Drug nanocarriers, the future of chemotherapy. *Eur J Pharm Biopharm.* 2015; 93: 52–79.

Pistolis G, Malliaris A. Study of poly(propylene imine) dendrimers in water, by exciplex formation. *Langmuir.* 2002; 18: 246–51.

Prieto MJ, Temprana CF, del Río Zabala NE, Marotta CH, Alonso SV. Optimization and *in vitro* toxicity evaluation of G4 PAMAM dendrimer-risperidone complexes. *Eur J Med Chem.* 2011; 46(3): 845–50.

Rajananthanan P, Attard GS, Sheikh NA, Morrow WJ. Evaluation of novel aggregate structures as adjuvants: Composition, toxicity studies and humoral responses. *Vaccine.* 1999; 17: 715–30.

Roberts JC, Bhalgat MK, Zera RT. Preliminary biological evaluation of polyamidoamine (PAMAM) Starburst dendrimers. *J Biomedical Mat Res.* 1996; 30: 53–65.

Sadekar S, Thiagarajan G, Bartlett K, Hubbard D, Ray A, McGill LD, Ghandehari H. Poly(amido amine) dendrimers as absorption enhancers for oral delivery of camptothecin. *Int J Pharm.* 2013; 456: 175–85.

Scutaru AM, Wenzel M, Scheffler H, Wolber G, Gust R. Optimization of the N-lost drugs melphalan and bendamustine: Synthesis and cytotoxicity of a new set of dendrimer-drug conjugates as tumor therapeutic agents. *Bioconjug.* 2010; 21: 1728–43.

Shen W, Van Dongen MA, Han Y, Yu M, Li Y, Liu G *et al.* The role of caveolin-1 and syndecan-4 in the internalization of PEGylated PAMAM dendrimer polyplexes into myoblast and hepatic cells. *Eur J Pharm Biopharm.* 2014; 88: 658–63.

Shi C, He Y, Feng X, Fu D. ε-Polylysine and next-generation dendrigraft poly-l-lysine: Chemistry, activity, and applications in biopharmaceuticals. *J Biomater Sci Polym Ed.* 2015; 26: 1343–56.

Shukla R, Thomas TP, Desai AM, Kotlyar A, Park SJ, Baker JR. HER2 specific delivery of methotrexate by dendrimer conjugated anti-HER2 mAb. *Nanotechnology.* 2008; 19: 295102.

Sideratou Z, Tsiourvas D, Paleos CM. Solubilization and release properties of PEGylated diaminobutane poly(propylene imine) dendrimers. *J Colloid Interface Sci.* 2001; 242: 272–6.

Spangler BD. Inventor biosensors utilizing dendrimer immobilized ligands and there use thereof patent. US7138121; 2006.

Stasko NA, Johnson CB, Schoenfisch MH, Johnson TA, Holmuhamedov EL. Cytotoxicity of polypropylenimine dendrimer conjugates on cultured endothelial cells. *Biomacromolecules.* 2007; 8: 3853–9.

Sun Y, Jiao Y, Wang Y, Lu D, Yang W. The strategy to improve gene transfection efficiency and biocompatibility of hyperbranched PAMAM with the cooperation of PEGylated hyperbranched PAMAM. *In J Pharm.* 2014; 465: 112–9.

Svenson S. Dendrimers as versatile platform in drug delivery applications. *Eur J Pharm Biopharm.* 2009; 71: 445–62.

Svenson S, Tomalia DA. Dendrimers in biomedical applications reflections on the field. *Adv Drug Deliv Rev.* 2005; 57: 2106–29.

Thiagarajan G, Greish K, Ghandehari H. Charge affects the oral toxicity of poly(amidoamine) dendrimers. *Eur J Pharm Biopharm.* 2013a; 84: 330–4.

Thiagarajan G, Sadekar S, Greish K, Ray A, Ghandehari H. Evidence of oral translocation of anionic G6.5 dendrimers in mice. *Mol Pharm.* 2013b; 10: 988–98.

Tolia GT, Choi HH. The role of dendrimers in topical drug delivery. *Pharm Tech.* 2008; 32: 88–98.

Tomalia DA. Birth of a new macromolecular architecture: Dendrimers as quantized building blocks for nanoscale synthetic polymer chemistry. *Prog Polym Sci.* 2005; 30: 294–324.

Tomalia DA, Baker H, Dewald JR, Hall M, Kallos G, Martin S *et al.* A new class of polymers: Starburst-dendritic macromolecules. *Polym.* 1985; 17: 117–32.

Tomalia DA, Baker H, Dewald JR, Hall M, Kallos G, Martin S *et al.* Dendrimers II: Architecture, nanostructure and supramolecular chemistry. *Macromolecules.* 1986; 19: 2466–8.

Tomalia DA, Dewald JR, Hall MR, Martin SJ, Smith PB. Reprints of the 1st SPSJ International Polymer Conference, *Society of Polymer Science Japan, Kyoto*, 1984; 65.

Tomalia DA, Frechet JM. Discovery of dendrimers and dendritic polymers: A brief historical perspective. *J Polym Sci A Polym Chem.* 2002; 40: 2719–28.

Tomalia DA, Reyna LA, Svenson S. Dendrimers as multi-purpose nanodevices for oncology drug delivery and diagnostic imaging. *Biochem Soc Trans.* 2007; 35: 61–7.

Tomalia DA, Rookmaker M. *Poly (Propylene Imine) Dendrimers. Polymer Data Handbook.* New York: Oxford University Press; 2009.

Twyman LJ, Beezer AE, Esfand R, Dardy MJ, Michell JC. The synthesis of water soluble dendrimers and their application as possible drug delivery systems. *Tetrahedron Lett.* 1999; 40: 1743–6.

Vandamme TF, Brobeck L. Poly(amidoamine) dendrimers as ophthalmic vehicles for ocular delivery of pilocarpine nitrate and tropicamide. *J Control Release.* 2005; 20: 23–38.

Vannucci L, Fiserová A, Sadalapure K, Lindhorst TK, Kuldová M, Rossmann P *et al.* Effects of N-acetyl-glucosamine-coated glycodendrimers as biological modulators in the B16F10 melanoma model *in vivo. Int J Oncol.* 2003; 23(2): 285–96.

Vutukuri DR, Basu S, Thayumanavan S. Dendrimers with both polar and a polar nanocontainer characteristics. *J Am Chem Soc.* 2004; 126: 15636–7.

Wang P, Zhao XH, Wang ZY, Meng M, Li X, Ning Q. Generation 4 polyamidoamine dendrimers is a novel candidate of nano-carrier for gene delivery agents in breast cancer treatment. *Cancer Lett.* 2010; 298: 34–49.

Wang Y, Bai Y, Price C, Boros P, Qin L, Bielinska AU *et al.* Combination of electroporation and DNA/dendrimer complexes enhanced gene transfer into murine cardiac transplants. *Ame J Transplant.* 2001; 1: 334–8.

Wang ZX, Yoshiaki I, Yoshifumi H. Mechanism of enhancement of effect of dendrimers on transdermal permeation through polyhydroxyalkanoate matrix. *J Bioscience Bioeng.* 2003; 96: 537–40.

Wei T, Chen C, Liu J, Liu C, Posocco P, Liu X et al. Anticancer drug nanomicelles formed by self-assembling amphiphilic dendrimer to combat cancer drug resistance. *PNAS* 2015; 112: 2978–83.

Wolinsky JB, Grinstaff MW. Therapeutic and diagnostic applications of dendrimers for cancer treatment. *Adv Drug Deliv Rev.* 2008; 60: 1037–55.

Wu P, Feldman AK, Nugent AK, Hawker CJ, Scheel A, Voit B *et al.* Efficiency and fidelity in a click chemistry route to triazole dendrimers by the copper(I)-catalyzed ligation of azides and alkynes. *Angew Chem Int Ed.* 2004; 43: 3928–32.

Yang CH, Lin YT, Hung YH, Liao JW, Peir JJ, Liu HM *et al.* Autoradiographic and histopathological studies of boric acid-mediated BNCT in hepatic VX2 tumor-bearing rabbits: Specific boron retention and damage in tumor and tumor vessels. *Appl Radiat Isot.* 2015 Dec; 106: 176–80.

Yang H, Morris JJ, Lopina ST. Synthesis and characterization of Au core-Au-Ag shell nanoparticles from gold seeds: Impacts of glycine concentration and pH. *J Colloid Inter Sci.* 2004; 273: 145–54.

Yavuz B, Pehlivan SB, Vural İ, Ünlü N. *In vitro/in vivo* evaluation of dexamethasone-PAMAM dendrimer complexes for retinal drug delivery. *J Pharm Sci.* 2015; 104: 3814–823.

Yin H, Liao L, Fang J. Enhanced permeability and retention (epr) effect based tumor targeting: The concept, application and prospect. *JSM Clin Oncol Res.* 2014; 2: 1–5.

Yiyun C, Tongwen X, Rongqiang F. Polyamidoamine dendrimers used as solubility enhancers of ketoprofen. *Eur J Med Chem.* 2005; 40: 1390–3.

Zhang L, Zhu S, Qian L, Pei Y, Qui Y, Jiang Y. RGD-modified PEG–PAMAM–DOX conjugates: In vitro and in vivo studies for glioma. *Eur J Pharm Biopharm.* 2011; 79: 232–40.

Ziemba B, Janaszewska A, Ciepluch K, Krotewicz M, Fogel WA, Appelhans D, Voit B, Bryszewska M, Klajnert B. *In vivo* toxicity of poly(propyleneimine) dendrimers. *J Biomed Mater Res A.* 2011; 99: 261–68.

# 4 Designing Nanocargos for Multi-Drug Resistant Cancerous Cells: Strategies and Applications

*Shivani Rai Paliwal,* Pawan Tekchandani, Balak Das Kurmi, and Rishi Paliwal*

**CONTENTS**



* Corresponding author.

## 4.1 INTRODUCTION

Cancer diagnosis and therapy have improved over the last few years, yet cancer is the major cause of mortalities worldwide (Fitzmaurice *et al.*, 2017). Multi-drug resistance (MDR) is the foremost concern regarding failure of chemotherapeutics drugs. This phenomenon occurs in about 50% of individuals going through chemotherapy and accounts for a vast majority of deaths due to cancer. Most cancerous cells, including lung, breast, ovarian, and colon, develop resistance to chemotherapeutics involving mechanisms of MDR. Cancerous tissue contains a diverse population of cancerous cells including drug-sensitive and drug-resistant cells. Chemotherapy effectively fights against sensitive cells, leaving behind the resistant cells, which grow further and result in tumor progression. Thus, it can be said that MDR is the major hurdle to successful chemotherapy. Chemotherapeutic drug resistance is a combined characteristic of the type of anti-neoplastic agent, cancerous cells against which it is employed, and the specific host. Drug resistance in cancer can be broadly categorized as natural resistance and acquired resistance (Harris and Hochhauser, 1992). Prevalence of MDR was first reported in microbes such as bacteria and viruses. Further studies demonstrate the development of MDR in cancerous cells is a complex phenomenon involving one or more mechanisms working together (Lopes-Rodrigues *et al.*, 2017). Some known mechanisms of MDR can be categorized as (i) increased drug efflux by efflux pumps; (ii) decreased influx; (iii) increased DNA repair; (iv) increased concentration of cytochrome p450-metabolizing enzymes, well-known for rapid metabolism and inactivation of internalized chemotherapeutic agent(s); and (v) termination of the apoptotic process. Resistant cells at different stages or with diverse mechanisms need different treatment concentration(s) and different drug exposure time. Since MDR is a complex phenomenon, overcoming a single mechanism offers an incomplete solution to combat all the mechanisms of cancer-cell resistance to chemotherapy. Over the years, many strategies have been employed for reversal of MDR, such as use of MDR modulators, but still, they suffer from a lot of drawbacks. Nanomedicine is a promising tool for chemotherapy, utilizing a nanosized drug-delivery system of cytotoxic agents for enhancement of anticancer efficiency and diminishing side effects to normal cells. Being tiny particles, nanocarriers offer high drug payload, surface modification, and selective drug targeting at different cellular and subcellular locations.

## 4.2 HISTORICAL BACKGROUND

Cells with the ability to efflux a variety of molecules outside the cytoplasm have been identified for decades. One such mechanism is tubular secretion of molecules by the kidney, first demonstrated by Rennick *et al.* (1947). Later, the first incidence of MDR in cancer was observed by Ling and Thompson (1974) in Chinese hamster ovary (CHO) cell lines. They demonstrated a stable colchicine-resistant cell clone, a derivative of CHO cells that prevented the entry of colchicine to the cytoplasm. Furthermore, cells were resistant to other drugs also, such as vinblastine, demecolcine, and actinomycin D. Unlike colchicine's uptake, which is a passive process, resistance in cells was found to be the active process as it was suppressed by the use of inhibitors such as cyanides, azides, and dinitrophenol (See *et al.*, 1974). Further studies found a specific correlation between efflux pump transporters of the cell membrane and a drug-resistant phenotype (Juliano and Ling, 1976). A variety of drug substances used in cancer chemotherapy were found to be specific substrates for the efflux pump, including colchicine, vinblastine, DOX (doxorubicin), etoposide, PTX (paclitaxel), and other small molecules. P-glycoprotein (P-gp) was isolated by Riordan and Ling (1979) and a couple of years later, its significance in pleiotropic drug resistance was demonstrated followed by the observation that transfer of genetic information from resistant cell lines was able to transform sensitive cells (Debenham *et al.*, 1982). The nonresistant cells acquire resistance correlated to overexpression of MDR proteins. In 1985, gene expressing P-gp, called MDR-1, was cloned. Thereafter, function of this protein as an energy-dependent pump involved in efflux of substrate molecules from intracellular compartments was demonstrated as a sequence homology with bacterial lysine protein transport and other investigations (Chen *et al.*, 1986). Further studies demonstrated existence of several MDR proteins apart from P-gp at the cellular level. One such instance is MRP expressed by lung cancer cell lines that were resistant to DOX and other chemotherapeutic agents (Cole *et al.*, 1992). These proteins are significant targets for effective chemotherapy of MDR cancer.

## 4.3 MDR: DEFINITION AND CONCEPT

The concept of MDR was first discussed by Biedler and Riehm (1970) as the insensitivity to treatment offered by tumor cells against an anticancer agent. Additionally, these cells showed resistance to other drugs as well, even when they have dissimilar chemical structures or different mechanisms of action. MDR is a critical issue as the cancerous cell become resistant to a whole spectrum of the drugs (Pauwels *et al.*, 2007). Although attempts have been made to overcome drug resistance, such as using combinations of drugs having different mechanisms of action in place of using a single drug (Hodge *et al.*, 2012; Inoue *et al.*, 2012; Slovin, 2012; Teoh and Secord, 2012), the combination drug therapy approach is ineffective in the case of MDR cancer, because these cells become resistant to previously used drugs along with other cancer chemotherapeutics with different anticancer mechanisms used for the treatment. In other words, it can be referred to as a condition of resilience against structurally and functionally dissimilar drugs (Harris and Hochhauser, 1992).

## 4.4 MECHANISM OF MDR

Natural resistance or primary resistance is the intrinsic insensitivity of tumor cells for a drug presented since the beginning of therapy. Cancer-initiating cells or stem cells are vital subsets of cancer cells and are notorious in character for intrinsic drug resistance (Singh *et al.*, 2017). These cells overexpress certain stem cell markers of resistance and also initiate a growth of new MDR tumor cells. Further, the resistance that develops after the primary encounter to the drug is said to be acquired resistance. The most important mechanisms of drug resistance include increased drug outflow by transporters and reduced entry of drugs, along with the enhanced DNA repair, detoxification, and inhibition of caspase-mediated cell death (i.e., apoptosis). Mechanistic classification of MDR and responsible factors is presented in Figure 4.1.

### 4.4.1 Efflux Pump-Mediated MDR

In normal physiology, efflux transporters are part of physiological defence against xenobiotics. These transporters are expressed majorly in brain, placenta, lung, and testis (Leslie *et al.*, 2005). They play an important role in altering the pharmacokinetics of drugs and other xenobiotics (Kalia, 2017). A majority of anticancer drugs act on intracellular targets. Therefore, success of chemotherapy using such intracellular acting agents requires their accumulation at the site of action in sufficient concentrations. A critical factor leading to failure of chemotherapy is overexpression of the efflux transporters on the cell membrane of cancer cells (Fletcher *et al.*, 2010), which causes lowering of intracellular concentrations of drugs by increasing their outflow. Efflux transporters join the ABC super-family, which accounts for most of the drug-efflux activity (Gillet *et al.*, 2007. The ABC transporter superfamily, comprising 48 members of transporter proteins (Foulkes *et al.*, 2017), is constituted of seven sub-families, ABC A–G. The well-known members are ABCB1 [P-gp, MDR protein 1 (MDR1), ABCC1-3 (MDR-associated protein 1–3), and ABCG2 (breast cancer resistance protein) (Arrigoni *et al.*, 2016). These proteins are helpful in efflux of toxic substances and other harmful substances out of the cell utilizing ATP as an energy source. Out of 48 ABC transporters currently identified, 13 have been recognized to contribute to MDR (Xue and Liang, 2012).

#### 4.4.1.1 Structure and Function of P-gp

P-gp, a peptide with molecular weight (MW) of 170 kDa, possesses diverse substrate specificity (Aller *et al.*, 2009). Generally, weak bases and neutral compounds have been observed to be the most susceptible

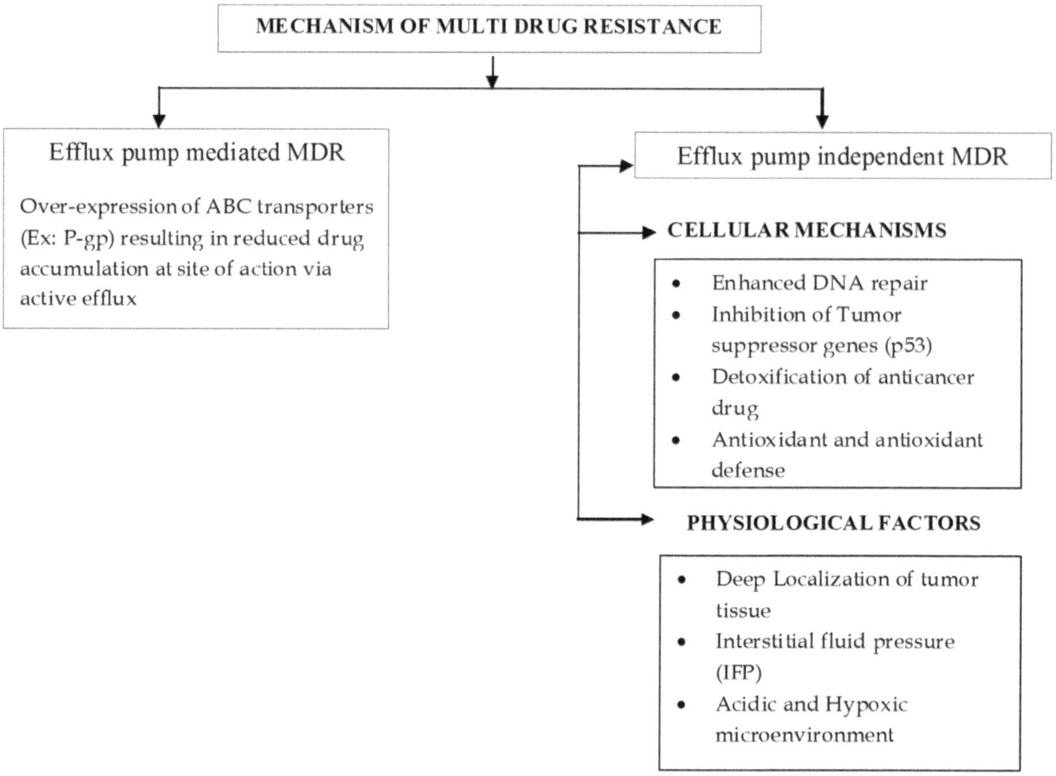


**FIGURE 4.1** Mechanistic classification of multi-drug resistance and responsible factors.

to these transporters. However, some studies suggest that even acidic and amphiphilic molecules are also vulnerable, but to a lesser extent (Seelig *et al.*, 2000; Loscher and Potschka, 2005). P-gp has been observed to interact with structurally and functionally dissimilar molecules such as various anticancer agents, immunosuppressants, HIV protease inhibitors, cardiac glycosides, analgesics, anthelmintics, steroids, antibiotics, lipids, small cytokines (such as interferon-$\gamma$, interleukin-2, and interleukin-4), and many more (Krishna and Mayer, 2001). The extent of P-gp expression depends upon the type of tumor. High P-gp expression is generally observed in carcinoma of the kidney, colon, adrenal gland, liver, and pancreas, while lower P-gp levels are seen in cancer of the stomach, lung, oesophagus, ovary, and breast and melanomas, lymphomas, and multiple myelomas. However, in the latter case, high P-gp levels may be observed after primary exposure to drugs owing to acquired drug resistance. Moreover, intermediate levels are observed in various soft tissue carcinomas, neuroblastomas, and haematological malignancies (Loo and Clarke, 2005). Typically, drugs are small molecules with low MW and are internalized via passive diffusion and pumped out from the cells by these types of drug-efflux transporters. Such a prerequisite can enable us to overcome pump-mediated MDR of cancer cells by changing the drug uptake mechanism, e.g., receptor-mediated endocytosis can be employed as an effective approach to bypass the efflux transporters.

### 4.4.2 Efflux-Independent MDR

In addition to efflux pump-mediated drug resistance, many other mechanisms are involved in MDR (Chaudhry and Asselin, 2009). In contrast to pump-mediated mechanisms, efflux pump-independent mechanisms generally do not hinder influx and accumulation of anticancer drugs in the cells. Instead, they lower the cytotoxic potential of these agents via processes such as increased drug metabolism, anti-apoptotic and antioxidant defences, enhanced DNA repair, mutations in tumor suppressor genes, etc. For better illustration, we have classified efflux-independent mechanisms into two broad categories, namely cellular mechanisms and physiological mechanisms.

#### 4.4.2.1 Cellular Mechanisms

Altered genetics and mutations in the tumor cells give rise to various cellular adaptations, which lead to the development of MDR. These efflux pump-independent cellular mechanisms include decreased drug influx, increased enzymatic degradation and inactivation of drug, enhanced DNA repair, inhibition of growth regulators, anti-apoptotic and antioxidant defence, etc. In spite of a lot of possible mechanisms, the clinically significant mechanisms are discussed in the following sections.

##### *4.4.2.1.1 Anti-Apoptotic and Antioxidant Defence*

Apoptosis, known as programmed cell death, is one of the most important steps in the life cycle of a cell. This process is represented by plasmolysis/cell shrinking, cell membrane blebs/protrusion of the plasma membrane, DNA injury by condensation and fragmentation, cell organelle integrity maintenance, and eventually, signalling and consequent removal by phagocytes (Kerr *et al.*, 1972; Hanahan and Weinberg, 2000). Inhibition of apoptosis is an important mechanism of efflux pump-independent cancer-cell drug resistance. Anticancer drugs cause damage to cancer cells which induces apoptosis. It is a well-regulated phenomenon by pro-apoptotic (Bax, Bad) and anti-apoptotic (Bcl-2, Bcl-XL) protein fractions. It frequently involves inputs of the tumor suppressors p53 and p63/73, as well as the inhibitors of programmed cell death (IAPs, inhibitors of apoptosis proteins) family (survivin, livin) (Lane, 1992; Reed *et al.*, 1998; Moll *et al.*, 2005). Cancer cells endure many adaptive changes to evade the caspase-mediated cell death and exhibit survival. Overexpression of anti-apoptotic proteins such as Bcl-2, Bcl-XL, and Mcl-1 is one such adaptation to prevent apoptosis (Mashima and Tsuruo, 2005; Viktorsson *et al.*, 2005). These proteins induce MDR by antagonizing the pro-apoptotic proteins. A strong correlation exists between the expression of efflux transporters and the up-regulation of Bcl-XL, signifying additional pro-/anti-apoptotic ability of efflux transporters (Lage, 2008). Other such adaptive changes include up-regulation of the PI3K/Akt and IAP pathways. The PI3K/Akt activation acts by promoting cell growth, predominantly through NFκB activation and Bad inactivation, and consequently increases survival (Cantley, 2002).

The antioxidant system of cells consists of cytochromes; catalase, superoxide dismutase, antioxidants (α-tocopherol, steroids, thyroxine, etc.), and glutathione peroxidase of cells also defend against oxygen radicals (Cheeseman, 1993; Irshad and Chaudhuri, 2002; Bounous and Molson, 2003). Such antioxidant defence mechanisms involve prevention, interception, repair, or replacement (Fernandez and Videla, 1996; Vera-Ramirez *et al.*, 2011). The whole antioxidant system prevents injury to the cancer cells by free radicals generated in response to the cancer chemotherapeutics.

##### *4.4.2.1.2 Detoxification of Anticancer Drugs*

Cancer cells possess various mechanisms for drug metabolism leading to inactivation and degradation. Exposure to drugs induces certain metabolizing enzymes in the cancer cells, such as UDP-transferases and glutathione-S, which leads to drug degradation followed by elimination by forming conjugates and consequently affects the therapeutic efficacy of drugs (Minko *et al.*, 2004). Dihydrodiol dehydrogenase is overexpressed in the case of primary ovarian tumors, which primarily accounts for the development of cellular resistance against cisplatin (Deng *et al.*, 2002). Another group of enzymes, termed as cytochrome P450 (CYP450), which is commonly found in the liver, is also responsible for drug inactivation by altering its chemical structure. Genetic alterations leading to structural aberrations in CYP450 cause defective pharmacokinetics and inactivation of drugs owing to altered enzymatic activity (Gottesman *et al.*, 2002).

##### *4.4.2.1.3 DNA Repair*

Normal cells have a well-developed mechanism to prevent and repair damage caused by harmful substances to genetic material. The same mechanism is utilized to overcome anticancer-drug-induced damage by cancer cells. In this way, they develop resistance against drugs which produce cytotoxic action by directly or indirectly damaging DNA, such as for antimetabolites (mercaptopurine), alkylating agents (busulfan, cyclophosphamide), anthracyclines (DOX), epipodophyllotoxins (etoposide), and platinum compounds (carboplatin, cisplatin). For example, a commonly used chemotherapeutic drug, DOX, kills cells by stabilizing the topoisomerase II complex. On the other hand, resistant tumor cells reactivate topoisomerases in response to DOX treatment and overcome the damage and thereby offer resistance against the drug. This example demonstrates that the DNA repair system would be a possible target for the tumor tissue to modify consecutively to provoke MDR.

##### *4.4.2.1.4 Tumor Suppressor Genes (p53)*

Genesis of abnormal cells is controlled by the cell-cycle regulator gene referred to as gene *TP53*. Tumor protein p53 is a cellular SV40 large T antigen-binding protein also referred to as phosphoprotein p53 or, simply, p53. It has been identified to play a vital role as a tumor suppressor and apoptosis inducer to prevent tumor growth and progression. As stated above, a variety of drugs act by directly or indirectly damaging DNA. Lack of p53 function facilitates the cells with injured DNA to proliferate by uncontrolled replication and become resistant which shows an altered growth pattern. Such types of genetic mutations provide genomic instability and apoptosis resistance towards the *TP53* gene. In most MDR cancers, the *TP53* gene mutation has been observed (Luqmani, 2005). Further, some genes encoding bcl-2/Bax and HRAS are reported to express abundantly in MDR. Disturbance in such types of check-points helps cancer cells become resistant by avoiding the process of apoptosis (Nelson *et al.*, 2004). These MDR mechanisms have been observed in pre-tumorigenic cells that express telomerase required for cell immortalization and tumor growth (Yague *et al.*, 2007). It was observed that pRb (retinoblastoma) mutations and p53 tumor suppressor protein cascades turn into chemotherapeutic cell resistance.

#### 4.4.2.2 Physiological Factors

The microenvironment of tumor tissue possesses various characteristics that distinguish it from normal cells. The physiological factors which need to be focused on concerning MDR are hypoxia, interstitial fluid pressure (IFP), and acidic pH. These factors are significant as they cause MDR and are generally interconnected with each other.

##### *4.4.2.2.1 Interstitial Fluid Pressure (IFP)*

IFP serves as a determining factor for MDR of several chemotherapeutic drugs. In normal tissue, IFP is maintained by the harmony of angiogenic factors and the lymphatic system. This results in a highly organized system of vessels which supplies nutrients and oxygen felicitously to the tissues. In solid tumors, due to unharmonized functioning of angiogenic factors, disturbed vasculature is characterized, having dilated and complicated vessels with several loops, blunt ends, and shunts (Fukumura and Jain, 2007). Added to these is the poorly developed fenestration which causes leaking and irregular blood flow within the tumor (Carmeliet and Jain, 2000; Tredan *et al.*, 2007). The synergistic effects of leaky vasculature and the absence of lymphatic system results in accumulation of proteins and fluids and a hike in the IFP up to 100 mm Hg as compared to normal IFP (which is equal to atmospheric pressure) (Cairns *et al.*, 2006).

Another consequence of uncontrolled growth of tumor cells is the compression of blood vessels which further aggravates the problem (Padera *et al.*, 2004). The IFP at the tumor center and at the tumor periphery (which resembles more like normal tissue) have different features (Milosevic *et al.*, 2004). As the blood vessels are irregular in nature in the tumor cells, there is an incongruous rate of cell proliferation, and areas of the tumor with the lowest rate of cell proliferation are in areas distant from the vessels (Hirst and Denekamp, 1979). Chemotherapeutic drugs, which are largely administrated through intravenous injection, are dependent on the systemic circulation to act on their destined targets. The haphazard, complicated tumor vasculature, along with high and low IFP in different areas of the tumor tissue, holds back a successful drug delivery (Heldin *et al.*, 2004).

This hinders transportation of giant anticancer molecules such as proteins, peptides, and antibodies to the tumor uniformly. Hence, MDR is observed in solid tumors due to their peculiar action and physiology.

#### *4.4.2.2.2 Localization of Tumor Tissue*

Tumor tissue localization provides a shield toward anticancer drugs because tumor cells reside up to 200 μm away from the blood vessels and create complexity for the drug molecules to penetrate deep inside and approach viable tumor cells (Tredan *et al.*, 2007). In general practice, it is possible to combat this situation by using low MW compounds with higher doses to maintain passive diffusion for transport, which is influenced by the concentration gradients. Further, the ECM (extracellular matrix), drug characteristics, and degradation of drug molecules in the tumor tissue are also considerable factors for drug delivery systems to be formulated to deliver a high concentration of drugs to deeper tissues of the tumor. The delivery of drug over an extended period is also an ideal strategy to potentiate the therapeutic benefit with the least possible side effects. Certain chemotherapeutic drugs show anticancer effects on rapidly-proliferating cells. Tumor cells located away from the blood capillaries are generally in the lag phase, hence these cells may experience a low drug concentration owing to limited drug penetration and distribution (Minchinton and Tannock, 2006). Such conditions also lead to drug resistance.

#### *4.4.2.2.3 Hypoxic Microenvironment*

About 50%–60% of locally advanced tumors are characterized by the presence of a hypoxic microenvironment. Irregular vasculature formed as a result of abrupt angiogenesis and lack of RBCs (red blood cells) lead to reduction in oxygen supply to the tumor tissue (Dewhirst *et al.*, 1996). Further, rapid proliferation and hence increased oxygen consumption by the tumor tissue creates hypoxic conditions. Hypoxia is an important phenomenon attributed to development of MDR in cancer cells (Brown, 1990). Oxygen is the prerequisite for effective treatment via radiation therapy as cytotoxicity is caused by DNA damage induced by a free-radical mechanism which requires oxygen (Brown and Wilson, 2004). Similar correlation is observed between MDR and hypoxia in the case of chemotherapeutic agents that rely on the oxygen for their cytotoxic activity, such as antibiotics (bleomycin), alkylating agents (melphalan), podophyllotoxin derivative (etoposide), and photosensitizers (porphyrin) (Freitas, 1985; Shannon *et al.*, 2003; Wilson and Hay, 2011). Moreover, hypoxia also decreases cell proliferation and greatly affects the activity of drugs that act on rapidly multiplying cells (Amellem and Pettersen, 1991).

Tumors are typically diverse in the context of hypoxic regions, but generally, they show two major regions: the external shell of rapidly dividing cells enclosing the internal hypoxic/necrotic core (Hockel and Vaupel, 2001). A labile transcriptional factor referred to as hypoxia-inducible factor 1α (HIF1α) is stabilized due to a hypoxic microenvironment. Further, it causes activation of numerous genes related to the hypoxia responsive element (HRE). The *MDR1* gene is one among such genes controlled by HRE (Comerford *et al.*, 2002). As stated earlier, up-regulation of the *MDR1* gene causes efflux pump-mediated MDR.

#### *4.4.2.2.4 Acidic Microenvironment*

Tumor extracellular matrix is characterized by low pH as demonstrated by Warburg in 1965, a phenomenon referred to as the 'Warburg effect'. Warburg observed a higher accumulation of lactic acid in tumor microenvironments via the glycolysis pathway (Warburg, 1956). In fact, glycolysis is more prominent in hypoxic cells instead of oxidative phosphorylation, owing to its better efficacy of producing $CO_2$ and carbonic acid (Denko, 2008). The ineffective lymphatic drainage leads to accumulation of the lactic acid produced by glycolysis metabolism which in turn results in low extracellular pH (Zhang *et al.*, 2010). On the contrary, intracellular pH is comparatively normal due to activity of the ion pumps such as the $Na^+/H^+$ exchanger, V-type $H^+$ ATPase (proton pump), and the sodium ($Na^+$)-dependent chloride ($Cl^-$)/bicarbonate exchanger. The reduction in pH may range from 5.8–7.2, whereas the pH of blood and tissue under normal physiology is about 7.4 (Lee *et al.*, 2008). The acidic extracellular environment adversely influences the activity of weakly basic drugs. As stated earlier, intracellular accumulation is the prerequisite for drugs acting on intracellular targets. The weakly basic drugs undergo ionization in acidic microenvironments, which affects their internalization via passive diffusion and eventually hampers their therapeutic efficacy (Raghunand and Gillies, 2000).

## 4.5 CLASSES OF NANO-DRUGS FOR CANCER THERAPY

Nanotechnology with a broad range of particle size (ranging from 1–200 nm), has been applied in medicine to circumvent several therapeutic limitations. Nanomedicine involves different applications of nanotechnology in medicine for diversified purposes of detection and diagnosis, cure and prevention, and finally treatment of the disease. This branch of drug delivery is widely investigated for anticancer therapies to improve drug delivery. A wide range of nanocarriers are well-established for drug delivery, that is dendrimers, quantum dots, liposomes, polymer-drug conjugates, solid lipid nanoparticles (SLNs), polymersomes, nanoparticles, micelles, nanoemulsions, nanospheres, nanoshells, CNTs (carbon nanotubes), etc. Nanocarriers are able to significantly control the pharmacokinetic and pharmacodynamic properties of the encapsulated therapeutics by either protecting them from the destructive bioenvironment or release modification at the target site (Kurmi *et al.*, 2015).

### 4.5.1 Lipid-Based Carriers

Lipid-based nanocarriers acquire significant notice as a drug delivery system due to their inherent properties like biocompatibility, biodegradability, low immunogenicity, and capacity to deliver hydrophilic as well as hydrophobic drugs. Liposomes are lipid-based, bilayered vesicles having the ability to carry drug load in an aqueous compartment in its lipid bilayer, depending on the physiochemical nature of the drug molecule. The delivery of drug-encapsulated liposomes towards cancer cells is regulated by passive targeting or by active targeting (i.e., attaching a targeting ligand). This may provide a pharmacologic advantage for liposomes to overcome drug resistance (Rai *et al.*, 2008; Paliwal *et al.*, 2012). A number of liposomal formulations have been developed for that purpose such as with DOX (Yang *et al.*, 2007; Torchilin, 2010). Liposomes can be easily modified for targeting and triggering purposes, and PEGylation also helps to protect from reticuloendothelial system (RES) uptake. Hence, increased circulation time and increased drug retention in the tumor is possible.

Solid lipid nanoparticles (SLNs) belong to a class of drug delivery system resembling liposomes, but differ in having a solid lipid core. They are somewhat closer to nanoemulsions, however they differ in the lipidic nature. SLNs are colloidal carriers capable of loading both hydrophilic and lipophilic drugs, usually depend on the method of preparation. They exhibit various advantages such as low cytotoxicity, stability over liposomal formulations, excellent reproducibility, and potential applications using various routes of administrations such as IV, oral, dermal, or topical (Paliwal *et al.*, 2011; Kurmi *et al.*, 2017). The certain limitations of SLNs include agglomeration or coagulation, less drug entrapment capacity, and rapid drug leaching during storage, which can be overcome by the modification of SLNs with triglycerides, glyceride mixtures, or waxes, the so-called nanostructured lipid carriers (NLCs). It is possible with SLNs and NLCs for anchoring with target-recognizing ligands to deliver drugs by targeting cancer cells with controlled drug release. This makes them suitable for application in cancer therapy.

### 4.5.2 Polymer-Based Carriers

Polymeric nanoparticulate carriers exhibit potential advantages regarding drug delivery in cancer therapeutics. These systems provide extend blood-retention time, higher drug-loading ability, and chemical modification or surface functionalization for various targeting purposes. Various polymers are useful as drug carriers for targeted drug delivery; some of them have been approved by the US FDA (Food and Drug Administration) for one or more applications. Surface and material properties of nanoparticle systems can be utilized for enhanced drug delivery to the tumor. For example, modification of the surfaces with hydrophilic molecules provides longer circulation, whereas positively charged surfaces may enhance endocytosis of polymeric nanoparticles (Kurmi *et al.*, 2010).

Polymer-drug conjugates can also be developed by the formation a biodegradable linkage between water-soluble polymers and an anticancer agent. Such formulations have the capacity to passively target tumor vasculature by the EPR (enhanced permeability and retention) effect. Various types of polymer-drug conjugates are under clinical investigation, such as poly (L-glutamic acid)-PTX (CT-2103, Xyotax®), a synthetic polymer-drug conjugate that has advanced to Phase III clinical trials (Beg *et al.*, 2016).

Polymeric micelles form monolayers spontaneously by self-assembling under certain conditions like CMC (critical micelle concentration) of amphiphilic surfactants, pH, temperature, and ionic strength with a hydrophobic core and hydrophilic shell in the nanometre range. The small size and hydrophilic shells are the properties which protect them from the mononuclear phagocyte system (MPS). Attachment of a targeting ligand further enhances the tumor accumulation, as well as cellular uptake in cancer cells via receptor-mediated endocytosis (Gothwal *et al.*, 2016; Amjad *et al.*, 2017).

Dendrimers are hyper-branched, monodisperse, highly water soluble, and unimolecular polymeric systems. The size, shape, MW, and composition of dendrimers can be controlled by optimization of the synthesis procedure (Kurmi *et al.*, 2011). All of these parameters make dendrimers suitable for applications in cancer therapeutics. The presence of various controlled terminal groups ($–NH_2$ in PPI, PAMAM) are helpful in anchoring with targeting ligands, imaging contrast agents, drugs, genes, and chemical sensors. (Singh *et al.*, 2016).

### 4.5.3 Miscellaneous Nanocarriers

Besides the above nanocarriers, several other types of systems are also useful for biomedical applications in cancer therapeutics due to their unique intrinsic physical and biological properties such as optical, electrochemical, and magnetic characteristics. These categories include various inorganic materials formulations such as CNTs, metallic nanoparticles, quantum dots, and various protein-based nanoparticles, aptamers etc.

CNTs are able to enter in cells using their needle-like penetration process and can deliver anticancer drug molecules into the cytoplasm of the cancer cell. Large surface availability provides them attachment sites for anchoring or conjugation with various epitopes (Kaur *et al.*, 2017). Metallic nanoparticles such as iron-based nanocarriers are flexible to formulate. Such carriers localize magnetically and treat with hyperthermia responsive delivery of drugs for treatment of cancer. Protein-based nanomedicine systems are also possible to formulate by various natural protein subunits with different drug molecules. Aptamers are known as chemical antibodies and provide a better platform to avoid the traditional antibodies due to low immunogenic potential, easy synthesis, structural modification flexibility, and good affinity, specificity, and stability (Powell *et al.*, 2017).

## 4.6 NANOCARGOS TO OVERCOME MDR

### 4.6.1 Combination Therapy Using Nanocarriers

Combination therapy represents the delivery of active agents (which have the ability to alter MDR) with an anticancer drug (responsible for the anticancer activity) to cancer cells. A literature review revealed that combination therapy with an MDR modulator and anticancer drugs incorporating nanocargos serves as an excellent platform for this purpose. These formulations localize actively or passively towards tumor sites and remain responsible for enhanced therapeutic activity with reduced

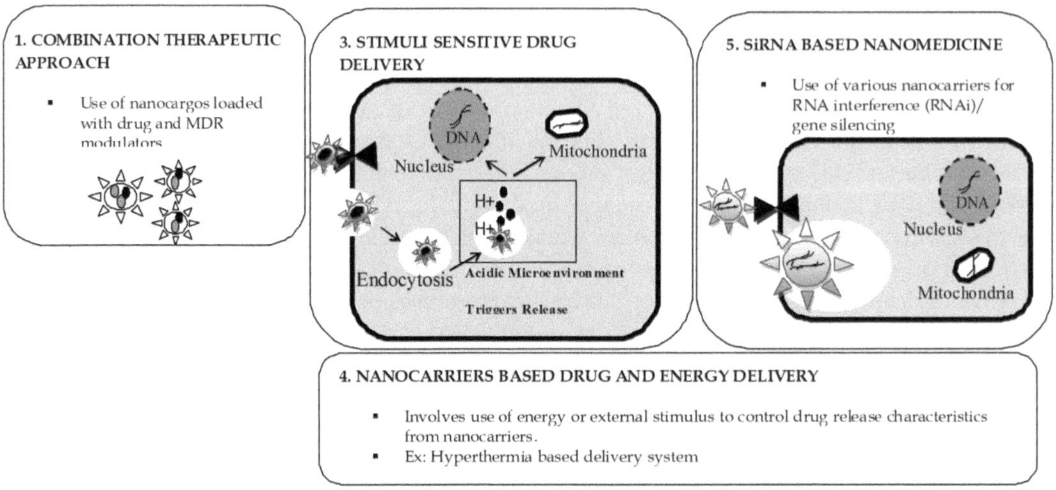


**FIGURE 4.2** Schematic presentation of various nanotechnological approaches to overcome multi-drug resistance.

associated systemic toxicity and potentially circumvent the problem of drug resistance. Numerous therapeutic agents have been under investigation for clinical studies to reverse MDR in cancer such as MDR modulators (chemosensitizers or MDR inhibitors, apoptosis modulators), and anticancer agents having MDR-overcoming properties. Co-administration of a chemotherapeutic agent with an MDR modulator is a common strategy for overcoming MDR in cancer (Alamolhodaei *et al.*, 2017).

Figure 4.2 shows a schematic presentation of various nanotechnological approaches to overcome MDR.

#### 4.6.1.1 Drug and Efflux Modulators

MDR modulators are agents with the ability to alter the MDR process or affect various efflux systems of MDR. These agents are categorized on the basis of their affinity for transporter proteins and their relative side effects as first-, second-, and third-generation MDR modulators (Prasad *et al.*, 2016). First-generation MDR modulators consist of various chemical classes of drugs like calcium-channel blockers, calmodulin antagonists, steroidal agents, antibiotics (e.g., erythromycin), protein kinase C inhibitors, and antimalarials (e.g., quinine). Cyclosporin-A, is a most effective first-generation MDR modulator commonly used as an immunosuppressant (Zhu *et al.*, 2016). Requirements of high dose or very high concentration of such agents result in the unacceptable high toxicity or lethal adverse effects, which is responsible for their limited clinical applicability as MDR modulators. To overcome such limitations, some structural analogues of first-generation modulators, such as dexverapamil and emopamil, are available which were developed synthetically with a high potent nature as well as considerably lower toxicity. Such advanced derivatives are categorized as second-generation modulators. Valspodar, biricodar, and elacridar, without immunosuppressive or nephrotoxicity, are second-generation modulators having 5- to 30-fold more potency than cyclosporin-A, but major problems with most of them are the competition for cytochrome P4503A4 along with the anticancer drugs. This may result in increased plasma concentrations of anticancer drugs (Saraswathy and Gong, 2013). Hence, continuous research aiming to minimize all of these problems lead to the growth of third-generation modulators. These are highly specific in nature with minimal cytochrome P4503A4 interaction and noncompetitive P-gp antagonist ability (Lin *et al.*, 2010). Co-delivery of the MDR modulator and anticancer drugs also increases the chemotherapeutic effect by blocking physiological anticancer drug efflux from cancer cells. Simultaneously, it is important to limit the exposure of the normal cells to the efflux inhibitor. Alteration of pharmacokinetics and tumor accumulation of the two co-delivered agents is the main problem with combinational delivery of anticancer drugs and efflux modulators. These two obstacles can be minimized or removed by using targeted nanosized delivery systems to co-encapsulate the anticancer drug and efflux modulators or chemosensitizers.

Various studies of MDR modulators and anticancer drugs via nanocarriers clearly suggest their effectiveness for drug delivery to overcome MDR with respect to efficacy, toxicity, targeting, MTD (maximum tolerated dose), etc. (Wang *et al.*, 2005; Pakunlu *et al.*, 2006; Wu *et al.*, 2007; Song *et al.*, 2009).

#### 4.6.1.2 Drug and Apoptosis Modulators

The MDR cells are basically characterized by high or low expression of pro-apoptotic factors and anti-apoptotic factors simultaneously. This results in a higher threshold for apoptosis and survival for cytotoxic drugs. Various therapeutic strategies attempt to overcome this unique ability of MDR cells by combining delivery of pro-apoptotic modulators or anti-apoptotic inhibitors (apoptosis modulators) with an anticancer agent. The rationale behind this approach is that apoptotic modulator lowers the apoptotic threshold of MDR cells and makes anticancer agents more effective as compared to single treatment of MDR cancer. For example, the naturally occurring plant product quercetin has a wide range of biological activities such as induction of apoptosis and inhibition of cell invasion, along with angiogenesis (Tan *et al.*, 2003). A number of studies revealed that quercetin may be used to overcome MDR in many types of resistant cells (Nabekura *et al.*, 2005; Borska *et al.*,2012). Emodin is another agent that has the ability to augment chemotherapeutic agent sensitivity against human tumor cells. Li *et al.*, 2009 investigated emodin in overcoming chemoresistance of A2780/PTX cells. After treatment with emodin alone and in combination with PTX, apoptosis was determined by flow cytometric analysis. They found that the emodin-induced apoptosis alone required a high concentration whetherPTX-induced apoptosis existed at a low concentration (Li *et al.*, 2009). Curcumin (diferuloylmethane) belongs to the natural product category, also having a history of traditional use because of its effective antitumor properties over the past decades (Joe and Lokesh, 1994; Aggarwal *et al.*, 2003). Different therapeutic strategies using curcumin have been developed with the aim of inhibiting DNA repair pathways and sensitizing chemoresistant tumors for apoptosis. Co-administration of ceramide (a pro-apoptotic modulator) with PTX by nanoparticulate formulation in an MDR human ovarian cancer cell line also increases the cytotoxicity of PTX in the MDR cells by three fold. The nanoparticle delivery system probably increased the EPR effect for successful co-delivery of both the drugs for treating MDR cancer (van Vlerken *et al.*, 2007).

### 4.6.2 Formulation Strategy to Overcome MDR

Over the past few years, different formulation strategies have been applied to combat MDR. The use of low MW inhibitors with more expectations has been applied, but these attempts have been unsuccessful due to low selectivity, inherent toxicity, and pharmacokinetic interactions with most of the anticancer drug molecules. Nanomedicine has also been extensively evaluated to overcome MDR (Table 4.1). In the last three decades, a significant emphasis has been placed on nanoparticle-prolonged blood-circulation properties and their ability to aid in tumor accumulation due to EPR effect or passive targeting approaches (Lammers *et al.*, 2008; Jain and Stylianopoulos, 2010). Generally, the design or development of a nanomedicine is an attempt to improve the bound or encapsulated chemotherapeutic agent with respect to biodistribution, accumulation at the target site, and ultimately maintain the balance between the therapeutic efficacy and toxicity. After efficient/enhanced delivery of chemotherapeutics to tumors *in vivo*, it is more important to understand the mechanism behind these carrier materials, which are taken up or accumulated by the respective cancer cells, as well as enhancement of retention time within cells. The cellular retention time of nanomedicine mainly depends on the equilibrium between drug uptake and drug efflux by the cells. As discussed elsewhere, the MDR cells have the ability to disturb this phenomenon and affect the concentration of chemotherapeutic agent within cancerous cells after passive diffusion. This is rapidly sensed by MDR proteins and such rapidly internalized small molecules are effluxed out of the cells (Ambudkar *et al.*, 2006, Tekchandani *et al.*, 2016).

**TABLE 4.1**
**Patents Published on Nanocarrier-Based MDR Treatment Reports**

| S. No. | Patent Title | Therapeutic Strategy/System | Reference |
|---|---|---|---|
| 1 | Production of immunogens by antigen conjugation to liposomes | Immunoliposome | Heath *et al.* (1986) |
| 2 | Magnetically-localizable, polymerized lipid vesicles and method of disrupting same | Stimuli responsive | Chang (1987) |
| 3 | Surface modified anticancer nanoparticles | Ligand targeted | Liversidge *et al.* (1995) |
| 4 | Liposome encapsulated taxol and a method of using the same | Passive targeted liposome | Rahman *et al.* (1997) |
| 5 | Methods for conferring multidrug resistance on a cell | Various strategies to overcome MDR | Deeley and Cole (1999) |
| 6 | Microfabricated particles and method for treating solid tumors | Treatment of solid tumors | Martin and Ferrari (2003) |
| 7 | Vinorelbine compositions and methods of use | Vinorelbine compositions for cancer treatment | Zhang and Ahmad (2004) |
| 8 | Anticancer drug aldehyde conjugate drugs with enhanced cytotoxicity compounds, compositions and methods | Drug conjugates | Taatjes *et al.* (2004) |
| 9 | Method of *in vivo* drug targeting to solid tumors via acoustically triggered drug delivery in polymeric micelles | Triggered targeted drug delivery | Rapoport (2005) |
| 10 | Use of P-gp inhibitor surfactants at the surface of a colloidal carrier | P-gp inhibitor surfactants | Benoit and Lamprecht (2006) |
| 11 | Nanoparticulate delivery systems for treating multi-drug resistance | Nanoparticulate carriers | Amiji *et al.* (2006) |
| 12 | Gene delivery mediated by liposome-DNA complex with cleavable PEG surface modification | Gene delivery via liposome | Huang and Zalipsky (2008) |
| 13 | Echogenic microbubbles and microemulsions for ultrasound-enhanced nanoparticle-mediated delivery of agents | Stimuli-responsive nanoparticulate drug delivery | Rapoport and Gao (2009) |
| 14 | PEG and targeting ligands on nanoparticle surface | Active targeting | Panyam *et al.*, (2010) |
| 15 | Transferrin-conjugated nanoparticles for increasing efficacy of a therapeutic agent | Ligand (Tf) targeting | Labhasetwar and Sahoo (2010) |
| 16 | Nanoparticles effective for internalization into cells | Endocytosis of nanoparticles | Drummond *et al.* (2010) |
| 17 | Drug conjugates with polyglycerols | Drug conjugates | Kratz *et al.* (2011) |
| 18 | pH-sensitive biodegradable polymeric particles for drug delivery | pH responsive | Radovic-moreno *et al.* (2011) |
| 19 | Polymer-surfactant nanoparticles for sustained release of compounds | Passive targeting | Panyam and Chavanpatil (2011) |
| 20 | Cationic polymer coated mesoporous silica nanoparticles and uses thereof | Charged based applications | Zink *et al.* (2012) |
| 21 | Complex drug delivery composition and method for treating cancer | Drug-polymer complex | Minko *et al.* (2012) |
| 22 | Method and composition for treating cancer | Various possible ways | Manganaro and Rockwell (2012) |
| 23 | Cytotoxic therapy by proton flux modulation | Proton flux-modulated therapy | Rosenberg and Hill (2013) |
| 24 | Targeted crosslinked multilamellar liposomes | Targeted liposome | Wang *et al.* (2014a) |

*(Continued)*

**TABLE 4.1 (*Continued*)**
**Patents Published on Nanocarrier-Based MDR Treatment Reports**

| S. No. | Patent Title | Therapeutic Strategy/System | Reference |
|---|---|---|---|
| 25 | Polymer-conjugated protein micelles | Conjugated micelles for passive targeting | Perumal *et al.* (2014) |
| 26 | Protocells and their use for targeted delivery of multicomponent cargos to cancer cells | Targeted delivery | Brinker *et al.* (2015) |
| 27 | Multifunctional self-assembling polymeric nanosystems | Polymeric micelles | Amiji and Iyer (2015) |
| 28 | Nano-hybrid delivery system for sequential utilization of passive and active targeting | Passive and active targeting | Hong and Liu (2015) |
| 29 | Polymeric nanoparticles useful in theranostics | Theranostically used nanoparticle | Wu *et al.* (2015) |
| 30 | Cancer cell targeting using nanoparticles | Different ways of targeting | Zale and Ali (2016) |
| 31 | Mixed micelles | Micellar systems | Gombotz *et al.* (2016) |
| 32 | On-demand drug release using magnetoelectric nanoparticles | Magnetic responsive system | Khizroev *et al.* (2016) |
| 33 | Targeted dendrimer-drug conjugates | Targeted dendrimers | El-sayed *et al.* (2016) |

#### 4.6.2.1 Conventional Nanomedicine

Most of the standard nanomedicine (micelles or liposomes) and their composition materials are inert in nature with lack of intrinsic pharmacological activity to alter MDR-related activity of cells. However, their cellular uptake may be vary from other low MW chemotherapeutic agents and other 'standard' nanomedicines, which are generally assumed to have significant inherent ability or potential to overcome MDR. The conventional dosage form is generally passively accumulated at the site of action effluxed out of the MDR cells, as discussed earlier. There are various mechanisms also involved in the entry of nanomedicine in cells, such as clathrin-mediated endocytosis, caveolae-mediated endocytosis, macropinocytosis, and other clathrin- and caveolae-independent endocytosis mechanisms. Nanocarriers are a better formulation strategy to facilitate endocytosis and overcome some undesirable properties of drug molecules and also can prevent drugs from premature *in vivo* degradation, which results in drug release over a long period of time systemically.

#### 4.6.2.2 Pluronic Nanomedicine

Pluronics are block copolymers of hydrophilic poly (ethylene oxide) (PEO) and hydrophobic and poly (propylene oxide (PPO), arranged in a PEO-PPO-PEO structure and also known as poloxamers. Pluronics are materials of high interest in cancer drug delivery to overcome MDR due to their potent drug encapsulating properties and having the intrinsic ability to modify biological/pharmacological responses of MDR cancer cells *in vitro* and *in vivo* (Batrakova and Kabanov, 2008). Hydrophilic-lipophilic balance (HLB) of the pluronics copolymers depends on the length of the blocks and spontaneously forms micelles above the critical micelle concentration (CMC). The micelle core possesses a hydrophobic nature due to the presence of PPO blocks which allows incorporation of hydrophobic drugs. Pluronics prevented the development of MDR in combination with various anticancer drugs (Alakhova and Kabanov, 2014). The MDR sensitization by pluronics mechanisms involves various complex and specific events which occur at different parts in the cell. The polymer-cell interaction begins at the cell membrane where most of the drug-efflux transporters remain present. Pluronics exert an inhibitory effect on ABC transporters (i.e., P-gp, MRP, and BCRP known as Breast Cancer Resistance Protein) by suppressing the transporters' ATPase activity and their interaction with the drug. The amphiphilic structure of pluronics might be responsible for alteration of the lipid microenvironment and can interact with the cell membrane, as well as alter its properties, which are crucial for proper function of ABC transporters. The pluronics copolymer is rapidly taken up by the cells via a caveolae-mediated endocytosis pathway and depletes the intracellular ATP levels in MDR cancer cells. The chemosensitization ability of the copolymers in MDR cells is strongly correlated with the ability to deplete cellular ATP levels (Batrakova *et al.*, 2001). It also stimulates the production of reactive oxygen species (ROS) and release of cytochrome *c*, which are the early signs of the mitochondrial apoptotic pathway. Some reports suggest that the alterations in gene expression profiles with copolymers in limited situations are also responsible for pluronics action on MDR cells (Alakhova *et al.*, 2010). Several studies have been published in this regard which confirm the anti-MDR effects of pluronics nanomedicine. Along with the unique biological activities of pluronics, the drug solubilization capability also make them very attractive polymers for drug delivery purposes. However, the biological response alteration property (i.e., ability to chemosensitize MDR cells) has been observed in some other polymers also (Werle, 2008; Cambon *et al.*, 2013). Hence, it can be concluded that the nanocargos developed using pluronics polymers have great attention to combat MDR with other anticancer agents.

### 4.6.3 Stimuli-Sensitive Drug Delivery

A usual limit of conventional nanoparticle-based drug delivery is partial and uncontrolled release of drug in the bioenvironment of choice. Nanoformulation, when accumulated in the cellular environment of either normal or cancerous cell, starts release of the loaded content or drug without recognition of cell type. Hence, it is necessary to customize nanomedicine to develop such stimuli-responsive systems that produce the biological response at the disease target to achieve improved delivery efficiency and therapeutic efficacy with minimal adverse effects. Stimuli-responsive systems have unique features to show a large ‘response’ to a small alteration in the chemical or physical conditions of the environment (Iyer *et al.*, 2013). They are effective carriers to achieve localized drug release. The stimuli-responsive concept is applied to release the drug by a stimulus such as pH and other physical forces, which include magnetic fields, ultrasound, hyperthermia, or light. This unique technology has been applied using nanomedicine to overcome MDR. Such delivery systems can be classified on the basis of the source of required stimuli, namely, externally regulated and self-regulated delivery systems, respectively (Gao *et al.*, 2012).

#### 4.6.3.1 pH-Sensitive Nanocarriers

The pH-sensitive nanocarriers respond to internal biological changes in the pH to modulate the drug release. As discussed previously, the microenvironment of a tumor is significantly different from that of normal tissue. The lower pH of a tumor cell has been useful for designing smart materials, one of which is sensitive to pH of the environment and also provides controlled drug delivery towards the tumor location (Gerweck and Seetharaman, 1996). MDR cells acquire low pH during the endocytosis process due to activation of V-ATPase in the endosomal membrane that causes an influx of protons during the maturation process of endosomes from early endosomes to late endosomes. This pH drop is continuing even further when late endosomes fuse with the lysosomes (Yamashiro and Maxfield, 1988). Recently, a large number of studies have been reported that are based on the pH-activated mechanism to overcome efflux-dependent MDR by disrupting the endosomal membrane and releasing drugs into the cytoplasm in a burst-like manner from cargos of NPs (nanoparticles) (Yadav *et al.*, 2009; Kievit *et al.*, 2011; Gu *et al.*, 2012; Paliwal *et al.*, 2015, 2016). The development of a pH-responsive nanocarriers system allows targeting of the lysosomal environment instead of endosomes to retard the undesired drug release in the tumor stroma. Hence, it is exceptionally important to design the pKb of polymers in such a manner which is sensitive to the lysosomal pH to obtain the precise pH-sensitive drug release. The stability of the nanocarrier in the slightly acidic conditions of extracellular tumor sarcoma is also a considerable thought in the development of MDR-reversal pH-sensitive nanocarriers. Although, such carriers should be very sensitive to structural transformations and solubility changes in the more acidic lysosomal compartments to trigger drug release. Occasionally, there is a situation when pH stimuli formulation fails to produce the desired therapeutic level at the subcellular organelles target even after being released in the endosome or lysosome due to drug escape from the endosomes. It may happen because the pH-responsive polymers caused alterations of endosomal membranes by membrane interactions and/or proton absorption, resulting osmotic swelling and rupturing of the membranes and hence creation of defects such as pores or channels in the lipid membranes (Lee *et al.*, 2005; Chen *et al.*, 2009). It was observed that DOX, in an acidic microenvironment of late endosomes/lysosomes, was subjected to rapid protonation, rendering it water soluble and positively charged, which hampered its diffusion through the endosomal/lysosomal membrane, trapping it (Larsen *et al.*, 2000). Hence, it is beneficial in such cases that nanocarriers run away from the endosome/lysosome by distorting their membranes in an intact form instead of only disintegrating in the endosome/lysosome (Shen *et al.*, 2009).

### 4.6.4 Nanocarrier-Based Drug Delivery

Various advanced nanotechnological methods are reported now that involve either the energy or any external stimulus for controlling the drug release from nanocarriers. These systems provide more precisely controlled delivery and release profiles depending upon the strength and duration of the applied external energy or stimuli. This is also a beneficial step towards targeting therapeutics for diseases like cancer. For such purposes, heat, light, ultrasound, and magnetic and electrical energy have been generally used to regulate the delivery of the drug at the site of interest.

#### 4.6.4.1 Ultrasound Delivery

The applications of ultrasound energy for the purpose of imaging and diagnostics are well investigated. In contrast to the diagnostic uses of ultrasound, with expectation to minimize energy deposition within the tissue which leads to reduce 'biological effect', the therapeutic use of ultrasound is supposed to be maximizing the energy deposition to generate its effect. The energy of ultrasound can be utilized for the purpose of the tumor-specific drug release from nanocarriers. Ultrasound energy is able to focus over a small area which leads to very high intensity with deeper penetration efficiency (Cheung and Neyzari, 1984). Ultrasound induces thermal effects that help nanocarrier systems to extravagate the tumor. The use of ultrasound techniques with other release mechanics could be highly beneficial for MDR cancers. Ultrasound energy for enhanced drug delivery with less residual toxicity concern is suitable for the development of nanocarriers for therapeutic application in MDR cancer cells.

#### 4.6.4.2 Hyperthermia-Based Delivery

Hyperthermia application in cancer therapy involves the localized heating of tumor between the temperatures of 41 and 46°C. Induction of hyperthermia is responsible for facilitating drug delivery from a thermosensitive nanocarrier. For this, it is necessary to explore potentially important aspects to achieve a controlled desired local temperature without damaging surrounding tissues. The thermo-responsive ability of polymers can be utilized for this purpose. The encapsulated anticancer drug is released from such thermosensitive material above body temperature by external heating techniques. Drug delivery using nanocarriers along with hyperthermia has also been useful to overcome MDR-related cancer. A recent report suggested that alteration in the vesicle permeability, perfusion, and interstitial fluid flow across the tumor also play an important role in efficiency improvement during local hyperthermia-mediated drug delivery (Li *et al.*, 2013). Hence, the application of a thermosensitive nanocarrier delivery system holds tremendous opportunities to regulate the cancer growth due to improved localization of drug in the tumor.

#### 4.6.4.3 Photodynamic Delivery

Photodynamic therapy (PDT) has wide range of applications in cancer therapeutics due to its specificity and selectivity. Light is used as external stimuli for this purpose. The administration of a photosensitizer followed by illumination towards the cancer site is required in PDT (Huang *et al.*, 2015). The PDT can be readily tuned and light exposure can be focused on the surface with easy control of its spatial and temporal exposure. To achieve this, light-sensitive materials have been extensively used in drug delivery. PDT has clinical approval with minimum invasiveness in therapeutic modality. It has been potentially useful for curative and palliative management of various premalignant (e.g., actinic keratosis, Barrett's oesophagus) and malignant (e.g., obstructive lung cancer and oesophageal cancer) diseases worldwide (Huang, 2005; Huang *et al.*, 2009). PDT is less affected by cancer resistance at the cellular and molecular level via drug efflux, hypoxia, levels of pigmentation, and damage reversal mechanisms. Serious considerations have been taken to overcome these defence mechanisms in order to make anticancer PDT more effective and acceptable. Skin photosensitivity is a major side effect of PDT which can be reduced by the application of nanotechnology-based photosensitizer drug delivery. The dual targeting ability of nanocarriers may be beneficial to localize drugs into tumors by conjugation with tumor-targeting ligands in addition to PDT. An interesting approach of plasmonic nanobubble-enhanced endosomal escape (PNBEE) also has been reported recently for fast, selective, and guided intracellular drug delivery (Lukianova-Hleb *et al.*, 2012).

#### 4.6.4.4 Magnetically Responsive Delivery

Some materials, which are responsive to applied magnetic fields such as magnetite can be potentially incorporated into a delivery vector or nanocarrier to design a magnetically regulated/responsive drug-delivery system. The basic principle behind this approach is that the therapeutic agent can be attached or encapsulated within micro- or nanoparticles. These systems are then directed towards targeting sites under the influence of magnetic fields. Beside the application of magnetically responsive delivery in cancer targeting, it is useful for targeting other sites also. In addition to facilitating targeted drug delivery, the unique advantage of magnetic targeting is rapid sedimentation of drug-encapsulated nanocarriers in tumor vasculature (McBain *et al.*, 2008). Alternatively, one can also attain controlled drug release from magnetically active delivery vehicles by applying magnetic field-induced hyperthermia (Moroz *et al.*, 2002).

### 4.6.5 Si-RNA Based Nanomedicine

At present, RNA interference (RNAi) gene silencing or inhibition has gained much attention for gene-specific therapeutics. There are some limitations with the use of low MW compounds for inhibition of activity of MDR genes to overcome MDR. Though, it is a feasible therapeutic approach but, practically, clinical success with this approach is limited. Hence, development of innovative and alternate strategies such as RNA interference (RNAi) to overcome MDR is desired. Generally, RNAi can occur in two different ways: the first is small interfering RNA (siRNAs) delivery to deliver the target cells, which is also called RNA-based approach and other is a DNA-based approach which involves the generation of effector siRNAs by the intracellular processing of RNA hairpin transcripts (Saraswathy and Gong, 2013). Besides the potent use of RNAi therapy for minimizing the MDR effect, questions arise regarding *in vivo* delivery of siRNAs to targeted tumor sites because of the rapid degradation of naked siRNAs due to presence of serum ribonucleases. Several approaches are available to fulfil various requirements of siRNA delivery to overcome MDR by using different types of nanocarriers such as polymeric nanoparticles, cationic liposomes, lipoplexes, dendrimers, and micelles (Gao *et al.*, 2012). Yhee *et al.* (2015) developed a novel tumor targeted glycol chitosan nanoparticle loaded with small siRNA for P-gp down-regulation for reversal of MDR in cancer. These nanoparticles were quite stable and provided efficient delivery of protected siRNA to adriamycin-resistant breast cancer cells (MCF-7/ADR) after intravenous administration. A combination therapeutic strategy is a standard approach for better treatment of MDR cells. In a recent study, co-delivery of docetaxel (DTX) and BCL-2 siRNA has been achieved using PEGylated liposome in a lung cancer model (A549). The *in vivo* anticancer study on A549 cells bearing a xenograft tumor model illustrated significant tumor suppression with the proposed delivery system as a consequence of the synergistic effect of DTX potency and MDR reversing the ability of BCL-2 siRNA in the tumor mass (Qu *et al.*, 2014). Shen *et al.*, (2014) investigated micelles consisting of a polycationic PEI-CyD shell to incorporate the siRNA and hydrophobic core to condense the antineoplastic agent, doxorubicin. *In vivo* studies confirmed that the dual cargo loaded micelles induced tumor-cell apoptosis and inhibited the growth of resistant tumor. Summary of recent research reports (2011–2016) related to the nanocarrier-based MDR treatment are presented in Table 4.2.

**TABLE 4.2**
**Recent Research Published on Nanocarrier-Based MDR Treatment**

| S. No. | Nanocarriers | Approaches | Cancer | Reference |
|---|---|---|---|---|
| 1 | mPEG-PCL-g-cellulose micellar nanoparticles | Overcoming P-gp efflux in MDR | MCF-7/WT and MCF-7/ADR | Chen *et al.* (2011) |
| 2 | pH-responsive mesoporous silica nanoparticles | Synergistic cell cycle arrest/ apoptosis-inducing chemosensitization effect of the surfactant CTAB | MCF-7/ADR | He *et al.* (2011) |
| 3 | Iron oxide nanoparticle | pH-sensitive release | C6-ADR | Kievit *et al.* (2011) |
| 4 | Long-circulatory liposomes | Co-delivery of the P-gp inhibitor and cytotoxic agents | SKOV-3 and SKOV-3TR | Patel *et al.* (2011) |
| 5 | Redox-responsive micellar system | Redox responsive system | MCF-7/ADR | Wang *et al.* (2011) |
| 6 | Dual-functional liposome | Drug-efflux transporter inhibition and targeting | HCC Huh-7 and MDCKII/BCRP | Zhang *et al.* (2012) |
| 7 | Pluronic P85-PEI/TPGS complex nanoparticles | Co-delivery of PTX and survivin shRNA | A549 and A549/T | Shen *et al.* (2012) |
| 8 | CD44-targeted hyaluronic acid (HA) nanoparticle (NP) | SiRNA-based down-regulation of anti-apoptotic genes overexpressed in cisplatin-resistant tumors | Human lung tumors that were sensitive and resistant to cisplatin | Ganesh *et al.* (2013) Shah *et al.* (2013) |
| 9 | PAMAM-based nanocarriers | Co-delivery of chemotherapy and siRNA agents | Human alveolar adenocarcinoma cell line, A549, mouse yolk sac embryo cells, c-166 | Biswas *et al.* (2013) |
| 10 | Photofrin II polymeric micelles | Photodynamic therapy | MCF-7/WT SKOV-3 | Lamch *et al.* (2014) |
| 11 | pH-responsive nanocarrier | Chemo- and photothermal therapy with pH responsiveness | MCF-7/WT and MCF-7/ADR | Feng *et al.* (2014) |
| 12 | Multifunctional nanoparticles | Receptor-mediated endocytosis | MCF-7 and MCF-7/ADR | Wang *et al.* (2014a,b) |
| 13 | Chitosan-g-TPGS nanoparticles | P-gp inhibition by TPGS | HepG2 and BEL-7402 | Guo *et al.* (2013) |
| 14 | Single-walled carbon nanotube (SWCNT) | Cholanic acid-derivatized hyaluronic acid act as sensitizer to influence drug uptake and induce apoptosis | OVCAR8 and OVCAR8/ADR | Bhirde *et al.* (2014) |
| 15 | Dual drug-loaded mixed micelles (PF-DP) | Pluronic-based synergistic effect | MCF-7/ADR tumor-bearing mice | Chen *et al.* (2015) |
| 16 | PEGylated polymeric micelles | Co-delivery with P-gp inhibitor and apoptosis inhibitor | MCF7/ADR cells | Wang *et al.* (2015) |
| 17 | Polyamidoamine dendrimer-modified selenium nanoparticles | Combined chemotherapy and gene therapy technology | A549/DDP cells | Zheng *et al.* (2015) |
| 18 | Chitosan nanoparticles | P-gp down-regulation by SiRNA | MCF-7/ADR | Yhee *et al.* (2015) |
| 19 | Exosomes (natural nanoparticles) | Co-localization of airway-delivered exosomes | $MDCK_{MDR1}$ (P-gp +) cells | Kim *et al.* (2016) |
| 20 | Yolk-shell magnetic mesoporous nanocomposites | Combined magnetic and ligand targeting | Hep-G2/MDR cancer cells | Wang *et al.* (2016) |

*(Continued)*

**TABLE 4.2 (*Continued*)**
**Recent Research Published on Nanocarrier-Based MDR Treatment**

| S. No. | Nanocarriers | Approaches | Cancer | Reference |
|---|---|---|---|---|
| 21 | Mesoporous silica nanoparticle | Ligand (MAb) targeting | PC9-DR | Cabral and Kataoka (2016) |
| 22 | Q-Graphene | Ligand (hyaluronic acid) targeting | Human lung epithelial tumor cell line A549 and human embryonic lung cell line MRC-5 | Luo *et al.* (2016) |
| 23 | Reduction-sensitive liposome | Co-delivery of (P-gp) inhibitor and anticancer agents and increased tumor accumulation | MCF-7/ADR tumor | Tang *et al.* (2016) |
| 24 | Polymeric micelles | Combination therapy with SiRNA | SKOV-3, MCF-7 and A549 | Riehle *et al.* (2016) |

## 4.7 FUTURE PERSPECTIVES

In this chapter, we covered several nanocarrier-mediated formulation development strategies with increased cytotoxicity to drug-resistant tumor cells in comparison to the treatment with either entity alone, prolonged release of encapsulated drug/inhibitor which leads to a sustained sensitization of resistant cells facilitating tumor-cell killing by overcoming MDR, and reduced systemic pharmacokinetic interactions between cytotoxic drugs and MDR inhibitors. The complicated mechanisms of MDR are yet to be fully understood, and this makes application of nanotechnologies for combating MDR more complicated. Hence, a basic need for future nanomedicine development is based on the research outcomes on physiology and pharmacology of MDR cancer cells to consider the nanocarrier's application in overcoming MDR. Numerous materials are available that can be integrated with the carrier to overcome MDR. They exhibit MDR modulation at even very low concentrations, and their clinical effectiveness has been approved by drug regulatory authorities for varied pharmaceutical applications. However, further in-depth characterization would be required to elucidate their mode of action and their interaction with plasma membranes both as nanocarriers and in free form. Nanocarriers for the delivery of anticancer drugs have made significant contributions to reverse MDR but the complications of manufacturing and scale-up of such nanocarriers must be addressed in the future. A rapidly increasing literature in the form of publications provides hope where latest techniques are being employed in the design of nanocarriers with different structure, but the reproducibility concern of these methods is still a question today. Further, the translation of nanocarriers for chemotherapeutics delivery towards MDR cancer from bench to bedside has to face many questions with several issues of their safety and toxicological manifestation. Another concern is the structural and chemical stability of nanocarriers during storage and the need of optimization for premature release of the drug. In totality, we can be optimistic but with cautions while applying nanocarrier-mediated drug delivery towards MDR cancer with a new hope to cancer patients.

## ACKNOWLEDGMENT

The authors acknowledge SERB-DST, New Delhi, for providing financial assistance in the form of young scientist project grant to SRP (Reg. No. SB/YS/LS333/2013).

## REFERENCES

Aggarwal BB, Kumar A, Bharti AC. Anticancer potential of curcumin: Preclinical and clinical studies. *Anticancer Res.* 2003; 23: 363–98.

Alakhova DY, Kabanov AV. Pluronics and MDR reversal: An update. *Mol Pharm.* 2014; 11: 2566–78.

Alakhova DY, Rapoport NY, Batrakova EV, Timoshin AA, Li S, Nicholls D *et al.* Differential metabolic responses to pluronic in MDR and non-MDR cells: A novel pathway for chemosensitization of drug resistant cancers. *J Control Release.* 2010; 142: 89–100.

Alamolhodaei NS, Tsatsakis AM, Ramezani M, Hayes AW, Karimi G. Resveratrol as MDR reversion molecule in breast cancer: An overview. *Food Chem Toxicol.* 2017; 103: 223–32.

Aller SG, Yu J, Ward A, Weng Y, Chittaboina S, Zhuo R *et al.* Structure of P-glycoprotein reveals a molecular basis for poly-specific drug binding. *Science.* 2009; 323: 1718–22.

Ambudkar SV, Kim IW, Sauna ZE. The power of the pump: Mechanisms of action of P-glycoprotein (ABCB1). *Eur J Pharm Sci.* 2006; 27: 392–400.

Amellem O, Pettersen EO. Cell inactivation and cell cycle inhibition as induced by extreme hypoxia: The possible role of cell cycle arrest as a protection against hypoxia-induced lethal damage. *Cell Prolif.* 1991; 24: 127–41.

Amiji MM, Iyer AK. Multifunctional self-assembling polymeric nanosystems. US Patent 9,173,840; 2015.

Amiji MM, Shenoy DB, Vlerken LV. Nanoparticulate delivery systems for treating multi-drug resistance. US Patent 20,060,257,493; 2006.

Amjad MW, Kesharwani P, Amin MCIM, Iyer AK. Recent advances in the design, development, and targeting mechanisms of polymeric micelles for delivery of siRNA in cancer therapy. *Prog Polym Sci.* 2017; 64: 154–81.

Arrigoni E, Galimberti S, Petrini M, Danesi R, Di Paolo A. ATP-binding cassette transmembrane transporters and their epigenetic control in cancer: An overview. *Expert Opin Drug Metab Toxicol.* 2016; 12: 1419–32.

Batrakova EV, Kabanov AV. Pluronic block copolymers: Evolution of drug delivery concept from inert nanocarriers to biological response modifiers. *J Control Release.* 2008; 130: 98–106.

Batrakova EV, Li S, Vinogradov SV, Alakhov VY, Miller DW, Kabanov AV. Mechanism of pluronic effect on P-glycoprotein efflux system in blood-brain barrier: Contributions of energy depletion and membrane fluidization. *J Pharmacol Exp Ther.* 2001; 299: 483–93.

Beg S, Katare O, Saini S, Garg B, Khurana RK, Singh B. Solid self-nanoemulsifying systems of olmesartan medoxomil: Formulation development, micromeritic characterization, *in vitro* and *in vivo* evaluation. *Powder Technology.* 2016; 294: 93–104.

Benoit J-p, Lamprecht A. Use of p-glycoprotein inhibitor surfactants at the surface of a colloidal carrier. US Patent 20,060,073,196; 2006.

Bhirde AA, Chikkaveeraiah BV, Srivatsan A, Niu G, Jin AJ, Kapoor A *et al.* Targeted therapeutic nanotubes influence the viscoelasticity of cancer cells to overcome drug resistance. *Acs Nano.* 2014; 8: 4177–89.

Biedler JL, Riehm H. Cellular resistance to actinomycin D in Chinese hamster cells *in vitro*: Cross-resistance, radioautographic, and cytogenetic studies. *Cancer Res.* 1970; 30: 1174–84.

Biswas S, Deshpande PP, Navarro G, Dodwadkar NS, Torchilin VP. Lipid modified triblock PAMAM-based nanocarriers for siRNA drug co-delivery. *Biomaterials.* 2013; 34: 1289–1301.

Borska S, Chmielewska M, Wysocka T, Drag-Zalesinska M, Zabel M, Dziegiel P. *In vitro* effect of quercetin on human gastric carcinoma: Targeting cancer cells death and MDR. *Food Chem Toxicol.* 2012; 50: 3375–83.

Bounous G, Molson JH. The antioxidant system. *Anticancer Res.* 2003; 23: 1411–5.

Brinker JC, Ashley CE, Jiang X, Liu J, Peabody DS, Wharton WR *et al.* Protocells and their use for targeted delivery of multicomponent cargos to cancer cells. US Patent 8,992,984; 2015.

Brown JM. Tumor hypoxia, drug resistance, and metastases. *J Natl Cancer Inst.* 1990; 82: 338–9.

Brown JM, Wilson WR. Exploiting tumour hypoxia in cancer treatment. *Nat Rev Cancer.* 2004; 4: 437–47.

Cabral H, Kataoka K. Multifunctional nanoassemblies of block copolymers for future cancer therapy. *Sci. Tech. Adv. Mater.* 2016.

Cairns R, Papandreou I, Denko N. Overcoming physiologic barriers to cancer treatment by molecularly targeting the tumor microenvironment. *Mol Cancer Res.* 2006; 4: 61–70.

Cambon A, Rey-Rico A, Barbosa S, Soltero JF, Yeates SG, Brea J *et al.* Poly(styrene oxide)-poly(ethylene oxide) block copolymers: From 'classical' chemotherapeutic nanocarriers to active cell-response inducers. *J Control Release.* 2013; 167: 68–75.

Cantley LC. The phosphoinositide 3-kinase pathway. *Science.* 2002; 296: 1655–7.

Carmeliet P, Jain RK. Angiogenesis in cancer and other diseases. *Nat.* 2000; 407: 249–57.

Chang EL. Magnetically-localizable, polymerized lipid vesicles and method of disrupting same. US Patent 4,652,257; 1987.

Chaudhry P, Asselin E. Resistance to chemotherapy and hormone therapy in endometrial cancer. *Endocr Relat Cancer.* 2009; 16: 363–80.

Cheeseman KH. Mechanisms and effects of lipid peroxidation. *Mol Aspects Med.* 1993; 14: 191–7.

Chen CH, Cuong NV, Chen YT, So RC, Liau I, Hsieh MF. Overcoming multidrug resistance of breast cancer cells by the micellar doxorubicin nanoparticles of mPEG-PCL-graft-cellulose. *J Nanosci Nanotechnol.* 2011; 11: 53–60.

Chen CJ, Chin JE, Ueda K, Clark DP, Pastan I, Gottesman MM *et al.* Internal duplication and homology with bacterial transport proteins in the mdr1 (P-glycoprotein) gene from multidrug-resistant human cells. *Cell.* 1986; 47: 381–9.

Chen R, Khormaee S, Eccleston ME, Slater NK. The role of hydrophobic amino acid grafts in the enhancement of membrane-disruptive activity of pH-responsive pseudo-peptides. *Biomaterials.* 2009; 30: 1954–61.

Chen Y, Zhang W, Huang Y, Gao F, Sha X, Fang X. Pluronic-based functional polymeric mixed micelles for co-delivery of doxorubicin and paclitaxel to multidrug resistant tumor. *Int J Pharm.* 2015; 488: 44–58.

Cheung AY, Neyzari A. Deep local hyperthermia for cancer therapy: External electromagnetic and ultrasound techniques. *Cancer Res.* 1984; 44: 4736s–44s.

Cole SP, Bhardwaj G, Gerlach JH, Mackie JE, Grant CE, Almquist KC *et al.* Overexpression of a transporter gene in a multidrug-resistant human lung cancer cell line. *Science.* 1992; 258: 1650–4.

Comerford KM, Wallace TJ, Karhausen J, Louis NA, Montalto MC, Colgan SP. Hypoxia-inducible factor-1-dependent regulation of the multidrug resistance (MDR1) gene. *Cancer Res.* 2002; 62: 3387–94.

Debenham PG, Kartner N, Siminovitch L, Riordan JR, Ling V. DNA-mediated transfer of multiple drug resistance and plasma membrane glycoprotein expression. *Mol Cell Biol.* 1982; 2: 881–9.

Deeley RG, Cole SP. Methods for conferring multidrug resistance on a cell. US Patent 5,891,724; 1999.

Deng HB, Parekh HK, Chow KC, Simpkins H. Increased expression of dihydrodiol dehydrogenase induces resistance to cisplatin in human ovarian carcinoma cells. *J Biol Chem.* 2002; 277: 15035–43.

Denko NC. Hypoxia, HIF1 and glucose metabolism in the solid tumour. *Nat Rev Cancer.* 2008; 8: 705–13.

Dewhirst MW, Kimura H, Rehmus SW, Braun RD, Papahadjopoulos D, Hong K *et al.* Microvascular studies on the origins of perfusion-limited hypoxia. *Br J Cancer Suppl.* 1996; 27: S247–51.

Drummond DC, Kirpotin DB, Marks JD, Zhou Y. Nanoparticles effective for internalization into cells. US Patent 20,100,008,978; 2010.

El-sayed ME, Ensminger W, Shewach D. Targeted dendrimer-drug conjugates. US Patent 9,345,781; 2016.

Feng L, Li K, Shi X, Gao M, Liu J, Liu Z. Smart pH-responsive nanocarriers based on nano-graphene oxide for combined chemo-and photothermal therapy overcoming drug resistance. *Adv Healthc Mater.* 2014; 3: 1261–71.

Fernandez V, Videla LA. Biochemical aspects of cellular antioxidant systems. *Biol Res.* 1996; 29: 177–82.

Fitzmaurice C, Allen C, Barber RM, Barregard L, Bhutta ZA, Brenner H *et al.* Global, regional, and national cancer incidence, mortality, years of life lost, years lived with disability, and disability-adjusted life-years for 32 cancer groups, 1990 to 2015: A systematic analysis for the global burden of disease study. *JAMA Oncology.* 2017; 3: 524–48.

Fletcher JI, Haber M, Henderson MJ, Norris MD. ABC transporters in cancer: More than just drug efflux pumps. *Nat Rev Cancer.* 2010; 10: 147–56.

Foulkes MJ, Henry KM, Rougeot J, Hooper-Greenhill E, Loynes CA, Jeffrey P *et al.* Expression and regulation of drug transporters in vertebrate neutrophils. *Sci Rep.* 2017; 7: 4967.

Freitas I. Role of hypoxia in photodynamic therapy of tumors. *Tumori.* 1985; 71: 251–9.

Fukumura D, Jain RK. Tumor microenvironment abnormalities: Causes, consequences, and strategies to normalize. *J Cell Biochem.* 2007; 101: 937–49.

Ganesh S, Iyer AK, Gattacceca F, Morrissey DV, Amiji MM. *In vivo* biodistribution of siRNA and cisplatin administered using CD44-targeted hyaluronic acid nanoparticles. *J Control Release.* 2013; 172: 699–706.

Gao Z, Zhang L, Sun Y. Nanotechnology applied to overcome tumor drug resistance. *J Control Release.* 2012; 162: 45–55.

Gerweck LE, Seetharaman K. Cellular pH gradient in tumor versus normal tissue: Potential exploitation for the treatment of cancer. *Cancer Res.* 1996; 56: 1194–8.

Gillet JP, Efferth T, Remacle J. Chemotherapy-induced resistance by ATP-binding cassette transporter genes. *Biochim Biophys Acta.* 2007; 1775: 237–62.

Gombotz WR, Pun S, Mount C, Kim TH, Dulken BW. Mixed micelles. US Patent 9,295,685; 2016.

Gothwal A, Khan I, Gupta U. Polymeric Micelles: Recent Advancements in the Delivery of Anticancer Drugs. *Pharm Res.* 2016; 33: 18–39.

Gottesman MM, Fojo T, Bates SE. Multidrug resistance in cancer: Role of ATP-dependent transporters. *Nat Rev Cancer.* 2002; 2: 48–58.

Gu YJ, Cheng J, Man CW, Wong WT, Cheng SH. Gold-doxorubicin nanoconjugates for overcoming multidrug resistance. *Nanomedicine.* 2012; 8: 204–11.

Guo Y, Chu M, Tan S, Zhao S, Liu H, Otieno BO *et al.* Chitosan-g-TPGS nanoparticles for anticancer drug delivery and overcoming multidrug resistance. *Mol Pharm.* 2013; 11: 59–70.

Hanahan D, Weinberg RA. The hallmarks of cancer. *Cell.* 2000; 100: 57–70.

Harris AL, Hochhauser D. Mechanisms of multidrug resistance in cancer treatment. *Acta Oncol.* 1992; 31: 205–13.

He Q, Gao Y, Zhang L, Zhang Z, Gao F, Ji X *et al.* A pH-responsive mesoporous silica nanoparticles-based multi-drug delivery system for overcoming multi-drug resistance. *Biomaterials.* 2011; 32: 7711–20.

Heath TD, Shek P, Papahadjopoulos D. Production of immunogens by antigen conjugation to liposomes. US Patent 4,565,696; 1986.

Heldin CH, Rubin K, Pietras K, Ostman A. High interstitial fluid pressure-an obstacle in cancer therapy. *Nat Rev Cancer.* 2004; 4: 806–13.

Hirst DG, Denekamp J. Tumour cell proliferation in relation to the vasculature. *Cell Tissue Kinet.* 1979; 12: 31–42.

Hockel M, Vaupel P. Tumor hypoxia: Definitions and current clinical, biologic, and molecular aspects. *J Natl Cancer Inst.* 2001; 93: 266–76.

Hodge JW, Ardiani A, Farsaci B, Kwilas AR, Gameiro SR. The tipping point for combination therapy: Cancer vaccines with radiation, chemotherapy, or targeted small molecule inhibitors. *Semin Oncol.* 2012; 39: 323–39.

Hong S, Liu Y. Nano-hybrid delivery system for sequential utilization of passive and active targeting. US Patent 9,168,225; 2015.

Huang S-k, Zalipsky S. Gene delivery mediated by liposome-DNA complex with cleavable peg surface modification. US Patent 7,390,780; 2008.

Huang Z. A review of progress in clinical photodynamic therapy. *Technol Cancer Res Treat.* 2005; 4: 283–93.

Huang Z, Hsu Y-C, Li L-B, Wang L-W, Song X-D, Yow CM *et al.* Photodynamic therapy of cancer—Challenges of multidrug resistance. *J Innov Opt Health Sci.* 2015; 8: 1530002.

Huang Z, Li L, Wang H, Wang X, Yuan K, Meyers A *et al.* Photodynamic therapy—An update on clinical applications. *J Innov Opt Health Sci.* 2009; 2: 73–92.

Inoue K, Narukawa M, Takeuchi M. Factors affecting efficacy and safety of add-on combination chemotherapy for non-small-cell lung cancer: A literature-based pooled analysis of randomized controlled trials. *Lung.* 2012; 190: 355–64.

Irshad M, Chaudhuri PS. Oxidant-antioxidant system: Role and significance in human body. *Indian J Exp Biol.* 2002; 40: 1233–9.

Iyer AK, Singh A, Ganta S, Amiji MM. Role of integrated cancer nanomedicine in overcoming drug resistance. *Adv Drug Deliv Rev.* 2013; 65: 1784–802.

Jain RK, Stylianopoulos T. Delivering nanomedicine to solid tumors. *Nat Rev Clin Oncol.* 2010; 7: 653–64.

Joe B, Lokesh BR. Role of capsaicin, curcumin and dietary n-3 fatty acids in lowering the generation of reactive oxygen species in rat peritoneal macrophages. *Biochim Biophys Acta.* 1994; 1224: 255–63.

Juliano RL, Ling V. A surface glycoprotein modulating drug permeability in Chinese hamster ovary cell mutants. *Biochim Biophys Acta.* 1976; 455: 152–62.

Kalia NP. Efflux-mediated drug resistance in *Staphylococcus aureus.* In: Arora G, Sajid A and Kalia VC, eds., *Drug Resistance in Bacteria, Fungi, Malaria, and Cancer* (ISBN: 978-3-319-48683-3). Springer, 2017; pp. 307–23.

Kaur S, Mehra NK, Jain K, Jain NK. Development and evaluation of targeting ligand-anchored CNTs as prospective targeted drug delivery system. *Artif Cells Nanomed Biotechnol.* 2017; 45: 242–50.

Kerr JF, Wyllie AH, Currie AR. Apoptosis: A basic biological phenomenon with wide-ranging implications in tissue kinetics. *Br J Cancer.* 1972; 26: 239–57.

Khizroev S, Guduru R, Liang P. On-demand drug release using magneto-electric nanoparticles. US Patent 20,160,030,724; 2016.

Kievit FM, Wang FY, Fang C, Mok H, Wang K, Silber JR *et al.* Doxorubicin loaded iron oxide nanoparticles overcome multidrug resistance in cancer *in vitro. J Control Release.* 2011; 152: 76–83.

Kim MS, Haney MJ, Zhao Y, Mahajan V, Deygen I, Klyachko NL *et al.* Development of exosome-encapsulated paclitaxel to overcome MDR in cancer cells. *Nanomedicine.* 2016; 12: 655–64.

Kratz F, Haag R, Calderon M. Drug conjugates with polyglycerols. US Patent 20,110,117,009; 2011.

Krishna R, Mayer LD. Modulation of P-glycoprotein (PGP) mediated multidrug resistance (MDR) using chemosensitizers: Recent advances in the design of selective MDR modulators. *Curr Med Chem Anticancer Agents.* 2001; 1: 163–74.

Kurmi BD, Gajbhiye V, Kayat J, Jain NK. Lactoferrin-conjugated dendritic nanoconstructs for lung targeting of methotrexate. *J Pharm Sci.* 2011; 100: 2311–20.

Kurmi BD, Kayat J, Gajbhiye V, Tekade RK, Jain NK. Micro- and nanocarrier-mediated lung targeting. *Expert Opin Drug Deliv.* 2010; 7: 781–94.

Kurmi BD, Tekchandani P, Paliwal R, Paliwal SR. Nanocarriers in Improved Heparin Delivery: Recent Updates. *Curr Pharm Des.* 2015; 21: 4509–18.

Kurmi BD, Tekchandani P, Paliwal R, Paliwal SR. Transdermal Drug Delivery: Opportunities and Challenges for Controlled Delivery of Therapeutic Agents Using Nanocarriers. *Curr Drug Metab.* 2017; 18: 481–95.

Labhasetwar VD, Sahoo SK. Transferrin-conjugated nanoparticles for increasing efficacy of a therapeutic agent. US Patent 20,100,015,051; 2010.

Lage H. An overview of cancer multidrug resistance: A still unsolved problem. *Cell Mol Life Sci.* 2008; 65: 3145–67.

Lamch L, Bazylinska U, Kulbacka J, Pietkiewicz J, Biezunska-Kusiak K, Wilk KA. Polymeric micelles for enhanced Photofrin II (R) delivery, cytotoxicity and pro-apoptotic activity in human breast and ovarian cancer cells. *Photodiagnosis Photodyn Ther.* 2014; 11: 570–85.

Lammers T, Hennink WE, Storm G. Tumour-targeted nanomedicines: Principles and practice. *Br J Cancer.* 2008; 99: 392–7.

Lane DP. Cancer. p53, guardian of the genome. *Nature.* 1992; 358: 15–6.

Larsen AK, Escargueil AE, Skladanowski A. Resistance mechanisms associated with altered intracellular distribution of anticancer agents. *Pharmacol Ther.* 2000; 85: 217–29.

Lee ES, Gao Z, Bae YH. Recent progress in tumor pH targeting nanotechnology. *J Control Release.* 2008; 132: 164–70.

Lee ES, Na K, Bae YH. Doxorubicin loaded pH-sensitive polymeric micelles for reversal of resistant MCF-7 tumor. *J Control Release.* 2005; 103: 405–18.

Leslie EM, Deeley RG, Cole SP. Multidrug resistance proteins: Role of P-glycoprotein, MRP1, MRP2, and BCRP (ABCG2) in tissue defense. *Toxicol Appl Pharmacol.* 2005; 204: 216–37.

Li J, Liu P, Mao H, Wanga A, Zhang X. Emodin sensitizes paclitaxel-resistant human ovarian cancer cells to paclitaxel-induced apoptosis *in vitro*. *Oncol Rep.* 2009; 21: 1605–10.

Li L, ten Hagen TL, Bolkestein M, Gasselhuber A, Yatvin J, van Rhoon GC *et al.* Improved intratumoral nanoparticle extravasation and penetration by mild hyperthermia. *J Control Release.* 2013; 167: 130–7.

Lin JN, Lin VC, Rau KM, Shieh PC, Kuo DH, Shieh JC *et al.* Resveratrol modulates tumor cell proliferation and protein translation via SIRT1-dependent AMPK activation. *J Agric Food Chem.* 2010; 58: 1584–92.

Ling V, Thompson LH. Reduced permeability in CHO cells as a mechanism of resistance to colchicine. *J Cell Physiol.* 1974; 83: 103–16.

Liversidge GG, Liversidge E, Sarpotdar PP. Surface modified anticancer nanoparticles. US Patent 5,399,363; 1995.

Loo TW, Clarke DM. Do drug substrates enter the common drug-binding pocket of P-glycoprotein through 'gates'? *Biochem Biophys Res Commun.* 2005; 329: 419–22.

Lopes-Rodrigues V, Di Luca A, Mleczko J, Meleady P, Henry M, Pesic M *et al.* Identification of the metabolic alterations associated with the multidrug resistant phenotype in cancer and their intercellular transfer mediated by extracellular vesicles. *Sci Rep.* 2017; 7: 44541.

Loscher W, Potschka H. Blood-brain barrier active efflux transporters: ATP-binding cassette gene family. *NeuroRx.* 2005; 2: 86–98.

Lukianova-Hleb EY, Belyanin A, Kashinath S, Wu X, Lapotko DO. Plasmonic nanobubble-enhanced endosomal escape processes for selective and guided intracellular delivery of chemotherapy to drug-resistant cancer cells. *Biomaterials.* 2012; 33: 1821–26.

Luo Y, Cai X, Li H, Lin Y, Du D. Hyaluronic Acid-Modified Multifunctional Q-Graphene for Targeted Killing of Drug-Resistant Lung Cancer Cells. *ACS Appl Mater Interfaces.* 2016; 8: 4048–55.

Luqmani YA. Mechanisms of drug resistance in cancer chemotherapy. *Med Princ Pract.* 2005; 14(Suppl 1): 35–48.

Manganaro A, Rockwell K. Method and composition for treating cancer. US Patent 20,120,164,065; 2012.

Martin FJ, Ferrari M. Microfabricated particles and method for treating solid tumors. US Patent 20,030,114,366; 2003.

Mashima T, Tsuruo T. Defects of the apoptotic pathway as therapeutic target against cancer. *Drug Resist Updat.* 2005; 8: 339–43.

McBain SC, Yiu HH, Dobson J. Magnetic nanoparticles for gene and drug delivery. *Int J Nanomedicine.* 2008; 3: 169–80.

Milosevic M, Fyles A, Hedley D, Hill R. The human tumor microenvironment: Invasive (needle) measurement of oxygen and interstitial fluid pressure. *Semin Radiat Oncol.* 2004; 14: 249–58.

Minchinton AI, Tannock IF. Drug penetration in solid tumours. *Nat Rev Cancer.* 2006; 6: 583–92.

Minko T, Dharap SS, Pakunlu RI, Wang Y. Molecular targeting of drug delivery systems to cancer. *Curr Drug Targets.* 2004; 5: 389–406.

Minko T, Sinko PJ, Stein S. Complex drug delivery composition and method for treating cancer. US Patent 8,124,051; 2012.

Moll UM, Wolff S, Speidel D, Deppert W. Transcription-independent pro-apoptotic functions of p53. *Curr Opin Cell Biol.* 2005; 17: 631–6.

Moroz P, Jones SK, Gray BN. Magnetically mediated hyperthermia: Current status and future directions. *Int J Hyperthermia.* 2002; 18: 267–84.

Nabekura T, Kamiyama S, Kitagawa S. Effects of dietary chemopreventive phytochemicals on P-glycoprotein function. *Biochem Biophys Res Commun.* 2005; 327: 866–70.

Nelson DA, Tan TT, Rabson AB, Anderson D, Degenhardt K, White E. Hypoxia and defective apoptosis drive genomic instability and tumorigenesis. *Genes Dev.* 2004; 18: 2095–107.

Padera TP, Stoll BR, Tooredman JB, Capen D, di Tomaso E, Jain RK. Pathology: Cancer cells compress intratumour vessels. *Nature.* 2004; 427: 695.

Pakunlu RI, Wang Y, Saad M, Khandare JJ, Starovoytov V, Minko T. *In vitro* and *in vivo* intracellular liposomal delivery of antisense oligonucleotides and anticancer drug. *J Control Release.* 2006; 114: 153–62.

Paliwal R, Paliwal SR, Agrawal GP, Vyas SP. Biomimetic solid lipid nanoparticles for oral bioavailability enhancement of low molecular weight heparin and its lipid conjugates: *In vitro* and *in vivo* evaluation. *Mol Pharm.* 2011; 8: 1314–21.

Paliwal SR, Paliwal R, Agrawal GP, Vyas SP. Hyaluronic acid modified pH-sensitive liposomes for targeted intracellular delivery of doxorubicin. *J Liposome Res.* 2016; 1–12.

Paliwal SR, Paliwal R, Pal HC, Saxena AK, Sharma PR, Gupta PN *et al.* Estrogen-anchored pH-sensitive liposomes as nanomodule designed for site-specific delivery of doxorubicin in breast cancer therapy. *Mol Pharm.* 2012; 9: 176–86.

Paliwal SR, Paliwal R, Vyas SP. A review of mechanistic insight and application of pH-sensitive liposomes in drug delivery. *Drug Deliv.* 2015; 22: 231–42.

Panyam J, Chavanpatil MD. Polymer-surfactant nanoparticles for sustained release of compounds. US Patent 20,110,020,457; 2011.

Panyam J, Patil Y, Khdair A. PEG and targeting ligands on nanoparticle surface. US Patent 20,100,015,050; 2010.

Patel NR, Rathi A, Mongayt D, Torchilin VP. Reversal of multidrug resistance by co-delivery of tariquidar (XR9576) and paclitaxel using long-circulating liposomes. *Int J Pharm.* 2011; 416: 296–99.

Pauwels EK, Erba P, Mariani G, Gomes CM. Multidrug resistance in cancer: Its mechanism and its modulation. *Drug News Perspect.* 2007; 20: 371–7.

Perumal OP, Podaralla SK, Averineni RK. Polymer conjugated protein micelles. US Patent 8,697,098; 2014.

Powell D, Chandra S, Dodson K, Shaheen F, Wiltz K, Ireland S *et al.* Aptamer-functionalized hybrid nanoparticle for the treatment of breast cancer. *Eur J Pharm Biopharm.* 2017; 114: 108–18.

Prasad NR, Muthusamy G, Shanmugam M, Ambudkar SV. South Asian medicinal compounds as modulators of resistance to chemotherapy and radiotherapy. *Cancers.* 2016; 8: 32.

Qu M-H, Zeng R-F, Fang S, Dai Q-S, Li H-P, Long J-T. Liposome-based co-delivery of siRNA and docetaxel for the synergistic treatment of lung cancer. *Int J Pharm.* 2014; 474: 112–22.

Radovic-moreno AF, Gao W, Swami A, Golomb G, Langer RS, Farokhzad OC. pH sensitive biodegradable polymeric particles for drug delivery. US Patent 20,110,065,807; 2011.

Raghunand N, Gillies RJ. pH and drug resistance in tumors. *Drug Resist Updat.* 2000; 3: 39–47.

Rahman A, Rafaeloff R, Husain SR. Liposome encapsulated taxol and a method of using the same. US Patent 5,648,090; 1997.

Rai S, Paliwal R, Vaidya B, Khatri K, Goyal AK, Gupta PN *et al.* Targeted delivery of doxorubicin via estrone-appended liposomes. *J Drug Target.* 2008; 16: 455–63.

Rapoport N, Gao Z. Echogenic microbubbles and microemulsions for ultrasound-enhanced nanoparticle-mediated delivery of agents. US Patent 20,090,117,177; 2009.

Rapoport N. Method of *in vivo* drug targeting to solid tumors via acoustically triggered drug delivery in polymeric micelles. US Patent 20,050,003,008; 2005.

Reed JC, Jurgensmeier JM, Matsuyama S. Bcl-2 family proteins and mitochondria. *Biochim Biophys Acta.* 1998; 1366: 127–37.

Rennick BR, Moe GK, Lyons RH, Hoobler SW, Neligh R. Absorption and renal excretion of the tetraethylammonium ion. *J Pharmacol Exp Ther.* 1947; 91: 210–7.
Riehle R, Pattni B, Jhaveri A, Kulkarni A, Thakur G, Degterev A *et al.* Combination nanopreparations of a novel proapoptotic Drug–NCL-240, TRAIL and siRNA. *Pharm Res.* 2016; 1–15.
Riordan JR, Ling V. Purification of P-glycoprotein from plasma membrane vesicles of Chinese hamster ovary cell mutants with reduced colchicine permeability. *J Biol Chem.* 1979; 254: 12701–5.
Rosenberg MA, Hill JW. Cytotoxic therapy by proton flux modulation. US Patent 20,130,011,480; 2013.
Saraswathy M, Gong S. Different strategies to overcome multidrug resistance in cancer. *Biotechnol Adv.* 2013; 31: 1397–407.
See YP, Carlsen SA, Till JE, Ling V. Increased drug permeability in Chinese hamster ovary cells in the presence of cyanide. *Biochim Biophys Acta.* 1974; 373: 242–52.
Seelig A, Blatter XL, Wohnsland F. Substrate recognition by P-glycoprotein and the multidrug resistance-associated protein MRP1: A comparison. *Int J Clin Pharmacol Ther.* 2000; 38: 111–21.
Shah V, Taratula O, Garbuzenko OB, Taratula OR, Rodriguez-Rodriguez L, Minko T. Targeted nanomedicine for suppression of CD44 and simultaneous cell death induction in ovarian cancer: An optimal delivery of siRNA and anticancer drug. *Clin Cancer Res.* 2013; 19: 6193–204.
Shannon AM, Bouchier-Hayes DJ, Condron CM, Toomey D. Tumour hypoxia, chemotherapeutic resistance and hypoxia-related therapies. *Cancer Treat Rev.* 2003; 29: 297–307.
Shen J, Wang Q, Hu Q, Li Y, Tang G, Chu PK. Restoration of chemosensitivity by multifunctional micelles mediated by P-gp siRNA to reverse MDR. *Biomaterials.* 2014; 35: 8621–34.
Shen J, Yin Q, Chen L, Zhang Z, Li Y. Co-delivery of paclitaxel and survivin shRNA by pluronic P85-PEI/TPGS complex nanoparticles to overcome drug resistance in lung cancer. *Biomaterials.* 2012; 33: 8613–24.
Shen Y, Tang H, Zhan Y, Van Kirk EA, Murdoch WJ. Degradable poly(beta-amino ester) nanoparticles for cancer cytoplasmic drug delivery. *Nanomedicine.* 2009; 5: 192–201.
Singh D, Minz AP, Sahoo SK. Nanomedicine-mediated drug targeting of cancer stem cells. *Drug Discov Today.* 2017; 22: 952–9.
Singh J, Jain K, Mehra NK, Jain NK. Dendrimers in anticancer drug delivery: Mechanism of interaction of drug and dendrimers. *Artif Cells Nanomed Biotechnol.* 2016; 1–9.
Slovin S. Chemotherapy and immunotherapy combination in advanced prostate cancer. *Clin Adv Hematol Oncol.* 2012; 10: 90–100.
Song XR, Cai Z, Zheng Y, He G, Cui FY, Gong DQ *et al.* Reversion of multidrug resistance by co-encapsulation of vincristine and verapamil in PLGA nanoparticles. *Eur J Pharm Sci.* 2009; 37: 300–5.
Taatjes DJ, Fenick DJ, Koch TH. Anti-cancer drug aldehyde conjugate drugs with enhanced cytotoxicity compounds, compositions and methods. US Patent 6,677,309; 2004.
Tan W-f, Lin L-p, Li M-h, Zhang Y-X, Tong Y-g, Xiao D *et al.* Quercetin, a dietary-derived flavonoid, possesses antiangiogenic potential. *Eur J Pharmacol.* 2003; 459: 255–262.
Tang J, Zhang L, Gao H, Liu Y, Zhang Q, Ran R *et al.* Co-delivery of doxorubicin and P-gp inhibitor by a reduction-sensitive liposome to overcome multidrug resistance, enhance anti-tumor efficiency and reduce toxicity. *Drug delivery.* 2016; 23: 1130–43.
Tekchandani P, Kurmi BD, Paliwal SR. Nanomedicine to deal with cancer cell biology in multi-drug resistance. *Mini Rev Med Chem.* 2016.
Teoh D, Secord AA. Antiangiogenic agents in combination with chemotherapy for the treatment of epithelial ovarian cancer. *Int J Gynecol Cancer.* 2012; 22: 348–59.
Torchilin VP. Antinuclear antibodies with nucleosome-restricted specificity for targeted delivery of chemotherapeutic agents. *Ther Deliv.* 2010; 1: 257–72.
Tredan O, Galmarini CM, Patel K, Tannock IF. Drug resistance and the solid tumor microenvironment. *J Natl Cancer Inst.* 2007; 99: 1441–54.
van Vlerken LE, Duan Z, Seiden MV, Amiji MM. Modulation of intracellular ceramide using polymeric nanoparticles to overcome multidrug resistance in cancer. *Cancer Res.* 2007; 67: 4843–50.
Vera-Ramirez L, Sanchez-Rovira P, Ramirez-Tortosa MC, Ramirez-Tortosa CL, Granados-Principal S, Lorente JA *et al.* Free radicals in breast carcinogenesis, breast cancer progression and cancer stem cells. Biological bases to develop oxidative-based therapies. *Crit Rev Oncol Hematol.* 2011; 80: 347–68.
Viktorsson K, Lewensohn R, Zhivotovsky B. Apoptotic pathways and therapy resistance in human malignancies. *Adv Cancer Res.* 2005; 94: 143–96.
Wang J, Goh B, Lu W, Zhang Q, Chang A, Liu XY *et al. In vitro* cytotoxicity of stealth liposomes co-encapsulating doxorubicin and verapamil on doxorubicin-resistant tumor cells. *Biol Pharm Bull.* 2005; 28: 822–8.
Wang J, Ma W, Tu P. Synergistically improved anti-tumor efficacy by co-delivery doxorubicin and curcumin polymeric micelles. *Macromol Biosci.* 2015; 15: 1252–61.

Wang P, Wong MK, Gray A, Joo K-I, Liu Y. Targeted Crosslinked Multilamellar Liposomes. US Patent 20,140,356,414; 2014a.

Wang Y, Dou L, He H, Zhang Y, Shen Q. Multifunctional nanoparticles as nanocarrier for vincristine sulfate delivery to overcome tumor multidrug resistance. *Mol Pharm.* 2014b; 11: 885–94.

Wang Y, Zhao R, Wang S, Liu Z, Tang R. *In vivo* dual-targeted chemotherapy of drug resistant cancer by rationally designed nanocarrier. *Biomaterials.* 2016; 75: 71–81.

Wang YC, Wang F, Sun TM, Wang J. Redox-responsive nanoparticles from the single disulfide bond-bridged block copolymer as drug carriers for overcoming multidrug resistance in cancer cells. *Bioconjug Chem.* 2011; 22: 1939–45.

Warburg O. On the origin of cancer cells. *Science.* 1956; 123: 309–14.

Werle M. Natural and synthetic polymers as inhibitors of drug efflux pumps. *Pharm Res.* 2008; 25: 500–11.

Wilson WR, Hay MP. Targeting hypoxia in cancer therapy. *Nat Rev Cancer.* 2011; 11: 393–410.

Wu J, Lu Y, Lee A, Pan X, Yang X, Zhao X *et al.* Reversal of multidrug resistance by transferrin-conjugated liposomes co-encapsulating doxorubicin and verapamil. *J Pharm Pharm Sci.* 2007; 10: 350–7.

Wu XY, Shalviri A, Cai P. Polymeric nanoparticles useful in theranostics. US Patent 20,150,132,384; 2015.

Xue X, Liang XJ. Overcoming drug efflux-based multidrug resistance in cancer with nanotechnology. *Chin J Cancer.* 2012; 31: 100–9.

Yadav S, van Vlerken LE, Little SR, Amiji MM. Evaluations of combination MDR-1 gene silencing and paclitaxel administration in biodegradable polymeric nanoparticle formulations to overcome multidrug resistance in cancer cells. *Cancer Chemother Pharmacol.* 2009; 63: 711–22.

Yague E, Arance A, Kubitza L, O'Hare M, Jat P, Ogilvie CM *et al.* Ability to acquire drug resistance arises early during the tumorigenesis process. *Cancer Res.* 2007; 67: 1130–7.

Yamashiro DJ, Maxfield FR. Regulation of endocytic processes by pH. *Trends Pharmacol Sci.* 1988; 9: 190–3.

Yang T, Choi MK, Cui FD, Lee SJ, Chung SJ, Shim CK *et al.* Antitumor effect of paclitaxel-loaded PEGylated immunoliposomes against human breast cancer cells. *Pharm Res.* 2007; 24: 2402–11.

Yhee JY, Song S, Lee SJ, Park SG, Kim KS, Kim MG *et al.* Cancer-targeted MDR-1 siRNA delivery using self-cross-linked glycol chitosan nanoparticles to overcome drug resistance. *J Control Release.* 2015; 198: 1–9.

Zale SE, Ali MM. Cancer cell targeting using nanoparticles. US Patent 9,295,727; 2016.

Zhang JA, Ahmad I. Vinorelbine compositions and methods of use. US20040228911 A1; 2004.

Zhang X, Guo S, Fan R, Yu M, Li F, Zhu C *et al.* Dual-functional liposome for tumor targeting and overcoming multidrug resistance in hepatocellular carcinoma cells. *Biomaterials.* 2012; 33: 7103–14.

Zhang X, Lin Y, Gillies RJ. Tumor pH and its measurement. *J Nucl Med.* 2010; 51: 1167–70.

Zheng W, Cao C, Liu Y, Yu Q, Zheng C, Sun D *et al.* Multifunctional polyamidoamine-modified selenium nanoparticles dual-delivering siRNA and cisplatin to A549/DDP cells for reversal multidrug resistance. *Acta Biomater.* 2015; 11: 368–80.

Zhu J, Wang R, Lou L, Li W, Tang G, Bu X *et al.* Jatrophane diterpenoids as modulators of p-glycoprotein-dependent multidrug resistance (MDR): Advances of structure–activity relationships and discovery of promising MDR reversal agents. *J Med Chem.* 2016; 59: 6353–69.

Zink JI, Nel AE, Xia T, Ji Z, Meng H, Li Z *et al.* Cationic polymer coated mesoporous silica nanoparticles and uses thereof. US Patent 20,120,207,795; 2012.

# 5 Aptamers for Nano-Delivery

*Naznin Akhtar, Suman Chaudhary, Jayanth Surya Narayanan Shankara Narayanan, Afrah Jalil Abd, Joanna MacDonald, Sarah Shigdar, Rupinder Kaur Kanwar, and Jagat R. Kanwar**

## CONTENTS



* Corresponding author.

## 5.1 INTRODUCTION

Aptamers have received extensive interest since they were discovered about 25 years ago by Tuerk and Gold. Aptamers are short (generally 15–80 nucleotides) single-stranded molecules (ssDNA or ssRNA) possessing a capability to fold stably in three-dimensional structures, thus allowing interaction with target molecules through electrostatic interactions, base stacking, hydrogen bonding, Van der Waals forces, or through a combination of these mechanisms (Ma *et al.*, 2015). Currently, the adopted methodology for potential identifying therapeutic candidates in medical research and diagnostics utilizes the power of combinatorial chemistry that relies on the screening of large chemical libraries (Song *et al.*, 2012). Interestingly, identification of aptamers from synthetic ssDNA or ssRNA libraries is accomplished through the Systematic Evolution of Ligands by EXponential (SELEX) process (Asai *et al.*, 2003), which shows similarity to this high throughput screening process. However, aptamer libraries are considered to contain the largest diversity, with up to $10^{15}$ different molecules, which make them highly attractive tools for lead discovery. Aptamer libraries as they surpass the levels of diversity observed with the combinatorial small-molecule chemistry (Bauer *et al.*, 2016), phage display and ribosome display (Song *et al.*, 2015), among others.

So far, numerous studies on the functionality of aptamers have demonstrated that antibodies are not the only candidates for high-affinity binding to targets. However, aptamers should never be seen only as a rival of antibodies. Aptamers have exceptional advantages that outweigh antibodies, including (i) nucleic acid chemistry that allows unlimited amplification of single aptamers enzymatically, (ii) ease to synthesize in automated instruments, (iii) ease to perform controlled modification and (iv) highly flexibile structures. Aptamers are able to bind a plethora of targets including peptides, proteins, cells as well as other small molecules (Ma *et al.*, 2015), and their use as therapeutic agents or analytical tools are under intense study (Bruno, 2015). However, only a single aptamer has been approved for clinical use since their discovery, and the introduction of aptamers into the market has not been very successful (Burke, 2015). Although several aptamers are currently being assessed at different clinical phases, aptamer developments are facing skepticism, and private investors are approaching this area with caution. In addition, the unexpected adverse events observed in a phase III clinical trial of Regado Biosciences, Inc., Durham, NC, USA (REG1), a reversible anticoagulant aptamer targeting Factor IXa, may significantly slowdown the progression of new therapeutic aptamers and will make the business around aptamers more challenging (Rusconi *et al.*, 2002). However, setbacks can be overcome, as many other well-established technologies have done in the past. Exploitation of the therapeutic potential of monoclonal antibodies was envisioned since their discovery in 1975. Cho *et al.* (2013) had to overcome many failures and wait for 27 years until the approval of the first humanized antibody, daclizumab (Roche, Indianapolis, IN, USA).

Antibodies are among the most commonly used tools in the biological sciences and clinical diagnostics. In addition, interest in antibody-based therapies has grown incrementally over the last few decades. Antibodies have been shown to be very effective in their capturing capacity for other molecules, and it is clear that antibodies brought better biomarker assays to the clinic and better medicines to patients. However, the currently produced antibodies are available to detect only a small percentage of the human proteome, and the need for antibody validation has recently started being deliberated upon (Chushak and Stone, 2009). There is still a large opportunity for superior technologies that could detect a larger number of molecules with better specificity and novelty to the assays in the clinic and this demand can be readily addressed using aptamer technology.

## 5.2 APTAMER TECHNOLOGY DEVELOPMENT

There are several limitations that have been slowing down the progression of aptamer technology, with some of these quite well-articulated in other critical reviews (Cox and Ellington, 2001; Dassie *et al.*, 2009). Hurdles limiting the development of aptamer technology could be assigned to the other investment that companies allocate to monoclonal antibodies bioprocesses, thus reducing their interest on other technologies that threaten their current developments (Rozenblum *et al.*, 2016). While patent protection is a good instrument to promote innovation, sometimes the patent landscape becomes a minefield and investors want to be sure that their technology will not be blocked for commercializing their development by a third-party patent. However, investors are not the main culprits, and the major limitation resides on the actual aptamer discovery process, which has many components that add significant uncertainty to the success of the methodology, such as polymerase chain reaction (PCR) bias, PCR artifacts and background binders (Dittmer *et al.*, 2004; Douglas *et al.*, 2012). Most academic laboratories are still using SELEX as their primary strategy for aptamer discovery. Schütze *et al.* (2011) wrote, 'SELEX is like a black box with an ability to remain closed for many rounds of selections with the inherent risk of identifying only unspecific artifacts'.

Advancement of technological instrumentation has been the major driver of several biotechnological achievements. Next-generation sequencing (NGS) is a clear example of how the interaction of different disciplines managed to successfully develop an instrument that made a breakthrough in all the areas of genomics (Reis-Filho, 2009). NGS merged the knowledge provided by life-science researchers, engineers and material scientists. Taking the case of the successful story of Illumina – a US-based biotech company involved in sequencing and array-based solutions for genetic research – the molecular biologists and biochemists provided the foundations to bridge PCR and sequencing by synthesis, material scientists provided knowledge for the fabrication of ordered nanowell arrays and engineers designed the automation to assemble all the processes in a single instrument (Seo *et al.*, 2018). Without these interdisciplinary collaborations, NGS would not be possible. Therefore, the aptamer field should learn from the history of NGS, and if private investors are not showing interest in the aptamers at the moment, efforts between academic labs should be focused on generating interdisciplinary collaborations in order to foster a prototype of a solid technological platform for aptamer discovery. From our point of view, existing technologies that have been built for this specific purpose could easily be adapted for the sake of aptamer discovery.

This chapter summarizes the latest achievements with aptamers, highlighting the major advantages of aptamers over other technologies. It also discusses the major limitations of aptamer technology and proposes a roadmap for building a technological platform for aptamer development and for aptamer generation. Furthermore, it emphasizes the application of nanoparticle-functionalized aptamers for targeted cancer therapeutics, as well as in brain delivery, to target various neurological diseases.

## 5.3 APTAMER SELECTION APPROACHES

SELEX is utilized for almost all aptamer selections. This application demonstrates immense compatibility to the targets. Notably, SELEX (Figure 5.1) application has undergone countless modifications to make it more cost-effective and to further enhance aptamer stability. Well-established SELEX approaches can be described briefly, as follows.

### 5.3.1 Conventional SELEX

Established two decades ago, this approach requires a few weeks for aptamer preparation (Tuerk and Gold, 1990). The underlying principle of this application depends on introducing several selection steps, according to the sort of aptamer desired. The protocol of conventional SELEX includes incubating target molecules with the chemically synthesized oligonucleotide library. The released DNA sequences undergo amplification using PCR. However, additional steps are needed

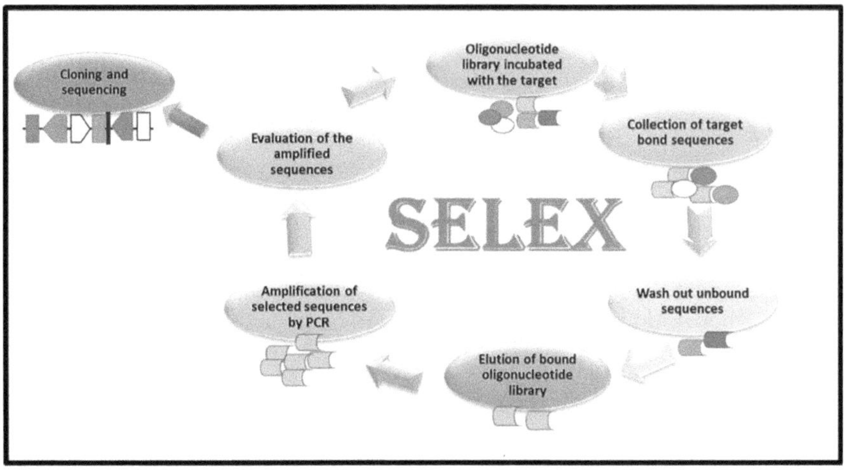


**FIGURE 5.1** Illustration of systematic evolution of ligands by exponential enrichment (SELEX).

for the synthesis of the RNA aptamer. This protocol is, though efficient, time-consuming, leading to exploration of alternative strategies.

### 5.3.2 Cell SELEX

The cell SELEX strategy do not require the prior purification of the target proteins expressed on the eukaryotic and prokaryotic cells (Chen et at., 2016). This technique is more functional, since the aptamer is adapted for the original configuration of the target *in vivo*. In this application, the target is incubated with the chemically synthesized DNA library, followed by aptamer elution and amplification using PCR. Mostly, it involves eight to ten selection rounds to obtain the aptamer. This technique is widely used against different cancer cell types (Ara *et al.*, 2012; Jiménez *et al.*, 2012; Ninomiya *et al.*, 2013), bacterial cells as well as viruses (Dwivedi *et al.*, 2013). The cell SELEX application could be considered as a promising tool for therapeutic applications.

### 5.3.3 Capillary Electrophoresis SELEX

This application has been introduced to reduce the number of rounds during the process of aptamer selection without affecting the aptamer-target affinity (Mendonsa and Bowser, 2004). This application is based on the principle that oligonucleotides bound with the target that possess reduced mobility in comparison to the free nucleotides. This application has undergone further modification by introducing micro free-flow electrophoresis (μFFE), which offers major yield of the cumulative complex *vis-à-vis* the capillary electrophoresis SELEX and less-diluted oligonucleotide sequences. However, this application requires a specific μFFE device to achieve the application (Jing and Bowser, 2011).

Non-equilibrium capillary electrophoresis of equilibrium mixtures (NECEEM), sweeping capillary electrophoresis (SweepCE), kinetic capillary electrophoresis (KCE) and equilibrium capillary electrophoresis of equilibrium mixtures (ECEEM) are additional techniques. These depend on the capillary electrophoresis, share the same principle of partitioning of the target and DNA library molecule mixture and utilize gel-free capillary electrophoresis. One such efficient and uncomplicated application for aptamer partitioning is NECEEM. The main feature of this technique is non-equilibrium production of the target and DNA, while ECEEM establishes various partitioning conditions compared to the previous one, and the complex of the protein-aptamer is maintained at equilibrium. This technique yields a smart aptamer that could be used for clinical applications, such as smart medications. Simultaneously, SweepCE has been reported to provide dissociation constant ($K_d$) and complex formation constant ($K_{on}$) in order to broaden the gathered information for interaction of aptamer-target complex (Okhonin *et al.*, 2004).

### 5.3.4 One-Round SELEX

Several investigations have been performed to reduce the total number of aptamer selection rounds. One-round SELEX involves several strategies, represented by MonoLEX, nanoselection, ASExp (Aptamer Selection Express) and selection on glass coverslips. The MonoLEX principle is based on incubation of the DNA library with a target bound to resin; the eluted aptamer is amplified using PCR. While the strategy of the nanoselection principle lies in incubation of protein with beads carrying fluorescent oligonucleotides and the immobilized target on a coverslip, the aptamer-protein complex is detected using fluorescent microscopy and extracted using atomic force microscopy (AFM). Similarly, it is amplified using PCR, and the obtained aptamer using nanoselection shows impressive outcomes.

The ASExp strategy depends on using microbeads in magnetic separation. In this regard, dsDNA needs to be amplified using PCR and incubated with the target, followed by supplementation of magnetic beads, which leads to aptamer isolation. This protocol is widely used against bacterial toxins. The simplest and the most cost-efficient approach was introduced by Arnold *et al.* (2012). The oligonucleotide library was incubated overnight with Kalikreine-related peptidase 6 (KLK6) after immobilization of the latter using a 96-well plate. Aptamer isolation and elution were performed using a higher concentration salt solution. This protocol is relatively effortless and inexpensive in comparison to the other methods of SELEX.

### 5.3.5 Magnetic Beads SELEX

The principle of this approach involves target immobilization on magnetic beads, which are incubated with oligonucleotides. The resultant complex of aptamer-target is obtained by magnetic separation and evaluated using epifluorescence microscopy. This protocol was subjected to further modification by introducing FluMag-SELEX (Stoltenburg *et al.*, 2005), capture SELEX by FluMag being considered as more suitable for capturing small targets that cannot be immobilized. Magnetic beads SELEX has introduced additional strategies, as is the case with microfluidic-based SELEX technology involving continuous-flow magnetic-activated, chip-based separation (CMACS) (Lou *et al.*, 2009). This protocol is undesirable due to handling difficulties and low purity of the obtained aptamer. Micro-magnetic separation chip (MMS) was later introduced to overcome the shortcomings with CMACS (Oh *et al.*, 2011b).

### 5.3.6 Additional Strategies for Aptamer Selection

The vast interest in the aptamer approach stimulates researchers to modify modern strategies to make the selection process more efficient. A succinct account of the recent applications is presented as follows:

#### 5.3.6.1 Multiplexed Massively Parallel SELEX

This strategy involves modification of the SELEX-SAGE protocol; the principle includes incubation of the oligonucleotide library (five nucleotides in length) with the immobilized target on a 96-well plate, following the steps of washing, elution and amplification using PCR. This application is rated as more efficient for aptamer selection.

#### 5.3.6.2 Fluoresence Activated Cell Sorting (FACS)-SELEX

The fluorescent-labeled DNA library is incubated with different cell mixtures, either viable or dead. An indicator of cell viability is used for cell staining, which is employed for separation of dead and live cells. This step is important for the selection of aptamer bound to live cells later bound aptamer can be isolated, eluted and amplified using PCR (Mayer *et al.*, 2010).

#### 5.3.6.3 MAI-SELEX

In order to prepare higher specificity binding aptamer to various protein subunits, MAI-SELEX was introduced. The protocol includes two selection stages. The first one is the affinity module, which involves bead separation after incubation of the immobilized target on magnetic beads with the DNA library. The other stage is the specificity module, represented by incubation of the target with the aptamer pool, and subsequently elution of the bound protein. Unbound aptamer is incubated again with the protein, and another specific aptamer can be selected.

#### 5.3.6.4 Graphene Oxide-SELEX (GO-SELEX)

The principle of GO-SELEX is represented by incubation of the oligonucleotide library with the target, followed by adding the whole mixture to graphene oxide (GO). The bound aptamer presents in the solution while the unbound ones will be absorbed by GO. Bound aptamer is subjected to elution and amplification using PCR (WooáKim and BockáGu, 2012). This application is efficient, simple, less time consuming and cost-effective as compared to other SELEX approaches.

### 5.3.7 Domain-Targeted SELEX

The domain-targeted SELEX principle includes incubation of oligonucleotides with the immobilized target on agarose beads. This protocol has several limitations and can not be applied to most proteins.

#### 5.3.7.1 Rapid SELEX

The protocol of this application is similar to the conventional SELEX in several aspects but it needs less time for aptamer selection as compared to the conventional SELEX. The main feature of this application is that there is no amplification of the obtained aptamer using PCR (Szeto *et al.*, 2013).

#### 5.3.7.2 Particle Display Selection

The principle of this protocol involves incubation of both the primers, fixed on beads, and the oligonucleotide library. The target should be incubated with the bound beads. Following this, collection of bound sequences along with amplification through PCR is done, which is then evaluated (Wang *et al.*, 2014). This application is rapid and simple.

#### 5.3.7.3 MARAS

It involves incubation of the oligonucleotide library with the magnetic nanoparticles attached to the target. The bound DNA library is collected and amplified using PCR (Lai and Hong, 2014). Higher affinity aptamers could be obtained using this protocol.

#### 5.3.7.4 Epitope-Specific-SELEX (ES-SELEX)

The principle of this protocol involves incubation of the oligonucleotide with the protein target, followed by complex isolation using chromatography, aptamer elution and amplification using PCR and subjecting to the next round of selection. The total number of selection rounds are nine (Lao *et al.*, 2014).

#### 5.3.7.5 Artificially Expanded Genetic Information Systems (AEGIS-SELEX)

This application of SELEX library is composed of two artificial and one natural nucleotide, and it requires 12 selection rounds (Sefah *et al.*, 2014). High aptamer affinity is obtained using this approach.

## 5.4 ADVANTAGES OF APTAMERS

There are many reviews that highlight the definitive advantages of aptamers over existing technologies. Herein, the focal point is extending the most relevant aspects of aptamers.

### 5.4.1 Versatility

Current trends in searching for novel therapeutic drugs are mainly focused on preparing chemical libraries with sources of new potential leads for advanced drug discovery (Zhou *et al.*, 2000). Aptamers are selected against a target of interest from a huge library of combinatorial nucleic acids that conatin up to $10^{15}$ (1 peta) species of nucleic acids. This library contains a lot of structural diversity and is currently considered to be the biggest combinatorial library.

### 5.4.2 Ease of Selection and Synthesis

Aptamers can be generated against toxic molecules or non-immunogenic targets with high affinity and specificity because, their production does not require specific cell lines or any animals. On the other hand, production of antibodies is hooked on living organisms. Moreover, antibodies are highly sensitive to the pH and temperature changes, while aptamers are more resistant to the environment. Additionally, the selection condition can be reformed accordingly to the ultimate applicability of the aptamer (Thiel and Giangrande, 2009).

### 5.4.3 Low Toxicity and Immunogenicity

Aptamers have demonstrated low to no immunogenicity upon administration in preclinical doses that are 1000-fold greater than doses employed in human and animal therapeutic applications (Glökler *et al.*, 2010; Gold *et al.*, 2010). However, the halt of phase III clinical trial of REG1 in 2014 due to an excess of severe allergic reactions suggests that low immunogenicity and toxicity may not be generalized to all aptamers (Lincoff *et al.*, 2016).

### 5.4.4 Adjustable Bioavailability

Chemical synthesis of aptamers is performed using solid-phase technology that facilitates cost-effective large-scale production and ensures reproducibility between different batches in comparison to biological drugs (Groff *et al.*, 2015). Moreover, they preserve their activity during long-term shelf storage at room temperature. Pharmacokinetics as well as bioavailability of nucleic acid ligands, can be controlled using modified nucleotides, which are added during synthesis. Besides, antidote control provides a safe means to regulate drug action and minimize side effects. While developing antidotes for antibodies and small drugs is a challenging task, it is not so for aptamers. Although REG1 did not pass a phase III trial, the origin of its antigenicity remains unknown and is still a great testimony on how aptamers can be developed with a rationally designed antidote control.

### 5.4.5 Targeting and Delivery

Knockdown of specific genes, such as oncogenes by RNA interference (RNAi) is a promising therapeutic technique for treating diseases associated with protein overexpression. Small interfering RNAs (siRNAs) can reduce gene expression via the RNAi pathway (Lares *et al.*, 2010). However, a major challenge in using RNAi *in vivo* is the delivery of the molecule to the correct tissue or cells, as well as its technical difficulty in crossing cell plasma membranes (Kee *et al.*, 2013). The siRNA-linked aptamers that target specific cell surface proteins are easy to synthesize and possess great potential to make RNAi attractive for therapies. An RNA aptamer has been generated against the human prostate-specific membrane antigen (PSMA) protein linked to a siRNA, which targets tumor overexpressed genes *PLK1* and *BCL2*, which are associated with tumor survival (McNamara *et al.*, 2006). The aptamer portion of these chimeras induced cell internalization and delivered the siRNA cargo. These chimeras were able to bind selectively to PSMA-positive cells and reduce gene expression of targeted genes. In addition, the PSMA-BCL2 chimera substantially decreased tumor volume in a prostate cancer xenograft mouse model when injected intratumorally and intraperitoneally (McNamara *et al.*, 2006). More recently, this research group elaborated a cell-internalization SELEX to obtain aptamers that are internalized by specific cell types. This method allowed rising of an aptamer–siRNA chimera to human epidermal growth factor receptor 2 (HER2)+ cells and aptamers against human papillomavirus type 16-associated cancers without knowing their mechanism of entrance. The usage of aptamers as potent systemic transporters has been extended to many other targets including delivery of siRNA and suppression of HIV replication (Hoinka *et al.*, 2015).

### 5.4.6 Aptamer-Based Nano-Architecture and Assembly

DNA is an ideal molecule for the building of objects at a nanometric scale. The specificity of A-T and G-C base pairing and the physicochemical nature of DNA allow the assembly of three-dimensional DNA-based nanostructures in solution with high efficiency and precise architecture (Jayasena, 1999). The first attempt to construct a DNA scaffold, conjugated to an aptamer resulted in the assembly of DNA tiles into highly ordered DNA lattices containing anti-thrombin aptamers placed orderly in defined patterns to selectively position the proteins (Jing and Bowser, 2011). These types of developments could lead to novel biosensing applications. Yi Shu *et al.* (2011) exploited the properties of the bacterial virus phi29 DNA packaging RNA (pRNA) molecule to assemble a hexameric ring nanostructure with an ability to deliver numerous therapeutics in precise cells. The addition of different receptor-binding ligands to the pRNA double-stranded 5′/3′-end helical domain resulted in chimeric molecules, such as pRNA/aptamer. Some researchers successfully managed to reduce HIV infectivity in a cultured cell line by constructing a pRNA dimer fused to an HIV gp120-binding aptamer, and an siRNA targeting the expression of HIV tat/rev regulatory elements (Khaled *et al.*, 2005). While the anti-HIV gp120 aptamer targeted the chimeric pRNA to the surface of HIV-infected cells for its internalization, the siRNA element was necessary for the inhibition of HIV replication.

DNA nanostructures can also be manipulated to create molecular machines of different complexities. These molecular devices can be controlled to release different kinds of cargos at specific targets by conformational switching. This conformational change is induced by a molecular effector or by intermolecular hybridization with complementary nucleic acids. A device with a more complex nanoarchitecture was constructed by a study that consisted of a cage-like structure with the configuration of a barrel containing an aptamer-based lock (Rozenblum *et al.*, 2016). This barrel was able to hold, locked a variety of payloads. The aptamer lock was used to keep the cage in its closed conformation using a complementary DNA sequence, which in the presence of the aptamer-target such as platelet-derived growth factor (PDGF), would open the cage and deliver the payloads to cells due to the formation of an energetically favored aptamer–ligand complex (Rozenblum *et al.*, 2016). Recently, stimuli-responsive DNA microcapsules were developed. In this study, Liao *et al.* (2015) constructed spherical microcapsules with oligonucleotide layers that were cross-linked through anti-ATP aptamer sequences. ATP-aptamers resulted in DNA-shell disassembly upon ATP binding, thus releasing the encapsulated bioactive molecules. These kinds of developments open newer pathways for novel applications such as smart drugs and stimuli-responsive DNA-based nanocapsules that provide selective control of load delivery and release.

## 5.5 APTAMER-TAGGED NANOPARTICLES FOR CANCER THERAPY AND DIAGNOSIS

Understanding the acidic microenvironment and taking advantage of the hyperpermeability of the tumor vasculature, many researchers have successfully developed diagnostic, therapeutic or

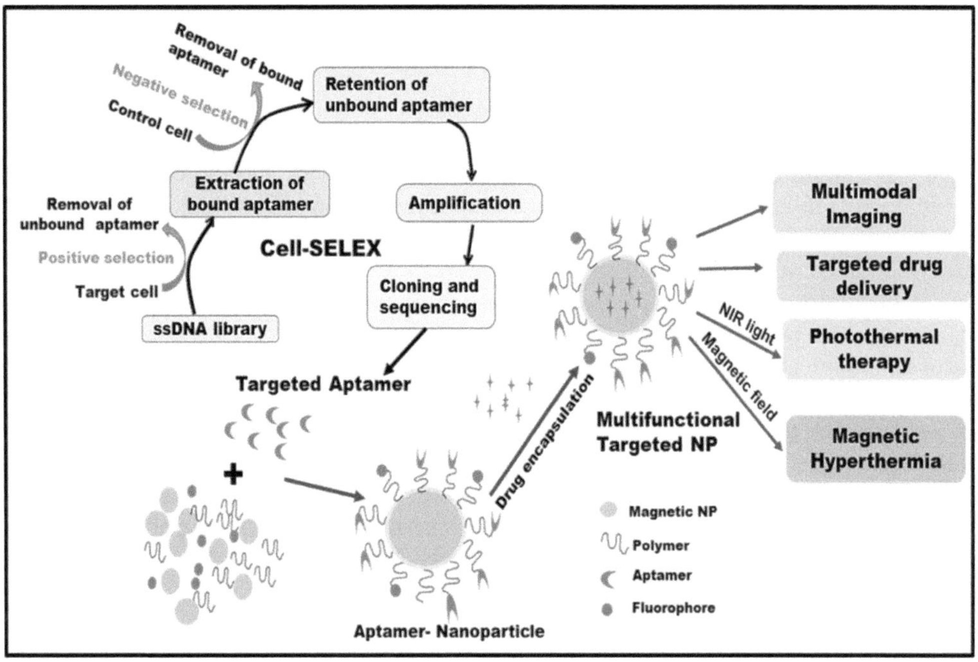


**FIGURE 5.2** Schematic illustration of aptamer selection and multifunctional nanoparticle formulation for targeted imaging and therapy of cancer cells.

theranostic devices for the controlled delivery of anticancer agents by entrapping them in 'sub-micronic colloidal systems' (i.e., nanoparticles). Until recently, numerous nanoparticles (NPs) such as carbon nanotubes (Taghdisi *et al.*, 2011), magnetic NPs (Leach *et al.*, 2016), metallic NPs (Malik *et al.*, 2015), liposomes (Moosavian *et al.*, 2016), lipid NPs (Zhang *et al.*, 2008), protein-based NPs (Pan *et al.*, 2016), polymeric NPs (Alibolandi *et al.*, 2015), micelles (Xu *et al.*, 2013), viral nanocarriers (Everts *et al.*, 2006) and dendrimers (Zhang *et al.*, 2015) have been developed for targeting cancer cells and the microenvironment.

However, precise targeting and site-specific internalization of the NPs is a prerequisite to overcome the challenges faced by current cancer therapeutics. In this milieu, targeting-specific biomarkers, which are over expressed on the surface of certain type of cancer cells, provide a lucrative option to upsurge anticancer efficacy while diminishing the systemic side effects of the existing mainstream cancer therapeutics. Ever since the discovery of aptamers possessing outstanding molecular recognition properties, these have attracted great attention for drug delivery to the cancer cell, as aptamers can bind to the specific biomarkers on these cells with very high affinity, thus providing a favourable way for targeted delivery of imaging and therapeutic agents (Figure 5.2). Many authors reported that the anticancer efficacy of NP-based drug delivery can be boosted by aptamer modification because of enhanced internalization in the targeted cells (Alibolandi *et al.*, 2015; Moosavian *et al.*, 2016).

## 5.6 DESIGN CONSIDERATIONS OF NPs FOR TARGETING THE CANCER MICROENVIRONMENT

For site-specific delivery and precise release of the anticancer agent from NP-aptamer bioconjugates, various parameters are needed to be optimized such as size, shape, zeta potential and surface properties due to the nonspecific accumulation of NPs in liver, spleen and other parts of the reticuloendothelial system (RES) which usually occurs, depending upon their size, physicochemical properties and administration route. In fact, nanoformulation systems exploit the abnormal vasculature and microenvironment of tumor tissues for site-specific targeting.

### 5.6.1 Size and Shape

Normal vasculature has highly organized continuous walls, with pores less than 1 nm (in brain and spinal cord) to around 15 nm (in kidney glomerulus) in non-sinusoidal capillaries. However, sinusoidal capillaries in the RES have larger pores, ranging from 5–60 nm (Sarin, 2010). On the contrary, tumors have an abnormal leaky vasculature, with pore size ranging between 200 nm and 1.2 μm, depending on the tumor type. Hence, to take the maximum benefit of highly fenestrated tumor vasculature, which provides enhanced penetration and retention effect (EPR effect), NPs in the size range of 20–50 nm are more effective, because, ultra-small (<10 nm) size NPs extravasate from the leaky tumor vessels (Tang *et al.*, 2014). Moreover, the hydrodynamic size of the NPs directly affects their concentration profile and clearance from circulation (Chavanpatil *et al.*, 2006). Effects of the size of targeted NPs have been studied extensively with most NPs being spherically shaped (Malik *et al.*, 2015; Leach *et al.*, 2016). However, Li and Lu (2015) recently described a DNA-modified gold NPs (DNA-AuNPs) onto hexagonal plate-like upconversion NPs (UCNPs) superstructure of around 100 nm size for targeted bimodal imaging of 4T1 cells using AS1411 aptamer.

### 5.6.2 Surface Properties

Surface charge and hydrophobicity of NPs play a key role in its biodistribution *in vivo* by regulating their interactions with serum proteins and other components of the vascular system, subsequently affecting their uptake and cytotoxicity (Lynch *et al.*, 2009). In case of systemic administration, hydrophilic surfaces provide 'stealth' properties protecting the NPs from opsonisation and removal by RES. On the contrary, hydrophobicity of NPs and their circulation times are inversely correlated due to adsorption of serum proteins, which leads to the formation of protein corona through adherence of proteins to the NP surface. This results in the hindrance of targeting ligands, leading eventually to the reduction in specificity to cancerous cells, and is thus ultimately detrimental for active-targeting strategies for NPs. Besides being small in size and having favorable surface physicochemistry, NPs should be biodegradable, have suitable topography, lower aggregation rate and high target specificity to eliminate or minimize the systemic side effects.

## 5.7 CATEGORIES OF NPs

Based on the building blocks used in the nanostructured formulation system, NPs can be classified into three major groups such as inorganic, organic and chimeric NPs. NP-aptamer conjugates currently under development for cancer therapy and imaging are presented in Table 5.1.

### 5.7.1 Inorganic NPs

Among the inorganic NPs, plasmonic NPs and, in particular, gold NPs (AuNPs) have great promise for bioimaging and photothermal cancer therapy (PTT) because of their biocompatibility, chemical inertness and amenability of structural modifications (Malik *et al.*, 2015). Moreover, nontoxicity, strong localized surface plasmon absorption (SPR), high absorption cross-section at near infrared (NIR) frequencies and stable photoluminescence (PL) make them highly attractive theranostic agents for multimodal imaging over organic dyes or toxic quantum dots, and also make them suitable for targeted therapy (Lee *et al.*, 2015). However, to specifically target cancer cells, AuNPs are needed to be conjugated with aptamers, because they do not inherently recognize the target cells. In fact, aptamer conjugation with AuNPs has a synergistic effect in targeted anticancer therapy, as aptamers protect AuNPs from aggregation (Gopinath *et al.*, 2014). At the same time, AuNPs also

**TABLE 5.1**
**NP-Aptamer Conjugates for Cancer Therapy and Imaging**

| Nanoparticles | Size | Aptamer/Target | Therapeutic Agent | Route and Dose | Status | Reference |
|---|---|---|---|---|---|---|
| **Inorganic Nanoparticles** | | | | | | |
| Dextran-coated nano-graphene oxide sheet | <200 nm | AS1411 aptamer, a ssDNAaptamer | Curcumin | – | *In vitro* cytotoxicity on the 4T1 and MCF-7 cells | Alibolandi *et al.* (2017) |
| Gold-coated super paramagnetic iron oxide nanoparticles (Au-SPION) | SPIO: 19 nm<br>Au-SPION: 24 nm | Thiol-modified MUC-1 aptamer/ MUC-1 | – | 100–500 μg/mL, NIR light (820 nm, 0.7 $WCm^{-2}$ for 2–8 min) | Real-time MR imaging and photothermal ablation of HT-29 cells | Azhdarzadeh *et al.* (2016) |
| Mesoporous silica nanoparticles (MSNs) | MSN: 159 ± 10 nm<br>MSN-Ap: 175 ± 5 nm<br>MSN-Ap-DOX: 181 ± 6 nm | EpCAM targeting aptamer | Doxorubicin (DOX) | 5–100 μg/mL | Targeted drug delivery in SW620 cells | Xie *et al.* (2016) |
| Super paramagnetic iron oxide nanoparticle | 10 nm | A10-3-J1/PSMA | Doxorubicin | 0.05 mg/mL | Targeted drug delivery in LNCap cells | Leach *et al.* (2016) |
| Gold nanospheres (AS1411-GNS) | Hydrodynamic diameter GNS: 6.4 ± 0.5 nm<br>GNSAp: 33.0 ± 3.4 nm | AS1411 aptamer/ nucleolin | – | 200–800 nM, intraperitoneal, 22 μg | Therapeutic efficacy *in vitro* on MCF-7 & MDA-MB-231 and *in vivo* ectopic xenograft mice model | Malik *et al.* (2015) |
| Mesoporous silica-coated $Cu_{1.8}S$ nanoparticles ($Cu_{1.8}S$-$mSiO_2$) | 17 nm | AS1411/nucleolin | Doxorubicin and curcumin | 125–1000 μg/mL + NIR light (980 nm, 5 min, 3 times) | *In vitro* photothermal ablation of MCF-7 cells | Zhang *et al.* (2015) |
| Gold nanoparticles (AuNPs) | 15.2 ± 0.7 nm | sgc8c aptamer/ PTK7 | Daunorubicin | 0.5 μM | *In vitro* pH-responsive drug release in Molt-4 cells | Danesh *et al.* (2015) |
| Silver nanoparticles (AgNPs) on the surfaces of mesoporous silica nanospheres | 90 nm | sgc8/PTK7 | Paclitaxel | 5.0 mM, intravenous, 1.0 mg $kg^{-1}$ | *In vivo* drug release in ectopic mice model | Liu *et al.* (2015) |
| Fluorescent silica nanoparticles (FSNPs) | 60–140 nm | DNA aptamer/ nucleolin | – | – | Targeted imaging of MCF-7 cells | Wang *et al.* (2015) |
| Gold nanostars (AuNS) | 40 nm | Anti-HER2 aptamer/ HER2 | – | 2.5 nM | Anticancer effects on SK-BR-3 cell | Lee *et al.* (2015) |

*(Continued)*

**TABLE 5.1 (*Continued*)**
**NP-Aptamer Conjugates for Cancer Therapy and Imaging**

| Nanoparticles | Size | Aptamer/Target | Therapeutic Agent | Route and Dose | Status | Reference |
|---|---|---|---|---|---|---|
| **Organic Nanoparticles** | | | | | | |
| Lipid-polymer hybrid nanoliposomes | 150–200 nm | Aptamer A6 ($NH_2$-Apt-6) | GAPDH siRNA | – | siRNA specifically target MDR gene which could be helpful to overcome chemoresistance of breast cancer cells | Powell *et al.* (2017) |
| DNA/RNA hybrid nanoparticles | 8.1 ± 1.5 nm | Endo28 thioaptamer | Doxorubicin | Intravenous, 100 μL of 10 μM nanoparticles | Annexin A2 aptamer anchored thio-DNA/2'F-RNA hybrid nanoparticles found to specifically target ovarian cancer cells | Pi *et al.* (2017) |
| DNA nanostructures | 6.1 & 7.4 nm | DNA Aptamer/ Nucleolin | Doxorubicin | Intravenous, 0.2 μM/kg, 10 mg/kg | Targeted drug delivery *in vitro* in MCF-7 cells and *in vivo* in BALB/c mice | Liu *et al.* (2016) |
| PEGylated-liposome | 110 nm | TSA14 aptamer/ HER2/erbB-2 | Doxorubicin | Intravenous, 15 mg/kg | Targeted drug delivery *in vitro* in TUBO cells and *in vivo* in ectopic xenograft mice model | Moosavian *et al.* (2016) |
| PLGA-b-PEG-COOH | 93.6 nm | Anti-PSMA aptamer | Docetaxel | Intravenous, 10 mg/kg, every other day for two weeks | Targeted drug delivery *in vitro* in LNCaP cells and *in vivo* in LNCaP tumor xenograft BALB/c mice | Chen *et al.* (2016) |
| Hyperbranched poly(2-((2(acryloyloxy)ethyl) disulfanyl)ethyl 4-cyano-4-(((propylthio)-carbonothioyl)-thio)-pentanoate-co-poly(ethylene glycol) methacrylate) (HPAEG) | Hydrodynamic diameter: 80.6 nm | AS1411/nucleolin | Doxorubicin | 0.1 – 10 μg/mL | Targeted drug delivery *in vitro* in MCF-7 cells | Zhuang *et al.* (2016) |

(*Continued*)

**TABLE 5.1 (*Continued*)**
**NP-Aptamer Conjugates for Cancer Therapy and Imaging**

| Nanoparticles | Size | Aptamer/Target | Therapeutic Agent | Route and Dose | Status | Reference |
|---|---|---|---|---|---|---|
| Polyethyleneimine | 151 ± 11 nm (TEM) 198 nm (DLS) | RNA aptamer/ EpCAM | siRNA | 0.1–3 μg/mL | Targeted gene delivery in MCF-7 and WERI-Rb1 cell | Subramanian *et al.* (2015) |
| PLGA-b-PEG nanopolymersome | Before conjugation: 124 ± 0.9 nm, after 136 ± 0.21 nm | EpCAM-fluoropyrimidine RNA Aptamer/EpCAM | Doxorubicin | Intravenous, 5 mg/kg | Targeted drug delivery *in vitro* in SK-MES-1 & A549 cells and *in vivo* in C57BL/6 nude mice | Alibolandi *et al.* (2015) |
| Polylactidenanoconjugate | 96 nm | A10 aptamer/ PSMA | Doxorubicin | Intravenous, 50 mg/kg, two times | Targeted drug delivery *in vivo* SB-HSA tumors in SCID/beige mice | Tang *et al.* (2015a,b) |
| Multifunctional DNA Dendrimer | – | sgc8 aptamer/ PTK7 | Doxorubicin | 1–9 μM | Bioimaging and drug delivery in CCRF-CEM and HeLa cells | Zhang *et al.* (2015) |
| PLL-alkyl-PEI copolymer | Copolymer:117–128 nm Aptamer-copolymer: 168–183 nm | AS1411 aptamer/ nucleolin | shRNA and plasmid | – | *In vitro* gene delivery efficacy and down-regulation of Bcl-XL gene in A549 cells | Askarian *et al.* (2015) |
| Poly(lactide-co-glycolide) functionalized with a targeting ligand and imaging agent (quantum dots) | Apt-Nut-NPs: 292 ± 10 nm QD-Nut-NPs: 257 ± 27 nm | EpCAM aptamer/ EpCAM | nutlin-3a | 0.01–1 μg/mL | *In vitro* targeted therapy and Imaging of MCF-7, SKOV3, and ZR751 cells | Das *et al.* (2015) |
| Dendrimer encapsulated Micro-RNA-34a (miR-34a) (PAM-Ap/pMiR-34a NPs) | 100–200 nm | S6 aptamer/ against A549 lung cancer cells | miR-34a | 2 μg/mL | *In vitro* therapeutic effect on A549 cells | Wang *et al.* (2015) |
| Tryptophan–phenylalanine dipeptide nanoparticles (DNPs) | 160 nm | MUC1 aptamer/ MUC1 | Doxorubicin | 25 and 100 nM | Targeted imaging and real time drug release monitoring in A549 and NIH-3T3 cells | Fan *et al.* (2016) |

(*Continued*)

**TABLE 5.1 (*Continued*)**
**NP-Aptamer Conjugates for Cancer Therapy and Imaging**

| Nanoparticles | Size | Aptamer/Target | Therapeutic Agent | Route and Dose | Status | Reference |
|---|---|---|---|---|---|---|
| **Chimeric Nanoparticles** | | | | | | |
| PEG-coated gold nanorods | – | DNA aptamer (KW16–13)/ surface marker on MCF10CA1 h cells | – | 100 μL, NIR irradiation | Photothermal killing of MCF10CA1h cells | Chandrasekaran *et al.* (2016) |
| DNAzyme-Nanocomposite probe equipped with built-in $Zn^{2+}$ arsenal (GNPs-ES-Dox/ ZnO) | 17.6 ± 1.3 nm | AS1411 aptamer/ nucleolin | Anti-miR-21 and doxorubicin | 0.5–1 μg $mL^{-1}$ | *In vitro* gene regulation and drug delivery in HeLa cells | He *et al.* (2016) |
| Alginate-coated chitosan-calcium phosphate-Fe3O4-bLf NCs (AEC-CP-Fe3O4-bLf NCs) | Before alginate coating: 90–120 nm<br>After alginate coating: 250–300 nm | LAN-modified aptamer/ EpCAM and Nucleolin | Iron saturated bovine lactoferrin | Oral; 12 g Fe-bLf/kg of the diet for 45 days | Multimodal imaging and therapy *in vivo* in ectopic xenograft mice model | Roy *et al.* (2015) |
| Hybrid nanoparticle 1-NPs-QDs-hAPT | 1-NPs-QDs: 35.5 ± 6.2 nm | MUC1-targeting aptamer/MUC1 | Doxorubicin | 0.62–10 μM | *In vitro* targeted cytotoxicity and imaging of MCF-7 cells | Du *et al.* (2015) |
| DNA-AuNPs/Lipo-UCNP superstructures | ~100 nm | AS1411/nucleolin | – | 100 μg/mL | Targeted dual-modality imaging of 4T1 cells | Li and Lu (2015) |
| Graphene quantum dots nanocomposite-capped fluorescent mesoporous silica nanoparticles (FAG-FMSNs-TA-Dox-PEG) | 50 nm | AS1411 and ATP aptamer/nucleolin and ATP | Doxorubicin | 2.5–20 μg/mL | Targeted drug delivery in HeLa cells | Zheng *et al.* (2015) |
| Graphene oxide wrapped doxorubicin (Dox)-loaded mesoporous silica nanoparticles (MSN-Dox@GO-Apt) | 76–111 nm | Cy5.5-AS1411 aptamer/ nucleolin | Doxorubicin | 2.4 μg $mL^{-1}$, NIR (808 nm, 0.25 and 0.5 W $cm^{-2}$, 10 min) | Targeted drug delivery in MCF-7 cells | Tang *et al.* (2015a,b) |
| Gold nanoparticle-hybridized graphene oxide (AuNP – GO) | AuNPs: ~13 nm | MUC1-binding-aptamer/MUC1 | – | 0.0174–1.74 μg/mL + NIR illumination | Photothermal therapeutic efficacy in MCF-7 cells | Yang *et al.* (2015) |

could safeguard aptamers from enzymatic degradation and their subsequent removal (Chang *et al.*, 2013). More recently, pH-tunable release of daunorubicin (dau) has been reported in Molt-4 cells from sgc8c aptamer-conjugated AuNPs (Danesh *et al.*, 2015). They reported an approximately a four-fold increase of dau release at acidic extracellular environment of cancer cells (pH 5.5) (Danesh *et al.*, 2015). Furthermore, Chandrasekaran *et al.* (2016) reported highly target-specific photothermal killing of cancer cells (tumor cell death of >96% and <1% in normal cells) upon NIR irradiation using KW16–13 aptamer-targeted gold nanorods.

The potential of superparamagnetic iron oxide NPs (SPIONs) revolutionized not only the possibilities of multimodal imaging, but also magneto-photo-thermal ablation of cancer cells (Espinosa *et al.*, 2016). SPIONs, due to their unique physical properties, ability to shorten the T2 relaxation time of their surrounding water protons and biocompatibility are extensively investigated as magnetic resonance imaging (MRI) contrast agents and as vehicles for targeted drug delivery (Wang *et al.*, 2008a,b; Leach *et al.*, 2016). Recently, combination(s) of gold and ferrite heterostructured NPs were also prepared for real-time MR imaging and photothermal therapy of human colon cancer cell (Azhdarzadeh *et al.*, 2016).

Mesoporous silica nanospheres (MSNs) constitute another kind of widely investigated NP for cancer diagnosis and therapy because of their unique mesoporous structure, chemical stability, biocompatibility and tunable size. In an *in vitro* experiment with colon cancer cell, Xie *et al.* (2016) demonstrated a controlled-release behavior of doxorubicin from aptamer-MSN conjugate in acidic environment (pH 5.5) as compared to a neutral environment of pH 7.4. Zheng *et al.* (2015) reported selective drug release in the cytoplasm of cancer cells as well as simultaneous monitoring of drug release with the dual-targeting of AS1411 and ATP aptamer using fluorescent MSNs. Furthermore, Liu *et al.* (2015) described sgc8 aptamer- functionalized MSN-based silver NPs for targeted delivery of paclitaxel in an ectopic mouse model.

Aptamer-functionalized zinc oxide (ZnO) nanostructures also hold great promise in cancer cell-targeted drug delivery, photothermal therapy and imaging, owing to their semiconductor properties and ability to form divalent Zn ions ($Zn^{2+}$) in the acidic microenvironment of the tumor. Though ZnO NPs are slightly toxic to normal cells; after aptamer-functionalization, these can target cancer cells precisely generating reactive oxygen species (ROS) to eradicate the target cells (Han *et al.*, 2015). In an *in vitro* study, MUC1 aptamer (S2.2)-functionalized ZnO NPs demonstrated a higher rate of ablation of cancer cells because aptamer conjugation greatly increased the accumulation of NPs within MCF-7 cells (Han *et al.*, 2015).

### 5.7.2 Organic NPs

Polymeric NPs have attracted a myriad of interest for targeted cancer therapy due to their unique advantages, including physiological stability, surface engineering, prolonged circulation time and high accumulation at tumor site through the EPR effect (Han *et al.*, 2014; Wang *et al.*, 2014). Among polymeric NPs, liposomes are one of the most recognized nano-vehicles as they are more biocompatible and biodegradable and less toxic (Ara *et al.*, 2014). Moreover, aptamer-targeted PEGylated liposomes showed high targeting efficiency and tumor-specific drug delivery in both *in vitro* and *in vivo* tumor models (Moosavian *et al.*, 2016). In addition, polyethylenimine (PEI) and poly(l-lysine) (PLL) copolymers (PLL-alkyl-PEI) NPs showed remarkable efficiency in gene delivery and suppression of Bcl-XL (anti-apoptotic gene) in A549 cells (Askarian *et al.*, 2015). They also reported up to a 5-fold increase in gene transfection ability in lung cancer cells after AS1411 aptamer functionalization of the NPs (Askarian *et al.*, 2015). Moreover, aptamer-vector mediated delivery of small hairpin RNA (shRNA) plasmid reduced the expression of mRNA and protein levels Bcl-XL in A549 cells. Besides these, many other organic nanostructures, such as aptamer-conjugated multifunctional DNA dendrimers, di-block copolymers of poly(lactic-co-glycolic acid) (PLGA)-b-PEG-COOH and polyethyleneimines, showed considerable efficiency for bioimaging and targeted drug delivery, in both *in vitro* and *in vivo* tumor models (Subramanian *et al.*, 2015; Zhang *et al.*, 2015; Chen *et al.*, 2016).

### 5.7.3 Chimeric NPs

Chimeric NPs provides versatile nanotherapeutic system for the transport of imaging and therapeutic agents to the tumor site (Tian *et al.*, 2016). Combination of organic and inorganic components for nanoformulation has been intensively studied for decades to triumph the drawbacks of conventional cancer therapy as well as diagnosis. Numerous chimeric NPs have been synthesized with optimum surface area-to-volume ratio, high stability and signal intensity with an ability to ordain manifold and concurrent targeting, imaging and therapy. For instance, He *et al.* (2016), developed a DNAzyme-nanocomposite probe containing a built-in $Zn^{2+}$ arsenal (GNPs-ES-Dox/ZnO), which showed promising ability to co-deliver antisense miRNA and anti-tumor drug for *in vitro* gene regulation as well as targeted therapy. This pH-sensitive chimeric nanostructure demonstrated enhanced therapeutic efficacy by downregulating the levels of endogenous miR-21 and also induced apoptosis in targeted cells. Roy *et al.* (2015) developed an orally deliverable locked nucleic acid (LNA)-aptamer decorated chimeric $Fe_3O_4$ nanocarrier (NC), which showed complete reduction of tumor in a xenograft cancer stem cell mouse model after 45 days of feeding an NC diet, with tumor recurrence in only 10% of mice fed in the targeted NCs treatment compared with 30% in the non-targeted NC feed group. Recently, GO-based NPs have emerged as prospective vehicles for targeted delivery of drugs owing to their excellent structural properties and high loading capacity (Tang *et al.*, 2015a,b).

## 5.8 NANOSENSORS

In our opinion, aptamer-based biosensor technology is the fastest-growing aptamer field and promises to continuously emerge with novel clinical or companion drug diagnostic and personalized theranostic assays. Aptamer-based biosensors have been recently reviewed and the increased interest in the application of aptamers to biosensors resides mainly in the physicochemical properties of aptamers (Bruno, 2015). Many academic laboratories exploit the capacity of aptamers to alter their conformation upon binding to ligand in order to synthesize label-free analytical assays. As mentioned above, aptamer activity can be modulated by competitive interaction with a target molecule or hybridization with a complementary nucleotidic sequence; aptamers are modified with ease and can be labeled with dyes and functional groups either to obtain a signal or for immobilization on solid supports. Nanosensors comprising of aptamers conjugated to nanoparticles offer an efficient hybrid system with dual properties of highly sensitive molecular recognition, along with a strong and diverse

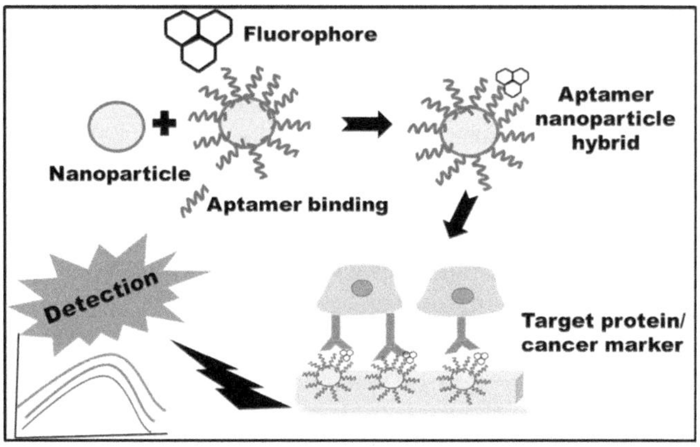


**FIGURE 5.3** Schematic illustration of aptamer-functionalized nanoparticle hybrid and its application as a nanosensor for highly specific and sensitive detection of targets, such as cancer cell markers. This can then be detected through photoluminescence, colorimetric detection system (Cha T-G *et al.*, *ACS Nano.* 2011; 5: 4236–44), chemiluminescence or a combination of both colorimetric analysis and chemiluminescence (Shwetha N *et al.*, *Anal Biochem.* 2013; 441: 73–9).

signal transduction through nanoparticles (Figure 5.3) (Xing *et al.*, 2012). Nanosensors provide a fast and simple method for the detection of numerous compounds such as mercury and lead through a colorimetric detection system. Critical aspect nanotheranostics involves the development of hybrids with clinical diagnostic and therapeutic potential (Liu and Lu, 2004; Lee *et al.*, 2007; Wang *et al.*, 2008a,b). GO based nanoparticles have emerged as a biocompatible material with a fluorescence-quenching property and an ability to protect the carrier (peptide/DNA) from enzymatic degradation (Chung *et al.*, 2013). This property of GO was exploited in a study for the development of a novel aptamer-two-photon-dye GO fluorescent nanosensor, which was used in biological fluids and live cells, as well as zebrafish for molecular probing. The sensing strategy developed, exhibited efficient *in vitro* as well as *in vivo* detection of ATP (Yi *et al.*, 2014).

Use of fluorophore molecules for detection purposes has possible associated cytotoxicity issues. Additionally, the interaction of dyes with the intracellular environment can hinder the reliability of these dyes. However, the small size, faster response rate and the mass transfer occurring over a small distance impart accuracy and precision in aptamer-based nanosensors. Furthermore, they require low voltage and power and have low cost as well (Chiu and Huang, 2009). Wang *et al.* (2012) demonstrated this using novel aptamer-based silver nanoparticles for individual as well as simultaneous detection of multiple proteins. The hybrid containing dye-labeled aptamers and silver nanoparticles (AgNPs) presented high sensitivity in the target protein detection. The limit of detection for these hybrid nanosensors was found to be around 0.4 fmol with a linear response ranging from 0.8 fmol to 0.5 pmol.

### 5.8.1 Clinical Diagnosis

The aptamer-conjugated nanosensor hybrids offer wide applications in clinical diagnosis (Farokhzad and Langer, 2006). A single quantum dot-based aptameric sensor formed by combining the high specificity of aptamers with single-molecule detection, nanotechnology and fluorescence resonance energy transfer (FRET) between cyanine 5 (Cy5) and 605 emitting quantum dot (605QD) and Iowa Black RQ was able to detect cocaine presence with extremely low consumption of sample through both on and off signal modes, thus offering a highly sensitive sensor that requires simple sample preparation technique (Zhang and Johnson, 2009). Further, the potential of single-walled carbon nanotubes (SWNTs) conjugated to DNA aptamer for non-invasive real-time detection of cell-signalling molecules was demonstrated in a study. The recognition molecule used was insulin-binding aptamer, which has the ability to detect insulin, and the photoluminescence emission upon selective binding of the aptamer conjugated SWNTs was observed. Results indicated that the photoluminescence of SWNTs quenches upon targeted recognition of insulin. Additionally, embedding of SWNT sensors in the extracellular matrix (ECM) detected insulin in glucose influx stimulated pancreatic INS-1 cells (Cha *et al.*, 2011). Bamrungsap *et al.* (2011) exploited the phenomenon of magnetic relaxation switch (MRSw) for lysozyme detection. This was attained by conjugating iron oxide-based nanoparticles with lysozyme aptamers, which have the ability to hybridize with the aptamer extension leading to the formation of clusters. When the lysozyme was added, it caused the binding of the aptamer to the target followed by disassembly of clusters and increase in spin-spin relaxation time ($T_2$). The high detection limit obtained demonstrated their immense potential in detection of clinical cancer samples. Key progress in structural DNA nanotechnology is the development of DNA origami (Seeman, 2010). This versatile technique has the potential to create nanoscale devices. It can detect the potential inhibitors to enhance the chemotherapy and enzymatic DNA repair activity of Human $O^6$-alkylguanine-DNA alkyltransferase, thus illustrating the applicability of this single-molecule recognition platform as nanosensors for a variety of complexes and proteins recognition (Tintoré *et al.*, 2013). An earlier study has reported nanosensors based on quantum dots with an ability to detect of a large number of molecules. These nanosensors consist of a thrombin binding aptamer with gold nanoparticle as the quencher. It was observed that both $Pb^{2+}$ and $K^+$ were able to bind to the TBA causing conformational change that led to the Au quencher closer to quantum dot. Further, there was a decrease in the photoluminescence observed which was a result of binding and increase in the concentrations of either $Pb^{2+}$ or $K^+$. It was capable of detecting the changes in the concentrations of ions making it an effective tool for diagnosis (Meshik *et al.*, 2014).

### 5.8.2 Cancer-Specific Delivery and Imaging

Cancer biomarkers include proteins overexpressed on the cancer cell surface or in blood and serum that can be used for the diagnosis of cancer. Most of these biomarkers exist in very low concentrations during the initial stages of cancer and are thus difficult to detect. However, nanomedicine-derived devices have emerged as effective biosensors or nanosensors with abilities to detect and monitor receptors or proteins overexpressed in cancer cells with high sensitivity and specificity (Perfézou *et al.*, 2012). Ultrasensitive and multiplexed detection of cancer cell biomarkers can be done through magnetic biomarkers (MNPs). High specificity was observed as the magnetic nanosensors were capable of interacting strongly with the target cells and were able to recognize and differentiate between cancer cells and normal cells (Bamrungsap *et al.*, 2012). Diagnosis, as well as prognosis of circulating tumor cells (CTCs), is imperative to achieve therapeutic efficacy. This was achieved in a study in which polydimethylsiloxane-based flat channel microfluidic devices were developed which were conjugated with LNA-modified epithelial cell adhesion molecule (EpCAM) and nucleolin-targeting aptamers. Results demonstrated >90% of capture efficiency with up to six times reusability. Additionally, a capture specificity and sensitivity of 100% and 92%, respectively, was detected in blood samples that were spiked with colon cancer cells. Further, an average of 5 $\pm$ 3 CTCs/mL was observed to be captured by the functionalized chip when analyzed on blood samples of 25 patients with head and neck cancer. Overall, this unique microfluidic system offered a quick, robust, sustainable, reusable and highly sensitive platform for low-cost cancer diagnosis and therapeutics. The tumor targeting potential of multi-modular nanocarriers was recently established using $Fe_3O_4$-saturated alginate enclosed chitosan coated calcium phosphate nanocarriers (NCs) functionalized using LNA-modified EpCAM and nucleolin aptamers. About 70% of the mice exhibited complete tumor regression when fed with the non-targeted nanocarriers. Further, the anticancer mechanism was observed to be mediated through the TNF-related apoptosis-inducing ligand (TRAIL) and Fas and Fas-associated protein with death domain (FADD)-induced p53 phosphorylation. Additionally, highly specific tumor-targeting abilities of these aptamer-conjugated NCs was validated using multi-modular imaging through NIR imaging, computerized tomographic (CT) techniques and MRI. The theranostic ability of these nanocapsules saturated with $Fe_3O_4$-encapsulating bovine lactoferrin was also demonstrated in a breast cancer model (claudin-low and triple-negative). These superparamagnetic nanocapsules internalized into the cancer cells through endocytosis, which was receptor-mediated. Additionally, they led to the apoptosis induction, which was mediated by the down-regulation of livin and survivin proteins. The real-time imaging designated selective localization of the nanocapsules to the tumor site along with tumor reduction, which was found to be time-dependent. This indicated remarkable nanotheranostic potential of these nanocapsules, which can be exploited for effective monitoring of anticancer drugs (Kanwar *et al.*, 2016). Application of aptamer-conjugated nanosensors has also been evaluated to detect a tumor suppressor protein, known as p53, using a colorimetric biosensing technique. AuNPs were conjugated with the RNA aptamer, which desorbed from the surface of the AuNPs due to its interactions with the target, leading to a change of color from red to purple, with the colorimetric detection limit observed as 0.2 ng/mL. Moreover, chemiluminescence reaction enhanced the catalytic activity of the aggregated nanoparticles, which was 10-fold higher in comparison to the AuNPs-based colorimetric method, indicating the potential of a combination of chemiluminescence and colorimetric-based aptasensor as a highly sensitive target-detecting system (Shwetha *et al.*, 2013). Extensive studies have demonstrated remarkable imaging and therapeutic potential of AuNPs for the cancer cells as fluorescent quantum dots, yet targeting is observed to be quite time-consuming for selective cancer cell detection and therapy (Lewis *et al.*, 2002). However, Chen *et al.* (2015) developed a targeting-free, fast screening, as well as selective cancer cell killing approach through Ag/Au nanoshells-modified glucose-responsive/-sensitive glucose oxidase.

Dark field scattering imaging indicated immense potential of these nanoplasmonic theranostic systems for fast screening along with visual differentiation of cancer cells from normal cells. Further, exposure to 808 nm laser presented ultrasensitive cancer cells targeting and their ablation.

## 5.9 BLOOD–BRAIN BARRIER: MOST CHALLENGING BARRIER

In order to function optimally, the human brain needs a precise microenvironment that is homeostatically maintained via the blood–brain barrier (BBB). This barrier has a key role in segregating brain tissue from the bloodborne toxins, thus restricting the access of most drugs and microorganisms (Abbott, 2013). The BBB contains endothelial cells lining the brain vasculature, coupled with surrounding astrocytes and pericytes (Figure 5.4). These brain endothelial cells are phenotypically dissimilar to those of the peripheral circulation, as they are characterized by highly polarized tight junctions connecting them to adjacent cells (Bernacki *et al.*, 2008). It is this junction that limits the free transport of molecules into the brain, along with lack of pinocytic vacuoles (Daneman, 2012). In addition, the highly polarized nature of the BBB endothelial cells generated by the proteins present within the tight junctions produces unique membrane compositions, which further restrict the transport of molecules into the brain (Daneman, 2012). This all combines to restrict the free movement of molecules with molecular weight greater than 500 daltons into the brain, thus preventing a reported 98% of small molecules from crossing the BBB (Gabathuler, 2010). However, while the BBB prevents small molecules and toxic substances from crossing into the brain, it also impedes the influx of hydrophilic substances essential for the maintenance of brain homeostasis from the blood to the brain. It is the presence of carrier-and receptor-mediated transporter systems that allows the movement of essential molecules required for cerebral function into the brain systems (Figure 5.5) (Svetlana *et al.*, 2011). The transportation of molecules across the BBB is also influenced by the presence of protein efflux transporters expressed on both the luminal and abluminal surfaces of the brain endothelial cells. These transporters have a dual role: first, to protect the CNS from the influx of neurotoxins, and second, to restrict the therapeutic drug access to the brain (Miller, 2010). This greatly hinders the treatment of brain disorders, thus limiting the effect of any treatment not given intracranially.

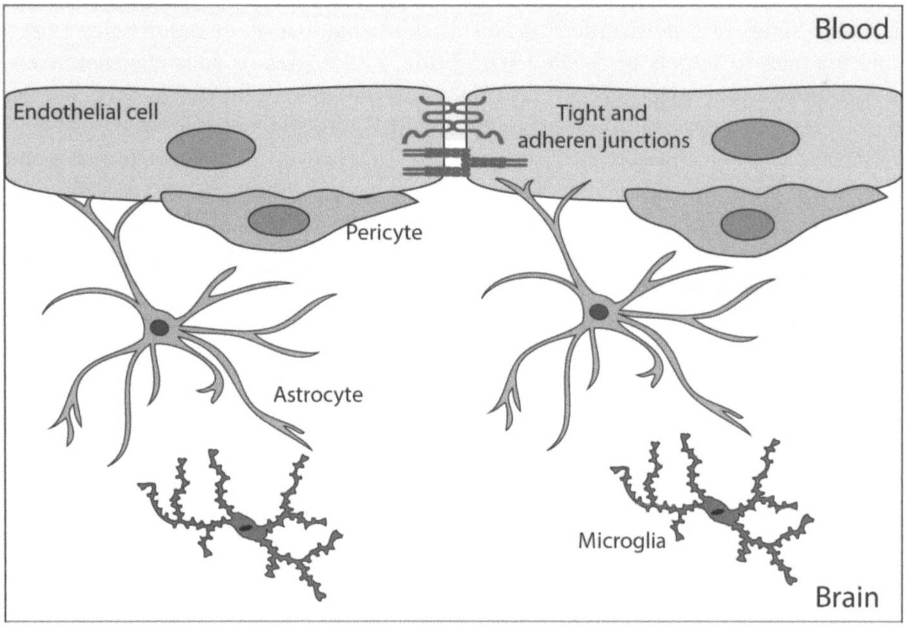


**FIGURE 5.4** Structure of the blood–brain barrier (BBB).

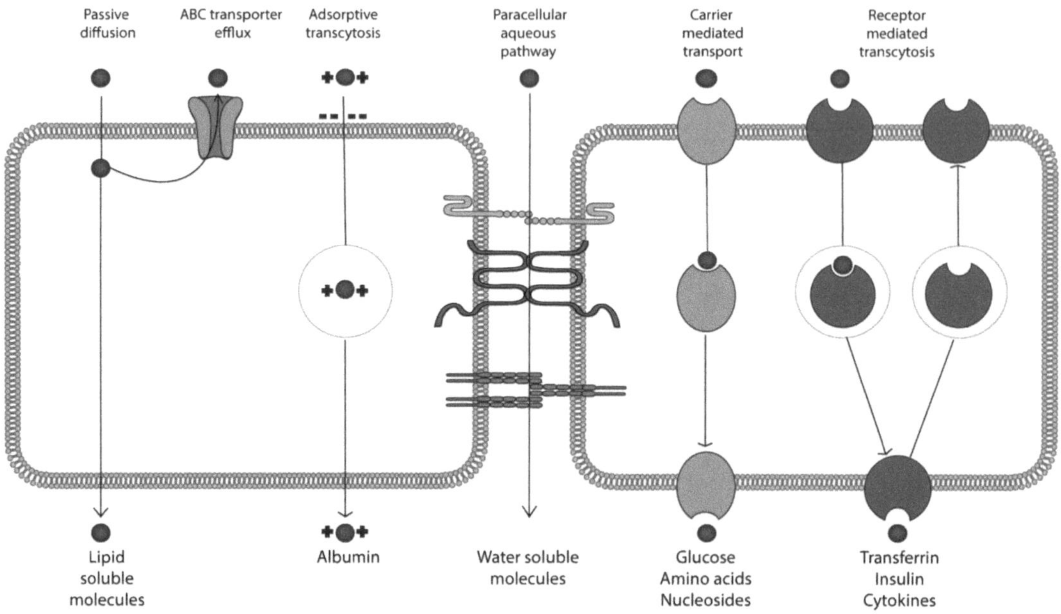


**FIGURE 5.5** Routes of transport across the blood–brain barrier (BBB). (Adapted from Abbott NJ. *J Anat.* 2002; 200: 629–38; Gabathuler R. *Neurobiol Dis.* 2010; 37: 48–57.)

A number of methods have been proposed for delivering therapeutics into the brain more effectively, such as nasal delivery, osmotic opening of the BBB, transporter vectors, viral vectors, or NPs (Cheng *et al.*, 2013). Some of these suggest widespread disruption of the BBB to allow therapeutics to enter the brain. However, this approach can be quite costly, requires hospitalisation and anaesthesia, and may also lead to permanently damaged neurons (Gabathuler, 2010). This has led to the search for advanced modalities that can actively enter the brain tissue without widespread disruption of the BBB.

Initial work to target CNS disorders has resulted in a number of aptamers being selected, in the traditional manner, to targets associated with brain pathologies or neurodegenerative disorders. Advances in the aptamer selection protocols have seen a number of advances that are beneficial for targeting of brain disorders. While a majority of SELEX techniques has been performed against protein targets, a new technique published in 2006 opened up the possibility of potential new targets with no prior knowledge of the disease target required (Shangguan *et al.*, 2006). This novel technique, cell SELEX, allowed selection against complex targets, such as cancer cells. Cerchia *et al.* (2009) generated aptamers against the U87MG cell line, using the non-tumorogenic T98G cell line as a negative control line during selection. The authors characterised eight aptamers from the selection, demonstrating high specificity and selectivity. Two of these aptamers were further studied by Aptekar *et al.* (2015), who demonstrated that one of these aptamers, SA43 (GL43), was able to recognize all four grades of glioblastoma in a histological study, suggesting the potential of this aptamer to be used for diagnostic applications. There is also high potential for this aptamer for being used as a therapeutic agent. Several other aptamers have also been developed for drug delivery, *in vivo* imaging and other diagnostic applications, which have been reviewed by Delac *et al.* (2015). While space precludes the discussion of all aptamers that have been selected for neurological disorders, Table 5.2 lists some of the more interesting aptamers that have been generated to date.

One of the new intriguing adaptations of the traditional SELEX procedure was development of an *in vivo* SELEX protocol. The first of such protocols involved selecting aptamers *in vivo* to intra-hepatic colon carcinoma-bearing mice using a single-stranded and chemically-modified RNA library (Mi *et al.*, 2010). This demonstrated that aptamer selection can be accomplished in complex

**TABLE 5.2**
**List of Various Aptamers Used for Targeted Drug Delivery in Brain**

| Disease | Aptamer | Target | Tested *In Vivo*? | Reference |
|---|---|---|---|---|
| Brain cancer metastases | SYL3C, EpA, EpB, EpC, TEPP, TENP, TEPN, TENN | Transferrin receptor and epithelial cell adhesion molecule (EpCAM) | Yes | Macdonald *et al.* (2017) |
| Alzheimer's disease | A1 aptamer | Beta-secretase-1 | Yes | Liang *et al.* (2015) |
| Alzheimer's disease | β55 RNA aptamer | Amyloid-β | Yes | Farrar *et al.* (2014) |
| Multiple sclerosis | Anti-myelin aptamer 3064 | Murine myelin | Yes | Nastasijevic *et al.* (2012) |
| Parkinson's disease | M5-15 aptamer | A-synuclein monomers | No | Tsukakoshi *et al.* (2010) |

environments and opened up the possibility for more complex selection strategies. Not only did this study represent the first published example of *in vivo* aptamer selection, but it also confirmed the findings from a number of *in vivo* biodistribution studies that intravenous delivery of aptamers leads to low distribution in the brain. Cheng *et al.* (2013) investigated the ability of a randomized pool of aptamer species to reach the brain in wild-type mice. Following injection via the tail vein, the brain tissue was harvested and the purified RNA was re-injected into subsequent animals. SELEX was performed for 22 rounds, and aptamers that were selected were further shown to be internalized into brain endothelial cells. However, given the number of rounds required, the authors commented that this might not be a suitable procedure for rapid aptamer generation.

Demonstrating that aptamers can cross the BBB, if specifically selected, suggests that additional routes may also be utilized. A recent development has been the utilization of the active transportation mechanisms to transport agents into the brain. Designing drugs and transport modalities for these transporters first allows them to pass through the BBB, but additionally increases drug delivery as a result of circumventing the efflux capacity of the BBB. One such example was demonstrated when Yu *et al.* (2011) showed that a transferring-receptor antibody was able to transcytose into the brain, if the binding affinity was tuned toward a low Kd value. While this showed proof-of-principle, the antibody produced side effects. However, building on this principle, the development of aptamers against the transferrin receptor is now being explored. Chen *et al.* (2008) reported the generation of DNA and RNA aptamers, with an ability to selectively recognize the extracellular domain of the mouse transferrin receptor. Originally 64 nucleotides long, these authors truncated the GS24 aptamer to 50 nucleotides with no influence on selectivity. Following this, Porciani *et al.* (2014) further truncated the aptamer to 35 nucleotides, which resulted in significant loss in binding affinity. Introduction of mutations into the stem region of the aptamer resolved this problem and, in turn, produced an aptamer with superior properties compared to the original. To further enhance the therapeutic potential of this aptamer utilizing rational engineering of the predicted two-dimensional structures, Macdonald *et al.* (2017) have further truncated the aptamer to a mere 14 nucleotides. Transitioning from a 64-nucleotide aptamer to 14 nucleotides has significantly widened the therapeutic potential of the transferrin-receptor aptamer. Its small size now allows it to be synthesized for the transport of wide drug varities to treat multiple neurological disorders including cancer and Alzheimer's disease.

In line with this, nanocarriers for brain targeted delivery also present a lucrative option for targeted therapeutic applications. Sriramoju *et al.* (2015) successfully demonstrated the potential of the recombinant dominant negative survivin mutant (SurR9-C84A)-conjugated biodegradable PLGA nanocarriers in reducing the expression of amyloid-β, along with enhancing the neuronal integrity markers. Also, it promoted the maintenance of endogenous survivin levels, which is imperative for restricting caspase activation and, thus, apoptosis. Therefore, a combinatorial approach of coupling high specificity of aptamers with site-specific smart 'nanosystems' can facilitate the therapy for numerous neurological diseases, thus overcoming the non-specific tissue distribution.

## 5.10 APTAMER LIMITATIONS

Original aptamer developments reported from the academic field have continuously been demonstrating immense potential of this technology, though it carries its own advantages and limitations as any early technology. While the advantages have been discussed above, two major limitations, in our opinion, which have been slowing down the progression of aptamer developments are (i) usage of PCR during SELEX, (ii) the rapid degradation of aptamers (especially RNA aptamers) by nucleases in blood, (iii) interaction of aptamers with intracellular targets, (iv) aptamer cross-reactivity and (v) reduced chemical information of unmodified nucleic acids (Lakhin *et al.*, 2013).

### 5.10.1 Inability to Use PCR

PCR is widely used in molecular biology, medical diagnostic assays and microbiology, where the template usually comprises a mixture of homologous nucleotide sequences. In these cases, it is assumed that the products after PCR are homogeneous. On the contrary, one of the numerous steps of SELEX, requires amplification of large and diverse DNA templates through PCR, that generates a reaction mixture containing complex chemical kinetics, which are hard to control. PCR amplification of an aptamer library usually produces various artifactual products, moreover, a bias is observed for some nucleotide sequences (Dittmer *et al.*, 2004; Lou *et al.*, 2009) that renders the aptamer selection procedure inefficient, time-consuming and tedious. Several factors might be generating these biased and artifactual PCR products: (i) not all the aptamer sequences are equally accessible to primer hybridization following denaturation, (ii) templates are not extended by polymerase with the same efficiency and (iii) single-stranded templates hybridize with each other acting as initiation primers. Evidence suggests that a DNA sequence with a rich content of guanosine and cytosine influences PCR amplification. Although template folding might also influence the bias in a PCR by hindering the primer priming regions or affecting polymerase progression, this is not yet well-documented. Hence, it is possible to assume that the result of the PCR is based on selection of PCR and aptamer, though the best binders may be lost during this procedure. There are two plausible approaches to decrease the effects of the PCR: (i) perform single-molecule PCR or (ii) decrease the SELEX cycle number as much as possible.

PCR can be significantly improved when performed on a single DNA template, which can be achieved by compartmentalization of the reaction in miniature reactors. Even more, NGS systems such as the Ion Torrent (Life Technologies, Carlsbad, CA, USA) and Illumina's sequencers use PCR-compartmentalization strategies for large genome sequencing. While the former uses emulsion PCR, the latter amplifies single-DNA templates in defined clusters by bridging PCR (Mc Namara *et al.*, 2006). Emulsion PCR is much easier to operate in an academic laboratory and has been applied for aptamer discovery to reduce by-product formation (Mendonsa and Bowser, 2004; Merriman *et al.*, 2012). Basically, emulsion PCR consists of generation of a water-in-oil emulsion containing all of the PCR elements, where a single DNA template is placed in each droplet by limited dilution. Thus, amplification occurs in the droplets, reducing PCR product-artifacts coming from inter-template hybridization.

Decreasing the number of SELEX rounds is another approach to reduce PCR bias and artifacts. Many processes are under development to make aptamer discovery more efficient, and these have been reviewed elsewhere (Niemeyer and Adler, 2002; Ni *et al.*, 2011). These processes include automated robotic SELEX platforms, capillary electrophoresis SELEX, microfluidics, GO separation, non-equilibrium capillary electrophoresis of equilibrium mixtures, capillary transient isotachophoresis, micro free-flow electrophoresis and high-throughput SELEX (Rusconi *et al.*, 2002; Oh *et al.*, 2011a; Merriman *et al.*, 2012; Riley *et al.*, 2014; Rozenblum *et al.*, 2016). While research efforts are being undertaken for resolving the hurdles in aptamer development, only a few aptamer-oriented commercial incubators have been raised, such as Somalogic (Boulder, CO, USA), Noxxon (Berlin, DE) and Apta Biosciences (Nanos, SG). Somalogic was a pioneer in translating the opportunities of modified nucleotides into its commercial Slow Off-rate Modified Aptamer (SOMAmer) platform

(Rusconi *et al.*, 2000). SOMAmers are aptamers containing nucleotides modified at the C5 position, which are used in SELEX. This allowed Somalogic to create a commercially efficient aptamer-discovery platform. Apta Biosciences has also exploited the benefits of novel nucleotide modifications and has created aptamers containing amino acid-like side chains (Schütze *et al.*, 2011). Clearly, chemical modifications introduce new functionalities to aptamers, increase affinity for their targets and enhance the aptamer discovery process. However, it has to be borne in the mind that how the addition of chemical complexity affects aptamer properties, such as denaturalization and antigenicity, is still elusive.

During the past 25 years, phenomenal progress in aptamer research has been made. Aptamers have the potential to supplant current antibody-based technologies and promise to introduce novel diagnostic assays and therapies. However, PCR bias and the intrinsic low chemical diversity of natural oligonucleotides still present hindrances that discourage the consolidation of aptamer-based companies. Nevertheless, a handful of companies have been established that provide different technological platforms for the discovery and commercialization of aptamers and services, including Base Pair Biotechnologies (Houston, TX, USA), Aptagen (Jacobus, PA, USA), Neoventures Biotechnology (London, ON, Canada), Somalogic (Boulder, CO, USA), Aptamer Sciences (Gyeonbuk, KR), Apta Matrix (Syracuse, NY, USA), AM Biotech (Houston, TX, USA), OTC Biotech (San Antonio, TX, USA), Apta Biosciences (Nanos, SG), Aptamer Group (York, UK) and Noxxon (Berlin, DE), among others. Interestingly, these companies claim that they have overcome some of the aptamer technology limitations using process automation, high-throughput screening, NGS and/or modified nucleotides. While several high-quality aptamers have been selected manually, the future of aptamer high-throughput screening in pharmaceutical industry may hinge on automation of SELEX (Shu *et al.*, 2011).

## 5.11 CONCLUSION

Aptamer-based therapeutics provides pharmaceuticals with small size, high affinity towards the targets and low immunogenicity. The aptamer-NPs bioconjugates have facilitated a new class of therapeutics with multi-modular tumor targeting and diagnostic properties. Along with this, the small size of aptamers and their ability to cross the BBB makes them ideal candidates for neurological therapeutics. Therefore, the researchers around the globe have focused towards developing various NPs with numerous valuable features during preclinical studies. However, only few of them find their way to clinics due to their intrinsic drawbacks in terms of hydrodynamic size, physicochemical properties, specificity, drug encapsulation efficiency, biodistribution after systemic delivery and pH-tunable drug release capacity. Despite the promising initial data generated during *in vitro* and *in vivo* studies with aptamer-functionalized NPs (Apt-NPs) for precisely targeting cancerous tissues, it still has a long way to go to diminish the gap between reverie and reality. Not only do the properties of NPs have to be optimized appropriately for effective translation of Apt-NPs to the clinic, but also target-specific aptamers need to be isolated, which can bind to a very specific fingerprint of malignant cells.

## REFERENCES

Abbott NJ. Astrocyte-endothelial interactions and blood-brain barrier permeability. *J Anat.* 2002; 200: 629–38.

Abbott NJ. Blood-brain barrier structure and function and the challenges for CNS drug delivery. *J Inherit Metab Dis.* 2013; 36: 437–49.

Alibolandi M, Mohammadi M, Taghdisi SM, Ramezani M, Abnous K. Fabrication of aptamer decorated dextran coated nano-graphene oxide for targeted drug delivery. *Carbohydr Polym.* 2017; 2: 218–29.

Alibolandi M, Ramezani M, Abnous K, Sadeghi F, Atyabi F, Asouri M, Ahmadi AA, Hadizadeh F. *In vitro* and *in vivo* evaluation of therapy targeting epithelial-cell adhesion-molecule aptamers for non-small cell lung cancer. *J Control Release.* 2015; 209: 88–100.

Aptekar S, Arora M, Lawrence CL, Lea RW, Ashton K, Dawson T, Alder JE, Shaw L. Selective targeting to glioma with nucleic acid aptamers. *PLOS One.* 2015; 10: 1–24.

Ara MN, Hyodo M, Ohga N, Hida K, Harashima H. Development of a novel DNA aptamer ligand targeting to primary cultured tumor endothelial cells by a cell-based SELEX method. *PLoS One*. 2012; 7(12): e50174.

Ara MN, Matsuda T, Hyodo M, Sakurai Y, Hatakeyama H, Ohga N, Hida K, Harashima H. An aptamer ligand based liposomal nanocarrier system that targets tumor endothelial cells. *Biomaterials*. 2014; 35: 7110–20.

Arnold S, Pampalakis G, Kantiotou K, Silva D, Cortez C, Missailidis S, Sotiropoulou G. One round of SELEX for the generation of DNA aptamers directed against KLK6. *Biol Chem*. 2012; 393: 343–53.

Asai R, Nishimura SI, Takahashi K. DNA aptamers that recognize fluorophore using on-chip screening in combination with an in silico evolution in.nucleic acids. *Symp Ser*. 2003; 3: 321–2.

Askarian S, Abnous K, Taghavi S, Oskuee RK, Ramezani M. Cellular delivery of shRNA using aptamer-conjugated PLL-alkyl-PEI nanoparticles. *Colloids Surf B*. 2015; 136: 355–64.

Azhdarzadeh M, Atyabi F, Saei AA, Varnamkhasti BS, Omidi Y, Fateh M, Ghavami M, Shanehsazzadeh S, Dinarvand R. Theranostic MUC-1 aptamer targeted gold coated superparamagnetic iron oxide nanoparticles for magnetic resonance imaging and photothermal therapy of colon cancer. *Colloids Surf B*. 2016; 143: 224–32.

Bamrungsap S, Chen T, Shukoor MI, Chen Z, Sefah K, Chen Y, Tan W. Pattern recognition of cancer cells using aptamer-conjugated magnetic nanoparticles. *ACS Nano*. 2012; 6: 3974–81.

Bamrungsap S, Shukoor MI, Chen T, Sefah K, Tan W. Detection of lysozyme magnetic relaxation switches based on aptamer-functionalized superparamagnetic nanoparticles. *Anal Chem*. 2011; 83: 7795–9.

Bauer M, Macdonald J, Henri J, Duan W, Shigdar S. The Application of Aptamers for Immunohistochemistry. *Nucl Acid Ther*. 2016; 26: 120–26.

Bernacki J, Dobrowolska A, Nierwińska K, Małecki A. Physiology and pharmacological role of the blood-brain barrier. *Pharmacol Rep: PR*. 2008; 60: 600–22.

Bruno JG. Predicting the uncertain future of aptamer-based diagnostics and therapeutics. *Molecules*. 2015; 20: 6866–87.

Burke PA. The ups and downs of clinical translation of new technologies: Déjà Vu All over again. *Mol Ther*. 2015; 23: 791.

Cerchia L, Esposito CL, Jacobs AH, Tavitian B, de Franciscis V. Differential SELEX in human glioma cell lines. *PLOS One*. 2009; 4: 1–10.

Cha T-G, Baker BA, Sauffer MD, Salgado J, Jaroch D, Rickus JL, Porterfield DM, Choi JH. Optical nanosensor architecture for cell-signaling molecules using DNA aptamer-coated carbon nanotubes. *ACS Nano*. 2011; 5: 4236–44.

Chandrasekaran R, Lee ASW, Yap LW, Jans DA, Wagstaff KM, Cheng W. Tumor cell-specific photothermal killing by SELEX-derived DNA aptamer-targeted gold nanorods. *Nanoscale*. 2016; 8: 187–96.

Chang C-C, Wei S-C, Wu T-H, Lee C-H, Lin C-W. Aptamer-based colorimetric detection of platelet-derived growth factor using unmodified goldnanoparticles. *Biosens Bioelectron*. 2013; 42: 119–23.

Chavanpatil MD, Khdair A, Panyam J. Nanoparticles for cellular drug delivery: Mechanisms and factors influencing delivery. *J Nanosci Nanotechnol*. 2006; 6: 2651–63.

Chen C, Dellamaggiore K, Ouellette C, Sedano C, Lizadjohry M, Chernis G, Gonzales M, Baltasar F, Fan A, Myerowitz R, Neufeld E. Aptamer-based endocytosis of a lysosomal enzyme. *Proc Natl Acad Sci USA*. 2008; 105: 15908–13.

Chen L, Li H, He H, Wu H, Jin Y. Smart plasmonic glucose nanosensors as generic theranostic agents for targeting-free cancer cell screening and killing. *Anal Chem*. 2015; 87: 6868–74.

Chen M, Yu Y, Jiang F, Zhou J, Li Y, Liang C, Dang L, Lu A, Zhang G. Development of Cell-SELEX Technology and Its Application in Cancer Diagnosis and Therapy. *Int. J. Mol. Sci*. 2016; 17: 2079.

Chen Z, Tai Z, Gu F, Hu C, Zhu Q, Gao S. Aptamer-mediated delivery of docetaxel to prostate cancer through polymeric nanoparticles for enhancement of antitumor efficacy. *Eur J Pharm Biopharm*. 2016; 107: 130–41.

Cheng C, Chen YH, Lennox KA, Behlke MA, Davidson BL. *In vivo* SELEX for identification of brain-penetrating aptamers. *Mol Ther Nucleic Acids*. 2013; 2: e67.

Chiu T-C, Huang C-C. Aptamer-functionalized nano-biosensors. *Sensors*. 2009; 9: 10356–88.

Cho M, Oh SS, Nie J, Stewart R, Eisenstein M, Chambers J, Marth JD, Walker F, Thomson JA, Soh HT. Quantitative selection and parallel characterization of aptamers. *Proc. Natl Acad Sci USA*. 2013; 110: 18460–5.

Chung C, Kim Y-K, Shin D, Ryoo S-R, Hong BH, Min D-H. Biomedical applications of graphene and graphene oxide. *Accounts Chem Res*. 2013; 46: 2211–24.

Chushak Y, Stone MO. In silico selection of RNA aptamers. *Nucleic Acids Res*. 2009; 1–9.

Cox JC, Ellington AD. Automated selection of anti-protein aptamers. *Bioorg Med Chem*. 2001; 9: 2525–31.

Daneman R. The blood-brain barrier in health and disease. *Ann Neurol*. 2012; 72: 648–72.

Danesh, NM, Lavaee P, Ramezani M, Abnous K, Taghdisi SM. Targeted and controlled release delivery of daunorubicin to T-cell acute lymphoblastic leukemia by aptamer-modified gold nanoparticles. *Int J Pharm.* 2015; 489: 311–7.

Das M, Duan W, Sahoo SK. Multifunctional nanoparticle–EpCAM aptamer bioconjugates: A paradigm for targeted drug delivery and imaging in cancer therapy. *Nanomedicine.* 2015; 11: 379–89.

Dassie JP, Liu X-y, Thomas GS, Whitaker RM, Thiel KW, Stockdale KR, Meyerholz DK, McCaffrey AP, McNamara II JO, Giangrande PH. Systemic administration of optimized aptamer-siRNA chimeras promotes regression of PSMA-expressing tumors. *Nat Biotechnol.* 2009; 27: 839–46.

Delac M, Motaln H, Ulrich H, Lah TT. Aptamer for imaging and therapeutic targeting of brain tumor glioblastoma. *Cytometry A.* 2015; 87: 806–16.

Dittmer WU, Reuter A, Simmel FC. A DNA-based machine that can cyclically bind and release thrombin. *Angew Chem Int Ed.* 2004; 43: 3550–3.

Douglas SM, Bachelet I, Church GM. A logic-gated nanorobot for targeted transport of molecular payloads. *Science.* 2012; 335: 831–4.

Du W, Yuan Y, Wang L, Cui Y, Wang H, Xu H, Liang G. Multifunctional bioconjugate for cancer cell-targeted theranostics. *Bioconjugate Chem.* 2015; 26: 2571–8.

Dwivedi HP, Smiley RD, Jaykus L-A. Selection of DNA aptamers for capture and detection of Salmonella Typhimurium using a whole-cell SELEX approach in conjunction with cell sorting. *Appl Microbiol Biotechnol.* 2013; 97: 3677–86.

Espinosa A, Di Corato R, Kolosnjaj-Tabi J, Flaud P, Pellegrino T, Wilhelm C. Duality of iron oxide nanoparticles in cancer therapy: Amplification of heating efficiency by magnetic hyperthermia and photothermal bimodal treatment. *ACS Nano.* 2016; 10: 2436–46.

Everts M, Saini V, Leddon JL, Kok RJ, Stoff-Khalili M, Preuss MA, Millican CL, Perkins G, Brown JM, Bagaria H. Covalently linked Au nanoparticles to a viral vector: Potential for combined photothermal and gene cancer therapy. *Nano Lett.* 2006; 6: 587–91.

Fan Z, Sun L, Huang Y, Wang Y, Zhang M. Bioinspired fluorescent dipeptide nanoparticles for targeted cancer cell imaging and real-time monitoring of drug release. *Nat Nanotechnol.* 2016; 11: 388–94.

Farokhzad OC, Langer R. Nanomedicine: Developing smarter therapeutic and diagnostic modalities. *Adv Drug Deliv Rev.* 2006; 58: 1456–9.

Farrar CT, William CM, Hudry E, Hashimoto T, Hyman BT. RNA aptamer probes as optical imaging agents for the detection of amyloid plaques. *PLOS One.* 2014; 9: 1–8.

Gabathuler R. Approaches to transport therapeutic drugs across the blood-brain barrier to treat brain diseases. *Neurobiol Dis.* 2010; 37: 48–57.

Glökler J, Schütze T, Konthur Z. Automation in the high-throughput selection of random combinatorial libraries—Different approaches for select applications. *Molecules.* 2010; 15: 2478–90.

Gold L, Ayers D, Bertino J, Bock C, Bock A, Brody EN, Carter J, Dalby AB, Eaton BE, Fitzwater T. Aptamer-based multiplexed proteomic technology for biomarker discovery. *PLOS One.* 2010; 5: 1–17.

Gopinath SC, Lakshmipriya T, Awazu K. Colorimetric detection of controlled assembly and disassembly of aptamers on unmodified gold nanoparticles. *Biosens Bioelectron.* 2014; 51: 115–23.

Groff K, Brown J, Clippinger MJ. Modern affinity reagents: Recombinant antibodies and aptamers. *Biotechnol Adv.* 2015; 33 (8): 1787–98.

Han Y, Zhang P, Chen Y, Sun J, Kong F. Co-delivery of plasmid DNA and doxorubicin by solid lipid nanoparticles for lung cancer therapy. *Int J Mol Med.* 2014; 34: 191–6.

Han Z, Wang X, Heng C, Han Q, Cai S, Li J, Qi C, Liang W, Yang R, Wang C. Synergistically enhanced photocatalytic and chemotherapeutic effects of aptamer-functionalized ZnO nanoparticles towards cancer cells. *Phys Chem Chem Phys.* 2015; 17: 21576–82.

He Z-M, Zhang P-H, Li X, Zhang J-R, Zhu J-J. A targeted DNAzyme-nanocomposite probe equipped with built-in $Zn2+$ arsenal for combined treatment of gene regulation and drug delivery. *Sci Rep.* 2016; 6: 1–10.

Hoinka J, Berezhnoy A, Dao P, Sauna ZE, Gilboa E, Przytycka TM. Large scale analysis of the mutational landscape in HT-SELEX improves aptamer discovery. *Nucleic Acids Res.* 2015; 43: 5699–707.

Jayasena SD. Aptamers: An emerging class of molecules that rival antibodies in diagnostics. *Clin Chem.* 1999; 45(9): 1628–50.

Jiménez E, Sefah K, López-Colón D, Van Simaeys D, Chen HW, Tockman MS, Tan W. Generation of lung adenocarcinoma DNA aptamers for cancer studies. *PLOS One.* 2012; 7: 1–7.

Jing M, Bowser MT. Isolation of DNA aptamers using micro free flow electrophoresis. *Lab Chip.* 2011; 11: 3703–9.

Kanwar JR, Kamalapuram SK, Krishnakumar S, Kanwar RK. Multimodal iron oxide ($Fe_3O_4$)-saturated lactoferrin nanocapsules as nanotheranostics for real-time imaging and breast cancer therapy of claudin-low, triple-negative (ER-/PR-/HER2-). *Nanomedicine.* 2016; 11: 249–68.

Kee ST, Madoff DC, Murthy R .Clinical Interventional Oncology E-Book. 2013.

Khaled A, Guo S, Li F, Guo P. Controllable self-assembly of nanoparticles for specific delivery of multiple therapeutic molecules to cancer cells using RNA nanotechnology. *Nano Lett.* 2005; 5: 1797–808.

Lai J-C, Hong C-Y. Magnetic-assisted rapid aptamer selection (MARAS) for generating high-affinity DNA aptamer using rotating magnetic fields. *ACS Comb Sci.* 2014; 16: 321–7.

Lakhin AV, Tarantul VZ, Gening LV. Aptamers: Problems, Solutions and Prospects. *Acta Naturae.* 2013; 5(4): 34–43.

Lao Y-H, Chiang H-Y, Yang D-K, Peck K, Chen L-C. Selection of aptamers targeting the sialic acid receptor of hemagglutinin by epitope-specific SELEX. *Chem Commun.* 2014; 50: 8719–22.

Lares MR, Rossi JJ, Ouellet DL. RNAi and small interfering RNAs in human disease therapeutic applications. *Trends Biotechnol.* 2010; 28(11): 570–9.

Leach JC, Wang A, YeK, Jin S. A RNA-DNA Hybrid Aptamer for nanoparticle-based prostate tumor targeted drug delivery. *Int J Mol Sci.* 2016; 17: 380.

Lee H, Dam DHM Ha, JW Yue J, Odom TW. Enhanced human epidermal growth factor receptor 2 degradation in breast cancer cells by lysosome-targeting gold nanoconstructs. *ACS Nano.* 2015; 9: 9859–67.

Lee JS, Han MS, Mirkin CA. Colorimetric detection of mercuric ion (Hg2+) in aqueous media using dna-functionalized gold nanoparticles. *Angew Chem Int Ed.* 2007; 46: 4093–6.

Lewis JS, Achilefu S, Garbow J, Laforest R, Welch M. Small animal imaging: Current technology and perspectives for oncological imaging. *Eur J Cancer.* 2002; 38: 2173–88.

Li L-L, Lu Y. Regiospecific hetero-assembly of DNA-functionalized plasmonic upconversion superstructures. *J Am Chem Soc.* 2015; 137: 5272–5.

Liang H, Shi Y, Kou Z, Peng Y, Chen W, Li X, Li S, Wang Y, Wang F, Zhang X. Inhibition of BACE1 activity by a DNA aptamer in an alzheimer's disease cell model. *PLOS One.* 2015; 10: 1–14.

Liao W-C, Lu C-H, Hartmann R, Wang F, Sohn YS, Parak WJ, Willner I. Adenosine triphosphate-triggered release of macromolecular and nanoparticle loads from aptamer/DNA-cross-linked microcapsules. *ACS Nano.* 2015; 9: 9078–86.

Lincoff AM, Mehran R, Povsic TJ, Zelenkofske SL, Huang Z, Armstrong PW, Steg PG, Bode C, Cohen MG, Buller C. Effect of the REG1 anticoagulation system versus bivalirudin on outcomes after percutaneous coronary intervention (REGULATE-PCI): A randomised clinical trial. *Lancet.* 2016; 387: 349–56.

Liu C, Zheng J, Deng L, Ma C, Li J, Li Y, Yang S, Yang J, Wang J, Yang R. Targeted intracellular controlled drug delivery and tumor therapy through *in situ* forming Ag nanogates on mesoporous silica nanocontainers. *ACS Appl Mater Interfaces.* 2015; 7: 11930–8.

Liu J, Lu Y. Accelerated color change of gold nanoparticles assembled by DNAzymes for simple and fast colorimetric Pb2+ detection. *J Am Chem Soc.* 2004; 126: 12298–305.

Liu J, Wei T, Zhao J, Huang Y, Deng H, Kumar A, Wang C, Liang Z, Ma X, Liang X-J. Multifunctional aptamer-based nanoparticles for targeted drug delivery to circumvent cancer resistance. *Biomaterials.* 2016; 91: 44–56.

Lou X, Qian J, Xiao Y, Viel L, Gerdon AE, Lagally ET, Atzberger P, Tarasow TM, Heeger AJ, Soh HT. Micromagnetic selection of aptamers in microfluidic channels. *Proc Natl Acad Sci U S A.* 2009; 106: 2989–94.

Lynch I, Salvati A, Dawson KA. Protein-nanoparticle interactions: What does the cell see. *Nat Nanotech.* 2009; 4: 546–7.

Ma H, Liu J, Ali MM, Mahmood MAI, Labanieh L, Lu M, Iqbal SM, Zhang Q, Zhao W, Wan Y. Nucleic acid aptamers in cancer research, diagnosis and therapy. *Chem Soc Rev.* 2015; 44: 1240–56.

Macdonald J, Henri J, Goodman L, Xiang D, Duan W, Shigdar S. Development of a bifunctional aptamer targeting the transferrin receptor and epithelial cell adhesion molecule (EpCAM) for the treatment of brain cancer metastases. *ACS Chem. Neurosci.* 2017; 84: 777–84.

Malik MT, O'Toole MG, Casson LK, Thomas SD, Bardi GT, Reyes-Reyes EM, Ng CK, Kang KA, Bates PJ. AS1411–conjugated gold nanospheres and their potential for breast cancer therapy. *Oncotarget.* 2015; 6: 22270.

Mayer G, Ahmed M-SL, Dolf A, Endl E, Knolle PA, Famulok M. Fluorescence-activated cell sorting for aptamer SELEX with cell mixtures. *Nat Protoc.* 2010; 5: 1993–2004.

McNamara JO, Andrechek ER, Wang Y, Viles KD, Rempel RE, Gilboa E, Sullenger BA, Giangrande PH. Cell type–specific delivery of siRNAs with aptamer-siRNA chimeras. *Nat Biotechnol.* 2006; 24: 1005–15.

Mendonsa SD, Bowser MT. *In vitro* evolution of functional DNA using capillary electrophoresis. *J Am Chem Soc*. 2004; 126: 20–1.

Merriman B, Torrent I, Rothberg JM, Team D. Progress in ion torrent semiconductor chip based sequencing. *Electrophoresis*. 2012; 33: 3397–417.

Meshik X, Xu K, Dutta M, Stroscio MA. Optical detection of lead and potassium ions using a quantum-dot-based aptamer nanosensor. *IEEE Trans Nanobioscience*. 2014; 13: 161–4.

Mi J, Liu Y, Rabbani ZN, Yang Z, Urban JH, Sullenger BA, Clary BM. *In vivo* selection of tumor-targeting RNA motifs. *Nat Chem Biol*. 2010; 6: 22–4.

Miller DS. Regulation of P-glycoprotein and other ABC drug transporters at the blood–brain barrier. *Trends Pharmacol Sci*. 2010; 31(6): 246–54.

Moosavian SA, Abnous K, Badiee A, Jaafari MR. Improvement in the drug delivery and anti-tumor efficacy of PEGylated liposomal doxorubicin by targeting RNA aptamers in mice bearing breast tumor model. *Colloids Surf B Biointerfaces*. 2016; 139: 228–36.

Nastasijevic B, Wright BR, Smestad J, Warrington AE, Rodriguez M, Maher LJ III. Remyelination induced by a DNA aptamer in a mouse model of multiple sclerosis. *PLOS One*. 2012; 7: 1–8.

Ni X, Castanares M, Mukherjee A, Lupold SE. Nucleic acid aptamers: Clinical applications and promising new horizons. *Curr Med Chem*. 2011; 18: 4206–14.

Niemeyer CM, Adler M. Nanomechanical devices based on DNA. *Angew Chem Int Ed*. 2002; 41: 3779–83.

Ninomiya K, Kaneda K, Kawashima S, Miyachi Y, Ogino C, Shimizu N. Cell-SELEX based selection and characterization of DNA aptamer recognizing human hepatocarcinoma. *Bioorg Med Chem Lett*. 2013; 23: 1797–802.

Oh SS, Ahmad KM, Cho M, Kim S, Xiao Y, Soh HT. Improving aptamer selection efficiency through volume dilution, magnetic concentration, and continuous washing in microfluidic channels. *Anal Chem*. 2011a; 83: 6883–9.

Oh SS, Qian J, Lou X, Zhang Y, Xiao Y, Soh HT. Generation of highly specific aptamers via micromagnetic selection. *Anal Chem*. 2011b; 83: 1866.

Okhonin V, Berezovski M, Krylov SN. Sweeping capillary electrophoresis: A non-stopped-flow method for measuring bimolecular rate constant of complex formation between protein and DNA. *J Am Chem Soc*. 2004; 126: 7166–7.

Pan UN, Khandelia R, Sanpui P, Das S, Paul A, Chattopadhyay A. Protein-based multifunctional nanocarriers for imaging, photothermal therapy, and anticancer drug delivery. *ACS Appl MaterInterfaces*. 2016.

Perfézou M, Turner A, Merkoçi A. Cancer detection using nanoparticle-based sensors. *Chem Soc Rev*. 2012; 41: 2606–22.

Pi F, Zhang H, Li H, Thiviyanathan V, Gorenstein DG, Sood AK, Guo P. RNA nanoparticles harboring annexin A2 aptamer can target ovarian cancer for tumor-specific doxorubicin delivery. *Nanomedicine*. 2017; 13: 1183–93.

Porciani D, Signore G, Marchetti L, Mereghetti P, Nifosi R, Beltram F. Two interconvertible folds modulate the activity of a DNA aptamer against transferrin receptor. *Mol Ther Nucleic Acids*. 2014; 3: 1–13.

Powell D, Chandra S, Dodson K, Shaheen F, Wiltz K, Ireland S, Syed M, Dash S, Wiese T, Mandal T, Kundu A. Aptamer-functionalized hybrid nanoparticle for the treatment of breast cancer. *Eur J Pharm Biopharm*. 2017; 114: 108–18.

Reis-Filho JS. Next-generation sequencing. *Breast Cancer Res*. 2009; 11: 1.

Riley KR, Saito S, Gagliano J, Colyer CL. Facilitating aptamer selection and collection by capillary transient isotachophoresis with laser-induced fluorescence detection. *J Chromatogr A*. 2014; 1368: 183–9.

Roy K, Kanwar RK, Kanwar JR. LNA aptamer based multi-modal, Fe3O4-saturated lactoferrin (Fe3O4-bLf) nanocarriers for triple positive (EpCAM, CD133, CD44) colon tumor targeting and NIR, MRI and CT imaging. *Biomaterials*. 2015; 71: 84–99.

Rozenblum GT, Lopez VG, Vitullo AD, Radrizzani M. Aptamers: Current challenges and future prospects. *Expert Opin Drug Discov*. 2016; 11: 127–35.

Rusconi CP, Scardino E, Layzer J, Pitoc GA, Ortel TL, Monroe D, Sullenger BA. RNA aptamers as reversible antagonists of coagulation factor IXa. *Nature*. 2002; 419: 90–4.

Sarin H. Physiologic upper limits of pore size of different blood capillary types and another perspective on the dual pore theory of microvascular permeability. *J Angiogenesis Res*. 2010; 2: 1.

Seeman NC. Structural DNA nanotechnology: growing along with nanoletters. *Nano Lett*. 2010; 10(6): 1971–8.

Sefah K, Yang Z, Bradley KM, Hoshika S, Jiménez E, Zhang L, Zhu G, Shanker S, Yu F, Turek D. *In vitro* selection with artificial expanded genetic information systems. *Procnatl Acad Sci USA*. 2014; 111: 1449–54.

Seo YT, Jeong S, Lee JK, Choi HS, Kim J, Lee HY. Innovations in biomedical nanoengineering: nanowell array biosensor. *Nano Convergence*. 2018; 5:9.

Shangguan D, Li Y, Tang Z, Cao ZC, Chen HW, Mallikaratchy P, Sefah K, Yang CJ, Tan W. Aptamers evolved from live cells as effective molecular probes for cancer study. *Proc Natl Acad Sci USA*. 2006; 103: 11838–43.

Shu Y, Cinier M, Shu D, Guo P. Assembly of multifunctional phi29 pRNA nanoparticles for specific delivery of siRNA and other therapeutics to targeted cells. *Methods*. 2011; 54: 204–14.

Shwetha N, Selvakumar L, Thakur, M. Aptamer–nanoparticle-based chemiluminescence for p53 protein. *Anal Biochem*. 2013; 441: 73–9.

Song K-M, Lee S, Ban C. Aptamers and their biological applications. *Sensors*. 2012; 12: 612–31.

Song Y, Zhang H, Zhu Z, Yang C. The clinical application of aptamers: Future challenges and prospects. In: *Aptamers Selected by Cell-SELEX for Theranostics*. Springer; 2015; pp. 339–52.

Sriramoju B, Neerati P, Kanwar RK, Kanwar JR. Brain targeted PLGA nanocarriers alleviating amyloid-B expression and preserving basal survivin in degenerating mice model. *Biochim BiophysActa*. 2015; 1852: 2423–31.

Stoltenburg R, Reinemann C, Strehlitz B. FluMag-SELEX as an advantageous method for DNA aptamer selection. *Analbioanal Chem*. 2005; 383: 83–91.

Subramanian N, Kanwar JR, Kumar Athalya P, Janakiraman N, Khetan V, Kanwar RK, Eluchuri S, Krishnakumar S. EpCAM aptamer mediated cancer cell specific delivery of EpCAM siRNA using polymeric nanocomplex. *J Biomed Sci*. 2015; 22: 1.

Svetlana MS, Nikola S, Richard FK, Anuska VA. Blood-brain barrier permeability: From bench to bedside. In: *Management of Epilepsy-Research, Results and Treatment*. 2011; pp. 113–140.

Szeto K, Latulippe DR, Ozer A, Pagano JM, White BS, Shalloway D, Lis JT, Craighead HG. RAPID-SELEX for RNA Aptamers. *PLOS One*. 2013; 8: 1–11.

Taghdisi SM, Lavaee P, Ramezani M, Abnous K. Reversible targeting and controlled release delivery of daunorubicin to cancer cells by aptamer-wrapped carbon nanotubes. *Eur J Pharm Biopharm*. 2011; 77: 200–6.

Tang L, Yang X, Yin Q, Cai K, Wang H, Chaudhurya I, Yao C, Zhou Q, Kwon M, Hartman JA, Dobrucki IT, Dobrucki LW, Borst LB, Lezmi S, Helferich WG, Ferguson AL, Fan TM, Cheng J. Investigating the optimal size of anticancer nanomedicine. *Proc Natl Acad Sci U S A*. 2014; 111(43): 15344–9.

Tang L, Tong R, Coyle VJ, Yin Q, Pondenis H, Borst LB, Cheng J, Fan TM. Targeting tumor vasculature with aptamer-functionalized doxorubicin–polylactide nanoconjugates for enhanced cancer therapy. *ACS Nano*. 2015a; 9: 5072–81.

Tang Y, Hu H, Zhang MG, Song J, Nie L, Wang S, Niu G, Huang P, Lu G, Chen X. An aptamer-targeting photoresponsive drug delivery system using 'off–on' graphene oxide wrapped mesoporous silica nanoparticles. *Nanoscale*. 2015b; 7: 6304–10.

Thiel KW, Giangrande PH. Therapeutic applications of DNA and RNA aptamers. *Oligonucleotides*. 2009; 19: 209–22.

Tian J, Chen J, Ge C, Liu X, He J, Ni P, Pan Y. Synthesis of PEGylated ferrocene nanoconjugates as the radiosensitizer of cancer Cells. *Bioconjugate Chem*. 2016; 26: 1518–24.

Tintoré M, Gállego I, Manning B, Eritja R, Fàbrega C. DNA origami as a DNA repair nanosensor at the single-molecule level. *Angew Chem Int Ed*. 2013; 52: 7747–50.

Tsukakoshi K, Harada R, Sode K, Ikebukuro K. Screening of DNA aptamer which binds to alpha-synuclein. *Biotechnol Lett*. 2010; 32: 643–8.

Tuerk C, Gold L. Systematic evolution of ligands by exponential enrichment: RNA ligands to bacteriophage T4 DNA polymerase. *Science*. 1990; 249: 505–10.

Wang AZ, Bagalkot V, Vasilliou CC, Gu F, Alexis F, Zhang L, Shaikh M, Yuet K, Cima MJ, Langer R. Superparamagnetic iron oxide nanoparticle–aptamer bioconjugates for combined prostate cancer imaging and therapy. *Chemmedchem*. 2008a; 3: 1311–5.

Wang H, Wang Y, Jin J, Yang R. Gold nanoparticle-based colorimetric and 'turn-on' fluorescent probe for mercury (II) ions in aqueous solution. *Anal Chem*. 2008b; 80: 9021–8.

Wang H, Zhao X, Guo C, Ren D, Zhao Y, Xiao W, Jiao W. Aptamer-dendrimer bioconjugates for targeted delivery of miR-34a expressing plasmid and antitumor effects in non-small cell lung cancer cells. *PLOS One*. 2015a; 10: 1–15.

Wang J, Gong Q, Maheshwari N, Eisenstein M, Arcila ML, Kosik KS, Soh HT. Particle display: A quantitative screening method for generating high-affinity aptamers. *Angew Chem Int Ed*. 2014; 53: 4796–801.

Wang X, Song P, Peng L, Tong A, Xiang Y. Aggregation-induced emission luminogen-embedded silica nanoparticles containing DNA aptamers for targeted cell imaging. *ACS Appl Mater Interfaces*. 2015b; 8: 609–16.

Wang Y, Xu D, Chen H-Y. Aptamer-based silver nanosensor for multiple protein detection. *Lab Chip*. 2012; 12: 3184–9.

WooáKim D, BockáGu M. Immobilization-free screening of aptamers assisted by graphene oxide. *Chem Commun*. 2012; 48: 2071–3.

Xie X, Li F, Zhang H, Lu Y, Lian S, Lin H, Gao Y, Jia L. EpCAM aptamer-functionalized mesoporous silica nanoparticles for efficient colon cancer cell-targeted drug delivery. *Eur J Pharm Sci*. 2016; 83: 28–35.

Xing H, Wong NY, Xiang Y, Lu Y. DNA aptamer functionalized nanomaterials for intracellular analysis, cancer cell imaging and drug delivery. *Curr Opin Chem Biol*. 2012; 16: 429–35.

Xu W, Siddiqui IA, Nihal M, Pilla S, Rosenthal K, Mukhtar H, Gong S. Aptamer-conjugated and doxorubicin-loaded unimolecular micelles for targeted therapy of prostate cancer. *Biomaterials*. 2013; 34: 5244–53.

Yang L, Tseng Y-T, Suo G, Chen L, Yu J, Chiu W-J, Huang C-C, Lin C-H. Photothermal Therapeutic Response of Cancer Cells to Aptamer–Gold Nanoparticle-Hybridized Graphene Oxide under NIR Illumination. *ACS Appl Mater Interfaces*. 2015; 7: 5097–106.

Yi M, Yang S, Peng Z, Liu C, Li J, Zhong W, Yang R, Tan W. Two-photon graphene oxide/aptamer nanosensing conjugate for *in vitro* or *in vivo* molecular probing. *Anal Chem*. 2014; 86: 3548–54.

Yu Y, Zhang Y, Kenrick M, Hoyte K, Luk W, Lu Y, Atwal J, Elliott J, Prabhu S, Watts R, Dennis M. Boosting brain uptake of a therapeutic antibody by reducing its affinity for a transcytosis target. *Sci Transl Med*. 2011; 3: 1–8.

Zhang C-y, Johnson LW. Single quantum-dot-based aptameric nanosensor for cocaine. *Anal Chem*. 2009; 81: 3051–5.

Zhang H, Ma Y, Xie Y, An Y, Huang Y, Zhu Z, Yang CJ. A controllable aptamer-based self-assembled DNA dendrimer for high affinity targeting, bioimaging and drug delivery. *Sci Rep*. 2015; 5: 10099.

Zhang L, Chan JM, Gu FX, Rhee J-W, Wang AZ, Radovic-Moreno AF, Alexis F, Langer R, Farokhzad OC. Self-assembled lipid – polymer hybrid nanoparticles: A robust drug delivery platform. *ACS Nano*. 2008; 2: 1696–702.

Zheng F-F, Zhang P-H, Xi Y, Chen J-J, Li L-L, Zhu J-J. Aptamer/graphene quantum dots nanocomposite capped fluorescent mesoporous silica nanoparticles for intracellular drug delivery and real-time monitoring of drug release. *Anal Chem*. 2015; 87: 11739–45.

Zhou W, Roland A, Jin Y, Iyer RP. Combinatorial synthesis using nucleic acid-based (NAB™) scaffold: parallel solid-phase synthesis of nucleotide libraries. *Tetrahedron Lett*. 2000; 41: 441–5.

Zhuang Y, Deng H, Su Y, He L, Wang R, Tong G, He D, Zhu X. Aptamer-functionalized and backbone redox-responsive hyperbranched polymer for targeted drug delivery in cancer therapy. *Biomacromolecules*. 2016; 17: 2050–62.

# 6 Multi-Dimensional Approaches for Targeted Drug Delivery Using Nanostructured Systems

*Charan Singh, Ranjot Kaur, Ripandeep Kaur, Raman Bahal,* and Bhupinder Singh**

## CONTENTS



## 6.1 INTRODUCTION

Nanotechnology unfolds umpteen avenues to explore its impact on various fields and possesses immense potential in the coming era of personalized medicine. The recent paradigm of using nanotechnology in drug delivery has now been become quite well-established over the past few decades. Development of nano-based drug delivery systems of already existing drugs, exhibiting poor and/or inconsistent bioavailability, is an appropriate option for augmenting their biopharmaceutical performance (Saini *et al.*, 2010). Compared to microparticles, nanoparticles can be more readily taken up by the tissues and circulated throughout the body into the cells and cell nuclei, and even across the placenta and blood–brain barrier. The hurdles invariably encountered with the conventional drug-delivery systems encompass nonspecific delivery of drug, poor bioavailability, reticular endothelial uptake, poor stability, and leading to systemic toxicity and incalculable side effects (Onoue *et al.*, 2014). The foremost challenge is to develop a delivery system that can act as a "magical

* Corresponding authors.

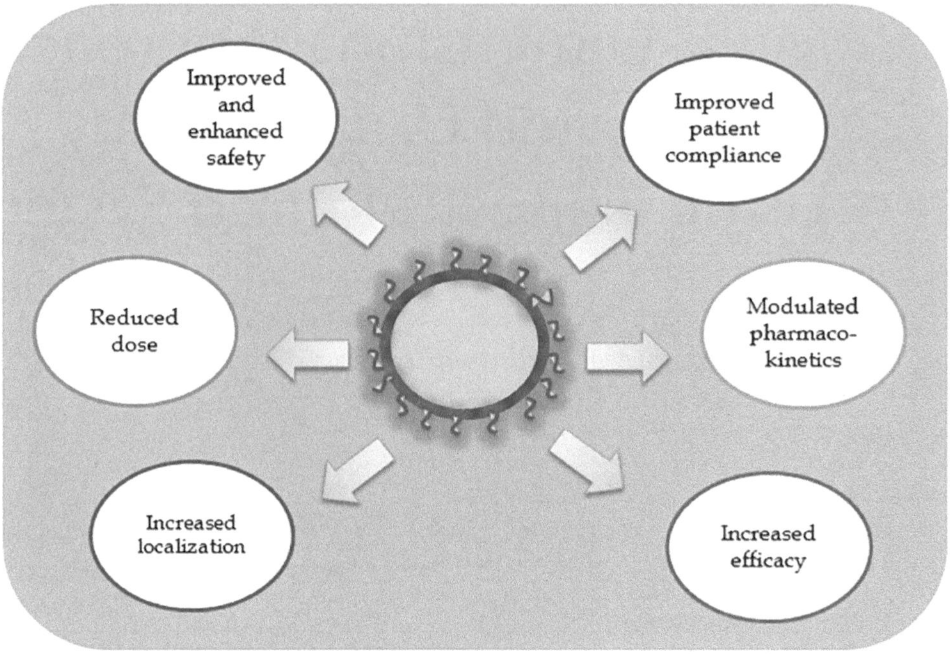


**FIGURE 6.1** Advantages of targeted drug-delivery systems.

cargo" not only for selectively targeting the affected pathological tissue but, more prominently, the organelle within the cell where the drug shows its action. In this context, tremendous progress has been made in the field of surface-engineered nanocarriers such as liposomes, dendrimers, solid lipid nanoparticles (SLNs), nanostructured lipid carriers (NLCs), carbon nanotubes (CNTs), and metallic nanoparticles, just to name a few. Such advances in the field of nano-drug delivery systems have created newer opportunities to transform the diagnosis and treatment of many diseases. The tailored nanosystems entirely possess novel therapeutic applications with better functionality, specificity, and efficiency due to their distinctive physicochemical and surface properties, as shown in Figure 6.1 (Ochekpe *et al.*, 2009). Surface-engineering of nanocarriers for site-specific delivery can be achieved by molecular identification of the diseased cells by various signature molecules overexpressed at the diseased site either via the ligand-receptor, antigen-antibody interactions, stimuli-sensitive approaches, or through aptamers (Jia *et al.*, 2013). In this context, this chapter entails to provide a holistic view over different approaches used to modify the surfaces of nanosystems.

Despite having various advantages over conventional drug-delivery approaches, nanotechnology-based systems still lag in reaching the target site and maintaining desired concentration, which often leads to off-target side effects. The objective of attaining desired drug levels at the target site can be effectively achieved by engineering the surface of nanostructured systems with surface-active agents. Surface engineering has emerged as one of the most novel approaches in drug delivery, as engineered nanocarriers have the ability to enhance the retention time of drug molecules in the blood and have the distinct potential to bind with ligand-specific receptor molecules (Li *et al.*, 2016). Moreover, surface-engineered nanoparticles are able to cross biological barriers and show improved pharmacokinetic profiles with minimal toxicity. Surface-engineered nanostructured systems can also be used to deliver therapeutic agents that are difficult to administer due to off-target side effects. For active or passive targeting, surface-modified systems represent promising avenues in drug delivery (Siafaka *et al.*, 2016) (Figure 6.2). Several research groups have reported various surface modifying agents for improving efficacy and targetability. Figure 6.3 presents an account of various surface-active agents used for engineering the surface of nanocarriers.

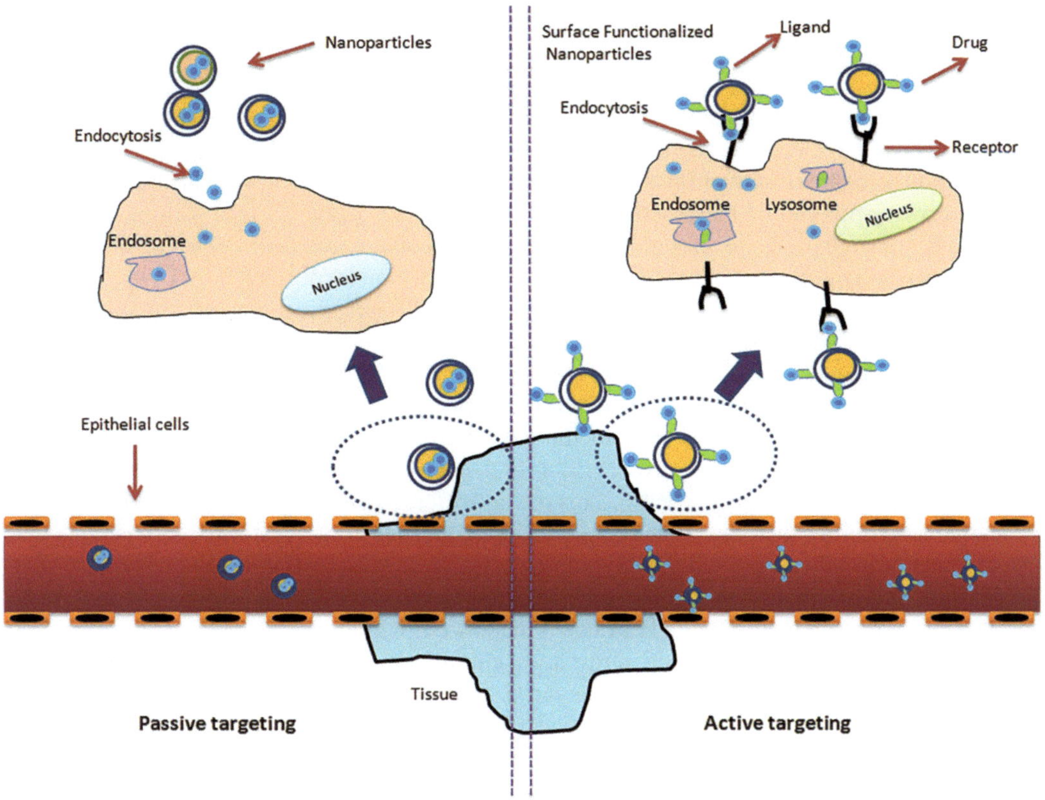


**FIGURE 6.2** Mechanism of drug uptake by nanoparticles via active and passive targeting.

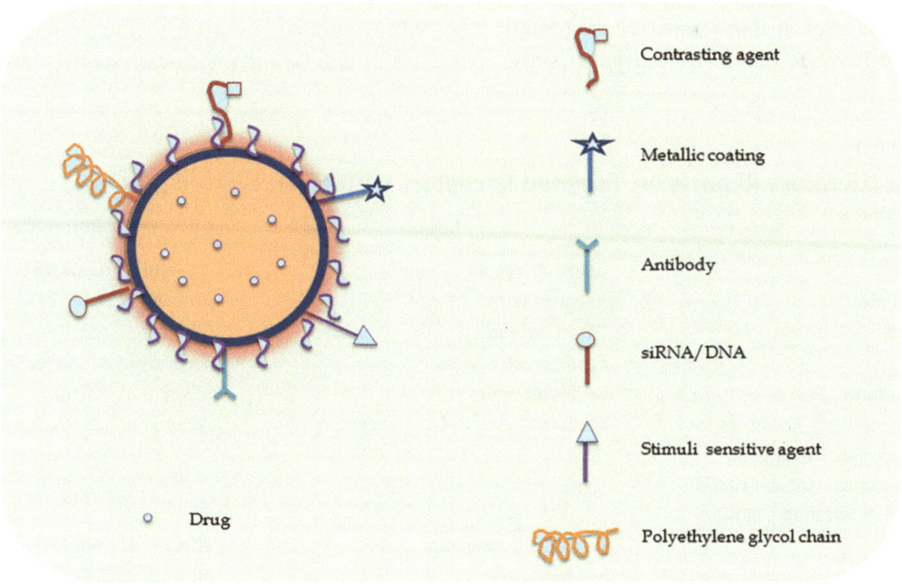


**FIGURE 6.3** Multifunctional nanoparticles scheme for molecular imaging, drug delivery, and therapy.

## 6.2 SURFACE ENGINEERING OF NANOSTRUCTURED SYSTEMS

### 6.2.1 Ligand-Based

Ligands are the molecules that are attached at the periphery of the nanosystems and produce signals when bound to the specific receptors that are overexpressed on the cells. Ligands have the ability to enhance the cellular uptake through endocytosis. Hence, the efficacy of drugs, biomolecules, and bioactive agents can be increased by modifying the nanoparticle surface with receptor-specific ligands *viz.* transferrin, folate, carbohydrates, proteins, peptides, etc. (Belhadj *et al.*, 2017). Ligand-based targeting is useful in intracellular delivery of macromolecules such as DNA, siRNA, and proteins, as it aids the nanostructured systems to cross the biological barriers through molecular recognition. Vital literature instances of various ligands used for targeting specific receptors have been listed in Table 6.1.

PEGylation is the most widely used method to improve the passive targeting of drug molecules. In this approach, surface modification of the nanostructured systems can be simply achieved by covalent grafting of polyethylene glycol (PEG) to the active functional moieties or adsorbing PEG chains of vivid lengths (Seo *et al.*, 2010). This method offers stealth properties to the nanosystems by providing a barrier for the adsorption of opsonin proteins, thus granting steric stabilization and long circulation of the nanocarriers in the blood. Zhang *et al.* (2015) successfully synthesized X-shaped, four-armed gemini-like PEGylated distearyl glycerol (Gemini-PEG2K-GCDS) nanomicellar carrier for delivery of doxorubicin (DOX). Pharmacokinetic studies of DOX-encapsulated Gemini-$PEG_{2K}$-GCDS nanomicelles in rats following intravenous administration confirmed longer retention of DOX in blood and larger values of the area under curve (AUC) compared with naïve DOX solutions.

Engineered nanosytems assist in the delivery of diverse therapeutics to the infected cells. The drugs that are difficult to administer due to adverse effects can be safely administered by attaching a receptor-specific moiety, thereby minimizing the toxic effects to the normal cells. Literature instances have reported the usage of different ligands for targeting various diseases such as cancer, microbial infections, neurodegenerative disorders, and many more. Some of the current research findings on ligand-specific targeting are represented in Table 6.2.

### 6.2.2 Stimuli-Responsive

The domain of stimuli-sensitive systems is a growing topic of research in the delivery of drugs and bioactives to the desired target tissues. Such stimuli can be intrinsic, such as pH, temperature,

**TABLE 6.1**
**Recent Literature Reports on Targeted Receptors with Ligand-Based Systems**

| Ligand | Target | Reference |
|---|---|---|
| Anisamide | Sigma receptors | Urandur *et al.* (2018) |
| Cyclic RGD | Integrin $\alpha v\beta 3$ | Belhadj *et al.* (2017) |
| Rituximab | CD20 antigen | Zhou *et al.* (2017) |
| Folate | Folate receptors | Esfandyari-Manesh *et al.* (2016) |
| Carbohydrates | Lectin receptors | Chen *et al.* (2016a) |
| EGF, transforming growth factor-α, Transtuzumab, Cetuximab, murine monoclonal anti-EGFR Ab, Cell-penetrating peptides. | Epidermal growth factor receptors | Yang *et al.* (2016), Xu *et al.* (2012) |
| Transferrin | Transferrin receptors | Zhao *et al.* (2014a) |
| RGD peptide | Integrin $\alpha \nu\beta 3$ | Su *et al.* (2012) |
| Galactosamine | Galactosamine receptors on hepatocytes | Shen *et al.* (2011) |

**TABLE 6.2**
**Recent Literature Instances on Usage of Ligands in Targeting Various Disorders**

| Disease | Nanosystems | Drug/Nucleic Acid/Biomolecule | Ligand | Target | Reference |
|---|---|---|---|---|---|
| Lung Cancer | Nanoliposomes | Triptolide | Anti-carbonic anhydrase IX antibody and cys-modified CPP33 peptide | Carbonic anhydrase expressing tumor cells | Lin *et al.* (2018) |
| Leishmaniasis | PLGA nanoparticles | Amphotericin B | Phosphatidyl-L-serine | Macrophage | Singh *et al.* (2018a,b) |
| Parkinson's disease | PEG-PLGA nanoparticles | Rotigotine | Lactoferrin | Lactoferrin receptors | Yan *et al.* (2018) |
| HIV | Gold nanoparticle | hexapeptide | TAT peptide | HIV integrase | Singh *et al.* (2018a,b) |
| Breast Cancer | Lipid-polymer hybrid nanoliposomes | GAPDH siRNA | Aptamer A6 ($NH_2$-Apt-6) | MDR gene | Powell *et al.* (2017) |
| Alveolar rhabdomyosarcoma | Lipid–protamine nanoparticles | siRNA | RGD peptide | PAX3-FOXO1 | Rengaswamy *et al.* (2016) |
| Cancer | Solid lipid nanoparticles | Docetaxel | Tetraiodothyroacetic acid, hyaluronan | Integrin $\alpha v\beta 3$/CD44- | Shi *et al.* (2016) |
| Cancer | PEG-folic acid nanoparticles | Paclitaxel | PEG-FA | Folate | Esfandyari-Manesh *et al.* (2016) |
| Brain tumor | Albumin nanoparticles | Doxorubicin | Mannose | Lectin | Byeon *et al.* (2016) |
| Lung tumor | Hyaluronan-based nanoparticles | – | Hyaluronan | CD44 receptor | Jeannot *et al.* (2016) |
| Lung cancer | Nanoparticles | Docetaxel | Peptide, LPLTPLP | Hsp90 | Jiang *et al.* (2016) |
| Ischemic stroke | Nanoparticles | Lexiscan | Chlorotoxin | Nogo-66 receptor | Han *et al.* (2016) |
| Chronic lymphocytic leukemia | Dendrimer-based nanoparticles | – | Maltotriose | BCR-signaling pathway | Franiak-Pietryga *et al.* (2016) |
| Periodontitis | PEG-poly(lactic acid) nanoparticles | Minocycline | RGD peptide | Periodontitis epithelial cells | Yao *et al.* (2015) |
| Rheumatoid arthritis | Nanoparticles | siRNA | Thiolated glycol chitosan | Notch1 pathway | Kim *et al.* (2015) |

enzymatic, or redox conditions, and/or extrinsic, such as magnetic field, ultrasound, or an apt irradiation. The working principle underlying these stimuli systems is their existence in solution (i.e., liquid or sol) form at the ambient temperature and their subsequent conversion into the gel form under the influence of various conditions of body milieu, be these physiological, chemical, or biochemical. These "smart" or "intelligent" systems are endowed with numerous merits such as ease of administration, sustained release, reduced frequent administration with improved patient compliance, and targeted and spatial delivery of a drug with reduced frequency of side effects (Singh *et al.*, 2017). Besides drug delivery, these smart systems have lately exhibited tremendous applications in tissue engineering, regenerative biomedicine, molecular imprinting, cancer therapy, gene delivery, theranostics, and even beyond. For the past two decades, the contribution of these systems has been growing exponentially in drug delivery, nanotechnology, and biotechnology. Usage of these multi-responsive systems has lately led to significant headway in the field of biomedical sciences. Some of the recent literature instances reported with *in situ* stimuli-responsive gelling systems are mentioned in Table 6.3.

### 6.2.3 Bioengineered Systems

Advances in the drug delivery area have led to the development of hybrid structures encompassing a biomolecule, that is, monoclonal antibodies, aptamers, peptides, immunoglobulins, hormones, etc., for site-specific targeting, thus paving the way to treat most deadly diseases (Kobsa and Saltzman, 2008). These are the potential agents that can facilitate recognizing and binding to cell-specific receptors for diagnostic and therapeutic purposes (Mansur *et al.*, 2016). These are attached to the nanostructured system surface, either covalently or non-covalently and offer higher drug-carrier capacity and improved specificity. Bioengineered systems have the potential to initiate signaling cascades that can potentiate the therapeutic response of the incorporated drug (Mansur *et al.*, 2014).

#### 6.2.3.1 Aptamers

Aptamers are single-stranded synthetic DNA or RNA oligonucleotides that are known to control gene expression and regulation. Aptamers are promising candidates that bind to target molecules owing to their low molecular weight, high specificity, and high affinity. Combination of aptamers with nanosystems has led to various targeted drug-delivery systems that offer theranostic applications. Aptamers have shown applications in the treatment of various disorders such as autoimmune diseases, cancer, HIV, and neurological malignancies, just to name a few (Nezlin, 2014). Taghdisi *et al.* (2016) has successfully demonstrated the use of gold nanoparticles decorated with polyvalent aptamers for targeting daunorubicin to cancer cells. The complex is effective in delivering and internalizing daunorubicin to target cells (human acute lymphoblastic leukemia cells). Furthermore, surface modification of the nanoparticles with antisense polyvalent aptamers has antagonized the cytotoxicity potential of the complex. In addition, aptamer-anchored nanoparticles (apt-NPs) containing docetaxel (DTX) demonstrated enhanced *in vitro* tumoricidal activity and cellular uptake over the conventional nanoparticles. In another promising study, the presence of aptamer was found to be responsible for mediating the antitumor effect against the prostate cancer (Chen *et al.*, 2016b).

#### 6.2.3.2 Monoclonal Antibodies

Surface functionalization of nanostructured systems with monoclonal antibodies is an area of immense significance in drug delivery and targeting. Monoclonal antibodies are ideal vehicles for delivering potent cytotoxic drugs to the target site. These are employed as therapeutic and targeting agents in various disorders such as cancer, rheumatoid arthritis, inflammatory bowel disease, and chronic inflammation (Kijanka *et al.*, 2016). Monoclonal antibodies have shown prominent ability to selectively localize at the target site. Loureiro *et al.* (2016) developed poly(lactic-co-glycolic acid) (PLGA) nanoparticles encapsulating i$A\beta_5$ peptide, surface-functionalized with anti-transferrin receptor monoclonal antibody (OX26) and anti-$A\beta$ (DE2B4) for targeting Alzheimer's disease (AD).

**TABLE 6.3**
**Some Literature Reports on Stimuli-Responsive Nanostructured Drug Delivery Systems**

| Stimuli | Carrier | Stimuli Agent | Drug/Biomolecule | Reference |
|---|---|---|---|---|
| **pH** | Nanovesicle | Cholesteryl hemisuccinate | Doxorubicin hydrochloride | Liang *et al.* (2018) |
| | Polymeric nanoparticle | poly(VI-co-HEI) and poly(VP-co-MVE) or poly(VP-co-MTOS) pseudo co-block polymer | Dexamethasone | Saldaña *et al.* (2018) |
| | Dendritic mesoporous silica nanoparticles | Polyethylenimine, nonionic hydrophilic polyethylenglycol | TNF-alpha | Kienzle *et al.* (2017) |
| | Mesoporous silica nanoparticles | Dioleoylphosphoethanolamine, cholesteryl hemisuccinate, distearoylphosphoethanolamine | Doxorubicin | Wang *et al.* (2016) |
| | Polysebacic anhydride nanocapsules | Poly (L-histidine) | Doxorubicin | Bagherifam *et al.* (2015) |
| | Gel | Sodium alginate, calcium chloride, sodium citrate, HPMC | Pregabalin | Madan *et al.* (2015) |
| | Nanogels | Dextrin | Doxorubicin | Manchun *et al.* (2014) |
| | Nanogels | Chitosan, poly(ethylene glycol) | 5-Florouracil | Zhou *et al.* (2013) |
| | Hydrogel | P(NIPAAm-co-PAA-co-BA) | bFGF | Garbern *et al.* (2011) |
| | Hydrogel | MAA and PEGMEMA | Diltiazem hydrochloride | Frutos *et al.* (2010) |
| | Nanoparticles | Dextran sulfate, chitosan | Insulin | Sarmento *et al.* (2006) |
| | Polyplexes | N-Ac-poly(L-histidine)-graft-poly(L-lysine) | siRNA | Benns *et al.* (2000) |
| **Redox** | Lipidic nanoparticle | Disulfide bond crosslinking and Cell penetrating peptide | Antigen protein ovalbumin | Wang *et al.* (2018) |
| | Mesoporous silica nanoparticles | Transferrin | Doxorubicin | Chen *et al.* (2017b) |
| | Nanoparticles | Hyaluronic acid | Luteinizing hormone | Lin *et al.* (2016) |
| | Mesoporous silica nanoparticles | d-α-tocopherol polyethylene glycol 1000 succinate | Doxorubicin | Han *et al.* (2015) |
| | Nanogels | Polyvinyl alcohol | Doxorubicin | Chen *et al.* (2015a) |
| | Nanogels | N,N-dimethylacrylamide | Doxorubicin | Qiao *et al.* (2013) |
| | Disulfide cross-linked micelles | Polyethylene glycol, cholic acids | Paclitaxel | Li *et al.* (2011) |
| | Micelles | Poly(ethyleneglycol)-b-poly(L-lysine)-b-poly(L-phenylalanine) | Methotrexate | Koo *et al.* (2008) |

*(Continued)*

**TABLE 6.3** (*Continued*)
**Some Literature Reports on Stimuli-Responsive Nanostructured Drug Delivery Systems**

| Stimuli | Carrier | Stimuli Agent | Drug/Biomolecule | Reference |
|---|---|---|---|---|
| **Ion activation** | Nanoparticles | Metal-ion-dependent DNAzyme/substrate | Doxorubicin | Chen *et al.* (2017a) |
| | Gels | Gellan gum, HPMC | Salbutamol sulphate | Salunke and Patil, (2016) |
| | Liposomes | Deacetylated gellan gum | Timolol maleate | Yu *et al.* (2015) |
| | Gels | Gellan | Clarithromycin | Bhimani *et al.* (2011) |
| | Gels | Sodium alginate | Clarithromycin, metronidazole benzoate | Patel *et al.* (2010) |
| | Gels | Sodium alginate | Ranitidine | Rohith *et al.* (2009) |
| | Gels | Gellan and sodium alginate | Metformin | Nagarwal *et al.* (2009) |
| | Gels | Xyloglucan and gellan | Paracetamol | Miyazaki *et al.* (2009) |
| **Temperature** | Silver-nanocomposite hydrogels | N-isopropylacrylamide | Green tea | Jayaramudu *et al.* (2018) |
| | Nanoparticles | Chitosan-gold copolymer hybrid | Erlotinib | Fathi *et al.* (2018) |
| | Mesoporous silica nanoparticles | Poly(ethylene glycol)/poly(epsilon-caprolactone) | Doxorubicin | Cho *et al.* (2017) |
| | Gels | Pluronic F127/Pluronic F68 | Ofloxacin | Al Khateb *et al.* (2016) |
| | Microcapsules and Hydrogels | Pluronic® | Doxorubicin and 5-fluorouracil | Kim *et al.* (2016) |
| | Nanoparticles | Hollow mesoporous copper sulfide | Doxorubicin | Feng *et al.* (2016) |
| | Gels | Thiolated poly(aspartic acid) | Sodium diclofenac | Horvát *et al.* (2015) |
| | Gels | Poloxamer 407 | Fluconazole | Gratieri *et al.* (2011) |
| | Hydrogels | Chitosan | Rutin | Tran *et al.* (2011) |
| | Gels | Poloxamer 407, Poloxamer P188 | Methazolamide | Qian *et al.* (2010) |
| | Hydrogels | MBCP-2 | hVEGF plasmid | Kwon *et al.* (2009) |
| | Gels | Xyloglucan | Paracetamol | Miyazaki *et al.* (2003) |
| **Photothermal** | Lipid based nanoconstructs | Silicon 2,3-naphthalocyanine bis(trihexylsilyloxide), phytantriol and 1-palmitoyl-2-oleoyl-sn-glycero-3-phosphoethanolamine mixed with cholesterol | – | Du *et al.* (2018) |
| | Liposomes | Gold nanoshell | Betulinic acid | Liu *et al.* (2017) |
| | Mesoporous Silica Nanocookies | Graphene oxide/carbon | Camptothecin | Chen *et al.* (2014b) |
| | Nanoconstructs | PEGylated graphene oxide | Resveratrol | Chen *et al.* (2014a) |
| | Nanosheets | Graphene oxide | Doxorubicin | Wang *et al.* (2013) |

The data demonstrated that functionalized nanoparticles show enhanced cellular uptake with controlled delivery as compared to plain nanoparticles. Similarly, antibody-conjugated mesoporous silica nanoparticles incorporating bevacizumab have been employed targeting ovarian cancer cells. The conjugate exhibited controlled release phenomenon at pH 7.4. Furthermore, antibody-conjugated nanosystems showed superior anticancer activity with profound apoptosis in OVCAR-5 cancer cells (Zhang *et al.*, 2015).

Both at the industrial and academic level, there has been increasing interest at arming monoclonal antibodies with drugs, toxins, cytokines, and radionuclides (Hess *et al.*, 2014). To date, Adcetris™ (Senter and Sievers, 2012) and Kadcyla™ (Ballantyne and Dhillon, 2013) are the two products that hold market authorization for the treatment of cancer, and more than 40 products are in their clinical development (Gebleux and Casi, 2016).

#### 6.2.3.3 Protein-Based Nanocages

Protein-based nanocages have gained a lot of attention as potential drug-delivery systems owing to their perfect symmetry, ideal physical properties, and versatile nature. Nanocages can be delivered both by passive or active targeting, depending upon their surface functionalization. Moreover, nanocages have the potential to minimize the premature loss of drug molecules, thereby improving the concentration of drug at the target site. Bioengineered protein-based nanocages can be used as effective tools to deliver therapeutic moieties in treatment and management of cancer, brain, and other infectious disorders (Alyautdin *et al.*, 1998). The research and usage of protein-based nanocages is not limited to the laboratory only, but have entered the preclinical and clinical stages (Table 6.4).

### 6.2.4 Metallic

Metals have been used as therapeutic agents in earlier nanomedicinal research (Zhang and Lippard, 2003). Decorating nanostructured systems with metals offers stellar advantages over their conventional counterparts as therapeutic, as well as diagnostic, agents. When conjugated with

**TABLE 6.4**
**Recent Literature Reports on Functionalized Protein-Based Nanocages for Specific Drug Targeting**

| Type of Nanocages | Drug | Functionalization | Target | Reference |
|---|---|---|---|---|
| Apoferritin | Oxaliplatin | Panitumumab | Epidermal growth factor receptor | Lin *et al.* (2018) |
| DNA octahedral nanocages | Doxorubicin | Biotin and folate | Folate receptor | Raniolo *et al.* (2018) |
| Apoferritin | Doxorubicin | – | Glioma cells | Chen *et al.* (2017a) |
| Apoferritin | Doxorubicin, docetaxel | – | Breast cancer | Ghosh *et al.* (2016) |
| Equine ferritin | Daunomycin | Hyaluronic acid | Lung cancer | Luo *et al.* (2015) |
| Human ferritin | TRAP peptide | MMP-2 cleavage peptide | Sepsis | Lee *et al.* (2015a,b) |
| Human ferritin | RFP Si-RNA | Protamine; HIV TAT peptide | Cancer | Lee *et al.* (2015a,b) |
| Human ferritin | Doxorubicin | Alexa546 | Cancer | Bellini *et al.* (2014) |
| Human ferritin | Doxorubicin | RGD4C peptide; ZW800 | Cancer | Zhen *et al.* (2013) |
| Pyrococcusfuriosus ferritin | Cy3-trastuzumab | Fc-binding peptide; Alexa 647 | Breast cancer | Kang *et al.* (2012) |

biomaterials, metals like silver, gold, and platinum demonstrate superior antimicrobial properties (Rai *et al.*, 2015). Functionalized metallic particles have been used to deliver drug to cellular and intracellular targets (Athar and Das, 2014). Metallic nanoparticles offer surface stability that allow surface engineering of these systems with biomolecules such as proteins, enzymes, and oligopeptides (Crespilho *et al.*, 2009); antibodies and antigens (Haes *et al.*, 2004; Pengo *et al.*, 2007); the biotin/streptavidin system (Zhu *et al.*, 2008); DNA, oligonucleotides, and aptamers (Nykypanchuk *et al.*, 2008); polymers like chitosan and fucoidan (Manivasagan *et al.*, 2016); polyvinylpyrrolidone (PVP) (Guo *et al.*, 2013); and PEG (Sarkar *et al.*, 2015).

Manivasagan and coworkers (2016) prepared the surface-modified, DOX-loaded, fucoidan-capped gold nanoparticles for drug delivery and photoacoustic imaging and further demonstrated the great potential of such for diagnostic and therapeutic applications.

Sarkar *et al.* (2015) reported PEG-stabilized, chitosan-*g*-polyacrylamide-modified silver nanoparticles for gene delivery. The study presented that functionalized silver nanoparticles may be used as efficient carriers with minimal cytotoxicity and enhanced therapeutic activity.

## 6.3 TARGETED-DELIVERY APPLICATIONS IN DISEASED STATES

This section is divided into parts that cover the various strategies of surface functionalization for targeted delivery of nanocarriers in various disease conditions such as cancer, HIV/AIDS, tuberculosis, malaria, AD and Parkinson's disease (PD). While the targeted delivery through surface functionalization approach needs comprehensive study of its own, the aim here is to provide an overview of the relevant strategies to combat delivery challenges of the biotherapeutics in the various prevalent diseases; some of the recent literature reports highlighting the role of functionalization are discussed.

### 6.3.1 Cancer

Recently, Powell *et al.* (2017) developed aptamer-labelled hybrid nanoparticles of protamine sulfate-condensed GAPDH siRNA for targeting breast cancer. Aptamer A6 has been used as a labelling agent because it binds with Her-2 receptors on breast cancer cells. The nanoparticles were prepared by a high-pressure homogenization technique using varied amounts of DOTAP, cholesterol, PLGA or PLGA-PEG, and Mal-PEG and then incubated with aptamer A6 for surface functionalization. The study demonstrated that cell transfection was improved significantly in Her-2 (+) SKBR-3 and 4T1-R cells, as compared to the cells in which Her-2 is poorly expressed. Labelling with aptamer further enhances the knockdown of P-gp. Therefore, the study concluded the potential of aptamer-labelled hybrid nanoparticles to deliver siRNA for targeting breast cancer.

In another recent study by Perillo *et al.* (2017), the multifunctional nanoparticles were synthesized for cancer theranostics. The nanoparticles were made up of a superparamagnetic iron oxide nanoparticulate (SPION) core, PEG 5000 coating, cyanine fluorescent dye, and the cell-penetrating peptide (CPP), that is, membranotropic peptide gH625. From a confocal spectral imaging study, it was observed that 3-fold higher uptake was obtained with SPIONs-PEG-CPP as compared with SPIONs-PEG only. Also, the presence of CPP impels the distribution of a higher NP fraction at subcellular levels, indicating endocytosis as the major mechanism for cellular uptake. Thus, the study provided a promising approach in terms of imaging, drug delivery, and hyperthermia in cancer theranostics.

He *et al.* (2017a,b) designed a multifunctional macrophage-targeting nano-based system for delivery of Cpg oligo deoxynucleotides (ODN) in cancer immunotherapy. The formulation includes excipients such as mannosylated carboxymethyl chitosan (for macrophage targeting), protamine sulphate (for improving ODN loading capacity), and calcium carbonate (for encapsulating ODN). The study demonstrated that the labelled nanoparticles delivered more ODN and significantly augmented the immune-stimulation capacity in comparison to lipofectamine 2000/ODN complexes. The ODN delivery regulated the NF-κB activity and enhanced the CD 80 expression, which in turn indicates the modulation in macrophage polarity. The study suggested the potential of functionalized targeted delivery in cancer immunotherapy (He *et al.*, 2017a,b).

Ravichandran *et al.* (2017) explored folic acid-engineered metallic nanoparticles of DOX for cancer theranostics. In this study, gold-coated magnetite core-shell nanoparticles were made using a seed-mediated technique. The iron core and gold shell work in a synergistic way to provide magnetic properties. The drug-loading efficiency of the optimized formulation was found to be 94.93% with the cumulative DOX release of 74.73% at pH 5.4. Under microwave-based hyperthermia, core-shell nanoparticles exhibited enough heat to induce both apoptosis and necrosis.

Okur *et al.* (2016) prepared CREKA-functionalized PEG nanoparticles loading DOX for targeting cancerous cells. The nanoparticles were prepared by the water-in-water emulsion technique. The study showed that CREKA-functionalized-nanoparticles demonstrate high cellular uptake and were more effective in inducing DOX-mediated apoptosis vis-à-vis nanoparticles conjugated with laminin peptide IKVAV. Furthermore, CREKA conjugation also played a role in increasing the fibrin-binding ability of PEG particles up to 94%. The results supported the notion that functionalizing the nanoparticles with tumor-specific agents could enhance the uptake and efficacy of the desired molecule.

Nag *et al.* (2016) synthesized transferrin-functionalized chitosan-PEG nanoparticles for delivering paclitaxel to cancerous cells. The nanoparticles were prepared by a solvent evaporation method, where the free amine groups of the chitosan nanoparticles were coupled with hydroxyl groups of PEG to provide stealth properties. Finally, functionalization was carried out by coupling transferrin carboxylic groups with hydroxyl groups of PEG. The ligand-anchored nanoparticles showed high cellular uptake with marked cytotoxicity towards human non-small-cell lung cancer cell lines over the non-ligand nanoparticles. Moreover, increased retention of drug was also observed in lungs and blood. The study provided proof of the concept for targeting nanoconstructs with ligands to improve therapeutic efficacy of drug molecules.

## 6.3.2 Neurodegenerative Diseases

### 6.3.2.1 Alzheimer's and Parkinson's Diseases

Lectins-modified polyethylene glycol-polylactide-polyglycolide (PEG-PLGA) nanoparticles have been prepared and evaluated via intranasal delivery for brain targeting to improve the spatial learning and memory of AD (Zhang *et al.*, 2014). In this study, basic fibroblast growth factor (bFGF) was loaded in nanoparticles conjugated with *Solanum tuberosum* lectin (STL). This conjugation allows selective binding to N-acetylglucosamine on the nasal epithelial membrane for its brain delivery. Functionalized nanoparticles (STL-bFGF-NP) showed a uniform particle size range and negative zeta potential. Further, bFGF concentration was higher in the olfactory bulb, cerebrum, and cerebellum of rats vis-à-vis unmodified nanoparticles after nose-to-brain delivery. Additionally, the study was extended to explore the neuroprotective effect using a mirror water maze task in rats. These results were consistent with choline acetyltransferase activity and histological observations of the rat's hippocampal region. Further, *in vivo* safety of STL-bFGF-NPs was confirmed by histopathology assays.

Currently, lipid nanodelivery has been gaining attention due to its biocompatible, biodegradable properties and its potential to entrap a variety of biotherapeutic molecules (Zhao *et al.*, 2014b). In this study, authors have shown that phospholipid-based gelatin nanoparticles encapsulating basic fibroblast growth factor (bFGF) can provide stability and improved therapeutic efficacy in 6-hydroxydopamine-induced hemiparkinsonian rats. The nanoparticles (particle size $143 \pm 1.14$ nm and Zeta potential $-38.2 \pm 1.2$ mV) were prepared by a water-in-water emulsion method, followed by freeze-drying for further analysis. Results of the studies revealed that nanoparticles could deliver the exogenous bFGF in the olfactory bulb and striatum and showed improved therapeutic effects without affecting the architecture of the nasal mucosa.

Markoutsa *et al.* (2014) prepared and evaluated mono- and dual-decorated (DUAL) liposomes (LIPs), which were stable, non-cytotoxic and having a particle size between 100 and 150 nm, by immobilization of MAb against transferrin (TfR[OX26 or RI7217]) and/or a peptide analogue of ApoE3 (APOe) to target low-density lipoprotein receptor (LPR). Owing to the additive effects of dual decoration, the LIP uptake and transport across the brain cells was significantly improved. Findings of the *in vivo* and *ex vivo* studies further suggested the increased targeting potential of MAb and DUAL-LIP *vis-à-vis* non-targeted PEG-LIPs.

Mourtas *et al.* (2014) synthesized and characterized a lipid-PEG-curcumin derivative and encapsulated it into multifunctional nanosized liposomes for targeting amyloid deposits in AD. Additionally, a lipid-PEG-curcumin derivative was decorated with a blood–brain barrier (BBB) transport mediator (i.e., an anti-transferrin antibody). The post-mortem brain sample analysis of AD patients revealed high concentrations of curcumin-derivative liposomes and curcumin-derivative anti-TrF liposomes. Further, it was shown that the brain targeting capability of the multifunctional curcumin nanoliposomes was very slightly affected after the attachment of antibody on their surface. Moreover, BBB cellular model studies also demonstrated that surface functionalization of the surface of nanoliposomes does not block deposition or prevention of A$\beta$ aggregation.

In another study by Yang *et al.* (2013), liposomes and cell-penetrating peptide (CPP)-modified liposomes loaded with rivastigmine were used to improve brain uptake and minimize side effects. Biodistribution studies exhibited higher concentration of rivastigmine via nose to brain delivery in comparison to free drug. It was also inferred that, while upon intranasal administration of free drug, the amount of rivastigmine was not quantifiable in the cerebellum. On the contrary, concentrations of rivastigmine remarkably improved upon intranasal administration. Based on the biodistribution studies, the authors concluded that pharmacodynamics and brain delivery of rivastigmine-loaded liposomes and modified liposomes significantly improved following nose-to-brain delivery.

### 6.3.3 Cardiovascular Diseases

Epshtein and Korin (2017) developed the micron-sized particles of 3–4 μm diameter, composed of several smaller nanoparticles (180–200 nm) prepared by a biocompatible and biodegradable polymer, poly(lactic-co-glycolyc acid) (PLGA). Once these microparticles reach the high shear region of occluded arteries, by means of shear, the nanoparticles release drug in the stenosis region.

Dong *et al.* (2017) prepared puerarin-loaded PEGylated lipid nanoparticles, which were surface-modified by a cyclic arginyl-glycylaspartic acid peptide (RGD). RGD has specific affinity for av$\beta_3$ integrin that is vastly expressed in angiogenic conditions on endothelial cells. Maximum biodistribution of the developed nanosystem was observed in the heart as compared to other organs.

Researchers from Colombia University Medical Center (CUMC), Brigham and Women's Hospital (BWH) and Harvard Medical School (HMS), developed Collagen IV (Col IV)–targeted nanoparticles containing Ac2-26 (Annexin/lipocortin 1-mimetic peptide) that have shown the potential to treat inflammation and fix plaques in fat-fed Ldlr−/− mice. Ac2-26 imitates Annexin A1 protein, which is responsible for mediating normal resolution against inflammatory responses. These were evaluated for their therapeutic effect on chronic, advanced atherosclerosis in fat-fed Ldlr−/− mice. Col IV–Ac2-26 nanoparticles selectively target the lesion region induced by plaques. The developed targeted nanoparticles enhanced the protective collagen layer overlying the lesions, control the oxidative stress, and reduce the plaque necrosis (Fredman *et al.*, 2015).

Holme and coworkers (2012) prepared a lentil-shaped vesicular system of 1,3-diaminophospholipid, which were observed to be shear-sensitive, as 80% of drug released from vesicles in the constricted model (simulating the obstructed arteries), whereas only 40% of drug released in the healthy model. The study revealed that the shear-susceptible characteristics of lentil-shaped vesicles have great scope for targeting the atherosclerotic region.

In another study by Dvir *et al.* (2011), the researchers explored the targeting potential of Gly-Gly-Gly-Gly-Asp-Arg-Val-Tyr-Ile-His-Pro-Phe peptide as a ligand against the AT1 receptors, which are overexpressed in the ischemic heart. They developed peptide-conjugated liposomes (AT1 liposomes). The specificity of developed liposomes against the injured cardiac cells was confirmed through confocal imaging. AT1 liposomes achieved 52% of targeting after 30 min of incubation with necrotic cardiac cells, while the nanoparticles with other random order of amino acids targeted 27% of the injured cardiac cells. The *in vitro* results were further confirmed in an *in vivo* mouse model, where the heart was extracted at different time intervals and imaging was carried out using IVIS Spectrum.

### 6.3.4 Infectious Disease

#### 6.3.4.1 Human Immunodeficiency Virus (HIV) and Tuberculosis (TB)

In a recent study, the mannan-conjugated rifampicin (RIF) and isoniazid (INH)-loaded spray-dried nano-embedded microparticles were developed and evaluated against experimental tuberculosis (TB). The nanoparticles were prepared by spray drying and subjected to physicochemical characterization, followed by *in vitro* and *in vivo* evaluation. Surface-functionalized nanoparticles exhibited selective uptake to macrophages due to the presence of a mannose moiety on the surface. *In vitro* release studies showed release of RIF and INH in almost 2 h and 30 min, respectively. This could be attributed to the difference in the intrinsic solubility of individual drug towards mannan. Cell-culture studies further indicated an almost 5-fold reduction of the number of bacilli as compared to the control group *vis-à-vis* surface-modified nanoparticles. *In vivo* studies exhibited prolonged residence at the local site of action and, consequently, boosting the therapeutic potential of the drug with an improved systemic toxicity profile (Goyal *et al.*, 2015).

Prados-Rosales and colleagues (2017) recently developed arabinomannan-protein conjugated vaccine for the treatment of *Mycobacterium tuberculosis* in a murine model. The authors studied the immunogenicity by ELISA and arabinomannan (AM) glycan microarrays and protection efficacy in mice. They reported that the conjugated vaccine elicited an AM-specific antibody response in the mice. Moreover, the bacteria count was remarkably lower in vital organs like lung and spleen *vis-à-vis* control mice. This study demonstrated the role of AM-binding antibodies in defense against Mtb infection and suggested that antibody-mediated immunity can make a vital contribution to the outcome of mycobacterial infection. Eventually, it was found to be evident that humoral immunity can contribute to protection against Mtb infections.

Kumar *et al.* (2015) encapsulated zidovudine (AZT) in lactoferrin nanoparticles for improved oral delivery of AZT to minimize the side effects. The nanoparticles were formulated within the size range of 50–60 nm and exhibited 67% encapsulation of AZT. Lactoferrin-functionalized nanoparticles showed constant release of AZT up to 96 h and showed approximately 2-fold lower toxicity *vis-à-vis* free drug counterpart. This study was extended to examine the stability of nanoparticles in simulated gastric and intestinal fluids. Further, it was observed that anti-HIV-1 activity of AZT remains unaltered and the concentration of AZT was higher in the plasma, followed by the liver and kidney. In this study, it was established that there was less or no significant organ toxicities after the surface engineering of the nanoparticles with lactoferrin.

Similarly, in another study, lactoferrin-functionalized triple drug combination of zidovudine-, efavirenz-, and lamivudine-loaded nanoparticles were fabricated and evaluated via the oral route of administration (Kumar *et al.*, 2017). The developed lactoferrin-functionalized nanoparticles showed a minimum burst release in the study and maximum payload at pH 5. Further, the nanoparticles exhibited negligible toxicity to the erythrocytes and remarkable anti-HIV activity after surface functionalization. It was also shown that there was >4-fold rise in AUC and AUMC, and >2-fold increase in half-life of each drug, respectively. Additionally, tissue distribution data suggested that nanoparticles have improved the bioavailability of all drugs with less tissue-related inflammation, as revealed through histopathological analysis.

#### 6.3.4.2 Malaria

During malarial infection, the human red blood cells (RBCs) overexpress the hexose transporter PfHT1 for the export of glucose as a survival source for the parasites residing inside the RBCs. Shafi and colleagues (2017) have used the concept of competitive delivery of dehydroascorbic acid against glucose. They have compared the targeting potential of dehydroascorbic acid by pre-derivitizing it with chitosan and post-surface derivatization of chitosan nanoparticles. Flow-cytometry results demonstrated that the uptake of pre-derivatized nanoparticles was superior in comparison to post-derivitized nanoparticles.

Marques *et al.* (2017) explored the use of glycosamine glycan, that is, heparin and chondroiton sulfate, for targeting parasite-infected RBCs (pRBCs). In this study, the authors formulated chitosan nanoparticles, which were further anchored by heparin through electrostatic interactions as well as through covalent bonding. The binding of heparin was confirmed by isothermal titration calorimetry (ITC) measurements. The nanoparticles covalently bound with heparin showed less anticoagulant activity *vis-à-vis* electrostatically bound heparin. The confocal images confirmed the binding of FITC-heparin to ookinetes of *P. berghei.* Besides this, specific binding potential of chondroitin sulfate to the *P. falciparum* CS2 strain was also confirmed by single-molecule force spectroscopy.

Moles *et al.* (2016) prepared lumefantrine-loaded liposomes, which were conjugated with polyclonal FCR3S1.2, R29, and PAvarO and monoclonal M17.1 antibodies against the NTS-DBL1α domain of PfEMP1 variants. Specific targeting and disruption of rosette was confirmed through confocal microscopic studies.

El Tahir and his team (2003) have shown that biotin-labelled human serum albumin has the ability to get direct access to iRBCs, as these are hypothesized to be essential for the parasite growth and act as great source of amino acids; also, albumin is supposed to export fatty acids to the pRBCs needed for the parasite's membrane biosynthesis. Taking into account the specific uptake features of serum albumin, Sidhaye and coworkers (2016) have developed albumin-based nanoparticles to target pRBCs actively. Selective uptake of the artemether-loaded albumin nanoparticles has been observed through confocal microscopy.

Liposomes decorated with monoclonal antibodies against the erythrocyte surface protein glycophorin A (anti-glycophorin A antibody) have been prepared and evaluated for selective delivery of chlorquine and primaquine by Moles and colleagues (2015).

Urbán *et al.* (2014) have observed through flow cytometry, confocal microscopy, and transmission electron microscope (TEM) that chlorquine and primaquine conjugated with AGMA1 ISA23 and ISA1 based poly (amidoamines) have shown favorable binding to iRBCs. Specific interaction with iRBCs was correlated with simulating characteristics of the polymer conjugated with peptides and proteins that are required for the development of the parasite.

## 6.4 PATENT MAPPING

The interest in nanostructured drug delivery systems on multidisciplinary fields for treatment and diagnosis is evidently reflected by the rising number of publications and issued patents every year. For illustration, the applications of surface-engineered nano-based drug-delivery systems in the treatment of infectious diseases, cancer, and neurodegenerative and cardiovascular disorders are compiled in Table 6.5.

## 6.5 CONCLUSION

Nanotechnology-mediated drug delivery holds immense potential in the treatment and diagnosis of several serious and even fatal diseases. Today, a growing number of literature reports have witnessed the remarkable promise of these systems in the field of medicine. Engineering the surface of nanosystems with appropriate linkers results in novel systems that have the potential to overcome biological barriers and reach the target site effectively. These multifunctional systems

**TABLE 6.5**
**Patent Mapping of Various Multifunctional Nanostructured Systems**

| Patent No. | Drug Delivery System (DDS) | Surface Modification | Purpose for Surface Modification | Disease |
|---|---|---|---|---|
| WO2009108822A1 | Nanoparticles | Polyamides | Minimize nonspecific binding, stabilization of the nanoparticles and reduced immunogenicity. | Microbial infection |
| US20150224212A1 | Nanoconjugates | Anti-Necl-5 monoclonal antibody (MAb Necl-5) | Target the Necl-5 glycoprotein, expressed by many epithelial carcinomas. | Prostate cancer |
| US9056129 B2 | Gold nanoparticles | Phosphonium salts (PS) and pH-cleavable ceramide derivatives | Enhanced permeation through plasma membrane and selective apoptosis of prostate cancer cells. | Prostate cancer |
| WO2009121631A2 | Immuno-nanoparticles | Trastuzumab and EP-CAM antibody, PEG (or hydrophilic polymer) | Inhibit the uptake of the nanoparticles by the RES and prolong systemic circulation. | Lung cancer |
| US20100111876 A1 | Metallic nanoparticles | Phosphonate, phosphonate monoester, phosphinate sulfonate, hydroxamate, siloxane, or catechalate. | Enhanced uptake especially by monocytes, macrophages, microglia in particular affected areas of the brain or amyloid plaques and to facilitate nanoparticle permeation through blood–brain barrier. | Alzheimer's disease |
| US8877207 B2 | Nanoparticles | Anti-β-amyloid antibody | Counteract neurite dystrophy and inhibiting disease progression. The antibody is specifically targeted against an amyloid-beta antigen associated with a neurodegenerative disease. | Alzheimer's disease |
| WO2006012201 A1 | Nanoparticle core is either a metal oxide or a doped semiconductor | Dextran sulfate, LDL, polyinosinic acid, fucoidan or silica | Target magnetic resonance imaging (MRI) contrast agent for *in vivo* imaging of early stage atherosclerosis. | Atherosclerosis |
| US8709472 B1 | Liposomes, polymersomes and polymer nanoparticles | Cysteine-arginine-proline-proline-arginine polypeptide | The peptide group has binding affinity for heart vein cells, heart capillary cells, heart artery cells, or heart endothelial cells. The invention useful for treating vascular disease. | Cardiovascular diseases |
| US8647673 B2 | Nanoparticles | Folic acid | Localized therapeutic action along with imaging properties. | Chlamydial infection |
| US8815295 B1 | Gold nanoparticles | Anti-respiratory syncytial virus (RSV) F | To prevent or inhibit infection by the respiratory syncytial virus. | Infections RSV and HIV |

are able to achieve manifold purposes in a concurrent manner, such as targeting to the specific site and simultaneous diagnosis along with therapeutics. Despite having enormous scope over their conventional counterparts, surface-modified systems still lack concerns on the toxicological aspects. There is an ardent need of the hour to explore these multipurpose systems at the cellular level, as such systems are still in their infancy. Growing understanding of this field will provide greater opportunities to the burgeoning researchers and scientists to harvest its amazing merits and address the unmet future demands.

## REFERENCES


Al Khateb K, Ozhmukhametova EK, Mussin MN, Seilkhanov SK, Rakhypbekov TK, Lau WM, Khutoryanskiy VV. *In situ* gelling systems based on Pluronic F127/Pluronic F68 formulations for ocular drug delivery. *Int J Pharm*. 2016; 502(1–2): 70–79.

Alyautdin RN, Tezikov EB, Ramge P, Kharkevich DA, Begley DJ, Kreuter J. Significant entry of tubocurarine into the brain of rats by adsorption to polysorbate 80-coated polybutylcyanoacrylate nanoparticles: An *in situ* brain perfusion study. *J Microencapsul*. 1998; 15(1): 67–74.

Athar M, Das AJ. Therapeutic nanoparticles: State-of-the-art of nanomedicine. *Adv Mater Rev*. 2014; 1(1): 25–37.

Bagherifam S, Skjeldal FM, Griffiths G, Maelandsmo GM, Engebraten O, Nystrom B, Hasirci V, Hasirci N. pH-responsive nano carriers for doxorubicin delivery. *Pharm Res*. 2015; 32(4): 1249–1263.

Ballantyne A, Dhillon S. Trastuzumab emtansine: First global approval. *Drugs*. 2013; 73(7): 755–765.

Belhadj Z, Zhan C, Ying M, Wei X, Xie C, Yan Z, Lu W. Multifunctional targeted liposomal drug delivery for efficient glioblastoma treatment. *Oncotarget*. 2017; 8(40): 66889–66900.

Bellini M, Mazzucchelli S, Galbiati E, Sommaruga S, Fiandra L, Truffi M *et al*. Protein nanocages for self-triggered nuclear delivery of DNA-targeted chemotherapeutics in cancer cells. *J Control Release*. 2014; 196: 184–196.

Benns JM, Choi JS, Mahato RI, Park JS, Kim SW. pH-sensitive cationic polymer gene delivery vehicle: N-Ac-poly(L-histidine)-graft-poly(L-lysine) comb shaped polymer. *Bioconjug Chem*. 2000; 11(5): 637–645.

Bhimani DR, Patel JK, Patel VP. Development and evaluation of floating *in situ* gelling system of clarithromycin. *Int J Pharm Res Dev*. 2011; 3(5): 32–40.

Byeon HJ, Thao le Q, Lee S, Min SY, Lee ES, Shin BS, Choi HG, Youn YS. Doxorubicin-loaded nanoparticles consisted of cationic- and mannose-modified-albumins for dual-targeting in brain tumors. *J Control Release*. 2016; 225: 301–313.

Chen J, Liu H, Zhao C, Qin G, Xi G, Li T, Wang X, Chen T. One-step reduction and PEGylation of graphene oxide for photothermally controlled drug delivery. *Biomaterials*. 2014a; 35(18): 4986–4995.

Chen J, Son H-N, Hill JJ, Srinivasan S, Su F-Y, Stayton PS, Convertine AJ, Ratner DM. Nanostructured glycopolymer augmented liposomes to elucidate carbohydrate-mediated targeting. *Nanomed Nanotech Biol Med*. 2016a; 12(7): 2031–2041.

Chen W, Achazi K, Schade B, Haag R. Charge-conversional and reduction-sensitive poly(vinyl alcohol) nanogels for enhanced cell uptake and efficient intracellular doxorubicin release. *J Control Release*. 2015a; 205: 15–24.

Chen W-H, Yu X, Cecconello A, Sohn YS, Nechushtai R, Willner I. Stimuli-responsive nucleic acid-functionalized metal-organic framework nanoparticles using pH- and metal-ion-dependent DNAzymes as locks. *Chem Sci*. 2017a; 8(8): 5769–5780.

Chen X, Sun H, Hu J, Han X, Liu H, Hu Y. Transferrin gated mesoporous silica nanoparticles for redox-responsive and targeted drug delivery. *Colloids Surf B*. 2017b; 152: 77–84.

Chen Y-W, Chen P-J, Hu S-H, Chen IW, Chen S-Y. NIR-triggered synergic photo-chemothermal therapy delivered by reduced graphene oxide/carbon/mesoporous silica nanocookies. *Adv Funct Mater*. 2014b; 24: 451–459.

Chen Z, Tai Z, Gu F, Hu C, Zhu Q, Gao S. Aptamer-mediated delivery of docetaxel to prostate cancer through polymeric nanoparticles for enhancement of antitumor efficacy. *Eur J Pharm Biopharm*. 2016b; S0939-6411(16): 30270–30273.

Chen Z, Zhai M, Xie X, Zhang Y, Ma S, Li Z *et al*. Apoferritin nanocage for brain targeted doxorubicin delivery. *Mol Pharm*. 2017c; 14(9): 3087–3097.

Cho I, Shim MK, Jung B, Janga EH, Park M, Kang HC. Heat shock responsive drug delivery system based on mesoporous silica nanoparticles coated with temperature sensitive gatekeeper. *Microporous Mesoporous Mater*. 2017; 253: 96–101.

Crespilho FN, Lima FCA, da Silva ArBF, Oliveira Jr ON, Zucolotto V. The origin of the molecular interaction between amino acids and gold nanoparticles: A theoretical and experimental investigation. *Chem Phys Lett.* 2009; 469(1–3): 186–190.

Dong Z, Guo J, Xing X, Zhang X, Du Y, Lu Q. RGD modified and PEGylated lipid nanoparticles loaded with puerarin: Formulation, characterization and protective effects on acute myocardial ischemia model. *Biomed Pharmacother.* 2017; 89: 297–304.

Du JD, Hong L, Tan A, Boyd BJ. Naphthalocyanine as a new photothermal actuator for lipid-based drug delivery systems. *J Phys Chem B.* 2018; 122: 1766–1770.

Dvir T, Bauer M, Schroeder A, Tsui JH, Anderson DG, Langer R, Liao R, Kohane DS. Nanoparticles for targeting the infarcted heart. *Nano Lett.* 2011; 11(10): 4411–4414.

El Tahir A, Malhotra P, Chauhan VS. Uptake of proteins and degradation of human serum albumin by Plasmodium falciparum–infected human erythrocytes. *Malar J.* 2003; 2(11): 1–8.

Epshtein M, Korin N. Shear targeted drug delivery to stenotic blood vessels. *J Biomech.* 2017; 50: 217–221.

Esfandyari-Manesh M, Darvishi B, Ishkuh FA, Shahmoradi E, Mohammadi A, Javanbakht M, Dinarvand R, Atyabi F. Paclitaxel molecularly imprinted polymer-PEG-folate nanoparticles for targeting anticancer delivery: Characterization and cellular cytotoxicity. *Mater Sci Eng C Mater Biol Appl.* 2016; 62: 626–633.

Fathi M, Zangabad P S, Barar J, Aghanejad A, Erfan-Niya H, Omidi Y. Thermo-sensitive chitosan copolymer-gold hybrid nanoparticles as a nanocarrier for delivery of erlotinib. *Int J Biol Macromol.* 2018; 106: 266–276.

Feng Q, Zhang Y, Zhang W, Shan X, Yuan Y, Zhang H, Hou L, Zhang Z. Tumor-targeted and multi-stimuli responsive drug delivery system for near-infrared light induced chemo-phototherapy and photoacoustic tomography. *Acta Biomater.* 2016; 38: 129–142.

Franiak-Pietryga I, Maciejewski H, Ostrowska K, Appelhans D, Voit B, Misiewicz M, Kowalczyk P, Bryszewska M, Borowiec M. Dendrimer-based nanoparticles for potential personalized therapy in chronic lymphocytic leukemia: Targeting the BCR-signaling pathway. *Int J Biol Macromol.* 2016; 88: 156–161.

Fredman G, Kamaly N, Spolitu S, Milton J, Ghorpade D, Chiasson R, Kuriakose G, Perretti M, Farokzhad O, Tabas I. Targeted nanoparticles containing the proresolving peptide Ac2-26 protect against advanced atherosclerosis in hypercholesterolemic mice. *Sci Transl Med.* 2015; 7(275): 275ra20.

Frutos G, Prior-Cabanillas A, Fau PR, Quijada-Garrido I. A novel controlled drug delivery system based on pH-responsive hydrogels included in soft gelatin capsules. *Acta Biomater.* 2010; 6(12): 4650–4656.

Garbern JC, Minami E, Stayton PS, Murry CE. Delivery of basic fibroblast growth factor with a pH-responsive, injectable hydrogel to improve angiogenesis in infarcted myocardium. *Biomaterials.* 2011; 32(9): 2407–2416.

Gebleux R, Casi G. Antibody-drug conjugates: Current status and future perspectives. *Pharmacol Ther.* 2016; 167: 48–59.

Ghosh S, Mohapatra S, Thomas A, Bhunia D, Saha A, Das G, Jana B. Apoferritin nanocage delivers combination of microtubule and nucleus targeting anticancer drugs. *ACS Appl Mater Interfaces.* 2016; 8(45): 30824–30832.

Goyal AK, Garg T, Rath G, Gupta UD, Gupta P. Development and characterization of nanoembedded microparticles for pulmonary delivery of antitubercular drugs against experimental tuberculosis. *Mol Pharm.* 2015; 12(11): 3839–3850.

Gratieri T, Gelfuso GM, de Freitas O, Rocha EM, Lopez RFV. Enhancing and sustaining the topical ocular delivery of fluconazole using chitosan solution and poloxamer/chitosan *in situ* forming gel. *Eur J Pharm Biopharm.* 2011; 79(2): 320–327.

Guo D, Zhu L, Huang Z, Zhou H, Ge Y, Ma W *et al.* Anti-leukemia activity of PVP-coated silver nanoparticles via generation of reactive oxygen species and release of silver ions. *Biomaterials.* 2013; 34(32): 7884–7894.

Haes AJ, Hall WP, Chang L, Klein WL, Van Duyne RP. A localized surface plasmon resonance biosensor: First steps toward an assay for Alzheimer's disease. *Nano Lett.* 2004; 4: 1029–1034.

Han L, Cai Q, Tian D, Kong DK, Gou X, Chen Z, Strittmatter SM, Wang Z, Sheth KN, Zhou J. Targeted drug delivery to ischemic stroke via chlorotoxin-anchored, lexiscan-loaded nanoparticles. *Nanomedicine.* 2016; 12(7): 1833–1842.

Han N, Zhao Q, Wan L, Wang Y, Gao Y, Wang P *et al.* Hybrid lipid capped mesoporous silica for stimuli-responsive drug release and overcoming multidrug resistance. *ACS Appl Mater Interfaces.* 2015; 7(5): 3342–3351.

He X-Y, Liu B-Y, Ai S-L, Xu L, Zhuo R-X, Cheng S-X. Functional polymer/inorganic hybrid nanoparticles for macrophage targeting delivery of oligodeoxynucleotides in cancer immunotherapy. *Mater Today Chem.* 2017a; 4: 106–116.

He X-Y, Liu B-Y, Ai S-L, Xu L, Zhuo R-X, Cheng S-X. Functional polymer/inorganic hybrid nanoparticles for macrophage targeting delivery of oligodeoxynucleotides in cancer immunotherapy. *Mater Today Chem.* 2017b; 4: 106–116.

Hess C, Venetza D, Neri D. Emerging classes of armed antibody therapeutics against cancer. *Med Chem Commun.* 2014; 5: 408–431.

Holme MN, Fedotenko IA, Abegg D, Althaus J, Babel L, Favarger F *et al.* Shear-stress sensitive lenticular vesicles for targeted drug delivery. *Nat Nano.* 2012; 7(8): 536–543.

Horvát G, Gyarmati B, Berkó S, Szabó-Révész P, Szilágyi BÁ, Szilágyi A *et al.* Thiolated poly(aspartic acid) as potential *in situ* gelling, ocular mucoadhesive drug delivery system. *Eur J Pharm Sci.* 2015; 67: 1–11.

Jayaramudu T, Varaprasad K, Sadiku ER, Kim HC, Kim J. Preparation of antibacterial temperature-sensitive silver-nanocomposite hydrogels from N-isopropylacrylamide with green tea. *J Appl Polym Sci.* 2018; 135(4): 1–9.

Jeannot V, Mazzaferro S, Lavaud J, Vanwonterghem L, Henry M, Arboleas M *et al.* Targeting CD44 receptor-positive lung tumors using polysaccharide-based nanocarriers: Influence of nanoparticle size and administration route. *Nanomedicine.* 2016; 12(4): 921–932.

Jia F, Liu X, Li L, Mallapragada S, Narasimhan B, Wang Q. Multifunctional nanoparticles for targeted delivery of immune activating and cancer therapeutic agents. *J Control Release.* 2013; 172(3): 1020–1034.

Jiang Y, Yang N, Zhang H, Sun B, Hou C, Ji C *et al.* Enhanced *in vivo* antitumor efficacy of dual-functional peptide-modified docetaxel nanoparticles through tumor targeting and Hsp90 inhibition. *J Control Release.* 2016; 221: 26–36.

Kang HJ, Kang YJ, Lee YM, Shin HH, Chung SJ, Kang S. Developing an antibody-binding protein cage as a molecular recognition drug modular nanoplatform. *Biomaterials.* 2012; 33(21): 5423–5430.

Kienzle AAO, Kurch S, Schloder J, Berges C, Ose R, Schupp J *et al.* Dendritic mesoporous silica nanoparticles for pH-stimuli-responsive drug delivery of TNF-alpha. *Adv Healthc Mater.* 2017; 6(13): 1–9.

Kijanka G, Bee JS, Bishop SM, Que I, Lowik C, Jiskoot W. Fate of multimeric oligomers, submicron, and micron size aggregates of monoclonal antibodies upon subcutaneous injection in mice. *J Pharm Sci.* 2016; 105(5): 1693–1704.

Kim DY, Kwon DY, Kwon JS, Park JH, Park SH, Oh HJ *et al.* Synergistic anti-tumor activity through combinational intratumoral injection of an *in-situ* injectable drug depot. *Biomaterials.* 2016; 85: 232–245.

Kim MJ, Park JS, Lee SJ, Jang J, Back SH, Bahn G *et al.* Notch1 targeting siRNA delivery nanoparticles for rheumatoid arthritis therapy. *J Control Release.* 2015; 216: 140–148.

Kobsa S, Saltzman WM. Bioengineering approaches to controlled protein delivery. *Pediatr Res.* 2008; 63(5): 513–519.

Koo AN, Lee HJ, Kim SE, Chang JH, Park C, Kim C *et al.* Disulfide-cross-linked PEG-poly(amino acid)s copolymer micelles for glutathione-mediated intracellular drug delivery. *Chem Commun Camb.* 2008; (48): 6570–6572.

Kumar P, Lakshmi YS, Bhaskar C, Golla K, Kondapi AK. Improved safety, bioavailability and pharmacokinetics of zidovudine through lactoferrin nanoparticles during oral administration in rats. *PLos ONE.* 2015; 10(10): e0140399.

Kumar P, Lakshmi Y, Kondapi A. Triple drug combination of zidovudine, efavirenz and lamivudine loaded lactoferrin nanoparticles: An effective nano first-line regimen for HIV therapy. *Pharm Res.* 2017; 34(2): 257–268.

Kwon JS, Park IK, Cho AS, Shin SM, Hong MH, Jeong SY *et al.* Enhanced angiogenesis mediated by vascular endothelial growth factor plasmid-loaded thermo-responsive amphiphilic polymer in a rat myocardial infarction model. *J Control Release.* 2009; 138(2): 168–176.

Lee EJ, Lee SJ, Kang YS, Ryu JH, Kwon KC, Jo E *et al.* Engineered proteinticles for targeted delivery of siRNA to cancer cells. *Adv Funct Mater.* 2015a; 25: 1279–1286.

Lee W, Seo J, Kwak S, Park EJ, Na DH, Kim S *et al.* A double-chambered protein nanocage loaded with thrombin receptor agonist peptide (TRAP) and gamma-carboxyglutamic acid of protein C (PC-Gla) for sepsis treatment. *Adv Mater.* 2015b; 27(42): 6637–6643.

Li J, Wu S, Wu C, Qiu L, Zhu G, Cui C *et al.* Versatile surface engineering of porous nanomaterials with bioinspired polyphenol coatings for targeted and controlled drug delivery. *Nanoscale.* 2016; 8(16): 8600–8606.

Li Y, Xiao K, Luo J, Xiao W, Lee JS, Gonik AM *et al.* Well-defined, reversible disulfide cross-linked micelles for on-demand paclitaxel delivery. *Biomaterials.* 2011; 32(27): 6633–6645.

Liang L, Fu J, Qiu L. Design of pH-sensitive nanovesicles via cholesterol analogue incorporation for improving *in vivo* delivery of chemotherapeutics. *ACS Appl Mater Interfaces.* 2018; 10(6): 5213–5226.

Lin C, Kuan C, Wang L, Wu H, Chen Y, Chang C *et al.* Integrated self-assembling drug delivery system possessing dual responsive and active targeting for orthotopic ovarian cancer theranostics. *Biomaterials.* 2016; 90: 12–26.

Lin C, Zhang X, Chen H, Bian Z, Zhang G, Riaz MK. Dual-ligand modified liposomes provide effective local targeted delivery of lung-cancer drug by antibody and tumor lineage-homing cell-penetrating peptide. *Drug Delv.* 2018; 25(1): 256–266.

Liu Y, Zhang X, Liu Z, Wang L, Luo L, Wang M *et al.* Gold nanoshell-based betulinic acid liposomes for synergistic chemo-photothermal therapy. *Nanomedicine.* 2017; 13(6): 1891–1900.

Loureiro JA, Gomes B, Fricker G, Coelho MA, Rocha S, Pereira MC. Cellular uptake of PLGA nanoparticles targeted with anti-amyloid and anti-transferrin receptor antibodies for Alzheimer's disease treatment. *Colloids Surf B.* 2016; 145: 8–13.

Luo Y, Wang X, Du D, Lin Y. Hyaluronic acid-conjugated apoferritin nanocages for lung cancer targeted drug delivery. *Biomater Sci.* 2015; 3(10): 1386–1394.

Madan J, Adokar B, Dua K. Development and evaluation of *in situ* gel of pregabalin. *Int J Pharm Investig.* 2015; 5(4): 226–233.

Manchun S, Cheewatanakornkool K, Dass CR, Sriamornsak P. Novel pH-responsive dextrin nanogels for doxorubicin delivery to cancer cells with reduced cytotoxicity to cardiomyocytes and stem cells. *Carbohydr Polym.* 2014; 114: 78–86.

Manivasagan P, Bharathiraja S, Bui NQ, Jang B, Oh YO, Lim IG *et al.* Doxorubicin-loaded fucoidan capped gold nanoparticles for drug delivery and photoacoustic imaging. *Int J Biol Macromol.* 2016; 91: 578–588.

Mansur AA, de Carvalho SM, Mansur HS. Bioengineered quantum dot/chitosan-tripeptide nanoconjugates for targeting the receptors of cancer cells. *Int J Biol Macromol.* 2016; 82: 780–789.

Mansur AA, Mansur HS, Soriano-Araújo A, Lobato ZI. Fluorescent nanohybrids based on quantum dot–Chitosan–antibody as potential cancer biomarkers. *ACS Appl Mater Interfaces.* 2014; 6(14): 11403–11412.

Markoutsa E, Papadia K, Giannou AD, Spella M, Cagnotto A, Salmona M *et al.* Mono and dually decorated nanoliposomes for brain targeting, *in vitro* and *in vivo* studies. *Pharm Res.* 2014; 31(5): 1275–1289.

Marques J, Valle-Delgado JJ, Urbán P, Baró E, Prohens R, Mayor A *et al.* Adaptation of targeted nanocarriers to changing requirements in antimalarial drug delivery. *Nanomed Nanotechnol Biol Med.* 2017; 13(2): 515–525.

Miyazaki S, Endo K, Kawasaki N, Kubo W, Watanabe H, Attwood D. Oral sustained delivery of paracetamol from *in situ* gelling xyloglucan formulations. *Drug Dev Ind Pharm.* 2003; 29(2): 113–119.

Miyazaki S, Takahashi A, Itoh K, Ishitani M, Dairaku M, Togashi M *et al.* Preparation and evaluation of gel formulations for oral sustained delivery to dysphagic patients. *Drug Dev Ind Pharm.* 2009; 35(7): 780–787.

Moles E, Moll K, Cheng JH, Parini P, Wahlgren M, Fernandez-Busquets X. Development of drug-loaded immunoliposomes for the selective targeting and elimination of rosetting *Plasmodium falciparum*-infected red blood cells. *J Control Release.* 2016; 241: 57–67.

Moles E, Urbán P, Jiménez-Díaz M, Viera-Morilla S, Angulo-Barturen I, Busquets M *et al.* Immunoliposome-mediated drug delivery to Plasmodium-infected and non-infected red blood cells as a dual therapeutic/prophylactic antimalarial strategy. *J Control Release.* 2015; 210: 217–229.

Mourtas S, Lazar AN, Markoutsa E, Duyckaerts C, Antimisiaris SG. Multifunctional nanoliposomes with curcumin-lipid derivative and brain targeting functionality with potential applications for Alzheimer disease. *Eur J Med Chem.* 2014; 80(80): 175–183.

Nag M, Gajbhiye V, Kesharwani P, Jain NK. Transferrin functionalized chitosan-PEG nanoparticles for targeted delivery of paclitaxel to cancer cells. *Colloids Surf B.* 2016; 148: 363–370.

Nagarwal R C, Srinatha A, Pandit JK. *In situ* forming formulation: Development, evaluation, and optimization using $3^3$ factorial design. *AAPS Pharmscitech.* 2009; 10(3): 977–983.

Nezlin R. Aptamers in immunological research. *Immunol Lett.* 2014; 162(2, Part B): 252–255.

Nykypanchuk D, Maye MM, van der Lelie D, Gang O. DNA-guided crystallization of colloidal nanoparticles. *Nat Biotechnol.* 2008; 451: 549–552.

Ochekpe NA, Olorunfemi PO, Ngwuluka NC. Nanotechnology and drug delivery part 2: Nanostructures for drug delivery. *Trop J Pharm Res.* 2009; 8(3): 275–287.

Okur AC, Erkoc P, Kizilel S. Targeting cancer cells via tumor-homing peptide CREKA functional PEG nanoparticles. *Colloids Surf B.* 2016; 147: 191–200.

Onoue S, Yamada S, Chan HK. Nanodrugs: Pharmacokinetics and safety. *Int J Nanomedicine.* 2014; 9: 1025–1037.

Patel RP, Dadhani B, Ladani R, Baria A, Patel J. Formulation, evaluation and optimization of stomach specific *in situ* gel of clarithromycin and metronidazole benzoate. *Int J Drug Deliv.* 2010; 2:141–153.

Pengo P, Baltzer L, Pasquato L, Scrimin P. Substrate modulation of the activity of an artificial nanoesterase made of peptide-functionalized gold nanoparticles. *Angew Chem Int Ed Engl.* 2007; 46(3): 400–406.

Perillo E, Herve-Aubert K, Allard-Vannier E, Falanga A, Galdiero S, Chourpa I. Synthesis and *in vitro* evaluation of fluorescent and magnetic nanoparticles functionalized with a cell penetrating peptide for cancer theranosis. *J Colloid Interface Sci.* 2017; 499: 209–217.

Powell D, Chandra S, Dodson K, Shaheen F, Wiltz K, Ireland S *et al.* Aptamer-functionalized hybrid nanoparticle for the treatment of breast cancer. *Eur J Pharm Biopharm.* 2017; 114: 108–118.

Prados-Rosales R, Carreño L, Cheng T, Blanc C, Weinrick B, Malek A *et al.* Enhanced control of Mycobacterium tuberculosis extrapulmonary dissemination in mice by an arabinomannan-protein conjugate vaccine. *PLoS Pathogens* 2017; 13(3): e1006250.

Qian Y, Wang F, Li R, Zhang Q, Xu Q. Preparation and evaluation of in situ gelling ophthalmic drug delivery system for methazolamide. *Drug Dev Ind Pharm.* 2010; 36(11): 1340–1347.

Qiao L, Wang X, Gao Y, Wei Q, Hu W, Wu L *et al.* Laccase-mediated formation of mesoporous silica nanoparticle based redox stimuli-responsive hybrid nanogels as a multifunctional nanotheranostic agent. *RSC Nanoscale.* 2013; (36): 1–8.

Rai M, Ingle A, Gupta I, Brandelli A. Bioactivity of noble metal nanoparticles decorated with biopolymers and their application in drug delivery. *Int J Pharm.* 2015; 496(2): 159–172.

Raniolo S, Giulia V, Ottaviani A, Unida V, Iacovelli F, Manetto A. Selective targeting and degradation of doxorubicin-loaded folate-functionalized DNA nanocages. *Nanomedicine.* 2018; 14(4): 1181–1190.

Ravichandran M, Oza G, Velumani S, Ramirez JT, Vera A, Leija L. Design and evaluation of surface functionalized superparamagneto-plasmonic nanoparticles for cancer therapeutics. *Int J Pharm.* 2017; 524(1): 16–29.

Rengaswamy V, Zimmer D, Suss R, Rossler J. RGD liposome-protamine-siRNA (LPR) nanoparticles targeting PAX3-FOXO1 for alveolar rhabdomyosarcoma therapy. *J Control Release.* 2016; 235: 319–327.

Rohith G, Sridhar BK, Srinatha A. Floating drug delivery of a locally acting H2-antagonist: An approach using an *in situ* gelling liquid formulation. *Acta Pharm.* 2009; 59(3): 345–354.

Saini R, Saini S, Sharma S. Nanotechnology: The future medicine. *J Cutan Aesthet Surg.* 2010; 3(1): 32–33.

Saldaña S, Palao-Suay R, Aguilar MR, Fernández L, Arévalo H, Trinidad A. pH-sensitive polymeric nanoparticles with antioxidant and anti-inflammatory properties against cisplatin-induced hearing loss. *J Control Release.* 2018; 270: 53–64.

Salunke SR, Patil SB. Ion activated *in situ* gel of gellan gum containing salbutamol sulphate for nasal administration. *Int J Biol Macromolec.* 2016; 87:41–47.

Sarkar K, Banerjee SL, Kundu PP, Madrasa G, Chatterjee K. Biofunctionalized surface-modified silver nanoparticles for gene delivery. *J Mater Chem B.* 2015; 3: 5266–52676.

Sarmento B, Ribeiro A, Veiga F, Ferreira D. Development and characterization of new insulin containing polysaccharide nanoparticles. *Colloids Surf B.* 2006; 53(2): 193–202.

Senter PD, Sievers EL. The discovery and development of brentuximab vedotin for use in relapsed Hodgkin lymphoma and systemic anaplastic large cell lymphoma. *Nat Biotechnol.* 2012; 30(7): 631–637.

Seo YG, Kim DW, Cho KH, Yousaf AM, Kim DS, Kim JH *et al.* Preparation and pharmaceutical evaluation of new tacrolimus-loaded solid self-emulsifying drug delivery system. *Mol Biol.* 2010; 624: 221–234.

Shafi H, Reddy DVS, Khan T, Ranjan R, Srivastava A, Vaishya S *et al.* Dehydroascorbate-derivatized chitosan particles for targeting antimalarial agents to infected erythrocytes. *Int J Pharm.* 2017; 524(1): 205–214.

Shen Z, Wei W, Tanaka H, Kohama K, Ma G, Dobashi T *et al.* A galactosamine-mediated drug delivery carrier for targeted liver cancer therapy. *Pharmacol Res.* 2011; 64(4): 410–419.

Shi S, Zhou M, Li X, Hu M, Li C, Li M *et al.* Synergistic active targeting of dually integrin alphavbeta3/ CD44-targeted nanoparticles to B16F10 tumors located at different sites of mouse bodies. *J Control Release.* 2016; 235: 1–13.

Siafaka PI, Okur NÜ, Karavas E, Bikiaris D. Surface modified multifunctional and stimuli responsive nanoparticles for drug targeting: Current status and uses. *Int J Mol Sci.* 2016; 17: 1440–1449.

Sidhaye A, Bhuran KC, Zambare S, Abubaker M, Niroshini N, Singh KK. Bio-inspired artemether-loaded human serum albumin nanoparticles for effective control of malaria-infected erythrocytes. *Nanomedicine.* 2016; 11(21): 1–11.

Singh B, Khurana RK, Garg B, Saini S, Kaur R. Stimuli-responsive systems with diverse drug delivery and biomedical applications: Recent updates and mechanistic pathways. *Crit Rev Ther Drug Carrier Syst.* 2017; 34(3): 209–255.

Singh L, Kruger HG, Maguire GEM, Govender T, Parboosing R. Development and evaluation of peptide-functionalized gold nanoparticles for HIV integrase inhibition. *Int J Pept Res Ther.* 2018a; 24: 1–12.

Singh PK, Jaiswal AK, Pawar VK, Raval K, Kumar A, Bora HK *et al.* Fabrication of 3-O-sn-Phosphatidyl-L-serine anchored PLGA nanoparticle bearing amphotericin B for macrophage targeting. *Pharm Res.* 2018b; 35(60): 1–12.

Su W, Wang H, Wang S, Liao Z, Kang S, Peng Y *et al.* PEG/RGD-modified magnetic polymeric liposomes for controlled drug release and tumor cell targeting. *Int J Pharm.* 2012; 426(1–2): 170–181.

Taghdisi SM, Danesh NM, Lavaee P, Emrani AS, Hassanabad KY, Ramezani M *et al.* Double targeting, controlled release and reversible delivery of daunorubicin to cancer cells by polyvalent aptamers-modified gold nanoparticles. *Mater Sci Eng C Mater Biol Appl.* 2016; 61: 753–761.

Tran NQ, Joung YK, Lih E, Park KD. *In situ* forming and rutin-releasing chitosan hydrogels as injectable dressings for dermal wound healing. *Biomac.* 2011; 12(8): 2872–2880.

Urandur S, Banala VT, Shukla RP, Mittapelly N, Pandey G, Kalleti N *et al.* Anisamide-anchored lyotropic nano-liquid crystalline particles with AIE effect: A smart optical beacon for tumor imaging and therapy. *ACS Appl Mater Interfaces.* 2018; 10(15): 12960–12974.

Urbán P, Valle-Delgado JJ, Mauro N, Marques J, Manfredi A, Rottmann M *et al.* Use of poly(amidoamine) drug conjugates for the delivery of antimalarials to Plasmodium. *J Control Release.* 2014; 177: 84–95.

Wang K, Yang Y, Xue W, Liu Z. Cell penetrating peptide-based redox-sensitive vaccine delivery system for subcutaneous vaccination. *Mol Pharm.* 2018; 15(3): 975–984.

Wang Y, Wang K, Zhao J, Liu X, Bu J, Yan X *et al.* Multifunctional mesoporous silica-coated graphene nanosheet used for chemo-photothermal synergistic targeted therapy of glioma. *J Am Chem Soc.* 2013; 135(12): 4799–4804.

Wang Z, Tian Y, Zhang H, Qin Y, Li D, Gan L *et al.* Using hyaluronic acid-functionalized pH stimuli-responsive mesoporous silica nanoparticles for targeted delivery to CD44-overexpressing cancer cells. *Int J Nanomedicine.* 2016; 11: 6485–6497.

Xu X, Chen S, Wu Q. Targeting epidermal growth factor receptor in tumors: From conventional monoclonal antibodies via heavy chain-only antibodies to nanobodies. *Eur J Pharm Sci.* 2012; 45(4): 399–407.

Yan X, Xu L, Bi C, Duan D, Chu L, Yu X *et al.* Lactoferrin-modified rotigotine nanoparticles for enhanced nose-to-brain delivery: LESA-MS/MS-based drug biodistribution, pharmacodynamics, and neuroprotective effects. *Int J Nanomedicine.* 2018; 13: 273–281.

Yang F, Ai W, Jiang F, Liu X, Huang Z, Ai S. Preclinical evaluation of an epidermal growth factor receptor–targeted doxorubicin–peptide conjugate: Toxicity, biodistribution, and efficacy in mice. *J Pharm Sci.* 2016; 105(2): 639–649.

Yang ZZ, Zhang YQ, Wang ZZ, Wu K, Lou JN, Qi XR. Enhanced brain distribution and pharmacodynamics of rivastigmine by liposomes following intranasal administration. *Int J Pharm.* 2013; 452(1–2): 344–354.

Yao W, Xu P, Zhao J, Ling L, Li X, Zhang B *et al.* RGD functionalized polymeric nanoparticles targeting periodontitis epithelial cells for the enhanced treatment of periodontitis in dogs. *J Colloid Interface Sci.* 2015; 458: 14–21.

Yu S, Wang QM, Wang X, Liu D, Zhang W, Ye T, Yang X, Pan W. Liposome incorporated ion sensitive *in situ* gels for opthalmic delivery of timolol maleate. *Int J Pharm.* 2015; 480(1–2): 128–136.

Zhang C, Chen J, Feng C, Shao X, Liu Q, Zhang Q *et al.* Intranasal nanoparticles of basic fibroblast growth factor for brain delivery to treat Alzheimer's disease. *Int J Pharm.* 2014; 461(1): 192–202.

Zhang CX, Lippard SJ. New metal complexes as potential therapeutics. *Curr Opin Chem Biol.* 2003; 7(4): 481–489.

Zhang Y, Guo J, Zhang XL, Li DP, Zhang TT, Gao FF *et al.* Antibody fragment-armed mesoporous silica nanoparticles for the targeted delivery of bevacizumab in ovarian cancer cells. *Int J Pharm.* 2015; 496(2): 1026–1033.

Zhao C, Liu X, Liu J, Yang Z, Rong X, Li M *et al.* Transferrin conjugated poly (γ-glutamic acid-maleimide-co-l-lactide)-1,2-dipalmitoylsn-glycero-3-phosphoethanolamine copolymer nanoparticles for targeting drug delivery. *Colloids Surf B.* 2014a; 123: 787–796.

Zhao YZ, Li X, Lu CT, Lin M, Chen LJ, Xiang Q *et al.* Gelatin nanostructured lipid carriers-mediated intranasal delivery of basic fibroblast growth factor enhances functional recovery in hemiparkinsonian rats. *Nanomedicine.* 2014b; 10(4): 755–764.

Zhen Z, Tang W, Chen H, Lin X, Todd T, Wang G *et al.* RGD-modified apoferritin nanoparticles for efficient drug delivery to tumors. *ACS Nano.* 2013; 7(6): 4830–4837.

Zhou S, Wu D, Yin X, Jin X, Zhang X, Zheng S. Intracellular pH-responsive and rituximab-conjugated mesoporous silica nanoparticles for targeted drug delivery to lymphoma B cells. *J Exp Clin Cancer Res.* 2017; 36(24): 1–17.

Zhou T, Xiao C, Fan J, Chen S, Shen J, Wua W *et al.* A nanogel of on-site tunable pH-response for efficient anticancer drug delivery. *Acta Biomater.* 2013; 9(1): 4546–4557.

Zhu SL, Li F, Du CL, Fu Y. A localized surface plasmon resonance nanosensor based on rhombic Ag nanoparticle array. *Senss Actuators B.* 2008; 134: 193–198.

# 7 Versatility of Poly(Lactic Acid) and Modified Poly(Lactic Acid) for Nanobioengineering Applications

*Chelladurai Karthikeyan Balavigneswaran, Kalyan Ramesh, Ranjot Kaur, and Nira Misra**

## CONTENTS



## 7.1 INTRODUCTION

Biomaterials are substances engineered to interact with any biological system for therapeutic (i.e., treat, augment, repair, or replace a tissue or cellular function of the body) or diagnostic purposes. These can be categorized in four classes: polymers, metals, ceramics, and composites. Synthetic polymers constitute a diverse class of biomaterials. They have been widely used in biomedical applications, as they can be synthesized with a wide variety of compositions and properties. They can also be fabricated into complex shapes and structures. In addition, their surface properties

* Corresponding author.

also can be readily modified physically or chemically. Polyesters, polyamides, polycarbonates, and polyurethane are the widely used synthetic polymers for biomedical applications. Among them, hydrophobic aliphatic poly(lactide) (PLA), poly(caprolactone) (PCL), and poly(glycolide) (PGA) are important polymers. These are biocompatible and biodegradable by various enzymes in the biological system under physiological conditions and have low immunogenicity and good mechanical properties. Importantly, they are US FDA-approved for clinical use. These unique properties have facilitated their use in biomedical applications as sutures, implants, drug-delivery carriers, and tissue-engineering scaffolds (Uhrich *et al.*, 1999; Oh, 2011). In view of these facts, PLA is a promising material, where lactic acid constitutes the chains of PLA, a byproduct of the glycolysis pathway in humans and mammals (Corneillie and Smet, 2015). Therefore, in this chapter, we intend to explain PLA properties, further modification of PLA with various hydrophilic polymers such as poly(ethylene glycol) (PEG), poly(N-isopropylacrylamide) (PNIPAAm), poly(vinyl pyrrolidone) (PNVP), polyacrylamide, poly(amino acids), and polysaccharides to prepare PLA-based synthetic polymers, followed by applications of such amphiphilic copolymers.

### 7.1.1 Chemistry of Lactic Acid and PLA

The chemistry of PLA is based on processing and polymerization of a lactic acid monomer. Lactic acid (2-hydroxypropionic acid) is a chiral molecule that exists as two enantiomers, L- and D-lactic acid (Figure 7.1). The L isomer rotates clockwise and the D isomer rotates anticlockwise on the plane of polarized light. The meso-form is an equimolar (racemic) mixture of D(−) and L(+) isomers and is optically inactive (Gupta *et al.*, 2007).

High molecular weight PLA can be synthesized in two ways (Figure 7.2):

1. Direct condensation, which involves solvents under high vacuum.
2. Formation of the cyclic dimer intermediate (lactide), which is solvent-free.

The direct condensation method was initially proposed by Carothers (1936) and the reaction is held at high vacuum and high temperatures for the removal of water produced during the condensation reaction (Mc Kenzie and Cowie, 1991). The resultant polymer molecular weight was observed at low-to-intermediate molecular weight (Lunt, 1998) due to some common side reactions (such as cyclization, incomplete reaction, the presence of water, etc.), and the polymer tended to be brittle, thus limiting its usage in industry. Further, it was found that high conversion is required to obtain a high degree of polymerization to further obtain the higher molecular weight polymer.

PLA with controlled molecular weight can be achieved by ring-opening polymerization (ROP), which is a solvent-free reaction. In addition, it is also possible to control the ratio and sequence of D- and L-lactic acid units by controlling residence time and temperature, in combination with catalyst type and concentration (Lunt, 1998; Lunt and Shafer, 2000; Radano *et al.*, 2000). ROP of lactides was first explained by Carothers in 1932, but he only obtained low molecular weight PLA due to impurities (Kaplan, 1998), and it was Dupont who improved the lactide purification techniques to obtain high molecular weight PLA (Lowe, 1954). Thus, ROP proceeding via a coordination–insertion pathway is efficient in terms of producing high molecular weight PLA (Kricheldorf and Sumbel, 1989; Kurcok *et al.*, 1995) (Figure 7.3). Another advantage of the coordination–insertion pathway is that it can be

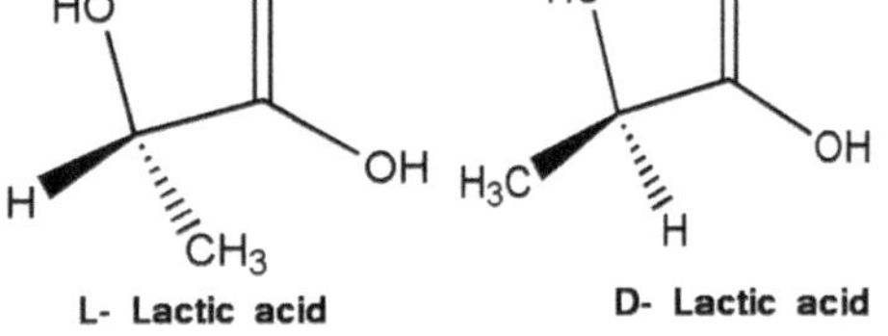


**FIGURE 7.1** Isomers of lactide.

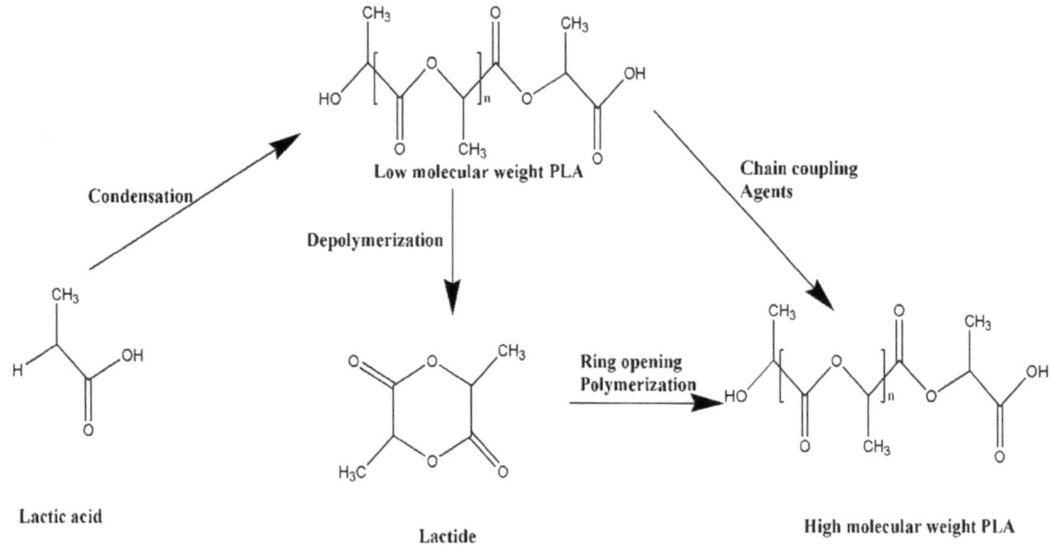


**FIGURE 7.2** Common routes to synthesis of high molecular weight PLA.

performed in bulk (whilst anionic and cationic ROP often require solvents). Many catalysts have been screened, and heavy metal-based catalysts such as tin, zinc, and aluminum were found to be the best to produce high molecular weight PLA, (Kricheldorf and Sumbel, 1989), which showed more than 90% conversion and less than 1% racemisation. Tin octoate [tin(II) 2-ethylhexanoate] is the catalyst most often used in industry because of its greater stability and lower toxicity (Ramesh *et al.*, 2012).

#### 7.1.1.1 Star Poly(Lactic Acid)

Star polymers are branched, multi-armed polymeric materials, where the branches extend from the core. Inter- and multi-disciplinary scientists have been showing their constant interest on star polymers to obtain interesting rheological and mechanical properties, such as increase of polymer strength. Star PLA can be categorized as discrete, polymeric, and dendritic/hyperbranched cores. These star polymers were first described by Schaefgen and Flory (1948), who synthesized four- and eight-arm star polymers of ε-caprolactam, and eventually made an efficient path in this area. Although star polymers were first reported in 1948, the first star PLA macromolecule was introduced in 1989 (Zhu *et al.*, 1989), and extensive research in this area has started in the last two decades due to their unique properties when compared to linear PLA.

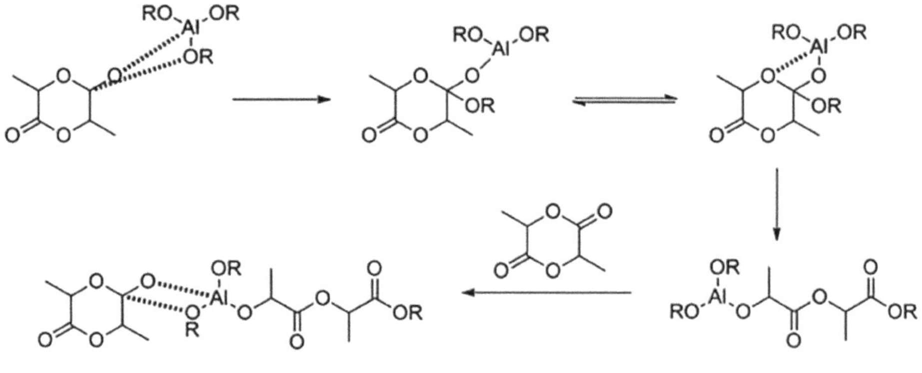

**FIGURE 7.3** Coordination-insertion mechanism using aluminium oxide. (Reproduced from Corneillie S, Smet M. *Polym Chem.* 2015; 6(6): 850–867. With permission.)

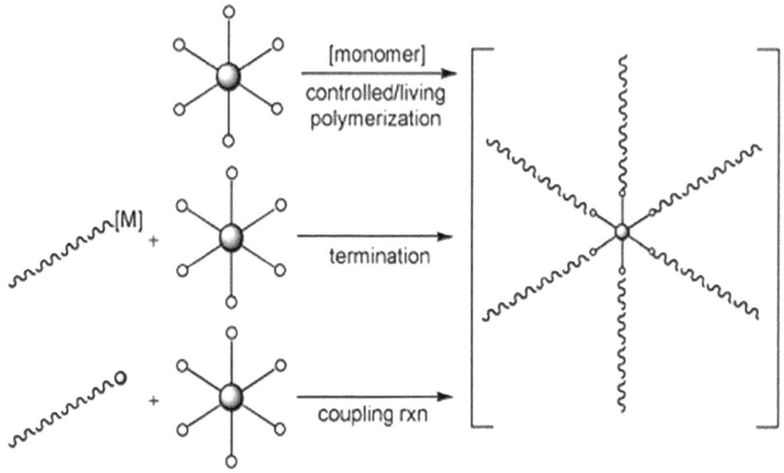


**FIGURE 7.4** Synthesis methodologies of star polymers. (Adapted from Cameron DJ, Shaver MP. *Chem Soc Rev*. 2011; 40(3): 1761–1776. With permission.)

In general, three methods have been generally used for star polymer synthesis (Figure 7.4) (Hatada *et al.*, 1997; Cameron and Shaver, 2011). The core-first method involves ring opening of the monomer by a multifunctional initiator and further living polymerization of a reactive monomer. In this method, the polymer chains are grown directly on the core. The arm-first method involves coupling linear polymer chains with a reactive core molecule. Multi-functional reagents will be used in this method to terminate linear living polymers. The other method can exploit the latent reactivity of telechelic linear polymers to chemically attach the polymer arms to the core.

Aliphatic polyesters are one of the important class of star polymers because the reactions are relatively easy, and readily available multifunctional alcohols to initiate the ROP of cyclic esters through a core-first approach polymer synthesis. The ROP reactions are the same as for linear polymers, whereas multifunctional alcohol will be used instead of monofunctional alcohol and catalyzed by a metal complex to form an active metal alkoxide that further follows a coordination–insertion mechanism to produce the star polymer. Most commonly used multifunctional alcohols for ring opening are listed in Figure 7.5.

### 7.1.2 Properties of Poly(Lactic Acid)

It is necessary to understand the properties of PLA before discussing the strategic improvement of linear and branched PLA.

#### 7.1.2.1 Crystallinity and Thermal Behavior of Poly(Lactic Acid)

Poly(l-lactic acid) (PLLA) and poly(d-lactic acid) (PDLA) are crystalline due to enantiomeric purity and stereo-regularity of the PLLA/PDLA chain. Conversely, PDLLA (equimolar of L- and D-lactic acid) is fully amorphous because of its irregular structure. In general, crystallization and thermal behavior of polymers depend on the polymer molecular weight, polymerization conditions, and purity of the polymer.

Pure PLLA and PDLA have almost identical properties, where the glass transition temperature ($T_g$) is around 50°C–70°C, the melting temperature is ($T_m$) between 170°C and 190°C, and the crystallinity is approximately 35% (Engelberg and Kohn, 1991; Celli and Scandola, 1992; Doi and Steinbüchel, 2002; Fambri *et al.*, 2002). In case of syndiotactic polylactide, lower glass transition temperature and melting temperature were observed as 34°C and 155°C, respectively, due to the different stereotactic configurations (Ovitt and Coates, 1999). It was further observed that increase in molecular weight leads to significant increase in $T_g/T_m$ and crystallinity for PLLA and PDLLA, respectively (Jamshidi *et al.*, 1988).

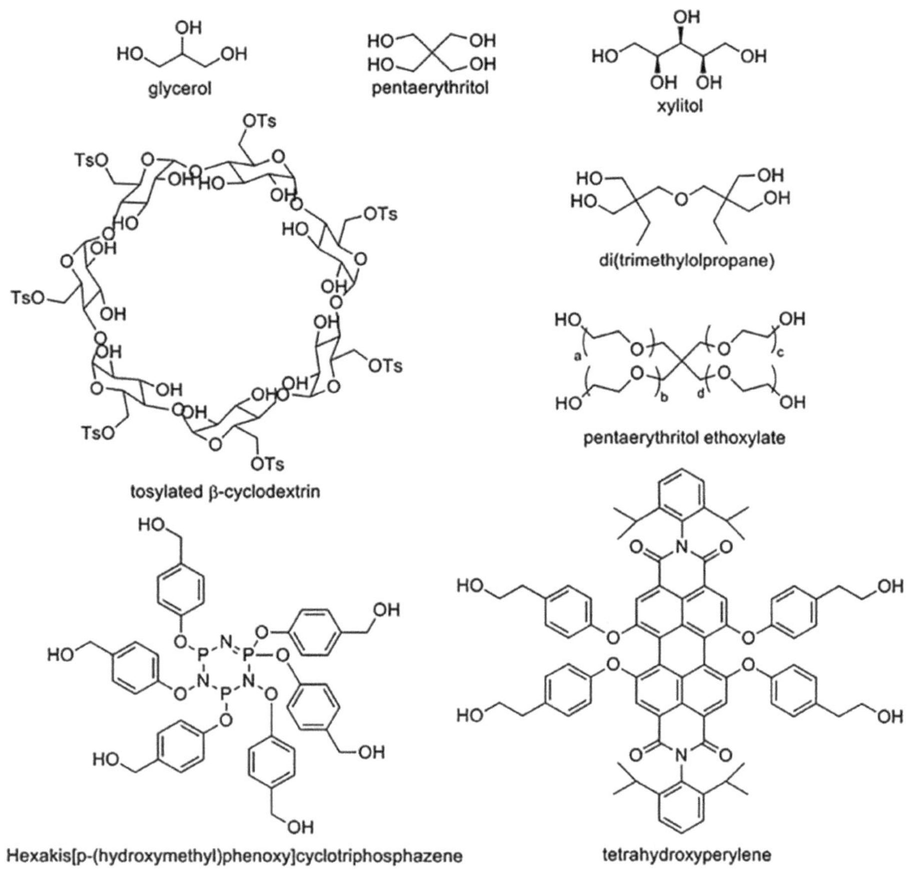


**FIGURE 7.5** Multifunctional alcohols used to synthesize multiarm PLA. (Reproduced from Corneillie S, Smet M. *Polym Chem.* 2015; 6(6): 850–867. With permission.)

Star PLA have shown lower $T_m$, $T_g$, crystallization temperature ($T_c$), and hydrodynamic volumes compared to linear PLA (Kim *et al.*, 2004; Stanford and Dove, 2010), but higher viscosity than linear PLA (Kim *et al.*, 2004). In addition, star-shaped PLA exhibits coiling nature (Long *et al.*, 1964). Lee *et al.* (2001) studied the thermal characterization of star PLA with different multifunctional alcohols. $T_g$ was closely dependent on molecular weight than the number of arms present in the star PLA and ranged from 49.1°C ($M_n = 12,700$) to 58.1°C ($M_n = 31,700$) for the star PLA (Ovitt and Coates, 1999). In addition, improper control in ROP resulted in lower $T_g$ values, indicating that polymer star uniformity plays an important role in controlling the thermal properties. Further, the crystallization behavior revealed the choice of initiator has a large effect on $T_c$, where pentaerythritol-based PLLAs have lower $T_c$ values compared to dipentaerythritol-based PLLAs (Han *et al.*, 2015; Tsuji *et al.*, 2016).

#### 7.1.2.2 Rheology and Mechanical Property

In general, increase in molecular weight usually leads into increase in strength, though contradictory results were observed by Perego *et al.* (1996). PLLA showed better behavior when compared with PDLLA, where tensile strength was 59 MPa ($M_n = 67,000$) and 44 MPa ($M_n = 114,000$) for PLLA and PDLLA, respectively, which can be explained due to the stereo-regularity of the polymer chain.

In case of viscosity, star-shaped PLLA showed higher viscosity and greater dependence of viscosity on temperature. It was observed that branched polymers have well-entangled arms, which results in higher viscosities due to the suppression of longitudinal motions (Long *et al.*, 1964; Kim *et al.*, 2004). Table 7.1 summarizes the rheological properties of the branched/star PLA (Corneillie and Smet, 2015).

## TABLE 7.1
## Properties of Branched PLA

| Structure | Properties | References |
|---|---|---|
| Star shaped 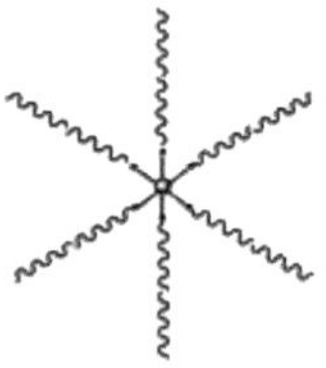 | • Higher solution viscosity<br>• More prominent shear-thinning behavior<br>• Higher storage and loss moduli<br>• Lower $T_g$, Tm and crystallinity | Kim *et al.* (2004) |
| Comb shaped  | • Higher degradability<br>• Lower $T_g$, Tm and crystallinity<br>• Increased hydrophilicity<br>• Lower intrinsic viscosity | Ohya *et al.* (1998b), Carlotti *et al.* (2001), Wan *et al.* (2008), Nouvel *et al.* (2009), Ma *et al.* (2013) |
| Hyper branched 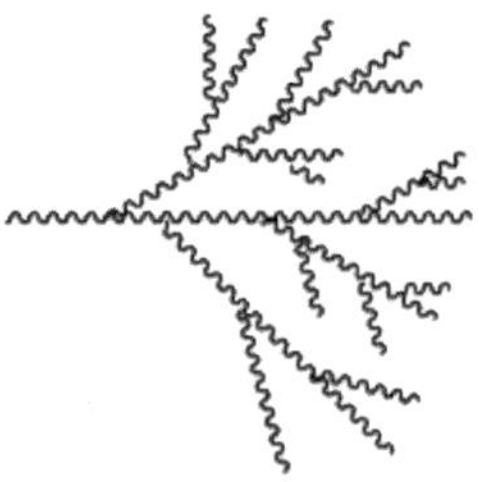 | • Lower $T_g$, Tm and melting enthalpy $\Delta H_m$<br>• Higher thermal stability | Wallach and Huang (2000), Tasaka *et al.* (2001), Cooper and Storey (2008), Fischer *et al.* (2012) |
| Long chain branched 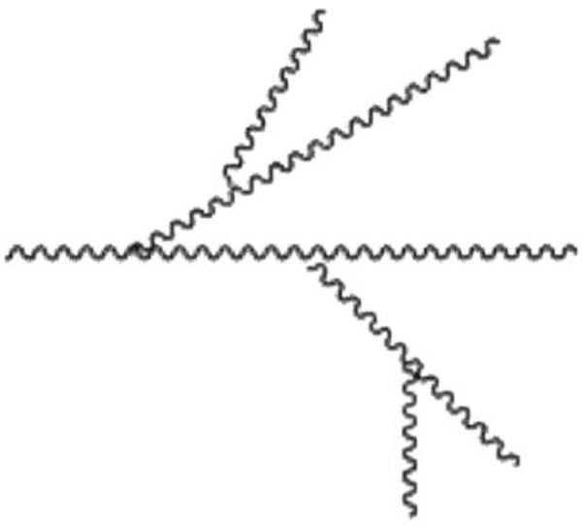 | • Lower crystallinity<br>• Lower contraction factor<br>• Lower intrinsic viscosity | Wallach and Huang (2000), Fischer *et al.* (2012) |
| Dumbell shaped 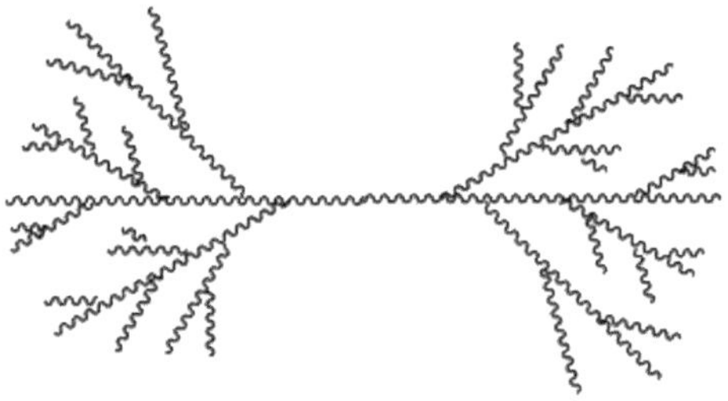 | • Lower $T_g$, $T_m$, $T_c$<br>• Lower crystallinity<br>• Decreased hydrophobicity | Han and Pan (2006), Gong *et al.* (2009) |

*Source:* Reproduced from Corneillie S, Smet M. *Polym Chem.* 2015; 6(6): 850–867. With permission.

## 7.2 NEED FOR THE MODIFICATION OF PLA

One serious challenge for PLA, when bringing it for biomedical applications, is associated with its solubility, as PLA is hydrophobic in nature. For instance, for hydrophobic drug-delivery applications, the hydrophobicity of PLA enhances hydrophobic drug loading into NPs, and, at the same time, because of the hydrophobicity, NPs rapidly get excreted from the blood circulation. This consequently decreases the drug efficiency *in vivo*. Functionalizing the PLA surface with any hydrophilic polymers is one of the promising approaches to achieve long circulation of PLA-based NPs in the blood, which has been facilitated by synthesis of well-controlled PLA-containing amphiphilic block copolymers (ABCs). These ABCs self-assemble in water to form micellar structures, where the hydrophobic core will contain the hydrophobic therapeutic molecule and the hydrophilic shell ensures water-solubility and biocompatibility of the NPs (Oh, 2011). In addition to the drug-delivery applications, PLA-based ABCs have been explored for the development of 3D scaffolds and cross-linked hydrogels for tissue-engineering applications and self-assembled metal-hybrid nanomaterials as a bioimaging platform. Another challenge with PLA is the need for bioconjugation with ligand biomolecules for targeting specific cells. This has also been achieved by PLA-based ABCs. The functional groups that have been explored include carboxylic acid (–COOH), amino (–$NH_2$), and hydroxyl (–OH) groups, which have been utilized for bioconjugation with targeting biomolecules to achieve active targeting of specific cells. The challenge of PLA-based biomaterials for tissue-engineering applications is that they do not have any integrin binding moieties to support the cell growth, though it is strong to mimic the mechanical properties of natural extracellular matrix (ECM) (Balavigneswaran *et al.*, 2017). This can be achieved by attaching some hydrophilic polymer into PLA so that protein adsorption will be increased; this increased protein adsorption property will help the cells to adhere into the matrix (Place *et al.*, 2009; Santoro *et al.*, 2016).

### 7.2.1 Poly(Ethylene Glycol) (PEG)

PEG is one of the FDA-approved polymers for clinical use and is mostly used as an effective method of improving the hydrophilic nature in biomedical science because of its prolonged blood circulation and diminished reticulo-endothelial system (RES) uptake (Oh, 2011). The common route of synthesis (Figure 7.6) is ROP of lactide by PEG to prepare PEG-containing PLA-based ABCs (Zhu *et al.*, 1990). Several other approaches have also been proposed for the facile synthesis of PLA-*b*-PEG-based ABCs, functionalized with various biomolecules. In one study, Shi *et al.* (2009) proposed to synthesize hemoglobin-functionalized PLA-*b*-PEG in which they synthesized triblock copolymer consisting of mPEG, pendent propargyl-containing polymer as a middle block, and PLA. These propargyl groups were further used for conjugation with azido-functionalized hemoglobin molecules through click-type Cu(I)-mediated Huisgen 1,3-dipolar cycloaddition as the artificial oxygen carrier.

### 7.2.2 Vinyl Polymers

#### 7.2.2.1 Poly(Vinylpyrrolidone) (PVP)

Poly(vinylpyrrolidone) (PVP) is a commercially available polymer in the brand name of Kollidon® from BASF. It has been used for tablet coating, as a binder, and as an excipient for the formulation

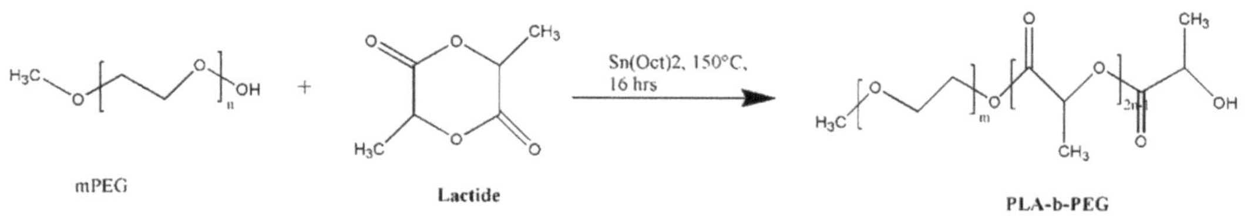


**FIGURE 7.6** General methods for the synthesis of PLA-b-PEG.

of poorly water-soluble drugs in the cosmetic and pharmaceutical industries. As a food additive, PVP has been used as a stabilizer (Torchilin, 1998). Possessing a highly hydrated structure, this polymer helps to increase the hydrophilicity of the other attached polymer (Stach *et al.*, 2008).

PVP can be synthesized by free-radical polymerization or by controlled radical polymerization (CRP) methods (Stach *et al.*, 2008). CRP methods follow the mechanism based on rapid dynamic equilibration between a small concentration of growing radicals and a majority of dormant species. Consequently, all the polymeric chains grow simultaneously and thus form the polymer with uniform distribution of molecular weights. These methods facilitate the synthesis of well-defined copolymers with narrow molecular weight distribution ($M_w/M_n < 1.5$) (Ray *et al.*, 2006), various chain architectures (block, gradient, graft, hyper-branched, and star), and facile functionalities. Examples of successful CRP methods include atom transfer radical polymerization (ATRP) (Wang and Matyjaszewski, 1995) and reversible addition-fragmentation chain transfer (RAFT) polymerization (Chiefari *et al.*, 1998; Luo *et al.*, 2004; Ray *et al.*, 2006). There are only a few reports on PLA-*b*-PVP (Le Garrec *et al.*, 2004; Luo *et al.*, 2004; Gaucher *et al.*, 2007, 2009; Ramesh *et al.*, 2012; Shin *et al.*, 2014). In general, PLA-*b*-PVP can be prepared in four steps (Figure 7.7): (i) ROP of lactide produces target molecular weight PLA, (ii) conversion of PLA–OH to the corresponding PLA–Br takes place, (iii) end group bromine gets converted to xanthate (PLA-X), and (iv) xanthate functionalized macro-chain transfer agent is used to polymerize vinylpyrrolidone and finally to

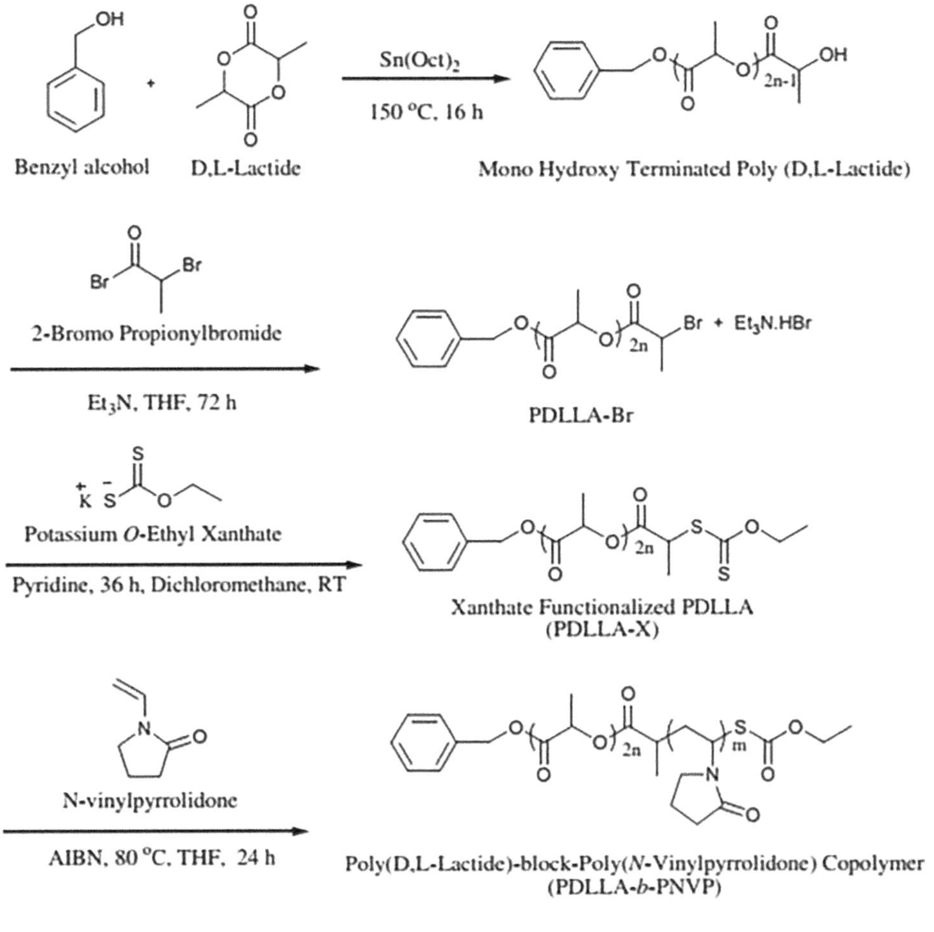


**FIGURE 7.7** Synthesis of PLA-b-PVP. (Reproduced from Ramesh *et al.*, *Polymer.* 2012; 53(25): 5743–5753. With permission.)

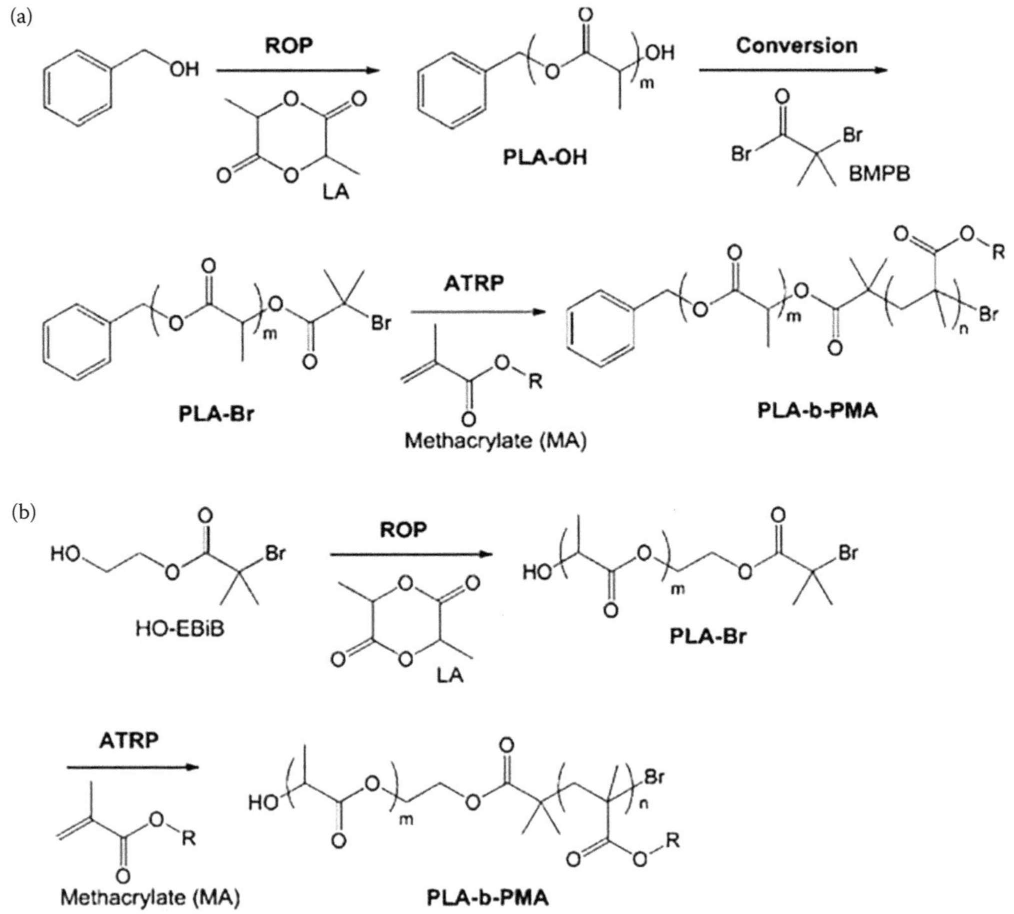


**FIGURE 7.8** A common method for the synthesis of PLA-b-PMA: (a) Synthesis involved ROP of PLA; conversion to PLA-Br, and Atom transfer radical polymerization (ATRP). (b) Synthesis involved the use of 2-hydroxyethyl-2-bromoisobutyrate (HO-EBiB). (Reproduced from Oh JK. *Soft Matter.* 2011; 7(11): 5096–5108. With permission.)

produce PLA-*b*-PVP polymer. Recently, one pot synthesis of PLA-b-PVP also has been reported (Shin *et al.*, 2014). The hydroxyl-functionalized N,N-diphenyldithiocarbamate and 2-hydroxyethyl 2-(N,N-diphenylcarbamothioylthio) propanoate (HDPCP) were used as the reversible addition–fragmentation chain transfer (RAFT) agent and dual initiator for RAFT polymerization, respectively, and ROP was achieved in a single-step procedure at 60°C.

### 7.2.2.2 Poly(Meth)Acrylates

Poly(meth)acrylates are often known for being nonimmunogenic, highly protein-resistant and non-cytotoxic (You *et al.*, 2007). Hence, it could be used as a hydrophilic molecule to modify PLA. CRP is a common polymerization technique to prepare PLA-based ABCs comprising hydrophilic poly(meth)acrylates (Davis and Matyjaszewski, 2002; Matyjaszewski and Davis, 2002; Oh, 2011). ATRP has been extensively explored for chain extension of PLA with hydrophilic polymers such as poly(meth)acrylate because of their applicability for the synthesis of well-controlled PLA-containing ABCs. A common three-step synthesis procedure, as explained in Figure 7.8, includes (i) ROP of lactide (LA) to produce PLA–OH, (ii) conversion of PLA–OH to the –Br terminated PLA (ATRP macroinitiator), and (iii) final ATRP reaction of hydrophilic monomers in the presence of –Br terminated PLA.

### 7.2.3 Poly(Amino Acids)

Poly(glutamic acid) (PGA), poly(lysine) (PLL), poly(cysteine), and poly(hydroxyethyl-l-asparagine) (PHEA), and poly(hydroxyethyl-l-glutamine) (PHEG) are the most commonly used poly(amino acids) for various drug-delivery applications (Romberg *et al.*, 2007; Oh, 2011). PGA, PLL, and PHEG are biodegradable polymers, and the degraded products are metabolized further by physiological pathways *in vivo*. The *in vitro* degradation kinetics show complete decomposition into single amino acids or degradation to oligomers with 4–9 repeating units in the presence of different enzymes (Pytela *et al.*, 1989; Romberg *et al.*, 2007). Biodegradability is the main advantage of poly(amino acids) being used as a surface-functionalizing moiety, and this enables prolonged blood circulation.

Pendant functional groups of poly(amino acids) have been found to be carboxylic acid (−COOH) for aspartic acid and glutamic acid, thiol (−SH) for cysteine, and amine (−$NH_2$) for lysine and are useful for functionalizing with the other biocompatible polymer, either by bioconjugation or shell-cross-linking. This unique property made them useful as a hydrophilic coating to hydrophobic PLA NPs. Synthesis of PLA-based poly(α-amino acids) always require the terminal protected α-amino acids to avoid the side reactions of −$NH_2$ and −COOH groups (Oh, 2011).

A general synthesis method involves three steps: (i) synthesis of $NH_2$–PLA, (ii) ROP of α-amino acid N-carboxy anhydride (NCA), and (iii) removal of NCA groups via hydrolysis or catalytic hydrogenation. As an example, the synthesis of PLA-*b*-poly(aspartic acid) (PLA-*b*-PAsp) is illustrated in Figure 7.9 (Arimura *et al.*, 2004). Ouchi *et al.* (2003b) reported the synthesis of PLA-*b*-PAsp, and the size of micelles they produced from the synthesized polymer was about 50 nm. PLA and poly(glutamic acid)-based ABCs have also been synthesized by Deng *et al.* (2005), in which they prepared triblock polymer of mPEG-*b*-PLA-*b*-poly(L-glutamic acid). In addition, Deng *et al.* (2007a, used the same approach to synthesize mPEG-*b*-PLA-*b*-poly(L-glutamic acid), and they then modified the triblock copolymer using the pendant −COOH group to further functionalize it with

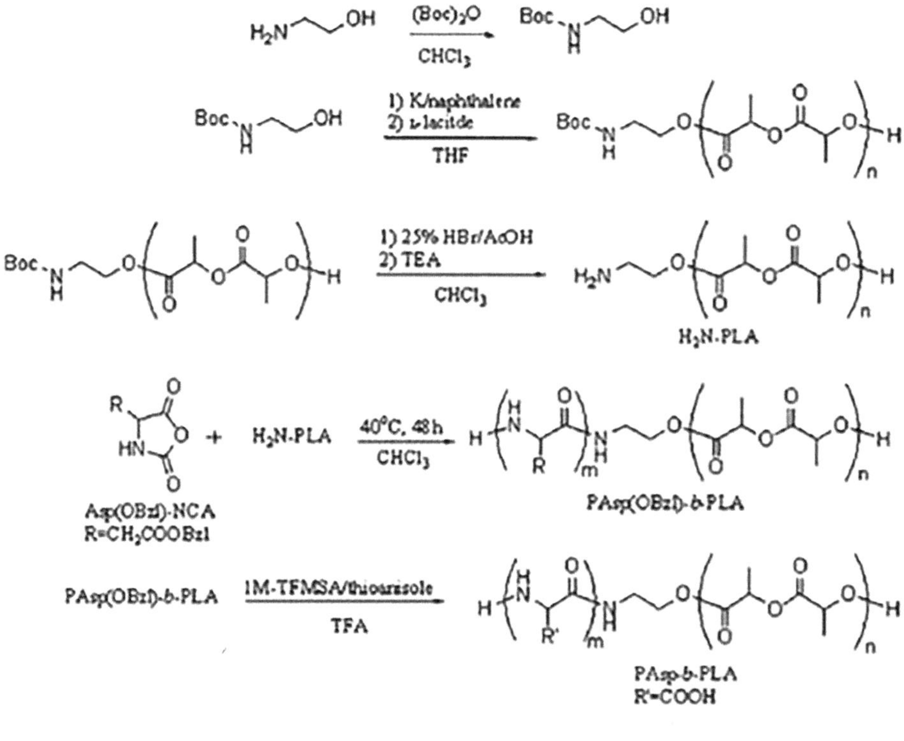


**FIGURE 7.9** Synthesis of PLA-b-PAsp. (Adapted from Arimura *et al.*, *Macromol Rapid Commun.* 2004; 25(6): 743–747. With permission.)

Arginylglycylaspartic acid (RGD) peptide to make it useful for targeted drug-delivery applications. Synthesis of mPEG-*b*-PLA-*b*-poly(L-lysine) (Deng *et al.* (2007b)) and PLA-*b*-poly(L-cysteine) (Sun *et al.*, 2008) has also been reported.

### 7.2.4 Polysaccharides

Polysaccharides such as cellulose are natural biopolymers that are abundant in nature; intrinsically, they are renewable and biodegradable. Amphiphilicity of PLA can be achieved by these polysaccharides, either by chemical grafting of polysaccharides into PLA using ethyl cellulose (Yuan *et al.*, 2007), cellulose (Mayumi *et al.*, 2006), cellulose fiber (Lönnberg *et al.*, 2006), hydroxypropyl cellulose (Östmark *et al.*, 2008), pullulan (Ohya *et al.*, 1998a ), amylose (Ohya *et al.*, 1998b), dextran (Ouchi *et al.*, 2003a), or hyaluronic acid (HA) (Pitarresi *et al.*, 2010) or by simple coating using any surfactant. Grafting of polysaccharides into PLA can be accomplished through three steps: (i) silylation of hydroxyl groups, (ii) ROP of LA using the hydroxyl groups, and finally (iii) silylether deprotection. These ABCs have been used as biodegradable films and fibers for biomedical applications (Raynaud *et al.*, 2008; Nouvel *et al.*, 2009).

## 7.3 FUNCTIONAL PLA FOR TISSUE-ENGINEERING APPLICATIONS

Materials-based tissue-engineering strategies consists of three components: a 3D scaffold that mimics the natural ECM, cells, and biological growth factors (this helps in the growth of cells inside the matrix). In the tissue regeneration process, 3D scaffolds play an important role in performing cellular processes (cell adhesion, proliferation, migration, and differentiation) by serving as a 3D template. Natural cellular processes are duly explained in Figure 7.10. Therefore, designing scaffolds should consider the characteristics of (i) biocompatible and biodegradable, which further does not need surgery after implantation at the targeted site, (ii) a proper surface chemistry to facilitate cell-material interactions, (iii) a 3D architecture which accommodates cell proliferation and further ECM deposition, (iv) an inter-connected pore network for exchanging nutrients and other wastes, and (v) an appropriate mechanical strength to support tissue growth (Ma, 2008). Recently, researchers are mostly focusing on the design and functionalization of scaffolds that can uphold the desirable cell-material interactions to guide neogenesis of tissue (Kohn, 2004; Langer and Tirrell, 2004; Lutolf and Hubbell, 2005).

Thermally-induced phase separation (TIPS), freeze drying, and electrospinning are the most common techniques used for the fabrication of 3D scaffolds. In the TIPS technique, polymers are initially dissolved in a suitable solvent and kept at a lower temperature to make them thermodynamically unstable, and thus the polymer spontaneously will be separated into two phases and the polymer-rich phase forms the nanofibrous matrix. Further, the polymer-lean phase is extracted under the proper conditions. This would produce 3D-nanofibrous structures (Ma and Zhang, 1999). Freeze drying involves freezing the polymer in a proper mold and further drying it in $-50°C$ to $-80°C$ to produce the scaffolds. Electrospinning is one of the well-established techniques used to fabricate fibrous scaffolds, whose pore size and fiber diameter can be easily modified by changing voltage, flow rate, concentration of polymer, and distance between substrate and nozzle. However, individual properties cannot be independently tuned (Pham *et al.*, 2006).

Researchers have studied the pendant –COOH-functionalized PLLA and found that the functionalization increased the hydrophilicity of the surface; thus the cells seeded on –COOH-modified PLLA scaffolds had higher adhesion and proliferation rates and spread well-over a larger area compared to unmodified scaffolds (Park *et al.*, 2007). Functionalizing scaffolds with the integrin-binding moieties will increase the cell-material interactions, since these bioactive molecules are the constituents of natural ECM proteins. Often, gelatin, which is a denatured form of collagen, has been used to alter the surfaces of 3D PLLA scaffolds (Liu *et al.*, 2005a,b; Liu *et al.*, 2006). The gelatin-modified scaffolds greatly increase the adhesion, proliferation, and spreading of MC3T3-E1 cells

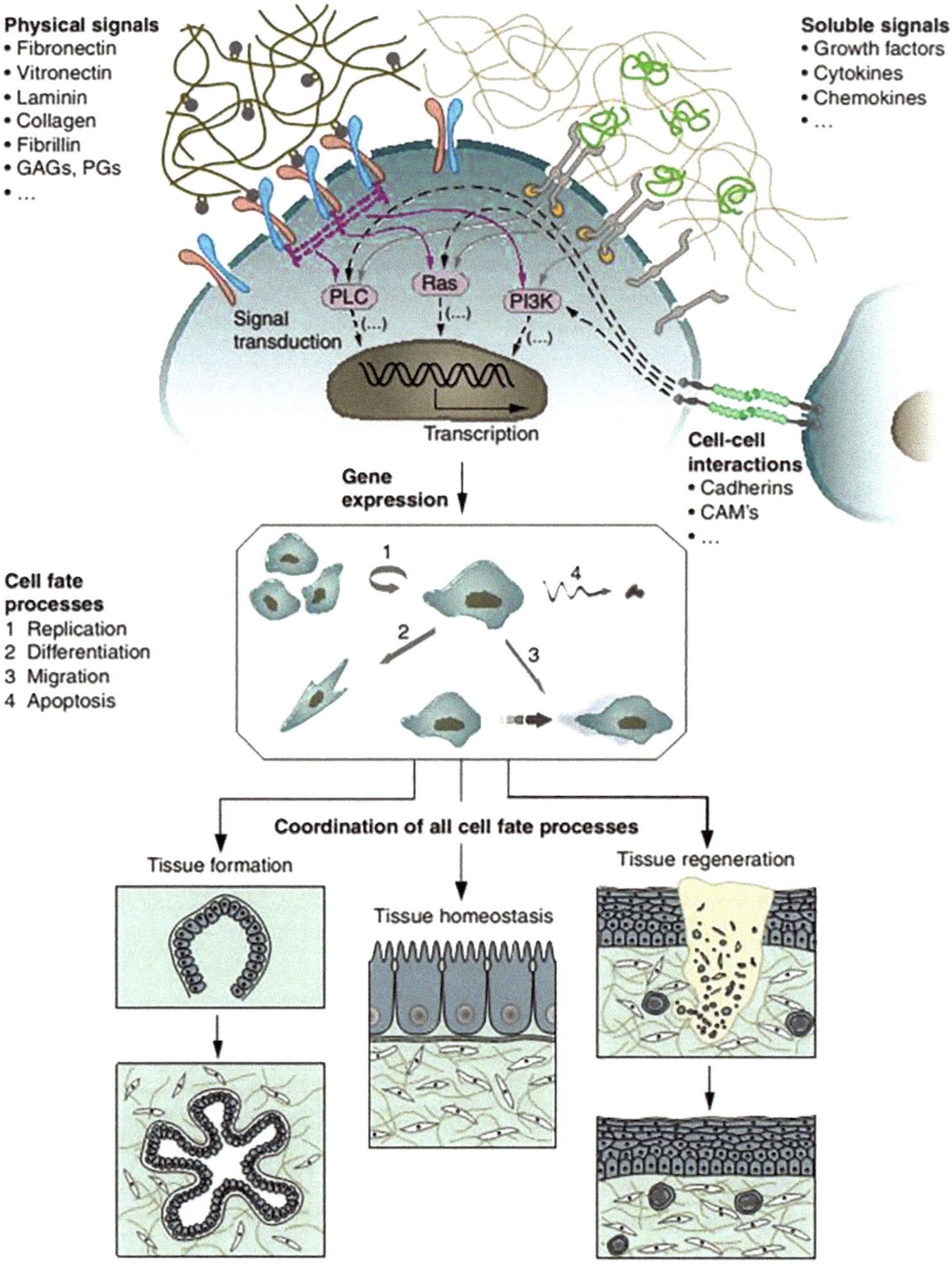


**FIGURE 7.10** Dynamicity of the natural cells in the tissue regeneration process. (Adapted from Lutolf M, Hubbell J. *Nat Biotechnol*. 2005; 23(1): 47–55. With permission.)

and collagen deposition than on the control scaffolds (Liu *et al.*, 2005b). Recently, we have shown the improved cell proliferation on the synthesized gelatin-grafted star-shaped PDLLA compared to unmodified star-shaped PDLLA (Figure 7.11) (Balavigneswaran *et al.*, 2017). Hydroxyappetite (HA)-incorporated PLLA/HA scaffolds showed significantly increased osteoblast adhesion and proliferation compared to unmodified PLLA scaffolds (Ma *et al.*, 2001). In addition to adhesion and proliferation, functionalized scaffolds have also been used to direct cell differentiation. Liu *et al.* (2009) have reported incorporation of HA on gelatin scaffolds for differentiation of MC3T3-E1 preosteoblasts

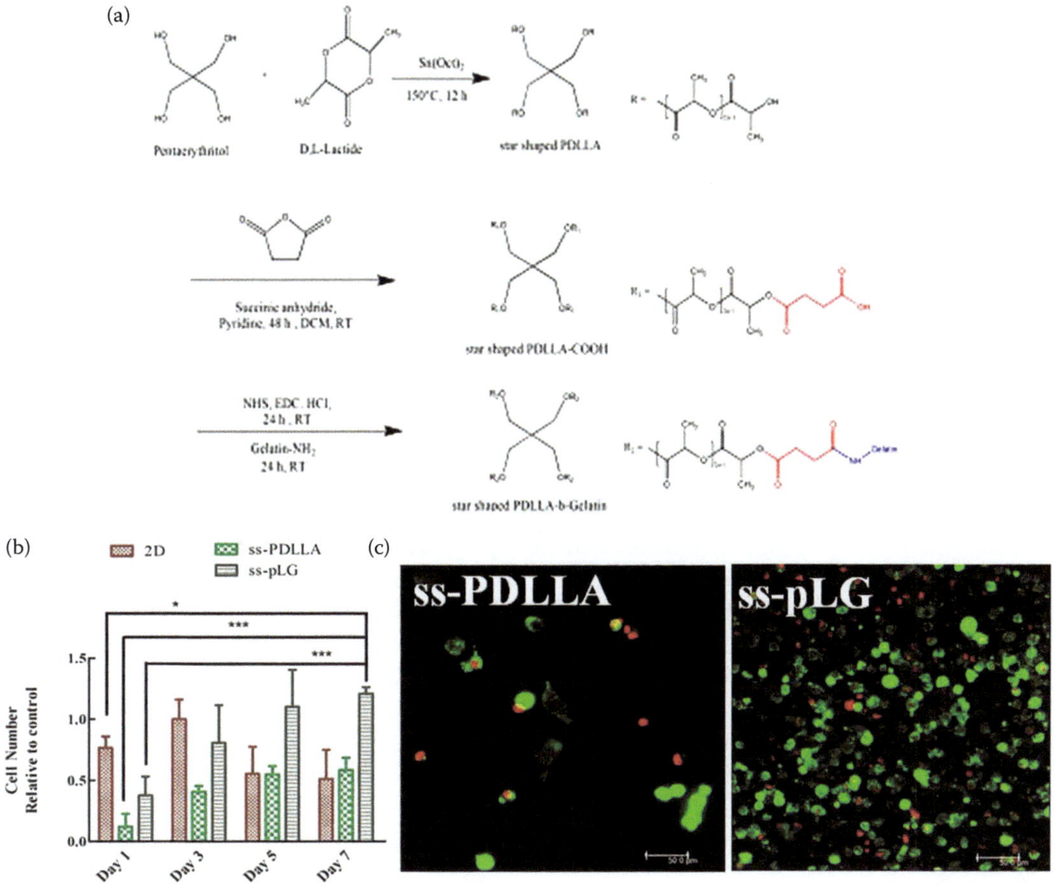


**FIGURE 7.11** (a) Synthesis scheme of gelatin-grafted star-shaped PDLLA; (b) 3T3-L1 cells proliferation within the scaffolds; (c) Live/dead cell staining of 3T3-L1 cells within the scaffolds after 7 days. (Reproduced from Balavigneswaran *et al.*, *Bioconjugate Chem.* 2017; 28(4): 1236–1250. With permission.)

into osteoblasts. Incorporation of HA not only helped in cell differentiation and tissue formation but also enhanced the mechanical strength of the scaffold. More frequently, specific peptide fragments also have been used to tailor cell adhesion and growth, where the RGD peptide is the vital integrin-binding sequence, which is ubiquitous in the ECM (Pierschbacher and Ruoslahti, 1983). Change of cell fate of human mesenchymal stem cells (MSCs) by different functional groups of hydrogel has been shown by Benoit *et al.* (2008). It has been found that carboxylic acid groups increase the expression of aggrecan, which is an indicator of chondrogenic differentiation. Phosphate groups have been found important for mineralized tissue formation, which increases expression of CBFA1, a marker for osteogenic differentiation. The t-butyl groups, mimicking lipid-filled adipocytes, increase the expression of peroxisome-proliferating antigen-receptor gamma (PPARG), which is a measure of adipogenic differentiation.

In general, cell differentiation can be achieved by delivering growth factors from scaffolds where these growth factors induce tissue neogenesis by guiding the cell fate via their cell-matrix interactions, followed by cell differentiation. Various methods have been employed to incorporate growth factors into scaffolds (Richardson *et al.*, 2001; Willerth *et al.*, 2008; Thomopoulos *et al.*, 2016). In a study, recombinant human bone morphogenic protein 7 (rhBMP-7) was encapsulated into PLGA microspheres, and the microspheres were incorporated into a nano-fibrous PLLA scaffold (Thomopoulos *et al.*, 2016). The rhBMP-7 microsphere-incorporated PLLA scaffolds significantly increased dentin regeneration. Ability of these small molecules to induce differentiation is an easier and safer way to guide the fate of stem cells (Benoit *et al.*, 2008; Thomopoulos *et al.*, 2016). In another recent study, epidermal growth factor (EGF) was encapsulated into PLLA scaffolds and used for the regeneration of intestine (Liu *et al.*, 2016).

Natural polysaccharide-grafted PLLA-based hydrogels have also been prepared and studied for the biomedical applications. De Jong *et al.* (2001) prepared nanogels of dextran-*g*-PLLA and dextran-*g*-PDLA with a diameter of 70 nm in water. They found that the degree of polymerization and substitution of oligo-LA affected the mechanical strength of the hydrogels.

## 7.4 PLA-BASED SMART BIOMATERIALS

### 7.4.1 Stimuli-Responsive PLA

Living creatures have the art of responding to various external stimuli naturally. For example, with the leaves of *Mimosa pudica,* nature is changed immediately when touched; Venus 20 flytrap closes fast enough to catch its prey; sea cucumbers change their stiffness in the face of danger; chameleons change colors according to the nature of the environment; and sunflowers follow the movement of the sun (Zhai, 2013). In the above-mentioned view, polymeric materials that can respond to changes with respect to the stimulated parameters are of particular importance for drug- and gene-delivery applications, and they are referred to as smart polymers or stimuli-responsive polymers (Marek, 2011). These polymers can respond to stimuli in several ways by physical (light, temperature) or chemical (pH, organic molecules, biomolecules) change. The degree of response of such polymers can be triggered and controlled by the intensity of applied stimuli (Brunsveld *et al.*, 2001; Liu and Urban, 2010). PLLA/PDLLA can be made stimuli responsive by chemically coupling with copolymers like PEG, poly(amino acids), PNVP and PNIPAM.

### 7.4.2 pH-Responsive PLA

The extracellular pH of tumor cells is more acidic (pH 6.5–7.2) than normal cells. pH-sensitive polymeric carriers have been widely used for tumor-targeted drug delivery based on the intrinsic pH difference (Engin *et al.*, 1995). Therefore, these pH-sensitive drug carriers are unique strategies, which trigger the release of drug/biomolecule at the targeted site.

In general, well-designed, pH-responsive polymeric nanocarriers should have high drug-loading efficiency, long-term stable circulation in the blood, respond rapidly to an acidic pH, and subsequently release the drugs/small molecules at the targeted site. For example, poly(β-amino esters) are soluble in water below pH 6.5 and insoluble above pH 6.5 because the tertiary amine's negative base-10 logarithm of the base dissociation constant ($pK_b$) is 6.5. Synthesis of pH-sensitive poly(β-amino esters) has been reported via Michael-type polymerization, and their degradability and cytotoxicity have been investigated (Lynn and Langer 2000; Potineni *et al.*, 2003). Later, self-assembling pH-sensitive poly(ethylene glycol)-*b*-poly(β-amino esters) were synthesized and evaluated for the camptothecin (CPT) and doxorubicin (DOX) release from the pH-responsive carrier (Ko *et al.*, 2007; Min *et al.*, 2010). However, these carriers showed the drawback in the lower drug-loading content immediate release of the drug, which needed further modification. Therefore, Zhang *et al.* (2012) introduced PLA block into the poly(β-amino esters) group to control the release the hydrophobic drug, DOX, by the change of pH.

Polyelectrolyte-coupled PLA also has been shown to be a pH-responsive carrier. Li *et al.* (2015) synthesized the stereo-mixtures D:L (1:1) of poly(N,N-dimethyl aminoethyl methacrylate) (PDMAEMA)-*b*-poly(D-lactic acid)-*b*-poly(N,N-dimethyl aminoethyl methacrylate) conjugated with poly(ethylene glycol). They showed increase in the size of micelles when the solution pH was reduced. Increase in size was observed due to the protonation of the PDMAEMA segments, which further pushes the osmotic pressure within the micelle. In other work by Hira *et al.* (2015), PVP

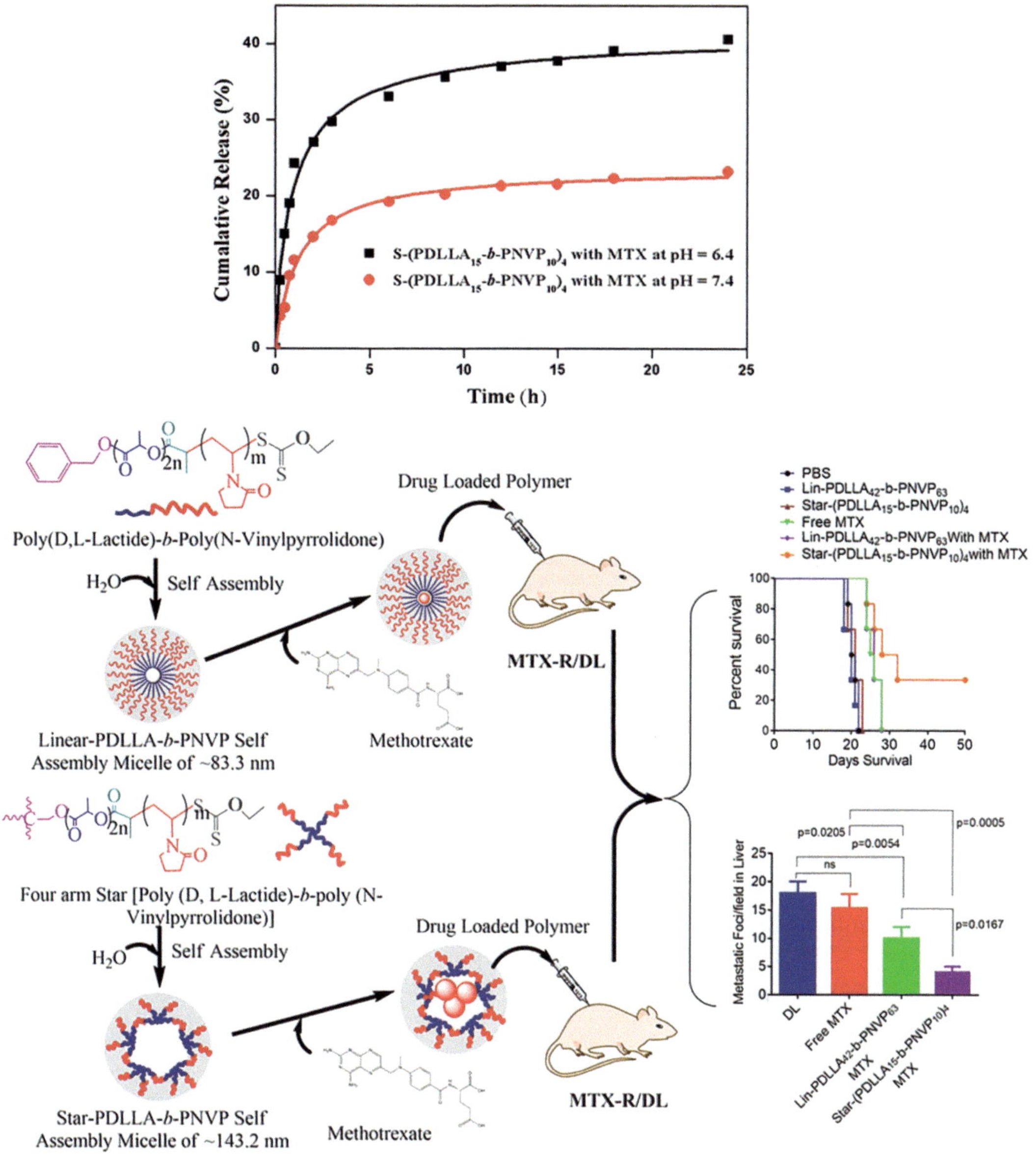


**FIGURE 7.12** MTX release from star PDLLA-b-PNVP and the efficacy of formulation *in vivo*. (Reproduced from Hira *et al.*, *ACS Appl Mater Interfaces*. 2015; 7(36): 20021–20033. With permission.)

was used as the hydrophilic group along with PLA blocks and used as the pH-responsive release of methotrexate (MTX) self-assembled micelles. Importantly, star-shaped PDLLA-*b*-PNVP showed the controlled release of MTX over a linear polymer system. *In vivo* studies also showed better drug efficiency from the star-based system (Figure 7.12).

### 7.4.3 Temperature-Sensitive PLA

In terms of thermo-sensitive materials, two types of carrier systems have been studied: (i) thermo-responsive liposomes (Weinstein *et al.*, 1979) and (ii) thermo-responsive polymeric micelles (Chung *et al.*, 2000). Polymeric thermo-responsive micelles are advantageous because they can incorporate water-insoluble drugs and many potent drugs that are insoluble in water, whereas the thermo-responsive liposomes are not suitable for the incorporation of water-insoluble drugs. Thermo-responsive polymers will contain a thermo-sensitive hydrophilic block and hydrophobic blocks, which will self-assemble and form the micelles. Drugs/small molecules will be incorporated into the hydrophobic inner core and the thermo-responsive character is possessed by the outer shell of the polymeric micelles. The micelles get formed when the temperature is below the phase-transition temperature of the thermo-responsive block, and when the temperature exceeds the phase-transition temperature, the drug gets released. Formation of micelles was based on the critical micelle concentration and the strength of the shrunken outer shell layer's hydrophobic interactions (Nakayama *et al.*, 2006).

Poly (N-isopropylacrylamide) (PNIPAAm) is the most widely studied synthetic temperature-responsive polymer and exhibits a thermally reversible phase transition at 32°C. This transition temperature is called as lower critical solution temperature (LCST). Below LCST, the polymer chains of PNIPAAm hydrate and expand, but above LCST, they dehydrate to form a compact, insoluble conformation (Yoshida *et al.*, 1995; Kanazawa *et al.*, 1996). This hydration/dehydration changes of PNIPAAm polymer chain groups can be utilized for stimuli-responsive drug-delivery systems. These hydrophilic/hydrophobic changes of NPs can be also used for passive drug targeting (Ayano *et al.*, 2012).

For the appropriate drug targeting in physiological temperature, the phase-transition temperature of PNIPAAm was modified by introducing a hydrophilic co-monomer such as N,N-dimethylacrylamide (DMAAm). In this view, scientists have synthesized poly(N-isopropylacrylamide-co-N, N-dimethylacrylamide)-*b*-poly(DL-lactide) (Kohori *et al.*, 1999; Chung *et al.*, 2000; Kohori *et al.*, 2002). The release property was further improved by hydrophilic betamethasone disodium 21-phosphate (BP)-encapsulated NPs formed from a blend of PLA homopolymers and PNIPAAm–PLA block copolymers (Ayano *et al.*, 2012). Dual pH- and temperature-responsive NPs also were studied for drug-delivery applications. Herein, the researchers have covalently linked the thermo-responsive and pH-sensitive block along with the biodegradable block PLA. In such work, mPEG-*b*-P(HPMALac-co-His) was synthesized and studied for their dual responsiveness of the NPs, made up from the respective polymer in which histidine was used as the pH-sensitive block and methacrylamide was used as the temperature-sensitive block (Chen *et al.*, 2012). In another work, Mao *et al.* (2011) have used poly(2-(dimethylamino)ethyl methacrylate) (PDMAEMA) as both pH- and temperature-sensitive block. They synthesized PLLA-b-PDMAEMA and studied the size changes at varying pH and temperature.

### 7.4.4 Smart Hydrogels for Tissue-Engineering Applications

In advanced therapeutic technology, an improved system to the conventional technology is anticipated, which could be used for tissue-engineering applications. Smart hydrogel system is one of the advanced technologies, where the polymer will form gel by the stimulus, such as temperature, light, etc.

Photo-induced hydrogels are produced using a light-sensitive cross-linker, which covalently couples with other monomers, producing a hydrogel. Dimethacrylates are the common photo-linkers, which are used to couple with PLA-based ABCs. Dimethacrylate derivatives of PLA-based ABCs can be physically gelled in aqueous solution through either sol–gel transition or stereo-complexation. The resulting physical gels form permanent cross-linking through photo-polymerization and increase the toughness of hydrogels (Hiemstra *et al.*, 2007; Censi *et al.*, 2010). On the other hand, the dimethacrylate derivatives have also been used as chemical cross-linkers to copolymerize with other monomers like NIPAM, whereas PNIPAM-based scaffolds have been widely used for tissue-engineering applications (Galperin *et al.*, 2010). PLA-based dimethacrylate and a biotinylated PEG methacrylate have also been reported to enhance cell-material interactions (Clapper *et al.*, 2008).

An injectable cell-delivery system is one of the advanced technologies in which polymer solutions that contain live cells could be injected at the defect site to form a temporary 3D scaffold for cell growth and tissue repair/regeneration. Thermo-responsive biodegradable polymers are one of the types, and the property of sol-to-gel transition point between room temperature and physiological temperature can be used for biomedical applications, as the polymer solution would be in a sol state at room temperature, and it would turn into gel *in situ*, once exposed to physiological temperature, without any toxic cross-linkers. An ideal injectable polymer system should exhibit immediate gelation in response to physiological temperature and should have high mechanical strength, which should match the injection site tissues. Also, cells entrapped in the hydrogel should not be released before gel formation. Specific control of the 3D-network structure and porosity inside the hydrogel are the required crucial parameters for the injectable hydrogels (sponges/foams like scaffolds) to guide the cellular functions. Importantly, such a thermo-gelling polymer will also have the ability to trap water-soluble pharmaceuticals or bioactive agents, such as proteins, and exhibit sustained release after injection at a target site. Probable applications of such a technology would be for the treatment of congenital diseases, as these patients cannot typically undergo complex surgical procedures (Lubiatowski *et al.*, 2006; Nagahama *et al.*, 2008).

Agrawal *et al.* (2008) proposed the *in situ* formation of cross-linked hydrogels in water (>20 wt%), where ABCs contained PEG and PLA. Such a hydrogel system has been explored for tissue-engineering applications. Nagahama *et al.* (2008) synthesized 8-Arm PEG-*b*-PLLA and conjugated cholesterol with 8-Arm PEG-*b*-PLLA and studied the formation of self-assembled temperature-induced hydrogels as injectable scaffolds for tissue engineering. Molecular weight and composition of block copolymers are the key parameters that affect sol–gel transition temperature and gelation (Lee *et al.*, 2006). Also, dual-responsive (pH and temperature) P(CL-co-LA)-b-PEG-b-P(CL-coLA) terminated with sulfamethazine end group-based hydrogels have also been produced and studied for their gelation behavior (Shim *et al.*, 2006).

## 7.5 SUMMARY AND CONCLUSION

Hyrophobic aliphatic polyester polylactide (PLA) is a vital synthetic biomaterial since its approval for clinical use by the FDA, being biocompatible and biodegradable by enzymes and hydrolysis under physiological conditions and having low immunogenicity and good mechanical properties. Most importantly, it is bio-based, where lactic acid is the building block of PLA, a byproduct of the glycolysis pathway in humans and mammals. In case of drug-delivery applications, one serious challenge for PLA is associated with their solubility. Hydrophobicity of PLA enhances hydrophobic drug loading onto NPs, but such NPs get rapidly excreted from the blood circulation because of their hydrophobicity, thus decreasing drug efficiency *in vivo*. Surface coating or covalent coupling of hydrophilic polymers onto the hydrophobic surface is a promising approach to achieve long circulation of NPs in the blood stream. Hence, the common methods for synthesis of PLA-based ABCs with PEG, PNIPAM, PMMA, PVP, polysaccharides, and polypeptides have been discussed and further explained for their stimuli-responsive drug release. In case of tissue-engineering applications, PLA-based biomaterials face the major problem of cell attachment. Therefore, various approaches have been discussed for modifying PLA to make it usable for tissue-engineering applications. Examples for PLA modifiers include dextran, gelatin, chitosan, and RGD-based moieties. Physical and chemical cross-linking methods have also been elaborated upon including temperature-induced sol–gel transition, ionic interactions, and photo-induced cross-linking polymerization to fabricate 3D-cross-linked hydrogels for tissue-engineering applications.

## ACKNOWLEDGMENTS

The authors gratefully acknowledge the financial support from the Ministry of Human Resource and Development (MHRD), New Delhi, India, in the form of a research fellowship. The authors also acknowledge the Indian Institute of Technology (BHU) for providing the infrastructure facility for our nanoparticulate research.

## REFERENCES


Agrawal SK, Sanabria-DeLong N, Tew GN, Bhatia SR. Structural characterization of PLA-PEO-PLA solutions and hydrogels: Crystalline *vs* amorphous PLA domains. *Macromol.* 2008; 41(5): 1774–1784.

Arimura H, Ohya Y, Ouchi T. The formation of biodegradable polymeric micelles from newly synthesized poly (aspartic acid)-block-polylactide AB-type diblock copolymers. *Macromol Rapid Commun.* 2004; 25(6): 743–747.

Ayano E, Karaki M, Ishihara T, Kanazawa H, Okano T. Poly (N-isopropylacrylamide)–PLA and PLA blend nanoparticles for temperature-controllable drug release and intracellular uptake. *Colloids Surf B.* 2012; 99: 67–73.

Balavigneswaran CK, Mahto SK, Subia B, Prabhakar A, Mitra K, Rao V *et al.* Tailored chemical properties of 4-arm star shaped poly (d, l-lactide) as cell adhesive three-dimensional scaffolds. *Bioconjugate Chem.* 2017; 28(4): 1236–1250.

Benoit DS, Schwartz MP, Durney AR, Anseth KS. Small functional groups for controlled differentiation of hydrogel-encapsulated human mesenchymal stem cells. *Nat Mater.* 2008; 7(10): 816–823.

Brunsveld L, Folmer B, Meijer E, Sijbesma R. Supramolecular polymers. *Chem Rev.* 2001; 101(12): 4071–4098.

Cameron DJ, Shaver MP. Aliphatic polyester polymer stars: Synthesis, properties and applications in biomedicine and nanotechnology. *Chem Soc Rev.* 2011; 40(3): 1761–1776.

Carlotti SJ, Giani-Beaune O, Schué F. Characterization and mechanical properties of water-soluble poly (vinyl alcohol) grafted with lactic acid and glycolic acid. *J App Polym Sci.* 2001; 80(2): 142–147.

Carothers WH. Polymers and polyfunctionality. *Trans Faraday Soc.* 1936; 32: 39–49.

Celli A, Scandola M. Thermal properties and physical ageing of poly (L-lactic acid). *Polymer.* 1992; 33(13): 2699–2703.

Censi R, Vermonden T, Deschout H, Braeckmans K, di Martino P, De Smedt SC *et al.* Photopolymerized thermosensitive poly (HPMAlactate)-PEG-based hydrogels: Effect of network design on mechanical properties, degradation, and release behavior. *Biomacromolecules.* 2010; 11(8): 2143–2151.

Chen YC, Liao LC, Lu PL, Lo CL, Tsai HC, Huang CY *et al.* The accumulation of dual pH and temperature responsive micelles in tumors. *Biomaterials.* 2012; 33(18): 4576–4588.

Chiefari J, Chong Y, Ercole F, Krstina J, Jeffery J, Le TP *et al.* Living free-radical polymerization by reversible addition-fragmentation chain transfer: The RAFT process. *Macromol.* 1998; 31(16): 5559–5562.

Chung JE, Yokoyama M, Okano T. Inner core segment design for drug delivery control of thermo-responsive polymeric micelles. *J Control Release.* 2000; 65(1): 93–103.

Clapper JD, Pearce ME, Guymon CA, Salem AK. Biotinylated biodegradable nanotemplated hydrogel networks for cell interactive applications. *Biomacromolecules.* 2008; 9(4): 1188–1194.

Cooper TR, Storey RF. Poly (lactic acid) and chain-extended poly (lactic acid)-polyurethane functionalized with pendent carboxylic acid groups. *Macromolecules.* 2008; 41(3): 655–662.

Corneillie S, Smet M. PLA architectures: The role of branching. *Polym Chem.* 2015; 6(6): 850–867.

Davis KA, Matyjaszewski K. *Statistical, gradient, block, and graft copolymers by controlled/living radical polymerizations.* Springer Berlin Heidelberg, USA, 2002 (ISBN: 9783540432449).

De Jong S, De Smedt S, Demeester J, Van Nostrum C, Kettenes-Van Den Bosch J, Hennink WE. Biodegradable hydrogels based on stereocomplex formation between lactic acid oligomers grafted to dextran. *J Control Release.* 2001; 72(1): 47–56.

Deng C, Chen X, Sun J, Lu T, Wang W, Jing X. RGD peptide grafted biodegradable amphiphilic triblock copolymer poly (glutamic acid)-b-poly (L-lactide)-b-poly (glutamic acid): Synthesis and self-assembly. *J Polym Sci A Polym Chem.* 2007a; 45(15): 3218–3230.

Deng C, Chen X, Yu H, Sun J, Lu T, Jing XA. Biodegradable triblock copolymer poly (ethylene glycol)-b-poly (l-lactide)-b-poly (l-lysine): Synthesis, self-assembly, and RGD peptide modification. *Polymer.* 2007b; 48(1): 139–149.

Deng C, Rong G, Tian H, Tang Z, Chen X, Jing X. Synthesis and characterization of poly (ethylene glycol)-b-poly (L-lactide)-b-poly (L-glutamic acid) triblock copolymer. *Polymer.* 2005; 46(3): 653–659.

Doi Y, Steinbüchel A. *Biopolymers, polyesters III-applications and commercial products.* Wiley-Blackwell, weinheim 2002 (ISBN: 3527302255, 9783527302253).

Engelberg I, Kohn J. Physico-mechanical properties of degradable polymers used in medical applications: A comparative study. *Biomaterials.* 1991; 12(3): 292–304.

Engin K, Leeper D, Cater J, Thistlethwaite A, Tupchong L, McFarlane J. Extracellular pH distribution in human tumours. *Int J Hyperthermia.* 1995; 11(2): 211–216.

Fambri L, Migliaresi C, Kesenci K, Piskin E. *Biodegradable polymers. Integrated biomaterials science*, Springer, US, 2002, pp. 119–187 (ISBN: 9781475776829).

Fischer AM, Wolf FK, Frey H. Long-chain branched poly (lactide)s based on polycondensation of AB2-type macromonomers. *Macromol Chem Phy.* 2012; 213(13): 1349–1358.

Galperin A, Long TJ, Ratner BD. Degradable, thermo-sensitive poly (N-isopropyl acrylamide)-based scaffolds with controlled porosity for tissue engineering applications. *Biomacromolecules.* 2010; 11(10): 2583–2592.

Gaucher G, Asahina K, Wang J, Leroux JC. Effect of poly (N-vinyl-pyrrolidone)-block-poly (D, L-lactide) as coating agent on the opsonization, phagocytosis, and pharmacokinetics of biodegradable nanoparticles. *Biomacromolecules.* 2009; 10(2): 408–416.

Gaucher G, Poreba M, Ravenelle F, Leroux JC. Poly (N-vinyl-pyrrolidone)-block-poly (D, L-lactide) as polymeric emulsifier for the preparation of biodegradable nanoparticles. *J Pharm Sci.* 2007; 96(7): 1763–1775.

Gong F, Cheng X, Wang S, Wang Y, Gao Y, Cheng S. Biodegradable comb-dendritic tri-block copolymers consisting of poly (ethylene glycol) and poly (l-lactide): Synthesis, characterizations, and regulation of surface morphology and cell responses. *Polymer.* 2009; 50(13): 2775–2785.

Gupta B, Revagade N, Hilborn J. Poly (lactic acid) fiber: An overview. *Prog Polym Sci.* 2007; 32(4): 455–482.

Han DH, Pan CY. Simple route for synthesis of H-shaped copolymers. *J Polym Sci A Polym Chem.* 2006; 44(9): 2794–2801.

Han L, Shan G, Bao Y, Pan P. Exclusive stereocomplex crystallization of linear and multiarm star-shaped high-molecular-weight stereo diblock poly-(lactic acid). *J Phys Chem B.* 2015; 119(44): 14270–14279.

Hatada EK, Kitatama T, Vogl O, Nakahama S. *Macromolecular design of polymeric materials.* Marcel Dekker, Inc., New York 1997 (ISBN: 9780824794651).

Hiemstra C, Zhou W, Zhong Z, Wouters M, Feijen J. Rapidly *in situ* forming biodegradable robust hydrogels by combining stereocomplexation and photopolymerization. *J Am Chem Soc.* 2007; 129(32): 9918–9926.

Hira SK, Ramesh K, Gupta U, Mitra K, Misra N, Ray B *et al.* Methotrexate-loaded four-arm star amphiphilic block copolymer elicits CD8+ T cell response against a highly aggressive and metastatic experimental lymphoma. *ACS Appl Mater Interfaces.* 2015; 7(36): 20021–20033.

Jamshidi K, Hyon SH, Ikada Y. Thermal characterization of polylactides. *Polymer.* 1988; 29(12): 2229–2234.

Kanazawa H, Yamamoto K, Matsushima Y, Takai N, Kikuchi A, Sakurai Y *et al.* Temperature-responsive chromatography using poly (N-isopropylacrylamide)-modified silica. *Analyt Chem.* 1996; 68(1): 100–105.

Kaplan DL. Introduction to biopolymers from renewable resources. DL Kaplan, ed, In: *Biopolymers from renewable resources.* Springer Berlin Heidelberg, New York 1998, pp. 1–29 (ISBN: 978-3-642-08341-9).

Kim ES, Kim BC, Kim SH. Structural effect of linear and star-shaped poly (L-lactic acid) on physical properties. *J Polym Sci Part B Polym Phys.* 2004; 42(6): 939–946.

Ko J, Park K, Kim YS, Kim MS, Han JK, Kim K *et al.* Tumoral acidic extracellular pH targeting of pH-responsive MPEG-poly (β-amino ester) block copolymer micelles for cancer therapy. *J Control Release.* 2007; 123(2): 109–115.

Kohn J. New approaches to biomaterials design. *Nat Mater.* 2004; 3(11): 745–747.

Kohori F, Sakai K, Aoyagi T, Yokoyama M, Yamato M, Sakurai Y *et al.* Control of adriamycin cytotoxic activity using thermally responsive polymeric micelles composed of poly (N-isopropylacrylamide-co-N, N-dimethylacrylamide)-b-poly (D, L-lactide). *Colloids Surf B.* 1999; 16(1): 195–205.

Kohori F, Yokoyama M, Sakai K, Okano T. Process design for efficient and controlled drug incorporation into polymeric micelle carrier systems. *J Control Release.* 2002; 78(1): 155–163.

Kricheldorf HR, Sumbel M. Polylactones—18. Polymerization of l, l-lactide with Sn (II) and Sn (IV) halogenides. *Eur Polym J.* 1989; 25(6): 585–591.

Kurcok P, Matuszowicz A, Jedliński Z, Kricheldorf HR, Dubois P, Jérôme R. Substituent effect in anionic polymerization of β-lactones initiated by alkali metal alkoxides. *Macromol Rapid Commun.* 1995; 16(7): 513–519.

Langer R, Tirrell DA. Designing materials for biology and medicine. *Nature.* 2004; 428(6982): 487–492.

Le Garrec D, Gori S, Luo L, Lessard D, Smith D, Yessine MA *et al.* Poly (N-vinylpyrrolidone)-block-poly (D, L-lactide) as a new polymeric solubilizer for hydrophobic anticancer drugs: *In vitro* and *in vivo* evaluation. *J Control Release.* 2004; 99(1): 83–101.

Lee J, Bae YH, Sohn YS, Jeong B. Thermogelling aqueous solutions of alternating multiblock copolymers of poly (L-lactic acid) and poly (ethylene glycol). *Biomacromolecules.* 2006; 7(6): 1729–1734.

Lee SH, Hyun Kim S, Han YK, Kim YH. Synthesis and degradation of end-group-functionalized polylactide. *J Polym Sci Polym Chem.* 2001; 39(7): 973–985.

Li Z, Yuan D, Fan X, Tan BH, He C. Poly (ethylene glycol) conjugated poly (lactide)-based polyelectrolytes: Synthesis and formation of stable self-assemblies induced by stereocomplexation. *Langmuir.* 2015; 31(8): 2321–2333.

Liu F, Urban MW. Recent advances and challenges in designing stimuli-responsive polymers. *Prog Polym Sci.* 2010; 35(1): 3–23.

Liu X, Smith LA, Hu J, Ma PX. Biomimetic nanofibrous gelatin/apatite composite scaffolds for bone tissue engineering. *Biomaterials*. 2009; 30(12): 2252–2258.

Liu X, Smith L, Wei G, Won Y, Ma PX. Surface engineering of nano-fibrous poly (L-lactic acid) scaffolds via self-assembly technique for bone tissue engineering. *J Biomed Nanotechnol*. 2005a; 1(1): 54–60.

Liu X, Won Y, Ma PX. Surface modification of interconnected porous scaffolds. *J Biomed Mater Res*. 2005b; 74(1): 84–91.

Liu X, Won Y, Ma PX. Porogen-induced surface modification of nano-fibrous poly (L-lactic acid) scaffolds for tissue engineering. *Biomaterials*. 2006; 27(21): 3980–3987.

Liu Y, Nelson T, Cromeens B, Rager T, Lannutti J, Johnson J *et al.* HB-EGF embedded in PGA/PLLA scaffolds via subcritical $CO_2$ augments the production of tissue engineered intestine. *Biomaterials*. 2016; 103: 150–159.

Long VC, Berry GC, Hobbs L. Solution and bulk properties of branched polyvinyl acetates IV—Melt viscosity. *Polymer*. 1964; 5: 517–524.

Lowe CE. Preparation of high molecular weight polyhydroxyacetic ester, US2668162 A; 1954.

Lubiatowski P, Kruczynski J, Gradys A, Trzeciak T, Jaroszewski J. Articular cartilage repair by means of biodegradable scaffolds. *Transplantation Proc*. 2006; 38(1): 320–322.

Lunt J. Large-scale production, properties and commercial applications of polylactic acid polymers. *Polym Degrad Stab*. 1998; 59(1): 145–152.

Lunt J, Shafer AL. Polylactic acid polymers from corn. Applications in the textiles industry. *J Ind Text*. 2000; 29(3): 191–205.

Luo L, Ranger M, Lessard DG, Le Garrec D, Gori S, Leroux JC *et al.* Novel amphiphilic diblock copolymer of low molecular weight poly (n-vinylpyrrolidone)-b lock-poly (d, l-lactide): Synthesis, characterization, and micellization. *Macromolecules*. 2004; 37(11): 4008–4013.

Lutolf M, Hubbell J. Synthetic biomaterials as instructive extracellular microenvironments for morphogenesis in tissue engineering. *Nat Biotechnol*. 2005; 23(1): 47–55.

Lynn DM, Langer R. Degradable poly (β-amino esters): Synthesis, characterization, and self-assembly with plasmid DNA. *J Am Chem Soc*. 2000; 122(44): 10761–10768.

Lönnberg H, Zhou Q, Brumer H, Teeri TT, Malmström E, Hult A. Grafting of cellulose fibers with pol(ε-caprolactone) and poly(L-lactic acid) via ring-opening polymerization. *Biomacromolecules*. 2006; 7(7): 2178–2185.

Ma FK, Li J, Kong M, Liu Y, An Y, Chen XG. Preparation and hydrolytic erosion of differently structured PLGA nanoparticles with chitosan modification. *Int J Biol Macromolec*. 2013; 54: 174–179.

Ma PX. Biomimetic materials for tissue engineering. *Adv Drug Deliv*. 2008; 60(2): 184–198.

Ma PX, Zhang R. Synthetic nano-scale fibrous extracellular matrix. *J Biomed Mater Res*. 1999; 46(1): 60–72.

Ma PX, Zhang R, Xiao G, Franceschi R. Engineering new bone tissue *in vitro* on highly porous poly (α-hydroxyl acids)/hydroxyapatite composite scaffolds. *J Biomed Mater Res*. 2001; 54(2): 284–293.

Mao J, Bo S, Ji X. pH/temperature-responsive behavior of amphiphilic block copolymer micelles prepared using two different methods. *Langmuir*. 2011; 27(12): 7385–7391.

Marek W. *Handbook of stimuli-responsive materials*. Wiley-VCH Verlag GmbH, weinheim 2011 (ISBN: 9783527327003).

Matyjaszewski K, Davis TP. *Handbook of radical polymerization*. Wiley Online Library, weinheim 2002 (ISBN: 978-0-471-39274-3).

Mayumi A, Kitaoka T, Wariishi H. Partial substitution of cellulose by ring-opening esterification of cyclic esters in a homogeneous system. *J Appl Polym Sci*. 2006; 102(5): 4358–4364.

Mc Kenzie J, Cowie G. *Polymers: Chemistry & physics of modern materials*. Nelson Thornes, UK, 1991 (ISBN 10: 0216929806).

Min KH, Kim JH, Bae SM, Shin H, Kim MS, Park S *et al.* Tumoral acidic pH-responsive MPEG-poly (β-amino ester) polymeric micelles for cancer targeting therapy. *J Control Release*. 2010; 144(2): 259–266.

Nagahama K, Ouchi T, Ohya Y. Temperature-Induced hydrogels through self-assembly of cholesterol-substituted star PEG-b-PLLA copolymers: An injectable scaffold for tissue engineering. *Adv Funct Mater*. 2008; 18(8): 1220–1231.

Nakayama M, Okano T, Miyazaki T, Kohori F, Sakai K, Yokoyama M. Molecular design of biodegradable polymeric micelles for temperature-responsive drug release. *J Control Release*. 2006; 115(1): 46–56.

Nouvel C, Raynaud J, Marie E, Dellacherie E, Six JL, Durand A. Biodegradable nanoparticles made from polylactide-grafted dextran copolymers. *J Colloid Interface Sci*. 2009; 330(2): 337–343.

Oh JK. Polylactide (PLA)-based amphiphilic block copolymers: Synthesis, self-assembly, and biomedical applications. *Soft Matter*. 2011; 7(11): 5096–5108.

Ohya Y, Maruhashi S, Ouch T. Graft polymerization of l-lactide on pullulan through the trimethylsilyl protection method and degradation of the graft copolymers. *Macromolecules.* 1998a; 31(14): 4662–4665.

Ohya Y, Maruhashi S, Ouchi T. Preparation of poly (lactic acid)-grafted amylose through the trimethylsilyl protection method and its biodegradation. *Macromol Chem Phy.* 1998b; 199(9): 2017–2022.

Ouchi T, Kontani T, Ohya Y. Modification of polylactide upon physical properties by solution-cast blends from polylactide and polylactide-grafted dextran. *Polymer.* 2003a; 44(14): 3927–3933.

Ouchi T, Uchida T, Arimura H, Ohya Y. Synthesis of poly (l-lactide) end-capped with lactose residue. *Biomacromolecules.* 2003b; 4(3): 477–480.

Ovitt TM, Coates GW. Stereoselective ring-opening polymerization of meso-lactide: Synthesis of syndiotactic poly (lactic acid). *J Am Chem Soc.* 1999; 121(16): 4072–4073.

Östmark E, Nyström D, Malmström E. Unimolecular nanocontainers prepared by ROP and subsequent ATRP from hydroxypropylcellulose. *Macromolecules.* 2008; 41(12): 4405–4415.

Park K, Ju YM, Son JS, Ahn KD, Han DK. Surface modification of biodegradable electrospun nanofiber scaffolds and their interaction with fibroblasts. *J Biomater Sci Polym Ed.* 2007; 18(4): 369–382.

Perego G, Cella GD, Bastioli C. Effect of molecular weight and crystallinity on poly(lactic acid) mechanical properties. *J Appl Polym Sci.* 1996; 59(1): 37–43.

Pham QP, Sharma U, Mikos AG. Electrospinning of polymeric nanofibers for tissue engineering applications: A review. *Tissue Eng.* 2006; 12(5): 1197–1211.

Pierschbacher MD, Ruoslahti E. Cell attachment activity of fibronectin can be duplicated by small synthetic fragments of the molecule. *Nature.* 1983; 309(5963): 30–33.

Pitarresi G, Palumbo FS, Albanese A, Fiorica C, Picone P, Giammona G. Self-assembled amphiphilic hyaluronic acid graft copolymers for targeted release of antitumoral drug. *J Drug Target.* 2010; 18(4): 264–276.

Place ES, Evans ND, Stevens MM. Complexity in biomaterials for tissue engineering. *Nature Mater.* 2009; 8(6): 457–470.

Potineni A, Lynn DM, Langer R, Amiji MM. Poly (ethylene oxide)-modified poly (β-amino ester) nanoparticles as a pH-sensitive biodegradable system for paclitaxel delivery. *J Control Release.* 2003; 86(2): 223–234.

Pytela J, Saudek V, Drobnik J, Rypáček F. Poly (N 5-hydroxyalkylglutamines). IV. Enzymatic degradation of n 5-(2-hydroxyethyl)-L-glutamine homopolymers and copolymers. *J Control Release.* 1989; 10(1): 17–25.

Radano CP, Baker GL, Smith MR. Stereoselective polymerization of a racemic monomer with a racemic catalyst: Direct preparation of the polylactic acid stereocomplex from racemic lactide. *J Am Chem Soc.* 2000; 122(7): 1552–1553.

Ramesh K, Mishra AK, Patel VK, Vishwakarma NK, Biswas CS, Paira TK *et al.* Synthesis of well-defined amphiphilic poly (D, L-lactide)-b-poly (N-vinylpyrrolidone) block copolymers using ROP and xanthate-mediated RAFT polymerization. *Polymer.* 2012; 53(25): 5743–5753.

Ray B, Kotani M, Yamago S. Highly controlled synthesis of poly (N-vinylpyrrolidone) and its block copolymers by organostibine-mediated living radical polymerization. *Macromolecules.* 2006; 39(16): 5259–5265.

Raynaud J, Choquenet B, Marie E, Dellacherie E, Nouvel C, Six JL *et al.* Emulsifying properties of biodegradable polylactide-grafted dextran copolymers. *Biomacromolecules.* 2008; 9(3): 1014–1021.

Richardson TP, Peters MC, Ennett AB, Mooney DJ. Polymeric system for dual growth factor delivery. *Nat Biotechnol.* 2001; 19(11): 1029–1034.

Romberg B, Metselaar JM, Baranyi L, Snel CJ, Bünger R, Hennink WE *et al.* Poly(amino acid)s: Promising enzymatically degradable stealth coatings for liposomes. *Int J Pharm.* 2007; 331(2): 186–189.

Santoro M, Shah SR, Walker JL, Mikos AG. Poly(lactic acid) nanofibrous scaffolds for tissue engineering. *Adv Drug Deliv Rev.* 2016; 107: 206–212.

Schaefgen JR, Flory PJ. Synthesis of multichain polymers and investigation of their viscosities1. *J Am Chem Rev.* 1948; 70(8): 2709–2718.

Shi Q, Huang Y, Chen X, Wu M, Sun J, Jing X. Hemoglobin conjugated micelles based on triblock biodegradable polymers as artificial oxygen carriers. *Biomaterials.* 2009; 30(28): 5077–5085.

Shim WS, Kim SW, Lee DS. Sulfonamide-based pH-and temperature-sensitive biodegradable block copolymer hydrogels. *Biomacromolecules.* 2006; 7(6): 1935–1941.

Shin SJ, Yu YC, Seo JD, Cho SJ, Youk JH. One-step synthesis of poly (N-vinylpyrrolidone)-b-poly(l-lactide) block copolymers using a dual initiator for RAFT polymerization and ROP. *J Poly Sci Polym Chem.* 2014; 52(11): 1607–1613.

Stach M, Lacik I, Chorvat Jr D, Buback M, Hesse P, Hutchinson RA *et al.* Propagation rate coefficient for radical polymerization of n-vinyl pyrrolidone in aqueous solution obtained by PLP – SEC. *Macromolecules.* 2008; 41(14): 5174–5185.

Stanford MJ, Dove AP. Stereocontrolled ring-opening polymerisation of lactide. *Chem Soc Rev.* 2010; 39(2): 486–494.

Sun J, Chen X, Lu T, Liu S, Tian H, Guo Z *et al.* Formation of reversible shell cross-linked micelles from the biodegradable amphiphilic diblock copolymer poly (L-cysteine)-block-poly (L-lactide). *Langmuir.* 2008; 24(18): 10099–10106.

Tasaka F, Ohya Y, Ouchi T. One-pot synthesis of novel branched polylactide through the copolymerization of lactide with mevalonolactone. *Macromol Rapid Commun.* 2001; 22(11): 820–824.

Thomopoulos S, Sakiyama-Elbert S, Silva M, Gelberman R, Xia Y, Schwartz A *et al.* Polymer nanofiber scaffold for a heparin/fibrin based growth factor delivery system. US8673323 B2; 2016.

Torchilin V. Polymer-coated long-circulating microparticulate pharmaceuticals. *J Microencapsul.* 1998; 15(1): 1–19.

Tsuji H, Ozawa R, Matsumura N. Effect of incorporated star-shaped four-armed stereo diblock poly (lactide) on the crystallization behavior of linear one-armed poly(l-lactide) or poly(d-lactide). *Polymer J.* 2016; 48(2): 209.

Uhrich KE, Cannizzaro SM, Langer RS, Shakesheff KM. Polymeric systems for controlled drug release. *Chem Rev.* 1999; 99(11): 3181–3198.

Wallach JA, Huang SJ. Copolymers of itaconic anhydride and methacrylate-terminated poly (lactic acid) macromonomers. *Biomacromolecules.* 2000; 1(2): 174–179.

Wan Y, Cao X, Zhang S, Wang S, Wu Q. Fibrous poly (chitosan-g-DL-lactic acid) scaffolds prepared via electro-wet-spinning. *Acta Biomater.* 2008; 4(4): 876–886.

Wang JS, Matyjaszewski K. Controlled/"living" radical polymerization. Atom transfer radical polymerization in the presence of transition-metal complexes. *J Am Chem Soc.* 1995; 117(20): 5614–5615.

Weinstein J, Magin R, Yatvin M, Zaharko D. Liposomes and local hyperthermia: Selective delivery of methotrexate to heated tumors. *Science.* 1979; 204(4389): 188–191.

Willerth SM, Rader A, Sakiyama-Elbert SE. The effect of controlled growth factor delivery on embryonic stem cell differentiation inside fibrin scaffolds. *Stem Cell Res.* 2008; 1(3): 205–218.

Yoshida R, Uchida K, Kaneko Y, Sakai K, Kikuchi A, Sakurai Y *et al.* Comb-type grafted hydrogels with rapid deswelling response to temperature changes. *Nature.* 1995; 374(6519): 240–242.

You ES, Jang HS, Ahn WS, Kang MI, Jun MG, Kim YC *et al. In vitro* biocompatibility of surface-modified poly(DL-lactide-co-glycolide) scaffolds with hydrophilic monomers. *J Ind Eng Chem.* 2007; 13(2): 219–224.

Yuan W, Yuan J, Zhang F, Xie X. Syntheses, characterization, and *in vitro* degradation of ethyl cellulose-graft-poly(ε-caprolactone)-block-poly(l-lactide) copolymers by sequential ring-opening polymerization. *Biomacromolecules.* 2007; 8(4): 1101–1108.

Zhai L. Stimuli-responsive polymer films. *Chem Soc Rev.* 2013; 42(17): 7148–7160.

Zhang CY, Yang YQ, Huang TX, Zhao B, Guo XD, Wang JF *et al.* Self-assembled pH-responsive MPEG-b-(PLA-co-PAE) block copolymer micelles for anticancer drug delivery. *Biomaterials.* 2012; 33(26): 6273–6283.

Zhu K, Bihai S, Shilin Y. Super microcapsules(SMC). I. Preparation and characterization of star polymethylene oxide (PEO) polylactide (PLA) copolymers. *J Polym Sci A Polym Chem.* 1989; 27(7): 2151–2159.

Zhu K, Xiangzhou L, Shilin Y. Preparation, characterization, and properties of polylactide (PLA)–poly (ethylene glycol)(PEG) copolymers: A potential drug carrier. *J Appl Polym Sci.* 1990; 39(1): 1–9.

# 8 Surface Functionalization and Antibacterial Characteristics of the Titanium-Based Metallic Biomaterials at Nanoscale

*Jalal Azadmanjiri, Wai Hong Wong, Jagat R. Kanwar,**
*Christopher C. Berndt, James Wang,*
*Vijay K. Srivastava, and Ajay Kapoor**

## CONTENTS



## 8.1 INTRODUCTION

It has been recognized that the osseointegration ability of titanium (Ti) implants is significantly affected by their surface characteristics, such as chemical compositions, surface topography and roughness (Pham *et al.*, 2012; He *et al.*, 2017). In this respect, scientists and engineers have been encouraged to investigate newer approaches of surface coatings (Yan *et al.*, 1997; Li *et al.*, 2008; Wang *et al.*, 2016). Therefore, developing new surfaces by functionalization via advanced techniques and capable coatings for enhancing the long-term mechanical stabilities, chemical stabilities with lesser corrosion and biological performances such as osseointegration, bioactivity, antibacterial are the current research field of interests in the dental and orthopaedic implant industries (Vetrone *et al.*, 2009; Tawfick *et al.*, 2012). Hence, the prime objective of this chapter is to explain functionalization of the surface of the Ti-based metallic biomaterials using highly bioactive elements and antibacterial agents.

* Corresponding authors.

Recent investigations show that tantalum (Ta) coating can improve the corrosion resistance and nanoporous titanium dioxide ($TiO_2$) coating can improve cell proliferation, differentiation and adhesion, because of their nano-topographic features and their large surface area. Hydroxyapatite (HA) coating also enhances bone growth and bonding between bone tissue and the surfaces of implants (Mandracci *et al.*, 2016; Parcharoen *et al.*, 2016). Biomaterial-related infection caused by bacterial contamination is one of the most serious problems in the field, and the approach to reduce the bacterial attachment is incorporating an antibacterial agent onto the implant surface. Studies show that silver (Ag) ion-implanted Ti and its alloys restrict the growth of *S. aureus* and *E. coli* on the surface, compared to a bare surface; this effect can conclude that Ag ion-implanted Ti is an excellent antibacterial structure. There are a wide range of implant devices built with polymeric biomaterials incorporated with drug therapies to enhance the therapeutic effects. Short half-lives of biomaterials is important for site minimal loss and minimize the toxicity of the system. Besides, the defects can be overcome by hybridizing organic compounds like carbon nanotubes (CNTs) and polymers. CNTs offer good for cell adhesion and their growth, as well as to avoid strong inflammatory responses. It also reinforces the fracture toughness and wear resistance of the HA coating. The natural polymer was grafted on a Ti surface because of their good biocompatibility, no or very less toxicity and ease of availability (Eleftheriou *et al.*, 2017). Synthetic water-soluble polymers have the ability to control the crystal growth of organic or inorganic sparingly soluble salts. Protein adsorption is a progressive process vital in determining the longevity of such implants.

## 8.2 FUNCTIONALIZATION OF THE SURFACE OF THE TITANIUM-BASED METALLIC BIOMATERIALS

### 8.2.1 Attachment of Highly Biocompatible Metal Layer on Titanium-Based Implants

Major issues confronted by the researchers and engineers, in the case of Ti-based implants, is chemical etching under the physiological environment and mechanical damage because of corrosion of the implant during its long-term implantation in the human body. Therefore, developing newer surface modification techniques is one of the current research fields, with initial studies showing that coating of a corrosion-resistant layer on top of the implant surface could resolve the problems. Xu *et al.* (2015) prepared two and six layers of Ta oxide ($TaO_2$) on the surface of a Ti substrate to study and compare their corrosion resistance and cytocompatibility with the bare Ti. Ta is an ideal choice to be coated because of its excellent biocompatibility and corrosion resistivity (Sieber and Schmuki, 2005). Hence, they employed the polymer-assisted deposition (PAD) technique to coat $TaO_2$ on the Ti substrate because of its fairly cheap precursor of tantalum chloride ($TaCl_5$). The concept of PAD involves coating of a solution made of a soluble polymer and a metal precursor onto a surface. The process, followed by elimination of organics with a high-temperature annealing process in an oxygen environment, results in the formation a metal oxide layer on the surface (Tsuchiya *et al.*, 2005; Xu *et al.*, 2015). The unwanted chemical reactions with metal ions are avoided, because of the encapsulation of metal ions by the polymer (Berger *et al.*, 2008). The factors that reveal the corrosion resistance are polarization resistance ($R_p$), corrosion current density ($I_{corr}$) and corrosion potential ($E_{corr}$). Materials with high $R_p$, low $I_{corr}$ and high $E_{corr}$ indicate good corrosion resistance (Wang *et al.*, 2011). Figure 8.1 exhibits the polarization curves obtained for all three samples belonging to bare Ti, Ti with two layers of Ta (Ta-2L), and Ti with six layers of Ta (Ta-6L). $E_{corr}$ values were −1.142 V/SCE, −0.969 V/SCE and −0.558 V/SCE for Ti, Ta-2L and Ta-6L, respectively. Tafel analysis was used to analyse polarization resistance ($R_p$) and corrosion current density ($I_{corr}$) from the potentiodynamic polarization curves, which are summarized in Table 8.1. Generally, Ta-6L indicates the best corrosion resistance and bare Ti shows the poorest corrosion resistance, based on the obtained results. Ta-6L presented the highest values of $E_{corr}$ and $R_p$, and the lowest ones of $I_{corr}$, compared to the bare Ti and Ta-2L, respectively (Xu *et al.*, 2015).

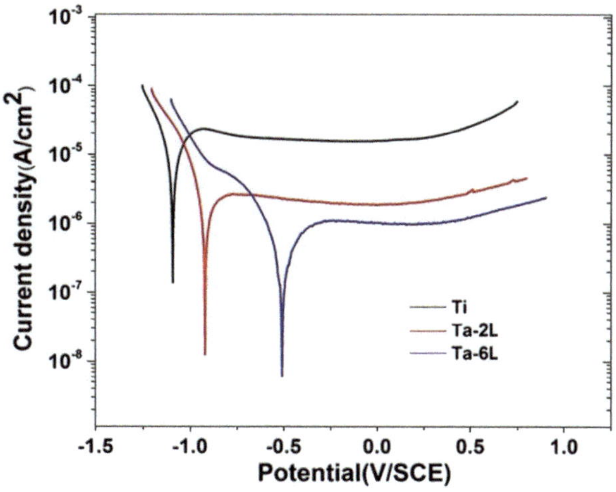


**FIGURE 8.1** Potentiodynamic polarization graphs of Ti, Ta-2L and Ta-6L. (From Xu GQ *et al. Surf Coat Tech.* 2015; 272: 58–65. With permission.)

**TABLE 8.1**
**Polarization Resistance ($R_p$), Corrosion Current Density ($I_{corr}$) and Corrosion Potential ($E_{corr}$) of Ti, Ta-2L and Ta-6L Samples**

| | Ti | Ta-2L | Ta-6L |
|---|---|---|---|
| $R_p$ (Ω) | $3.31 \times 10^3$ | $1.86 \times 10^4$ | $1.17 \times 10^5$ |
| $I_{corr}$ (A/cm²) | $2.82 \times 10^{-5}$ | $4.21 \times 10^{-6}$ | $3.68 \times 10^{-7}$ |
| $E_{corr}$ (V/SCE) | −1.142 | −0.969 | −0.558 |

*Source:* Xu GQ *et al. Surf Coat Tech.* 2015; 272: 58–65. With permission.

Surface modification improves the biological performances such as osseointegration by tailoring and controlling the physical and chemical surface characteristics, and the factors contributing to implant osseointegration are chemical composition, surface topography and roughness. Surface modification can be accomplished by coating bioactive materials and creating porous and rough surfaces (Lee *et al.*, 2011; Pham *et al.*, 2012). Micro- and nano-scale topographic modifications are vital methods that improve the osseointegration of Ti implants (Mendonca *et al.*, 2009; Pham *et al.*, 2012). Microporous $TiO_2$ using the micro-arc oxidation (MAO) technique shows better *in vivo* biocompatibility due to fixation and strong connection of *in vivo* bone to a Ti porous surface (Li *et al.*, 2004; Hong *et al.*, 2011; Pham *et al.*, 2012). Nanoporous $TiO_2$ created by an anodizing process could improve cell proliferation, differentiation and adhesion, because of the nano-topographic features and their significant surface area (Mendonca *et al.*, 2008). Pham *et al.* (2012) employed a novel way to produce nanoporous Ta-integrated Ti surfaces onto Ti substrates. They applied incredibly high negative substrate biases during sputtering of Ta on Ti in argon (Ar) gas. A nanoporous structure was created on the surface due to the considerable re-sputtering of deposited films when high energetic ions ($Ta^+$, $Ta^{2+}$, $Ar^+$) bombarded the surface of the substrate. They successfully produced a nanoporous surface of approximately 40–50 nm pore size and with a thickness of approximately 300 nm by employing a negative substrate bias voltage of 800 V. A nanostructured surface was also created using negative substrate bias voltages of 0, 200, 300 and 600 V, for comparison. The *in vitro* biocompatibility of substrates was examined using cell culture for 12 h, and the resultant confocal

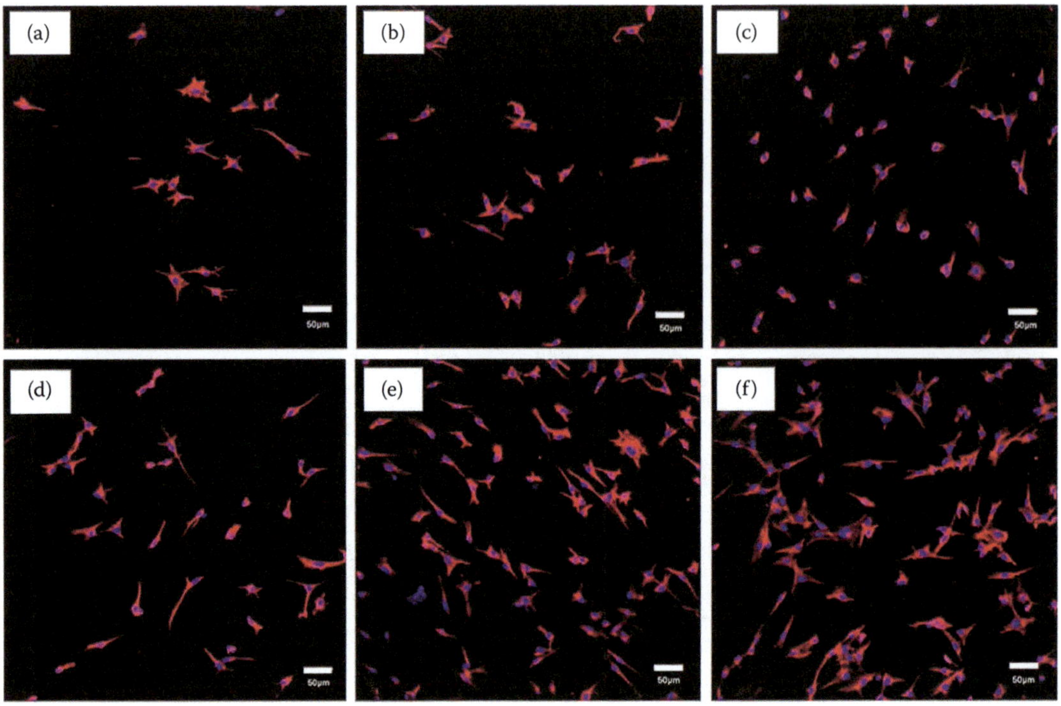


**FIGURE 8.2** CLSM images for (a) bare Ti substrate and nanostructured surfaces constructed using negative substrate bias voltages of (b) 0, (c) 200, (d) 300, (e) 600 and (f) 800 V. (Scale is 50 μm.) (From Pham VH *et al.* *J Mater Chem.* 2012; 22: 24798–24804. With permission.)

laser scanning microscopy (CLSM) images are shown in Figure 8.2a–f. All substrates showed good biocompatibility, in which a number of cells were attached and spread actively on the surfaces. The CLSM images can confirm when the negative substrate bias voltage increases, the number of cells tend to increase, notably due to enhanced surface roughness. The produced substrate using bias voltage of 800 V demonstrates the best cell attachment after 12 h. Figure 8.3a–d also shows the further examination by observing microfilament (actin) stress fibres (Figure 8.3a,b) and microtubules (tubulin) (Figure 8.3c,d) using CLSM on the bare Ti substrate (Figure 8.3a,c) and the Ti substrate with the nanoporous surface (Figure 8.3b,d), after 1 day of culturing (Pham *et al.*, 2012).

The nanoporous surface exhibited healthier organization of actin stress (Figure 8.3a,b) and higher expression of tubulin (Figure 8.3c,d) compared to bare Ti substrate. These results state that osteoblastic activity was enhanced by the nanoporous Ta-incorporated Ti surface. Figure 8.4 shows the degree of proliferation that was obtained from the MTS assay (i.e., cell proliferation assay) after cell culturing for 10 days and alkaline phosphatase (ALP) action of the cell after cell culturing for 15 days for both bare and nanoporous Ta-incorporated Ti substrates. The highly nanoporous surface performed better proliferation and ALP activity compared to blank Ti as shown by it exhibiting higher absorbance values (Pham *et al.*, 2012).

### 8.2.2 Hydroxyapatite Attachment on Titanium-Based Implants

HA with the formula [$Ca_{10}(PO_4)_6(OH)_2$] has been nicely identified due to its good biocompatibility and is an attractive material for hard-tissue implants. The reason for HA being so attractive is that it enhances quick bone development and bonding between the bone tissue and implant surfaces. This is due to a close resemblance to that of the chemical composition (Ca/P ratio) of teeth and bones. It also reduces the metal ion release to the body environment. HA is utilized medically as a packed

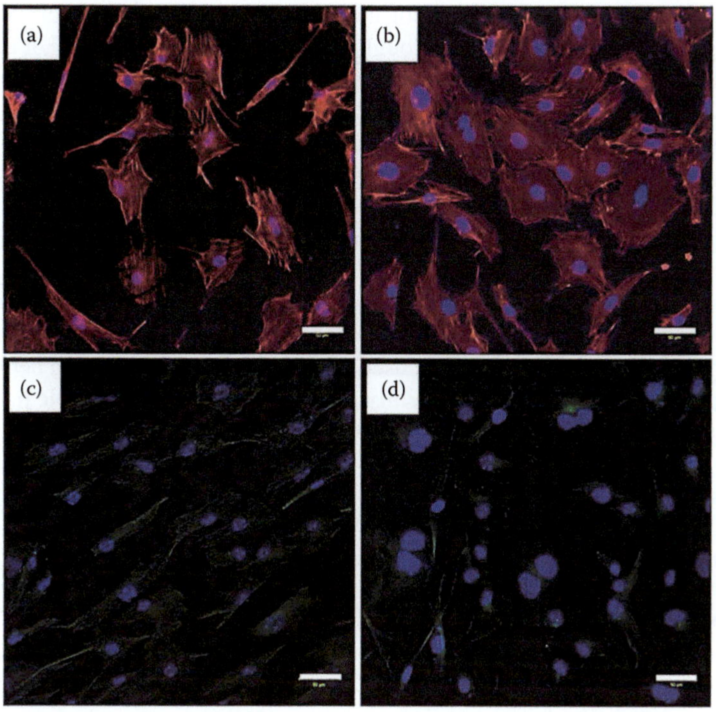


**FIGURE 8.3** CLSM images indicating actin stress fibres (a,b) and microtubules (c,d), on the bare Ti substrate (a,c) and the Ti substrate with the nanoporous surface (b,d). (From Pham VH *et al. J Mater Chem.* 2012; 22: 24798–24804. With permission.)

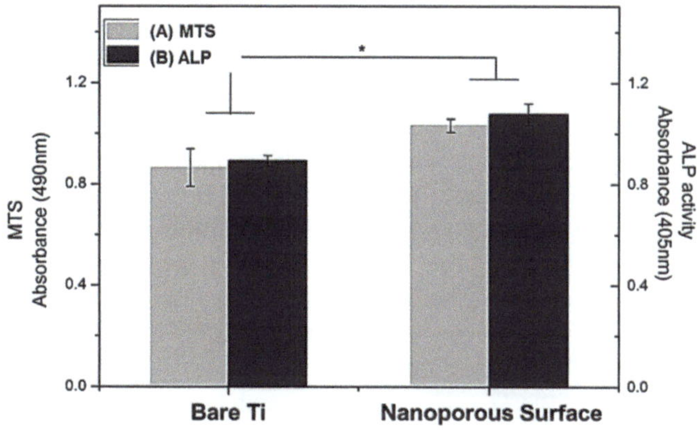


**FIGURE 8.4** (a) Proliferation and (b) ALP activity reading for bare Ti and nanoporous surface. (From Pham VH *et al. J Mater Chem.* 2012; 22: 24798–24804. With permission.)

and dense sintered material and coating on metallic implants (Hench, 1998; Dong *et al.*, 2003). Many surface deposition techniques can be used to deposit HA coating on metallic implants, which include high-velocity oxy-fuel (HVOF) spraying (Haman *et al.*, 1995), pulsed laser ablation (Singh *et al.*, 1996), electrodeposition (Dasarathy *et al.*, 1996), ion-beam sputtering (Ong *et al.*, 1991), radio frequency magnetron sputtering (Wolke *et al.*, 1994; Yamashita *et al.*, 1996), sol-gel (Haddow *et al.*, 1998), conventional ceramic processes (Hero *et al.*, 1994) and so forth.

Even though HA possesses excellent properties for implantation, it characterizes poor mechanical properties such as greater brittleness, lower tensile strength and lower impact-resistance compared to the human bone. HA can be observed to dissolve in the body fluids after long-term contact (Overgaard *et al.*, 1998; Negroiu *et al.*, 2008). Fracture toughness of HA being 1 $Mpa.m^{1/2}$, it is not applicable for high-load application, albeit the desired mechanical properties can be obtained by blending it with tough phase (Choi *et al.*, 1998; Kong *et al.*, 1999). Interfacial adhesion and reinforcement fillers are the vital factors among the mechanical properties of the HA-based composite.

Besides the bioactivity of the coating, the bond strength among the coating layer and metal substrate constitute important factors in characterizing the long-term stability of an implant. The detached fragments (on separation of coating) during actual application in the human body caused adverse effects on surrounding tissue (Wang *et al.*, 1996; Choi *et al.*, 2000). It has been reported that the bond strength between HA and Ti is relatively low with 20–30 MPa (Kim *et al.*, 2004), possibly due to the mismatch of the coefficients of thermal expansion (Heimann *et al.*, 1998; Zheng *et al.*, 2000). One of approaches that can be used to develop the bond strength between HA and Ti is forming a composite coating with mechanically strong bioinert, biocompatible metals (Zheng *et al.*, 2000). Investigations show that when 20 and 60 wt% of Ti is integrated into the HA coating, the bond strength of HA rose from 12.9 (for pure HA) to 14.5 and 17.3 MPa, respectively (Zheng *et al.*, 2000). Ion-beam-assisted deposition (IBAD) is one of the deposition techniques that can improve bond strength of the HA coating. It is reported that when ion-beam current increases, the bond strength increases because of widening of the atomic intermixed zone (Choi *et al.*, 2000).

#### 8.2.2.1 Growth of Hydroxyapatite from Simulated Body Fluid

The bioactive surface of a metallic biomedical implant can be improved by HA deposition on it. Accordingly, HA deposition has been widely applied on metallic implants (Oh *et al.*, 2005; Tsuchiya *et al.*, 2006; Kunze *et al.*, 2008). The fabricated $TiO_2$ nanotubes (NTs) by sodium hydroxide (NaOH) solution is one of metallic biomedical implants that has been used for the growth of HA through immersion in simulated body fluid (SBF) solution. Sodium titanate ($Na_2Ti_5O_{11}$ or $Na_2Ti_6O_{13}$) was fabricated after NaOH treatment, and HA was then created on the top of nanotubes upon soaking in SBF. The size of bone-like HA structure was about 25 nm (Oh *et al.*, 2005). The longer fabricated $TiO_2$ NTs develop a higher ratio of hydroxyapatite when soaking in SBF. For instance, $TiO_2$ NTs with 2 μm lengths promoted the growth of HA higher than $TiO_2$ NTs with 200 nm lengths. This is while the longer $TiO_2$ NTs took almost 2 days and short $TiO_2$ NTs took about 14 days. Furthermore, the annealing process shortened the growth time of HA for the short nanotubes to 2 days (Tsuchiya *et al.*, 2006). Annealed $TiO_2$ NTs with 1.66 μm length and 100 nm diameter exhibited higher growth rate of calcium (Ca, from 0.067 to 52.77) and phosphorus (P, from 0.058 to 36.27), upon extending the exposure time to SBF to 5 h. Despite the higher growth rate of HA on as-synthesized $TiO_2$ NTs, a homogeneity growth film exists on their surface (Kunze *et al.*, 2008). In addition, the nanotubular Ti exhibited a higher ratio of formation of the HA rather than the bare Ti. This effect is due to the high surface area in the nanotubular structure (Kunze *et al.*, 2008). The mechanism of HA formation from SBF on $TiO_2$ NTs is because of the dipolar nature of the $TiO_2$/liquid interface. Hydroxyl groups ($OH^-$) are adsorbed from SBF to form Ti–OH groups. Based on the pH of the media, the overall charge of the Ti–OH groups is created. In SBF, the pH is around 7.4 and therefore Ti–OH groups are negatively charged due to the existence of deprotonated acidic hydroxides. Calcium ions ($Ca^{2+}$) are then supported to adsorb from SBF as a result of such surface negative charges, so that $HOPO_4^{2-}$ or $H_2PO_4^-$ can easily react with the as-adsorbed $Ca^{2+}$ to finally produce calcium phosphates (Wang and Luo, 2011). Repetitive dipping and drying $TiO_2$ NTs in SBF enhanced the growth rate of HA as a consequence of filling the inner and outer areas of nanotubes, since capillary effect existed. Figure 8.5a and b shows fabrication of the dense HA layer after repetitive dipping for eight times, which was remarkably obvious, compared to two immersions (Wang and Luo, 2011).

#### 8.2.2.2 Deposition of Synthetic Hydroxyapatite Layer

Dong *et al.* (2003) prepared Ti-6Al-4V/HA composite layer with a plasma spraying method using composite powders of Ti-6Al-4V/HA. The powders were synthesized from pure HA and Ti-6Al-4V. A wet chemical method was used to produce HA raw material, that is, 0.3 M

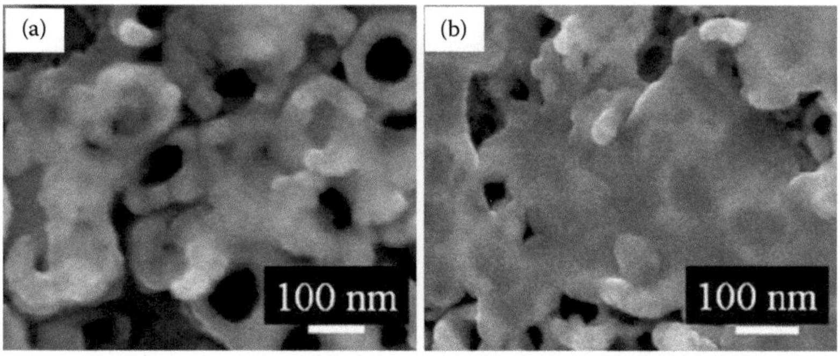


**FIGURE 8.5** FESEM micrographs of $TiO_2$ NTs dipped in SBF (a) twice and (b) 8 times. (From Wang LN, Luo JL. *Metall Mater Trans A-Phys Metall Mater Sci.* 2011; 42A: 3255–3264. With permission.)

orthophosphoric acid was reacted with 0.5 M calcium in a temperature control bath. The formed precipitate from the reactants is dried at 180°C and then heated at 800°C for 4 h for calcination. The used Ti-6Al-4V powders were 100 mesh, with a particle size range between 53 and 75 μm. Ti-6Al-4V/HA composite powders with particle size range from 75 to 106 μm formed, when Ti-6Al-4V granules were coated with slurry HA. The composite powders contained approximately 48 wt% of HA. The Ti alloy was sand-blasted with 24 mesh alumina ($Al_2O_3$) grit prior to the deposition of coating using plasma spraying with net energy of 12 kW, gun current of 600–1000 Å and powder federate of 0.01 kg/min. The process, followed by heat treatment, resulted in a coating with thickness of approximately 100 μm for 6 h at 600°C to discharge thermal residual stresses and accelerate the re-crystallization process. Figure 8.6a shows the scanning electron microscopy

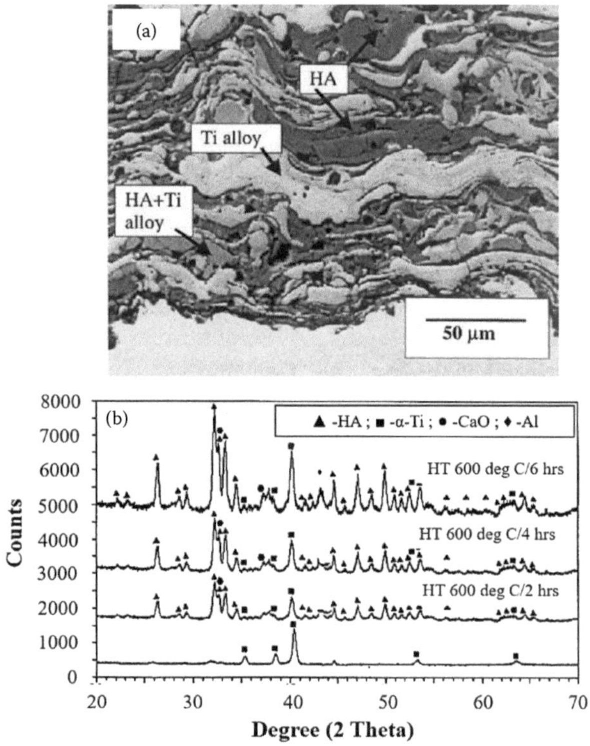


**FIGURE 8.6** (a) SEM image and (b) XRD graph of the as-sprayed and heat-treated composite of composite layer of the Ti-6Al-4V and HA. (From Dong ZL *et al. Biomater.* 2003; 24: 97–105. With permission.)

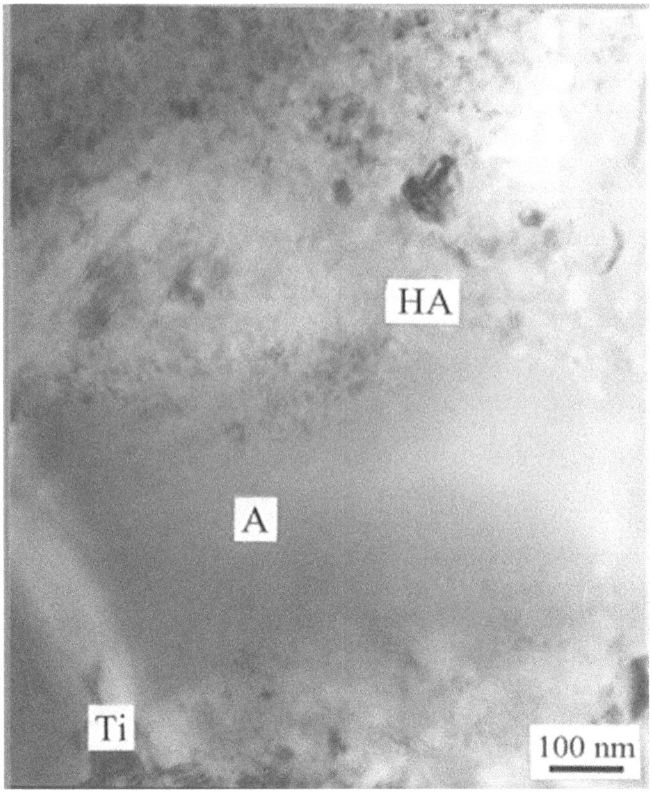


**FIGURE 8.7** TEM image of Ti/HA composite with fine grains and amorphous phase. (From Dong ZL *et al.* *Biomater.* 2003; 24: 97–105. With permission.)

(SEM) image of the composite layer of the coating, which clearly exhibits that the Ti-6Al-4V and HA phases are layered in an alternating structure. The x-ray diffraction (XRD) results (Figure 8.6b) show that the HA coating is the main crystalline phase and contains a huge amount of amorphous calcium phosphate phase.

Decomposition of HA in a high-temperature process caused the existence of CaO, $Ca_3(PO_4)_2$ (tricalcium phosphate, TCP) and $Ca_4P_2O_9$ (TTCP) in a small amount (Dong *et al.*, 2003). Transmission electron microscopy (TEM) analysis (Figure 8.7) demonstrates the coating structure consisting of fine HA crystalline material allocated as a result of an amorphous matrix and encapsulates Ti-6Al-4V crystals (Dong *et al.*, 2003). The amorphous calcium-phosphate phase can also be seen, as it was around almost ubiquitously in the coating. Study has also been done to the regions containing mainly Ti-6Al-4V as well as HA and the amorphous phase (Dong *et al.*, 2003). Amorphous phase can be mostly transformed into the crystalline HA phase through post-spray heat treatment. Because of the existence of ceramic and metallic phases in the starting powder, the microstructure of composite Ti-6Al-4V/HA coating is more complex.

HA coating is also fabricated using IBAD. A thin layer of HA was coated on Ti-6Al-4V alloy disc with 2 mm (Choi *et al.*, 2000). Commercial HA powder was sintered for 2 h in air at 1200°C as a deposition target. The substrate (Ti-6Al-4V) surface was grounded with silicon carbide (SiC) and then polished with diamond slurries, down to 1 μm before deposition. The chamber of a cryopump was set to a pressure of $10^{-7}$ Torr, and argon gas at a pressure of $10^{-4}$ Torr was introduced into the chamber. At the same time, the target was evaporated with an electron beam at 8.5 kV and about 0.1 A. An end-hall type ion gun at 130 V and current of 1 A was carried out to the substrate surface to support deposition. Investigations have been carried out to study the relationship between ion-bean

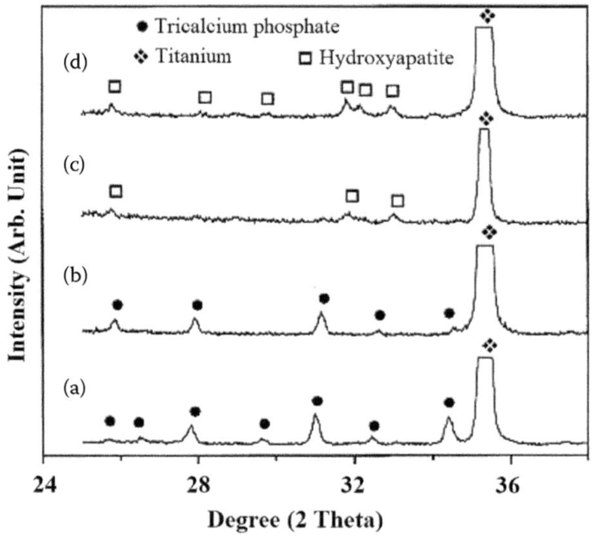


**FIGURE 8.8** XRD results obtained for coating after heat treatment. (From Choi JM *et al. Biomater.* 2000; 21: 469–473. With permission.)

intensity and layer-substrate bond strength, as well as the layer dissolution rate in a physiological saline solution. The layer phase and morphology of samples were also studied after heat treatment in a vacuum with a pressure of $3 \times 10^{-3}$ Torr for an hour at 630°C. Prior to heat treatment, the coating layer was amorphous with a thickness of around 700 nm (Choi *et al.*, 2000).

Figure 8.8 displays the obtained XRD results after heat treatment to estimate the composition of the coating. TCP formed after heat treatment without ion beam (Figure 8.8a) and ion beam with current of 0.6 A (Figure 8.8b). However, TCP disappeared and weak HA peaks (Figure 8.8c) were detected when the ion-beam current increased to 0.8 A. Stronger HA peaks are observed (Figure 8.8d) in the graph when the ion-beam current increased to 1.0 A. These XRD patterns indicate that ion-beam assistance plays a role in the composition of the coating layer. Properties of the deposited coating are affected by the difference in their composition.

Bond strength of the samples as a function of ion-beam current is displayed in Figure 8.9a. It increases with the increase of current before heat treatment, as ion bombardment widens the atomic intermixed area, which improves the layer-substrate adhesion strength (Cui *et al.*, 1997, 2000).

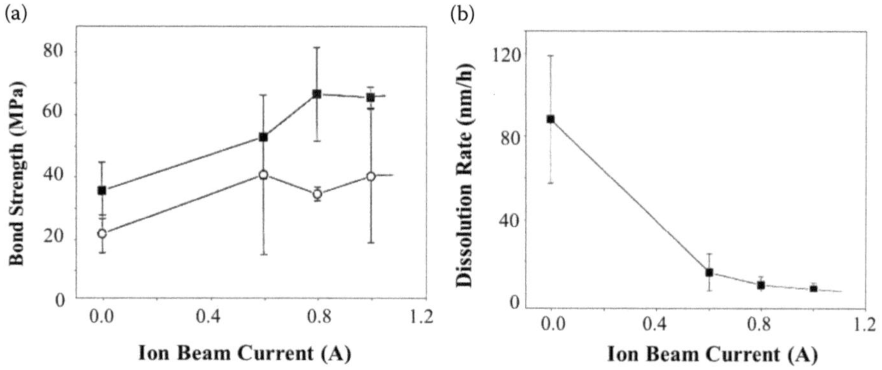


**FIGURE 8.9** (a) Bond strength between coating and substrates before (black) and after (white) heat treatment and (b) Dissolution rate of coating versus the ion-beam current. (From Choi JM *et al. Biomater.* 2000; 21: 469–473. With permission.)

However, bond strength decreases after the heating procedure due to severely cracked and slightly oxidized coating layer by OH, observed with a micrograph (Choi *et al.*, 1998, 2000).

Regardless of decreased in bond efficiency, the purpose of heat treatment has been carried out to crystallize the coating layer to reduce the dissolution rate of coating in a physiological solution for actual applications (Ji and Marquis, 1993; Chen *et al.*, 1994; Choi *et al.*, 2000). Dissolution rate of the coating was decreased by at least a factor of 10 (Figure 8.9b), with the assistance of ion beam during deposition without any heat treatment. The composition and structure of the coating layer became identical with crystalline HA, which has a low dissolution rate when ion beam assisted the deposition process. Hence, HA coating that has high bond strength and low dissolution rate at the same time can be deposited onto metal surface using the ion-beam assistance technique (Choi *et al.*, 2000).

### 8.2.3 Antibacterial Agent Incorporated on Titanium-Based Implants

Biomaterial-related infection, caused by bacterial contamination, is one of the most serious issues causing patient suffering from pain, financial burden or even fatalities in the biomedical field (Darouiche, 2004; Gao *et al.*, 2014). The sources of infection could be transfer of bacteria from infected tissues and hematogenous sources to the implant surface after surgery, insufficient pre-operation disinfection or a mistake during the process of surgery. Bacterial contamination is completely unavoidable, because they will accumulate in extracellular viscous polysaccharides secreted by them and form a biofilm even while applying strict sterilization and aseptic procedures. The bacteria are thus protected by the biofilm from the antibacterial agents, and host defence leads to persistent and chronic infections. Various strategies on surface modification have been developed to equip Ti-based implants with antibacterial ability to resist the bacterial attachment and formation of biofilms (Costerton *et al.*, 1999; Gao *et al.*, 2014). One of the methods is integrating antibacterial agents into the surface of an implant (Kingshott *et al.*, 2003; Harris *et al.*, 2004).

An easy and well-known approach to reduce the bacterial attachment to the implant surface is incorporating silver ions ($Ag^+$) onto the surface of a Ti-based implant (Rautray *et al.*, 2011; Huang *et al.*, 2014b). Investigations show that Ag-implanted Ti and its alloys restrict the growth of *S. aureus* and *E. coli* on the surface compared to the bare surface (Fiedler *et al.*, 2011; Huang *et al.*, 2014b), indicating it has excellent antibacterial activity. However, it causes a cytotoxic effect on osteoblasts and has low biocompatibility (Fiedler *et al.*, 2011; Huang *et al.*, 2014b). Researchers prepared Ca and Ag on nanograined Ti alloy, which enhanced the biological response of osteoblasts, as well as its antibacterial activity (Huang *et al.*, 2014b). Titania nanotubes ($TiO_2$-NTs) gained much attention in this field because of the unique topography and low elastic modulus (Crawford *et al.*, 2007; Park *et al.*, 2007; Mei *et al.*, 2014). Antibacterial ability of $TiO_2$-NTs is improved by storing the right number of antibacterial agents such as Ag in the $TiO_2$-NT structure which act as a reservoir (Ma *et al.*, 2011; Mei *et al.*, 2014).

#### 8.2.3.1 Attachment of Antibacterial Metallic Nanoparticles with Titanium Implants

The Ag nanoparticles are widely used as biomaterials and devices to avoid infection, for example, in burn ointments, vascular grafts and wound dressings (Gao *et al.*, 2014). No living bacterial cells were observed in $1 \times 10^{-5}$ dilution in the case of Ag nanoparticle-incorporated $TiO_2$ NTs. In contrast, bacterial cells remarkably thrived in the case of bare $TiO_2$ NTs. It was also noticed that biocompatibility increased by porous structure, while the bacterio-toxicity was significantly enhanced by Ag nanoparticle sensitization. The antibacterial role of Ag nanoparticles was explained either as a result of $Ag^+$ ions, which usually are capable to interact with the negatively charged bacterial cell wall membrane, thus destroying it, or due to interaction of $Ag^+$ ions with dissolved oxygen to generate activated oxygen (O*), which is able to decompose the bacterium (Petering, 1976; Schreurs and Rosenberg, 1982). Gao *et al.* (2014) integrated Ag nanoparticles into $TiO_2$-NTs arrays using a simple photo-reduction method, wherein the Ag nanoparticles adhered on the sidewalls of the nanotube. The NT-Ag arrays showed long-term antibacterial ability, but the fast release of Ag ions leads to cytotoxicity. Hence, they attempted to develop the new NT-Ag arrays with better cytocompatibility and controllable release of Ag ions. A new strategy was applied to fabricate the new NT-Ag array, with coated Ag to Ti and create Ti-Ag coating before anodization, instead of loading Ag into as-formed $TiO_2$-NTs arrays. The round Ti substrates (with 14 mm diameter × 2 mm thick) are polished to mirror surface and cleaned ultrasonically in acetone, ethanol and deionized water before coating the surface using pulsed DC magnetron sputtering with Ti and Ag targets at ambient temperature. The target power was set at 300 W, working pressure of 0.8 Pa and 80 V of substrate bias, as well as 3 h of deposition time.

The coated substrate is then subjected to electrochemical anodization to create an NT-$Ag_2O$ array with different content of Ag nanoparticles. The electrolyte used was ethylene glycol (EG) solution with 0.3 wt% $NH_4F$ and 2.0 vol% $H_2O$ (Gao *et al.*, 2014). Anodization was carried out for 4 h at room temperature with constant DC potential of 30 V on two electrodes. Figure 8.10a is the schematic diagram representing the formation process of NT-$Ag_2O$ arrays. Figure 8.10b depicts the TEM results of the tested specimens. Dark NPs are randomly dispersed on NT-$Ag_2O$ and no NPs are seen from $TiO_2$-NT under low-magnification images in Figure 8.10b (1–5). The observed NPs are well-embedded in amorphous $TiO_2$ substrate, which can be seen from HR-TEM images (local enlargement in each figure). The values of 0.246, 0.234 and 0.181 nm are the lattice spacing that matched with the (002), (011) and (012) crystallographic planes of hexagonal $Ag_2O$, respectively. As the concentrations of Ag increase in the coatings, the quantities of $Ag_2O$ also increase.

Figure 8.11 shows the SEM images of cell culture results that are done with MC3T3-E1 cells for 1 and 3 days. The SEM images show the cells with a generic polygonal osteoblastic shape. The cells have been spread uniformly on all samples of day 1 cell culture, under lower-magnification.

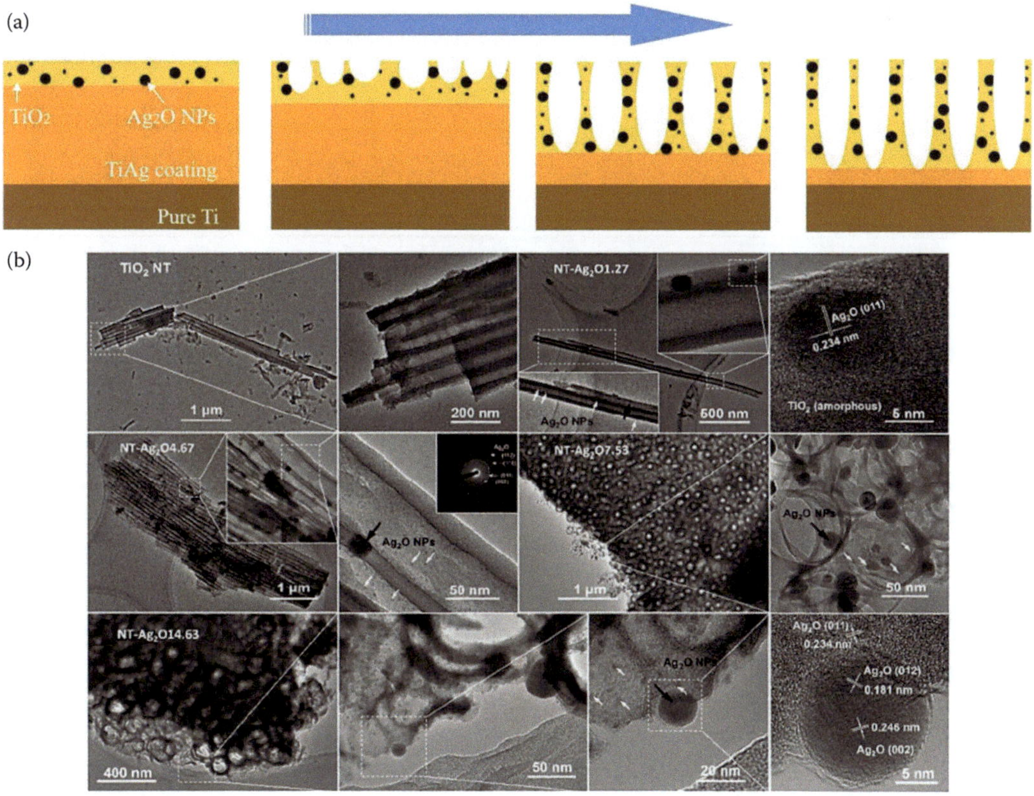


**FIGURE 8.10** (a) Schematic of formation process of the NT-$Ag_2O$ arrays and (b) TEM and HR-TEM images that acquired for $TiO_2$-NT and anodized NT-$Ag_2O$ arrays. (From Gao A *et al. Biomater.* 2014; 35: 4223–4235. With permission.)

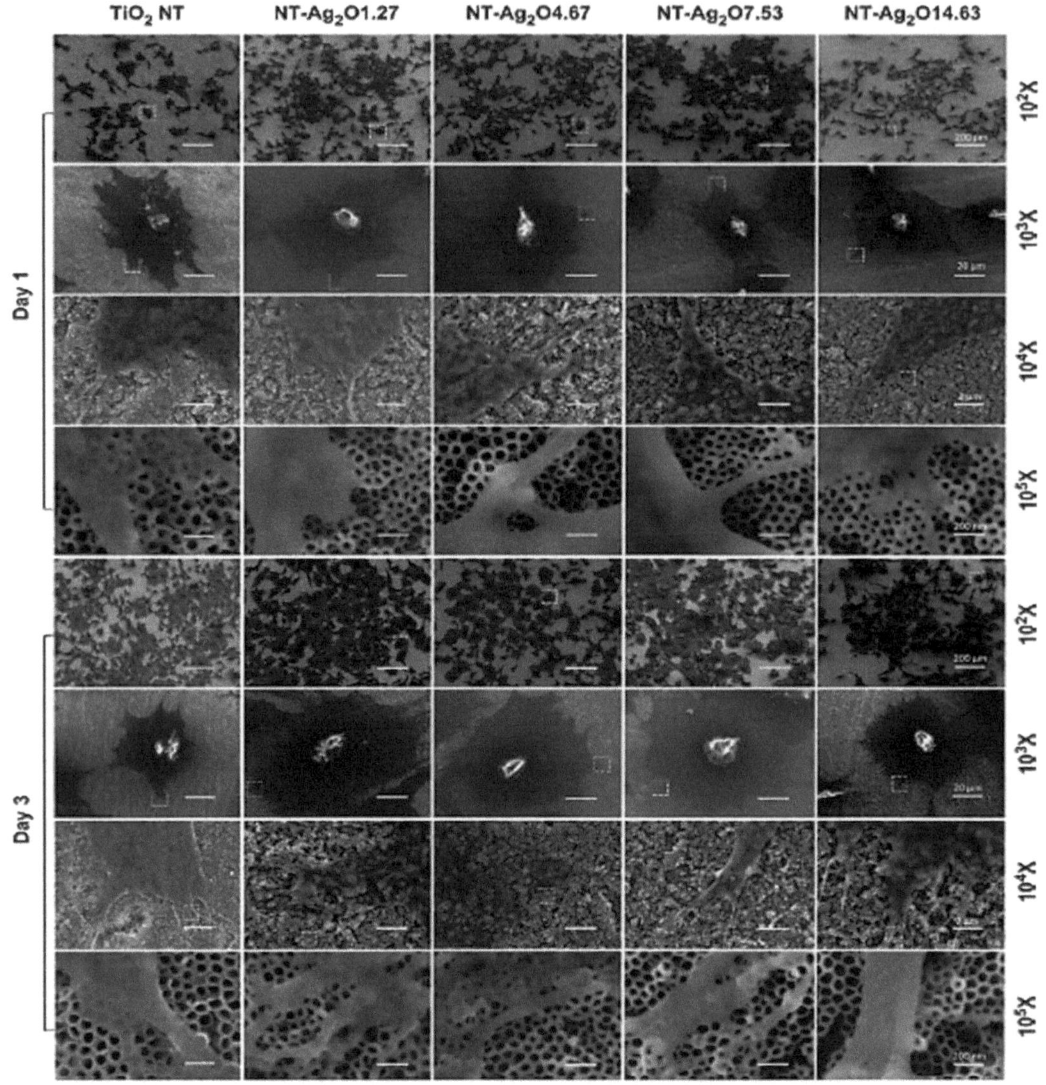


**FIGURE 8.11** SEM images of $TiO_2$-NT and NT-$Ag_2O$ arrays after cell cultured with MC3T3-E1 cells for 1 day and 3 days with different magnification. (From Gao A *et al. Biomater.* 2014; 35: 4223–4235. With permission.)

The finger-like protrusions and filopodia stretched out can be observed in the higher magnification images. The cells, on day 3, have been spread better compared to day 1, as well as the cells on NT-$Ag_2O$ are spread larger compared to $TiO_2$-NT.

Figure 8.12a displays the antibacterial activity of NT-$Ag_2O$ arrays against *S. aureus* and *E. coli* after 28 days of cell culture. A high antibacterial rate of 100% was displayed by all the NT-$Ag_2O$ on day 1 and the antibacterial rate doesn't decrease much at days 7 and 14.

The antibacterial rate decreases relatively fast for NT-$Ag_2O$ arrays with lower Ag concentration; however, the rate still remained at 97% on day 28. The long-term antibacterial activity of substrates is further confirmed with fluorescent staining (Figure 8.12b). The green colour and red stains represent the viable and the dead bacteria, respectively. The result has proven that there are more bacteria attached to $TiO_2$-NTs compared to NT-$Ag_2O$. The results also indicate that the NT-$Ag_2O$ not only can resist bacterial attachment, but also have the ability to kill attached bacteria (Gao *et al.*, 2014).

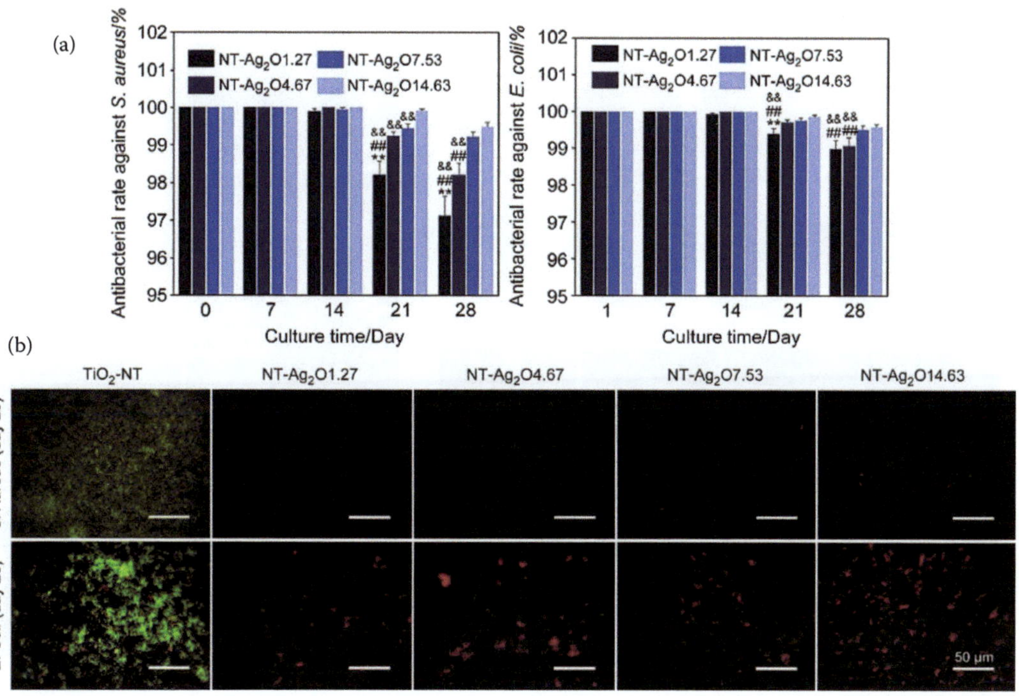


**FIGURE 8.12** (a) Antibacterial activity of NT-$Ag_2O$ arrays towards *S. aureus* and *E. coli*, and (b) Fluorescence images of adhered *S. aureus* and *E. coli* on $TiO_2$-NTs and NT-$Ag_2O$ arrays, with different atomic concentration of $Ag_2O$. (From Gao A *et al. Biomater.* 2014; 35: 4223–4235. With permission.)

Mei *et al.* (2014) used an anodization process to manufacture $TiO_2$-NTs onto a Ti substrate (10 × 10 × 1 mm), then a different depth of Ag was incorporated by low-voltage plasma immersion ion implantation (PIII) without destroying the structure, at 0.5 and 1.0 kV. The substrates are polished and follow by cleaning ultrasonically in acetone, ethanol and deionized water. There is untreated pure titanium (P-Ti) as a control and NT undergoes anodization for 30 min at a constant voltage of 20 V. Two anodized NT-Ag-0.5 and NT-Ag-1.0 subjected to Ag PIII on a plasma ion implanter for 90 min with the same pressure ($1.8 \times 10^{-4}$ Pa), pulse duration of 3000 μs and pulse frequency of 10 Hz and different working voltages of −0.5 kV and −1.0 kV. Bacterial culture with medium containing *Porphyromonas gingivalis* (Pg, Gram-positive bacteria) and *Actinobacillus actinomycetemcomitans* (Aa, Gram-negative bacteria) proceeded under standard anaerobic conditions (80% $N_2$, 10% $H_2$, 10% $CO_2$, at 37°C). Figure 8.13a displays the images of bacterial attachment on samples after one day of culturing. There are large numbers of bacteria attached to the P-Ti and NT but there are fewer on the NT-Ag-1.0 and even fewer on the NT-Ag-0.5.

The cultured samples were further evaluated using SYTO 9 green fluorescent for live bacterial and red fluorescent propidium iodide staining for dead bacteria. Figure 8.13b shows there are large numbers of live bacteria adhered to P-Ti and NT for both of the bacteria. This is while the live bacteria adhered on NT-Ag-0.5 and NT-Ag-1.0 are very few and most of the bacteria are dead on the surface. Fewer bacteria are observed on NT-Ag-1.0, possibly due to the hydrophobic surface; NT-Ag-0.5 has strongest sterilizing effect. The surface morphology of attached bacteria can be seen in Figure 8.13c. The shape of both bacteria Pg and Aa remain as their normal shape on P-Ti and NT samples, after one-day incubation. Most of the bacteria on NT-Ag-0.5 and NT-Ag-1.0 are inactivated as they are hardly intact (Mei *et al.*, 2014).

Besides that, antibacterial effects of the samples are also investigated on days 1, 4 and 7 of incubation. Figure 8.14 show that NT supported the growth of bacteria, possibly because of its great hydrophilicity and the nanotubular structure. In the meantime, a better antibacterial rate on Pa and

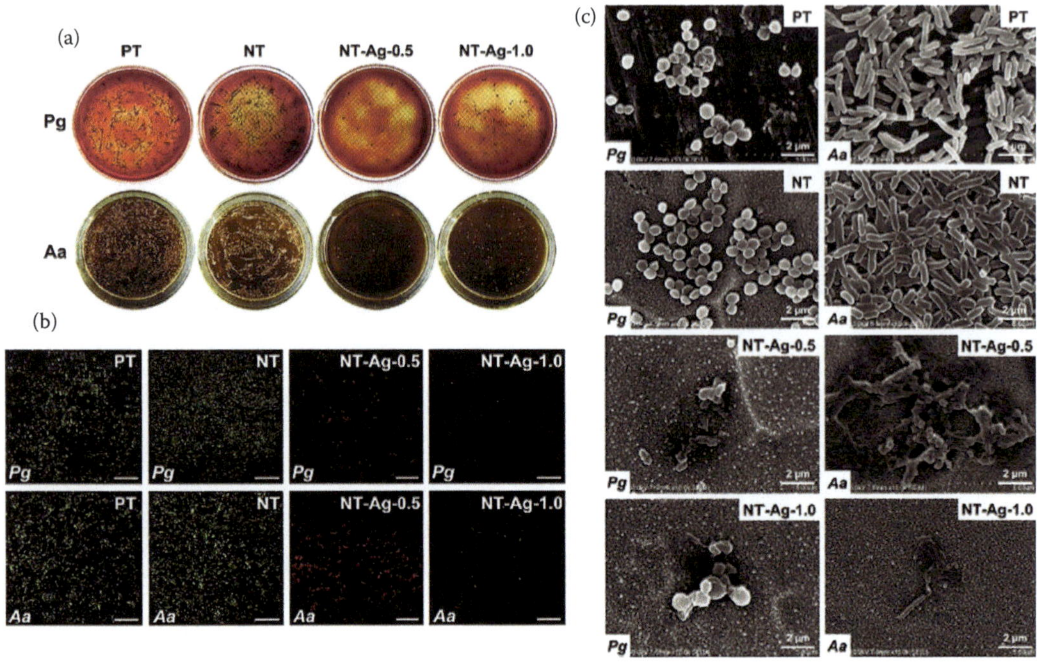


**FIGURE 8.13** (a) Bacterial attachment on samples after one day of culturing, (b) fluorescence and (c) SEM images of adhered Pg and Ag on pure P-Ti, NT and NT-Ag. (From Mei SL *et al. Biomater.* 2014; 35: 4255–4265. With permission.)

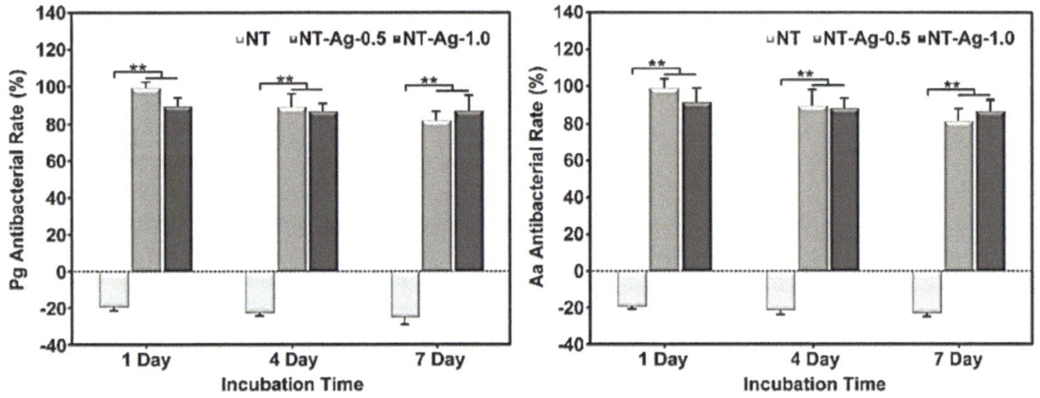


**FIGURE 8.14** Antibacterial rate of NT, NT-Ag-0.5 and NT-Ag-1.0 samples against Pg and Ag versus time of incubation. (From Mei SL *et al. Biomater.* 2014; 35: 4255–4265. With permission.)

Aa was observed in the samples that incorporated with Ag compared to NT. The NT-Ag-0.5 has better ability to fight bacteria compared to NT-Ag-1.0 after day 1 and day 4 of cell incubation, but the fighting rate decreases with increasing incubation time. However, the antibacterial rate of NT-Ag-1.0 remains nearly constant and has a higher antibacterial rate than NT-Ag-0.5 on day 7 (Mei *et al.*, 2014).

### 8.2.3.2 Combination of Drugs with Titanium Implants

The wide range of implant devices could be possibly built with polymeric biomaterial incorporated with drug therapies to enhance the therapeutic effects. A local drug-delivery system is created when drug-releasing agents are utilized on certain implant surfaces. Direct delivery of agents with short half-lives such as proteins and high localized drug concentrations with prolonged preservation at lower doses are allowed to site a minimal loss. Besides, the toxicity of the system can also be minimized (Lin and Metters, 2006; Choi *et al.*, 2009). The layer by layer (LbL) self-assembly method is used to fabricate the drug-delivery system because of its flexible choice of assembly components and easy process (Wood *et al.*, 2005; Zhang *et al.*, 2005; Ren *et al.*, 2006; Choi *et al.*, 2009). The drug-delivery system is used to keep up and control the release of integrated molecules (Wood *et al.*, 2005; Choi *et al.*, 2009). Bioactive molecules can be integrated onto a multilayered structure through submersion of a film into bioactive molecule's solution (Etienne *et al.*, 2005; Choi *et al.*, 2009). The LbL method is also one of the methods that used to incorporate bioactive agents within the multilayered structure (van den Beucken *et al.*, 2006; Choi *et al.*, 2009).

Choi *et al.* (2009) proposed that the factor that affects the controlled release of bioactive agents of a multilayered hydrogel is the location of a drug-loaded layer. They fabricated a multilayered phospholipid polymer hydrogel integrated with paclitaxel (PTX) as the bioactive agent on a Ti alloy surface to investigate the controlled drug release of the coating. An intermolecular reaction around two water-soluble polymers, phospholipid polymer (PMBV) and poly(vinyl alcohol) (PVA) solutions via LbL were used to create the coating. PMBV was first fabricated using 4-vinylphenylboronic acid units (VPBA), 2-methacryloyloxyethyl phosphorylcholine (MPC) and n-butyl methacrylate (BMA) (Sawada *et al.*, 2006; Choi *et al.*, 2009). PMBV and PVA were used to fabricate a PMBV/PVA multilayered hydrogel coating, because the phenylboronic acid of PMBV allows formation of a hydrogel with PVA (Konno and Ishihara, 2007; Choi *et al.*, 2009). A polar group, phosphorylcholine (PC) group is found in both MPC and living cell membranes. Therefore, biocompatibility of MPC polymer is better compared to another polymer (Ishihara *et al.*, 1992; Choi *et al.*, 2009). Additionally, formation of hydrophobic domains took place in aqueous solutions due to the present of a BMA unit in PMBV. Figure 8.15 illustrates the multilayered coating incorporated with PTX formed on the specimens that were fabricated by the LbL method (Wada *et al.*, 2007; Choi *et al.*, 2009).

Choi studied the release of PTX from two multilayered phospholipid polymer hydrogel coatings on a Ti substrate with different concentrations of 50 mg/mL PMBV and 15 mg/mL PVA (PMBV50/PVA15) and 25 mg/mL PMBV and 15 mg/mL PVA (PMBV25/PVA15). PTX was loaded with two layering schemes, top and bottom and samples submersed in PBS to study the release. Table 8.2 indicates the obtained results from the experimental work. The initial burst released all loaded multilayered hydrogels for the first 7 hours of immersion are ranging from 40% to 82%. The bottom scheme PMBV50/PVA15 coatings had the lowest burst release, which is 40% and there was a lag

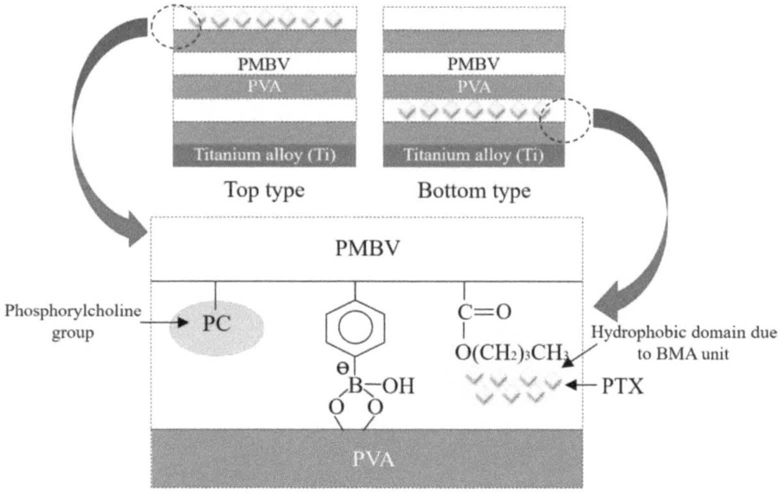


**FIGURE 8.15** A schematic representation of multi-layered PMBV and PVA integrated with PTX. (From Choi J *et al. Biomater.* 2009; 30: 5201–5208. With permission.)

**TABLE 8.2**
**The Release of PTX from Multilayered Hydrogel Coating**

| PTX-Loading Type | Coating Materials | Burst Release (0–7 h) (%) | Total Amount of PTX Released (μg) | Sustained Release (After 7 h) (%) |
|---|---|---|---|---|
| **Bottom type** | PMBV50/PVA15 | 40.0 | 0.40 | 37.5 |
| | PMBV25/PVA15 | 58.8 | 0.36 | 18.1 |
| **Top type** | PMBV50/PVA15 | 81.6 | 3.95 | 16.4 |
| | PMBV25/PVA15 | 45.5 | 3.48 | 24.9 |

*Source:* Choi J *et al. Biomater.* 2009; 30: 5201–5208. With permission.

time of 3 hours for PMBV50/PVA15 and 5 hours for PMBV25/PVA15 in the case of the bottom-layering scheme when starting release of PTX. Constant rate of release of the remaining PTX (16%–38%) was observed after the initial burst. The reason for burst time occurring is because early burst was reduced for bottom-type layering and top-type owning diffusion barrier (Choi *et al.*, 2009).

Bone loss is prevented by using a bisphosphonate (BP), which is a widely useful class of drug that restricts osteoclast-mediated bone resorption, impairs blood supply and improves osteoclast apoptosis (Brian *et al.*, 2005; de Oliveira *et al.*, 2015). Zoledronic acid (ZA) is a nitrogen-containing BP that is used in a range of anticancer protocols, together with angiogenic agents and glucocorticoids, to obtain a synergistic effect that could improve apoptosis (Bagan *et al.*, 2006; de Oliveira *et al.*, 2015). It is suggested that the low rates of osteonecrosis or failure linked with the implant placement, among individuals taking BPs, was evaluated to other invasive dental procedures (Madrid and Sanz, 2009; de Oliveira *et al.*, 2015). de Oliveira *et al.* (2015) conducted research investigating effects of the systemic use of ZA in conjunction with or without dexamethasone (DX) on osseointegration of Ti endosseous implants in male Wister rats; 27 Wister rats were divided into 3 groups and each rat is implanted with 2 sterile custom-fabricated Ti implants in the tibias. Group 1 was treated with ZA only. ZA and DX are used to treat group 2, and group 3 acts as a control receiving saline solution injections. Histological and histomorphometric analysis were carried out after selecting 3 rats from each group, on day 7, 14 and 28 of implantation.

Toluidine blue staining was carried out using light microscopy to perform histological analysis on the non-decalcified implants that were removed from the rats. All samples from all groups displayed existence of bone fragments in the medullary region, and cortical bones have close contact with implants after 7 days of implantation. The samples across the groups have identical bone cellularity and vascularity. Moreover, inflammatory, cell infiltrate, osteoclastic and osteoblastic activities were not presented. On day 14, new bone deposition around implants and the outer surface of tibia was observed, and the bone fragments were not noticed in all 3 groups. There was no remodelling of cortical bone in contact with implants for groups 1 and 2. However, new bone deposition along crucial bone area contacting with the implant was observed in group 3. The other thing observed on all samples is scarce inflammation. On day 28, a few samples from groups 1 and 2 and most of the samples from group 3 exhibited cortical bone remodelling. There is new bone deposition in the medullar area, and no inflammation or bone necrosis was observed.

### 8.2.4 Hybridization of the Organic Compound with Titanium-Based Implants

Ti-based alloy biomaterials being the favourite metal in oral and orthopaedic due to its good mechanical properties, but it is proven that there are negative effects of implantation which presented in various studies, such as inflammation because of the released toxic metallic ions (Hallab *et al.*, 2001; Adya *et al.*, 2005; Keegan *et al.*, 2007; Negroiu *et al.*, 2008). The defects can be overcome by hybridizing the organic compound with other materials to form a coating on the surface of the

Ti alloys. One of the novel methods for metallic surface treatment is to deposit a nanoparticle layer, including carbon nanoparticles-carbon nanotubes (CNTs), which show unique biological properties in application of medical therapies and diagnostics (Zhang *et al.*, 2010; Liao *et al.*, 2011; Dlugon *et al.*, 2015). CNTs can be utilized as a substrate for bone-tissue mineralization in the field of tissue engineering (Cho *et al.*, 2009; Dlugon *et al.*, 2015). Recent investigations also presented that CNTs are a perfect surface for cell adhesion and growth, as well as not causing strong inflammatory reactions such as necrosis in rat subcutaneous tissue (Yokoyama *et al.*, 2005; Aoki *et al.*, 2006; Inoue *et al.*, 2013). CNTs have been modified further to increase the range of application (Dlugon *et al.*, 2015). CNTs are incorporated with HA to reinforce the fracture toughness and wear resistance of the HA coating. The investigations show that multiwalled CNT incorporated with the HA coating develops the fracture toughness by 56% and crystallinity by 27% (Chen *et al.*, 2005; Rapacz-Kmita *et al.*, 2005; Balani *et al.*, 2007).

Various compounds are combined with HA to boost the biological and mechanical performances of the HA coating. Integration of polymers in the HA coating was widely investigated in their preparation, properties and application. Lack of toxicity, excellent biocompatibility and ease of availability are the reasons why natural polymers are used to integrate with HA (Kong *et al.*, 2006; Teng *et al.*, 2006; Wahl and Czernuszka, 2006). Some synthetic water-soluble polymers were also utilized and acted as crystallization promoters in HA/polymer composites. The polymers affect being assigned to the partial or total interfacial adsorption of polymer on the surface of growing crystals. Polyelectrolytes such as dermatan sulphate ($C_{14}H_{21}NO_{15}S_2$), heparin ($C_{12}H_{19}NO_{20}S_3$) and chondroitin sulfate ($C_{13}H_{21}NO_{15}S$) are the common polymers that integrated with HA (Bajpai and Singh, 2004). These polymers have the capability to control the crystal growth of organic or inorganic sparingly soluble salts (Dlugon *et al.*, 2015). Surface functionalization is another way of grafting the different chemical groups with covalent bonds onto the surface of a material to obtain desired biological responses (Dubruel *et al.*, 2006; Michiardi *et al.*, 2010). The cell attachment, activity and spreading of the materials can be modulated when polymers with appropriate chemical functionalities such as sulfonate and carboxylate group are grafted onto surfaces (Yammine *et al.*, 2005; Michiardi *et al.*, 2010). The investigations also show that grafting of styrene sodium sulfonate (NaSS) improved the osteoblast functionality and prevented the adhesion of bacteria (Helary *et al.*, 2009; Michiardi *et al.*, 2010).

#### 8.2.4.1 Interlock of Carbon Nanotubes (CNTs) on Titanium Implants

Bioactive ceramic materials can be fabricated with various methods such as plasma spray, hydrothermal methods and sol-gel to achieve desired properties for different biomaterials to be suited to different medical applications (Lu *et al.*, 2004; Dlugon *et al.*, 2015). Dlugon *et al.* (2015) reported the results of multi-walled CNTs that fabricated on the Ti surface using the electrophoretic deposition (EPD) process. Short CNTs with 1–2 μm width, 10–30 nm outer diameter and 5–10 nm inner diameter were functionalized and purified in a mixture of 65% nitric (V) acid and sulphuric (VI) acid at a ratio of 1:3. Nanotubes were rinsed with distilled water and centrifuged after submersion for 2 h in an acidic mixture at 70°C. Carboxylic (–COOH) and hydroxyl (–OH) groups formed on the surface of the CNTs after functionalization. Both sides of the Ti plates, 10 mm diameter, are then functionalized with CNTs. The Ti plate was etched in 5% hydrofluoric acid (HF) for 30 s and rinsed with distilled water prior to deposition of CNTs. The plate (anode) is placed in among two negatively charged electrodes immersed in CNT suspension with applied static voltage of 28 V and current ranging from 6 to 13 mA for 30 s. Figure 8.16 displays the atomic force microscopy (AFM) image of the deposited CNT layer on the surface of the Ti plate. The surface has isotropic morphology and it is uniform. The image also shows a fibrous mesh-like structure and bundles formed due to the aggregation of nanotubes. Corrosion resistance of specimens also are investigated by the linear polarization resistance (LPR) method.

"Surface-most layer of CNT" and "nanocomposites layer of CNT," built into Ti oxide, are the two main parts of the surface of Ti-CNT composites. It can be assumed that the fluctuation of

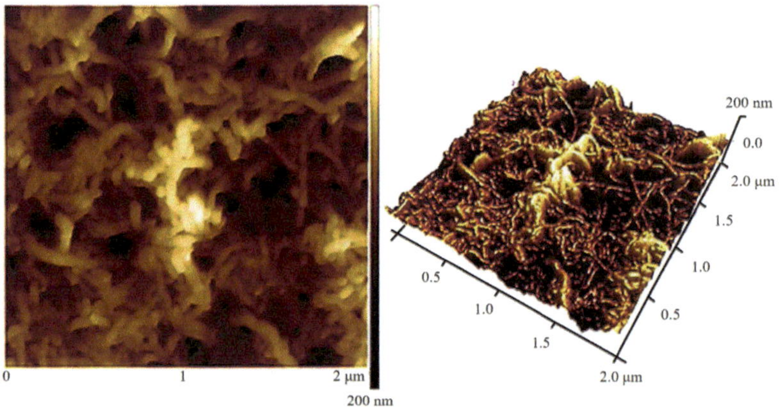


**FIGURE 8.16** An AFM image of deposited CNT layer on Ti. (From Dlugon E *et al. Bull Mater Sci.* 2015; 38: 1339–1344. With permission.)

corrosion potential to positive (0.270) is the responsibility of the nanocomposites part. The rise of corrosion current density and reduction of corrosion resistance are because of the porous fibrous and conductive CNT layer on the surface. These assumptions could be further confirmed with cyclic polarization curves (Dlugon *et al.*, 2015). The graph in Figure 8.17a is the normal human osteoblast (NHOst) cell viability after cell incubation. The cell viability of Ti and Ti-CNT is lower compared to the control (after 3 days of incubation); however, the Ti-CNT achieve best cell viability

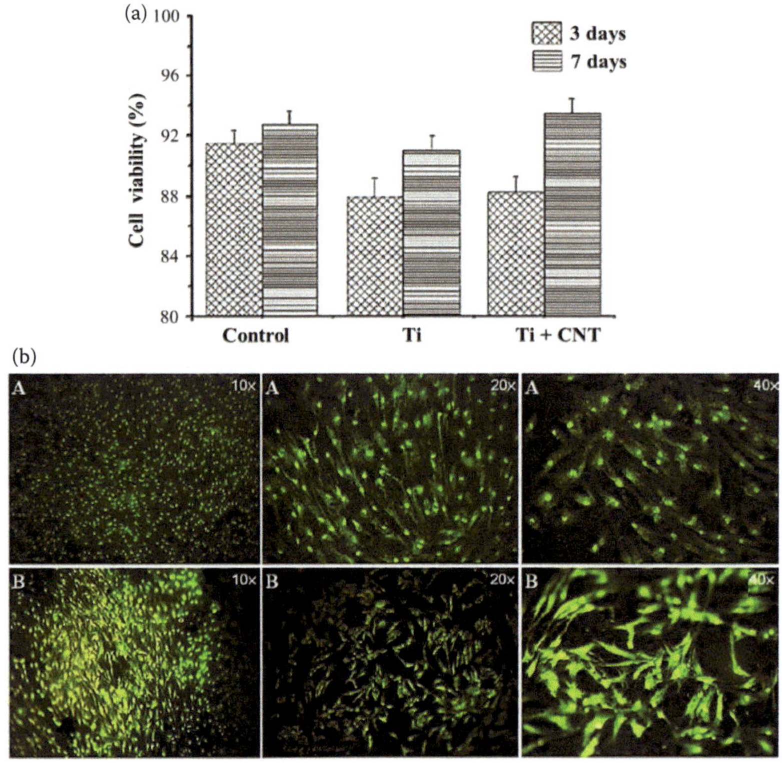


**FIGURE 8.17** (a) NHOst cells viability after cell incubation for 3 and 7 days and (b) NHOst cell morphology of Ti (A) and Ti-CNT (B) (with 10×, 20× and 40× magnifications) specimens. (From Dlugon E *et al. Bull Mater Sci.* 2015; 38: 1339–1344. With permission.)

after 7 days of incubation (Figure 8.17a). The best achievement in 7 days of incubation is because of the presence of nanotopography and functional groups on the surface of Ti-CNT, as well as the biomimetic shape. Based on the previous studies, nanotubes exhibited excellent bioactivity when contacting cells. Figure 8.17b demonstrates the cell morphology of the specimens after 7 days of cell incubation. The cytoplasm and nucleus are clearly displayed in this image. Ti cell (Figure 8.17b-A) displays a fibrous shape and CNT coating (Figure 8.17b-B) displays cells with a flattened shape that indicate better adhesion (Dlugon *et al.*, 2015).

Inoue *et al.* (2013) fabricated multiwalled carbon nanotubes (MWCNTs) on anodized Ti to investigate the behavior of *Sarcoma osteogenic* (SaOS2) cells and bone tissue on this coating. Ti plate and wire were polished with #600 and #1200 grit emery papers and cleaned ultrasonically with acetone and distilled water. The anode was connected with prepared Ti plate and wire while the cathode was connected with Ti plate with a thickness of 0.1 mm for the anodic oxidation. The electrolyte was an aqueous solution of 0.2 M calcium acetate ($C_4H_6CaO_4$) and 0.02 M DL-α-glycerophosphate disodium salt (DL-α-GP). The electrolyte was maintained at room temperature (25°C) with a current density of 30 mA/cm$^2$ and anodic voltage of 320 V. The Ti samples are anodized (denoted as Ano-Ti) after going through this process. The Ano-Ti samples were silanized with 3-aminopropyltriethexysilane ($C_9H_{23}NO_3Si$) and soaked in carboxylated MWCNT suspension for 12 hours at 100 ppm in order to form a CNT coating on the Ano-Ti (denoted as CNT-Ano-Ti) surface (Simka *et al.*, 2011). The surface of samples were seeded with 500 μl of cell suspension and incubated for 3 days in a 5% $CO_2$ atmosphere at 37°C. In the meantime, CNT-Ano-Ti wires were inserted into bone marrow space in femurs of rats and removed 2 weeks after implantation. The wires were planted into paraffin and stained with hematoxylin and eosin before further histological study was performed (Inoue *et al.*, 2013).

#### 8.2.4.2 Connective Polymer on Titanium Implants

There are new techniques that have been developed for functionalizing biomaterial surfaceds to alter the surface properties of the biomaterial. Michiardi *et al.* (2010) synthesized a new and easy method to graft polymers bearing sulfonate (styrene sodium sulfonate, NaSS) and carboxylate (methacrylic acid, MA) groups onto a Ti6Al4V surface. They aimed to improve anchorage of hip prostheses into bone by this investigation. The Ti6Al4V (disk or cylinder) substrates were polished, and both of the monomers NaSS and MA were purified by recrystallization in a mixture of ethanol and water and distillation, respectively. Oxidation of samples was carried out in a solution containing pure sulfuric acid ($H_2SO_4$) and hydrogen peroxide ($H_2O_2$) and oxygen-free environment for 1 or 3 min. Samples were rinsed after the oxidation process, then followed by submerging them into a monomer solution and radical polymerization for 15 h at temperature of 70°C (Helary *et al.*, 2009; Michiardi *et al.*, 2010). Some samples were submerged into 0.7 mol.l$^{-1}$ NaSS solution to form Ti alloy grafted with NaSS ($Ti_{NaSS}$) and some were submerged into MA: NaSS (80:20 mol) solution to form Ti alloy grafted with MA/NaSS copolymer ($Ti_{MA/NaSS}$). Untreated Ti6Al4V ($Ti_{alloy}$) acted as a control so that the obtained results could be compared.

Coluidine blue solution for 6 hours at 30°C was immersed for making complexes with surface grafted polymers. 4.2 ± 0.3 μg/cm$^2$ is the amount of grafter polyNaSS on the surface of Ti6Al4V that determines using toluidine blue colorimetric method, while the grafted MA/NaSS copolymer on the Ti6Al4V alloy is 1.3 ± 0.3 μg/cm$^2$. Even though there is a three times weight difference between these two films, both of the grafted films are similar in a molar basis. There are no significant differences on the amount of grafted polyNaSS or copolymer on Ti alloy that oxidized for 1 or 3 min. Cell culture characterization, with MG63 osteoblast-like cells was carried out and the cell adhesion was obtained using the equation (N1-N2)/$N_{cs}$. The results are demonstrated in Figure 8.18a. The higher percentage of the (N1-N2)/$N_{cs}$ gives higher number of cells detached from the sample surface, after applying 8 dyn.cm$^{-2}$ of shear force (Michiardi *et al.*, 2010).

The adhered cells on the $Ti_{NaSS}$ sample were more capable to resist detachment under the applying shear force. The percentage of untreated $Ti_{alloy}$ fell down between both of the grafted surfaces ($Ti_{NaSS}$ and $Ti_{MA/NaSS}$). The percentage of $Ti_{MA/NaSS}$ sample was the highest, which means the presence

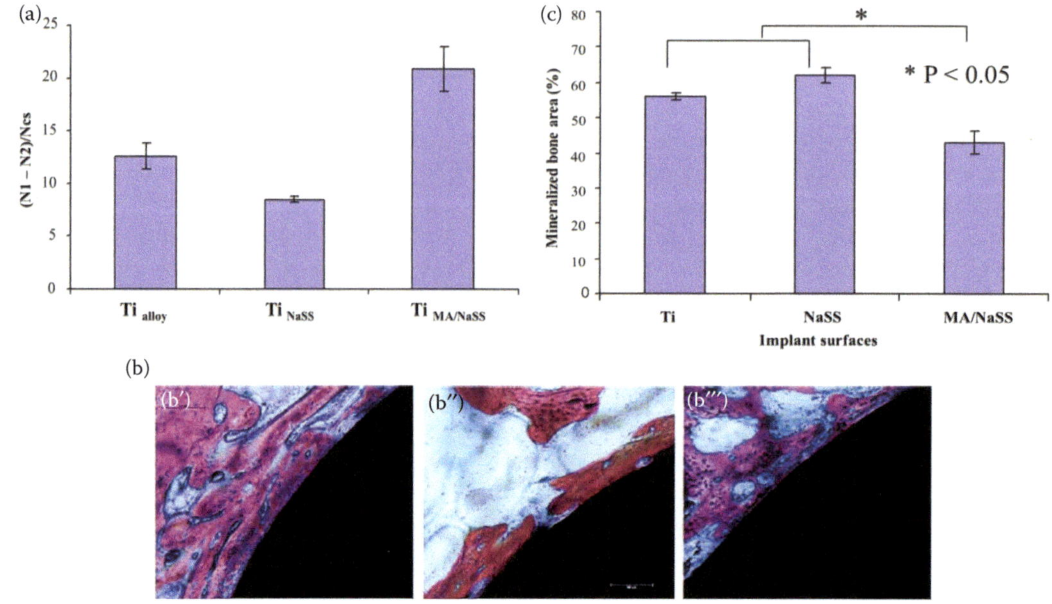


**FIGURE 8.18** (a) MG63 cell adhesion results using $(N_1\text{-}N_2)/N_{cs}$, (b) Micrograph (b′: Ti $_{alloy}$, b′: Ti $_{NaSS}$ and b″: Ti $_{MA/NaSS}$) of representative histology of the specimens after 4 weeks implantation (stained pink is representing the bone tissue that surrounded the surface of implants) and (c) Percentage of mineralized tissue on the implants that inserted into femoral rabbit model. (From Michiardi A *et al. Acta Biomater.* 2010; 6: 667–675. With permission.)

of NaSS and MA were allowing MG63 cell detachment under applied shear force. *In vivo* was carried out by inserting samples in a femoral rabbit for 4 weeks to evaluate the performance by histology and measurable histo-morphometry. The inserted implants were directly impact with the bone (Figure 8.18b), and no fibrous encapsulation of implants was perceived in any histological section (Michiardi *et al.*, 2010). Lamellar trabecular bone with wide haversian canals surrounded the grafted and control implants. Figure 8.18c shows the obtained percentage of mineralized tissue fraction from the *in vivo* study. The percentage of grafted polyNaSS (59% ± 4%) and untreated Ti alloy (57% ± 5%) are very similar; therefore, the grafted polyNaSS does not affect the mineralized content. 47% ± 5% is the percentage for MA/NaSS which shows a slight decline (Michiardi *et al.*, 2010).

Negroiu *et al.* (2008) fabricated nanocomposites of HA and sodium maleate ($C_4H_4Na_2O_5$) copolymer onto Ti substrates using matrix assisted pulsed laser evaporation (MAPLE) technique. Sodium maleate (NaM)-vinyl acetate (VA), with a ratio of 1:1, is the used copolymer and it was created using mild hydrolysis with diluted aqueous NaOH solution for 24 h at room temperature. Then, a solution was purified until its filtrate conductivity reached below 20 μS/cm. The NaM-VA copolymer was changed to denote MP, after it was recovered using freeze-drying (Chitanu *et al.*, 2005; Negroiu *et al.*, 2008). In order to fabricate HA-MP nanocomposites, a hydrothermal method is used by mixing soluble salts of calcium and ammonium phosphate ($H_{12}N_3O_4P$) aqueous solutions into an ammonia solution and MP. The nanocomposites powder is then hydrothermally synthesized for 3 hours at 100°C under pressure between 20 and 100 MPa. The process is then followed by filtering the precipitate, washing and oven drying (at 100°C). The dried targets, including 0.2% and 1% HA-MP powder containing 20% of NaM-VA copolymer, are used for the MAPLE technique to coat the Ti substrates at pressure of 13 Pa (under $N_2$ gas) and temperature of 30°C. (Note: The Ti coated with 0.2% and 1% HA-MP are denoted as HA-MP1 and HA-MP2, respectively.) The HA is also coated on Ti by PLD technique, as a third group, to investigate the difference between copolymer HA coatings (HA-MP1 and HA-MP2) and HA-Ti coating. The human embryonic kidney (HEK), human dermal fibroblast (HDF) and human mesenchymal stem (MSC) cells are used for 24 hours to investigate the biocompatibility of the obtained three groups of HA-MP1, HA-MP2 and HA, Ti coatings. The cells were grown on the all substrate at atmosphere of 19 air/1 $CO_2$ at 37°C (Negroiu *et al.*, 2008).

Cytoskeleton dynamics and nuclei morphology of both HA-MP nanocomposites surface have studied and the staining results of actin, microtubules and nuclei are shown on Figure 8.19. The red fluorescence is the actin long fibres, which form a network by extending all over the cell body for both HDF and MSC, while for the case of HEK cells is concentrated in the cell extremities. There are no visible differences of actin patterns between Ti-HA and Ti-HA-MP nanocomposites for cells MSC and HDF; however the MSC cells that attached to Ti-HA-MP2 are showing more intense cell growth compared to Ti-HA-MP1 (Negroiu *et al.*, 2008). HDF were positive for nuclear staining in some surface area of Ti-HA-MP and poorly stained for polymerized actin. The strong actin staining

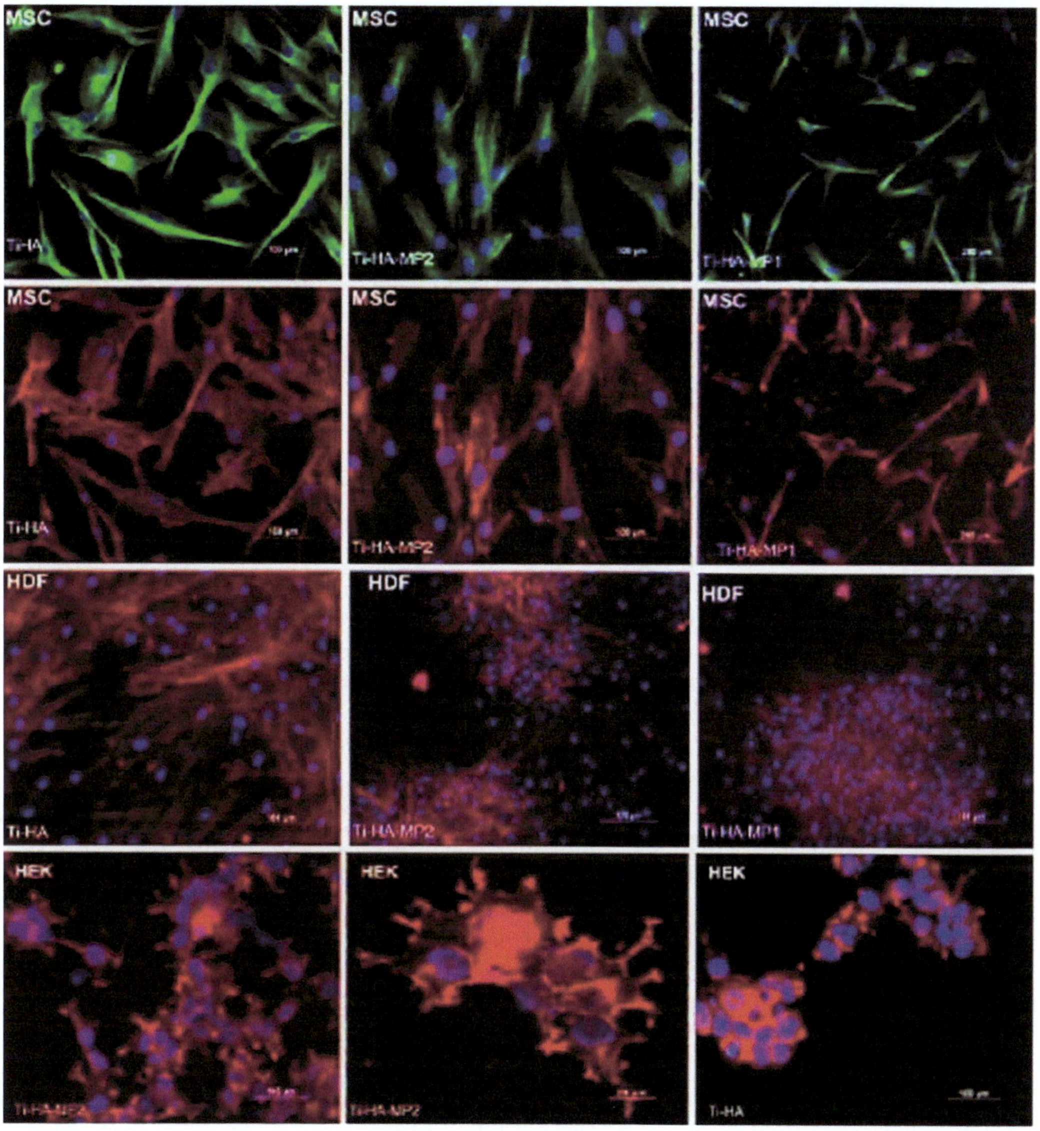


**FIGURE 8.19** Confocal images show cytoskeleton organization and nuclei morphology of MSC, HDF and HEK on various specimens that investigated. (From Negroiu G *et al. J Mater Sci-Mater Med.* 2008; 19: 1537–1544. With permission.)

of HDF displays a good spreading on the surface of Ti-HA-MP. The green fluorescence presents (Figure 8.19) the cortical microtubule network of MSC on Ti-HA or Ti-HA-MP which aligned with cell axis. This indicate that they are interphase and do no co-localize with actin. HEK cells exhibited small or fragmented nuclei and the other cells have uniformly stained nuclei with normal round shape morphology on Ti-HA or Ti-HA-MP (Negroiu *et al.*, 2008).

### 8.2.5 Adsorption of Protein on Titanium-Based Implants

Protein adsorption is a progressive method and the factors that drive this process are noncovalent hydrophobic interactions, van der Waals forces, electrostatic forces and hydrogen bonding (Andrade and Hlady, 1986; Kopf *et al.*, 2015). The noncovalent interactions are used to control the protein parameters such as secondary and tertiary structure of proteins and protein size (Rabe *et al.*, 2011; Kopf *et al.*, 2015). In addition, surface energy, roughness and chemistry of an implant also play a main role in influencing protein adsorption (Haynes *et al.*, 1994; Deligianni *et al.*, 2001; Rabe *et al.*, 2011; Kopf *et al.*, 2015). Hydrophilicity and hydrophobicity of biomaterials have a significant role in early bone response. It is reported that Ti-based implant with hydrophilic surface chemistry can improve cell response and enhance more quick osseointegration (Jimbo *et al.*, 2011; Hirakawa *et al.*, 2013; Kopf *et al.*, 2015). The blood-coagulation cascade is one of the mechanisms of adsorption of protein in biological processes (Rabe *et al.*, 2011; Kopf *et al.*, 2015). Fibrinogen and fibronectin are two proteins that are contained in blood plasma and play a significant role in blood coagulation. Fibrinogen is required for integrin-receptor-based cell adhesion and proliferation on surfaces, while fibronectin is an important component for extracellular matrix that prompt cell adhesion, as it is a cell-adhesion-promoting protein (Scotchford *et al.*, 2002; Laurens *et al.*, 2006; Kopf *et al.*, 2015).

Kopf *et al.* (2015) studied the adsorption levels of fibrinogen and fibronectin on Ti and TiZr surfaces with different surface properties such as wettability, (hydrophilic, $H_{phil}$) and (hydrophobic, $H_{phob}$) with or without nanostructures (NS). Six samples were fabricated, denoted as Ti SLA-$H_{phob}$, Ti SLA-$H_{phil}$, Ti SLA-$H_{phob}$ NS, Ti SLActive-$H_{phil}$ NS, TiZr SLA-$H_{phob}$ and TiZr SLActive-$H_{phil}$ NS, where the SLA term is used for "sandblasted with large grits and acid etched." The large grit sandblasting process produced a macro-roughness surface with a lateral feature size of approximately 20–40 μm. After that, acid etching created micro pits with a lateral feature size of about 2–4 μm. Protein-adsorption incubations in protein solutions, containing of fibrinogen (7 μg/mL) and fibronectin (3 μg/mL), were performed in the dark area for 10 min at room temperature.

β-type TiNbZr alloys gained much research attention for orthopaedic implants in the biomedical field due to the lower elastic modulus that is closer to human bone compared to pure Ti. β-type TiNbZr alloys can reduce the stress-shielding effect between bone and implant that can cause implant loosening and bone resorption (Jeong *et al.*, 2011; Nauth *et al.*, 2011; Ruther *et al.*, 2011; Huang *et al.*, 2014b). A porous structure on a surface a biocompatible material can enhance the biocompatibility of an implant, including protein adsorption on biomaterial surface, which may lead to enhancement of cell responses (adhesion, migration proliferation and mineralization) (Huang *et al.*, 2013; Huang *et al.*, 2014a). Huang *et al.* (2014b) developed a nanoporous surface on a β-type Ti25Nb25Zr alloy using simple electrochemical anodization methods to investigate the protein adsorption and cell responses. They hypothesized that the protein adsorption and cell responses would be enhanced with a nanoporous topographic surface structure on β-type Ti25Nb25Zr alloy.

The β-type Ti25Nb25Zr substrate discs were prepared using waterproof silicon carbide papers, followed by anodization using Galvano stat in a 5 M NaOH electrolyte for 10 min at temperature ~25°C and anodic current of 0.2 A (Huang *et al.*, 2014b). They used albumin and fibronectin proteins for their protein adsorption studies. The substrates were submerged in protein containing (5 mg/mL) phosphate-buffered saline (PBS) for 5 min and followed by washing with deionized water. The samples were then denoted as Ti25Nb25Zr-A for anodized sample and Ti25Nb25Zr-P for untreated specimen. Figure 8.20a shows the optical and FE-SEM micrographs of polished and ionized surfaces of specimens. The flat surface with light grey colour (Figure 8.20a-a′) is Ti25Nb25Zr-P. The sapphire

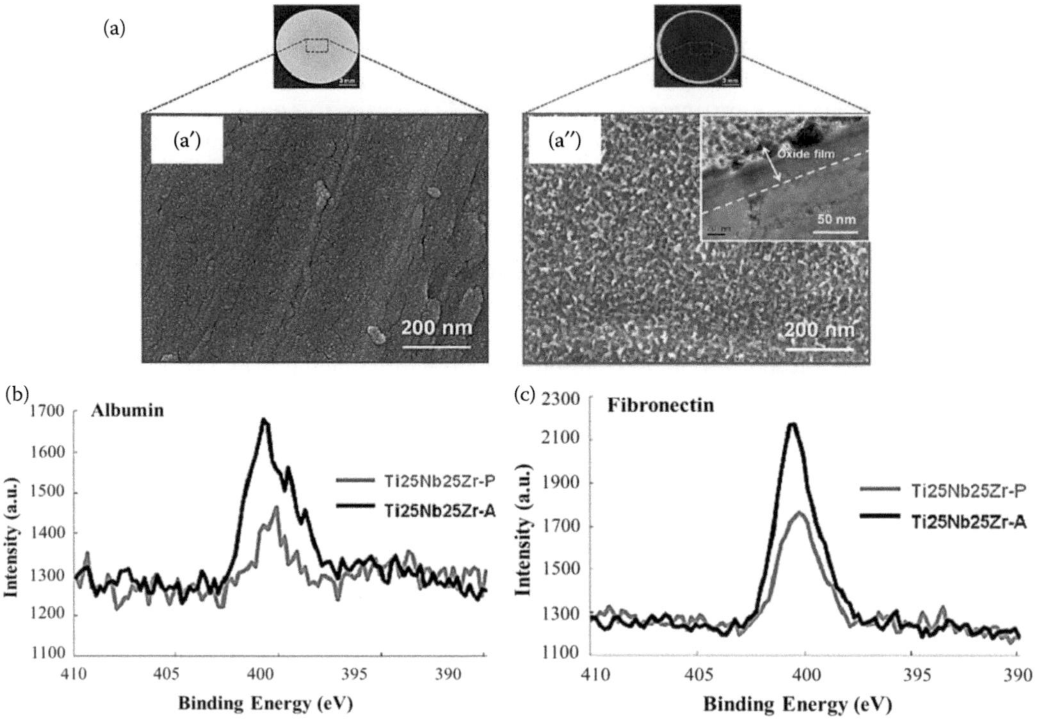


**FIGURE 8.20** (a) Optical and FE-SEM micrographs of polished surface and ionized surface of specimens (a′: before and a″: after anodization), and XPS analysis results of (b) albumin, (c) fibronectin proteins absorption after 5 min of immersion. (From Huang H-H *et al. Sur Coat Tech.* 2014a; 259(Part B): 206–212. With permission.)

blue colour (Figure 8.20a-a″) nanoporous topography with an average pore size of 8 nm and anodized layer thickness of 40 nm is Ti25Nb25Zr-A. The XPS analysis results, in terms of N1s spectra, of the specimens after 5 min of immersion in albumin- and fibronectin-containing solution, are presented in Figure 8.20b and c, respectively. The Ti25Nb25Zr-A specimen has better performance as it showed higher N1s intensity in both albumin (3.9-times) and fibronectin (1.8-times). This indicates that Ti25Nb25Zr-A has better ability in protein adsorption compared to Ti25Nb25Zr-P. This high level of protein adsorption, cell responses and modulating cell adhesion on the sample with nanoporous structure is due to the fact that nanoscale proteins are capable of interacting with a nanoscale surface topography (nanoporous) instead of untreated smooth surface (Song and Chen, 2007). This high adsorption of protein on nanoporous surface provides a larger number of sites for cellular bonding (Song and Chen, 2007; Huang *et al.*, 2014a).

## 8.3 CONCLUSIONS

The recent investigations show that a tantalum oxide coating can improve corrosion resistance and nanoporous $TiO_2$ coating can improve cell proliferation, differentiation and adhesion because of the nano-topographic features and their large surface area. HA is also coated on the implant surface because it enhances the bone growth and bonding between bone tissue and implants surfaces; however, the mechanical properties of HA are poor. In order to enhance the mechanical properties, HA is blended with tough phase Ti-6Al-4V and CNTs. The bond strength between HA and Ti is relatively low and researches are shown that it can be increased using ion-beam-assisted deposition methods. The researcher's studies demonstrate that the ion-beam current helped to increase the bond strength. Biomaterial-related infection caused by bacterial contamination is one of the most serious problems in the field, and the approach to reduce the bacterial attachment is incorporating an antibacterial agent onto the implant surface. Studies show that Ag ions implanted on Ti and its alloys restricted the growth of *S. aureus* and *E. coli* on the surface evaluated compared to the bare surface; this effect can conclude that Ag ions-implanted on Ti are excellent antibacterial structures.

Ag nanoparticles also are integrated into $TiO_2$-NTs arrays to equip Ti-based implants with antibacterial ability to resist bacterial attachment. It displays a long-term antibacterial ability; however, the fast release of Ag ions leads to cytotoxicity. Hence, development of new NT-Ag arrays with better cytocompatible and a controllable release of Ag ions were carried out. There are a wide range of implant devices built with polymeric biomaterial incorporated with drug therapies to enhance the therapeutic effects. Short half-lives is allowed to site a minimal loss and minimize the toxicity of the system. Besides, the defects can be overcome by hybridizing organic compounds such as CNTs and polymers. CNTs are an excellent surface for cell adhesion and growth, as well as not causing strong inflammatory responses. It also reinforces the fracture toughness and wear resistance of the HA coating. The natural polymer (i.e., dermatan sulphate and heparin) is grafted on the Ti surface because of the excellent biocompatibility, lack of toxicity and ease of availability. The synthetic water soluble polymers (which are a kind of natural polymer) have the ability to control the crystal growth of organic or inorganic sparingly soluble salts. Protein adsorption is a progressive process that plays vital role to determine the longevity of implants. Noncovalent interactions and surface properties of implants are the factors that influence protein absorption. Ti-based implants with hydrophilic surface chemistry can improve cell response and promote more rapid osseointegration.

## ACKNOWLEDGMENT

The authors acknowledge the financial support through the Australia-India Strategic Research Fund (AISRF) ST060048.

## REFERENCES

Adya N, Alam M, Ravindranath T, Mubeen A, Saluja B. Corrosion in titanium dental implants: Literature review. *J Ind Pros Soc.* 2005; 5: 126–131.

Andrade JD, Hlady V. Protein adsorption and materials biocompatibility- A tutorial review and suggested hypotheses. *Adv in Polyr Sci.* 1986; 79: 1–63.

Aoki N, Yokoyama A, Nodasaka Y, Akasaka T, Uo M, Sato Y *et al.* Strikingly extended morphology of cells grown on carbon nanotubes. *Chem Lett.* 2006; 35: 508–509.

Bagan JV, Jimenez Y, Murillo J, Hernandez S, Poveda R, Sanchis JM *et al.* Jaw osteonecrosis associated with bisphosphonates: Multiple exposed areas and its relationship to teeth extractions. Study of 20 cases. *Oral Onc.* 2006; 42: 327–329.

Bajpai AK, Singh R. Preparation and characterization of hydroxyapatite impregnated semi-interpenetrating polymer networks (IPNs) of polyvinyl alcohol and poly(acrylamide-co-acrylic acid). *J Macro Sci-Pure App Chem.* 2004; A41: 1135–1159.

Balani K, Anderson R, Laha T, Andara M, Tercero J, Crumpler E *et al.* Plasma-sprayed carbon nanotube reinforced hydroxyapatite coatings and their interaction with human osteoblasts *in vitro. Biomaterials.* 2007; 28: 618–624.

Berger S, Faltenbacher J, Bauer S, Schmuki P. Enhanced self-ordering of anodic $ZrO_2$ nanotubes in inorganic and organic electrolytes using two-step anodization. *Phys Stat Soli Rap Res Lett.* 2008; 2: 102–104.

Brian GM, Katz M, Crowley J. Osteonecrosis of the Jaw and Bisphosphonates. *New Eng J Med.* 2005; 353: 99–102.

Chen JY, Wolke JGC, Degroot K. Microstructure and crystallinity in hydroxyapatite coatings. *Biomaterials.* 1994; 15: 396–399.

Chen Y, Gan C, Zhang T, Yu G, Bai P, Kaplan A. Laser-surface-alloyed carbon nanotubes reinforced hydroxyapatite composite coatings. *App Phy Lett.* 2005; 86: 1–3.

Chitanu GC, Popescu I, Carpov A. Synthesis and characterization of maleic anhydride copolymers and their derivatives. 1. Addition polymerization–Literature survey. *Rev Roum De Chim.* 2005; 50: 589–599.

Cho J, Konopka K, Rozniatowski K, García-Lecina E, Shaffer MSP, Boccaccini AR. Characterisation of carbon nanotube films deposited by electrophoretic deposition. *Carbon.* 2009; 47: 58–67.

Choi J, Konno T, Takai M, Ishihara K. Controlled drug release from multilayered phospholipid polymer hydrogel on titanium alloy surface. *Biomaterials*. 2009; 30: 5201–5208.

Choi JM, Kim HE, Lee IS. Ion-beam-assisted deposition (IBAD) of hydroxyapatite coating layer on Ti-based metal substrate. *Biomaterials*. 2000; 21: 469–473.

Choi JW, Kong YM, Kim HE, Lee IS. Reinforcement of hydroxyapatite bioceramic by addition of Ni3Al and Al2O3. *J Amer Cera Soc*. 1998; 81: 1743–1748.

Costerton JW, Stewart PS, Greenberg EP. Bacterial biofilms: A common cause of persistent infections. *Sci*. 1999; 284: 1318–1322.

Crawford GA, Chawla N, Das K, Bose S, Bandyopadhyay A. Microstructure and deformation behavior of biocompatible TiO2 nanotubes on titanium substrate. *Acta Biomater*. 2007; 3: 359–367.

Cui FZ, Luo ZS, Feng QL. Highly adhesive hydroxyapatite coatings on titanium alloy formed by ion beam assisted deposition. *J Mater Sci-Mater Med*. 1997; 8: 403–405.

Darouiche RO. Treatment of Infections Associated with Surgical Implants. *New Eng J Med*. 2004; 350: 1422–1429.

Dasarathy H, Riley C, Coble HD, Lacefield WR, Maybee G. Hydroxyapatite/metal composite coatings formed by electrocodeposition. *J Biomed Mater Res*. 1996; 31: 81–89.

Deligianni DD, Katsala N, Ladas S, Sotiropoulou D, Amedee J, Missirlis YF. Effect of surface roughness of the titanium alloy Ti-6Al-4V on human bone marrow cell response and on protein adsorption. *Biomaterials*. 2001; 22: 1241–1251.

Dlugon E, Simka W, Fraczek-Szczypta A, Niemiec W, Markowski J, Szymanska M *et al*. Carbon nanotube-based coatings on titanium. *Bull Mater Sci*. 2015; 38: 1339–1344.

Dong ZL, Khor KA, Quek CH, White TJ, Cheang P. TEM and STEM analysis on heat-treated and *in vitro* plasma-sprayed hydroxyapatite/Ti-6Al-4V composite coatings. *Biomaterials*. 2003; 24: 97–105.

Dubruel P, Vanderleyden E, Bergada M, De Paepe I, Chen H, Kuypers S *et al*. Comparative study of silanisation reactions for the biofunctionalisation of Ti-surfaces. *Surf Sci*. 2006; 600: 2562–2571.

Eleftheriou CG, Zimmermann JB, Kjeldsen HD, David-Pur M, Hanein Y, Sernago E. Carbon nanotube electrodes for retinal implants: A study of structural and functional integration over time. *Biomaterials*. 2017; 112: 108e121.

Etienne O, Gasnier C, Taddei C, Voegel JC, Aunis D, Schaaf P *et al*. Antifungal coating by biofunctionalized polyelectrolyte multilayered films. *Biomaterials*. 2005; 26: 6704–6712.

Fiedler J, Kolitsch A, Kleffner B, Henke D, Stenger S, Brenner RE. Copper and silver ion implantation of aluminium oxide-blasted titanium surfaces: Proliferative response of osteoblasts and antibacterial effects. *Inter J Artif Organ*. 2011; 34: 882–888.

Gao A, Hang RQ, Huang XB, Zhao LZ, Zhang XY, Wang L *et al*. The effects of titania nanotubes with embedded silver oxide nanoparticles on bacteria and osteoblasts. *Biomaterials*. 2014; 35: 4223–4235.

Haddow DB, James PF, Van Noort R. Sol-gel derived calcium phosphate coatings for biomedical applications. *J Sol-Gel Sci Tech*. 1998; 13: 261–265.

Hallab NJ, Mikecz K, Vermes C, Skipor A, Jacobs JJ. Orthopaedic implant related metal toxicity in terms of human lymphocyte reactivity to metal-protein complexes produced from cobalt-base and titanium-base implant alloy degradation. *Molec Cell Biochem*. 2001; 222: 127–136.

Haman JD, Boulware AA, Lucas LC, Crawmer DE. High-velocity oxyfuel thermal spray coatings for biomedical applications. *J Ther Spr Tech*. 1995; 4: 179–184.

Harris LG, Tosatti S, Wieland M, Textor M, Richards RG. Staphylococcus aureus adhesion to titanium oxide surfaces coated with non-functionalized and peptide-functionalized poly(L-lysine)-grafted-poly(ethylene glycol) copolymers. *Biomaterials*. 2004; 25: 4135–4148.

Haynes CA, Sliwinsky E, Norde W. Structural and electrostatic properties of globular proteins at a polystyrene-water interface. *J Coll Inter Sci*. 1994; 164: 394–409.

He X, Zhang X, Wang X, Qin L. Review of antibacterial activity of titanium-based implants' surfaces fabricated by micro-arc oxidation. *Coatings*. 2017; 7: 45.

Heimann RB, Kurzweg H, Vu TA. Hydroxyapatite-bond coat systems for improved mechanical and biological performance of hip implants. In Thermal Spray: Meeting the Challenges of the 21st Century, C. Coddet, ed., ASM International: France, 1998, pp. 999–1005.

Helary G, Noirclere F, Mayingi J, Migonney V. A new approach to graft bioactive polymer on titanium implants: Improvement of MG 63 cell differentiation onto this coating. *Acta Biomater*. 2009; 5: 124–133.

Hench LL. Bioceramics. *J Am Ceram Soc*. 1998; 81(7): 1705–1728.

Hero H, Wie H, Jorgensen RB, Ruyter IE. Hydroxyapatite coatings on Ti produced by hot isostatic pressing. *J Biomed Mater Res*. 1994; 28: 343–348.

Hirakawa Y, Jimbo R, Shibata Y, Watanabe I, Wennerberg A, Sawase T. Accelerated bone formation on photo-induced hydrophilic titanium implants: An experimental study in the dog mandible. *Clin Oral Impl Res.* 2013; 24: 139–144.

Hong MH, Lee DH, Kim KM, Lee YK. Study on bioactivity and bonding strength between Ti alloy substrate and TiO2 film by micro-arc oxidation. *Thin Solid Films.* 2011; 519: 7065–7070.

Huang H-H, Wu C-P, Sun Y-S, Yang W-E, Lin M-C, Lee T-H. Surface nanoporosity of β-type Ti–25Nb–25Zr alloy for the enhancement of protein adsorption and cell response. *Sur Coat Tech.* 2014a; 259(Part B): 206–212.

Huang HH, Wu CP, Sun YS, Huang HM, Lee TH. Enhanced corrosion resistance and biocompatibility of beta-type Ti-25Nb-25Zr alloy by electrochemical anodization. *Thin Solid Films.* 2013; 549: 87–92.

Huang R, Han Y, Lu SM. Enhanced osteoblast functions and bactericidal effect of Ca and Ag dual-ion implanted surface layers on nanograined titanium alloys. *J Mater Chem B.* 2014b; 2: 4531–4543.

Inoue S, Uo M, Sakairi M, Hirata E, Lee MH, Bae TS *et al.* The effects of the coating of anodized titanium with multi-walled carbon nanotubes on bone formation. *Bioceramics.* 2013; 24: 529–530. Ishikawa K, Iwamoto Y. Stafa-Zurich, Trans Tech Publications Ltd., 621–624.

Ishihara K, Oshida H, Endo Y, Ueda T, Watanabe A, Nakabayashi N. Hemocompatibility of human whole-blood on polymers with a phospholipid polar group and its mechanism. *J Biomed Mater Res.* 1992; 26: 1543–1552.

Jeong YH, Choe HC, Eun SW. Hydroxyapatite coating on the Ti-35Nb-xZr alloy by electron beam-physical vapor deposition. *Thin Solid Films.* 2011; 519: 7050–7056.

Ji HX, Marquis PM. Effect of heat-treatment on the microstructure of plasma-sprayed hyroxyapatite coating. *Biomaterials.* 1993; 14: 64–68.

Jimbo R, Ono D, Hirakawa Y, Odatsu T, Tanaka T, Sawase T. Accelerated photo-induced hydrophilicity promotes osseointegration: An animal study. *Clin Imp Dent Rel Res.* 2011; 13: 79–85.

Keegan GM, Learmonth ID, Case CP. Orthopaedic metals and their potential toxicity in the arthroplasty patient–Review of current knowledge and future strategies. *J Bone and Joint Surg-Bri Vol.* 2007; 89B: 567–573.

Kim HW, Koh YH, Li LH, Lee S, Kim HE. Hydroxyapatite coating on titanium substrate with titania buffer layer processed by sol-gel method. *Biomaterials.* 2004; 25: 2533–2538.

Kingshott P, Wei J, Bagge-Ravn D, Gadegaard N, Gram L. Covalent attachment of poly(ethylene glycol) to surfaces, critical for reducing bacterial adhesion. *Langmuir.* 2003; 19(17): 6912–6921.

Kong LJ, Gao Y, Lu GY, Gong YD, Zhao NM, Zhang XF. A study on the bioactivity of chitosan/nano-hydroxyapatite composite scaffolds for bone tissue engineering. *Europ Poly J.* 2006; 42: 3171–3179.

Kong YM, Kim S, Kim HE, Lee IS. Reinforcement of hydroxyapatite bioceramic by addition of ZrO2 coated with Al2O3. *J Amer Cera Soc.* 1999; 82: 2963–2968.

Konno T, Ishihara K. Temporal and spatially controllable cell encapsulation using a water-soluble phospholipid polymer with phenylboronic acid moiety. *Biomaterials.* 2007; 28: 1770–1777.

Kopf BS, Ruch S, Berner S, Spencer ND, Maniura-Weber K. The role of nanostructures and hydrophilicity in osseointegration: *In-vitro* protein-adsorption and blook-interaction studies. *J Biomed Mater Res A.* 2015; 103(8): 2661–2672.

Kunze J, Muller L, Macak JM, Greil P, Schmuki P, Muller FA. Time-dependent growth of biomimetic apatite on anodic TiO2 nanotubes. *Electroch Acta.* 2008; 53: 6995–7003.

Laurens N, Koolwijk P, De Maat MPM. Fibrin structure and wound healing. *J Throm and Haem.* 2006; 4: 932–939.

Lee JH, Kim HE, Shin KH, Koh YH. Electrodeposition of biodegradable sol-gel derived silica onto nanoporous TiO2 surface formed on Ti substrate. *Mater Lett.* 2011; 65: 1519–1521.

Li LH, Kong YM, Kim HW, Kim YW, Kim HE, Heo SJ *et al.* Improved biological performance of Ti implants due to surface modification by micro-arc oxidation. *Biomaterials.* 2004; 25: 2867–2875.

Li Y, Lee IS, Cui FZ, Choi SH. The biocompatibility of nanostructured calcium phosphate coated on micro-arc oxidized titanium. *Biomater.* 2008; 29: 2025–2032.

Liao H, Paratala B, Sitharaman B, Wang Y. Applications of carbon nanotubes in biomedical studies. *Meth Molec Bio (Clifton, NJ).* 2011; 726: 223–241.

Lin CC, Metters AT. Hydrogels in controlled release formulations: Network design and mathematical modeling. *Adv Dru Del Rev.* 2006; 58: 1379–1408.

Lu YP, Li MS, Li ST, Wang ZG, Zhu RF. Plasma-sprayed hydroxyapatite plus titania composite bond coat for hydroxyapatite coating on titanium substrate. *Biomaterials.* 2004; 25: 4393–4403.

Ma QL, Mei SL, Ji K, Zhang YM, Chu PK. Immobilization of Ag nanoparticles/FGF-2 on a modified titanium implant surface and improved human gingival fibroblasts behavior. *J Biomed Mater Res Part A.* 2011; 98A: 274–286.

Madrid C, Sanz M. What impact do systemically administrated bisphosphonates have on oral implant therapy? A systematic review. *Clin Oral Impl Res.* 2009; 20: 87–95.

Mandracci P, Mussano F, Rivolo P, Carossa S. Surface treatments and functional coatings for biocompatibility improvement and bacterial adhesion reduction in dental implantology. *Coatings.* 2016; 6(7): 1–22.

Mei SL, Wang HY, Wang W, Tong LP, Pan HB, Ruan CS *et al.* Antibacterial effects and biocompatibility of titanium surfaces with graded silver incorporation in titania nanotubes. *Biomaterials.* 2014; 35: 4255–4265.

Mendonca G, Mendonca DBS, Aragao FJL, Cooper LF. Advancing dental implant surface technology–From micron- to nanotopography. *Biomaterials.* 2008; 29: 3822–3835.

Mendonca G, Mendonca DBS, Simoes LGP, Araujo AL, Leite ER, Duarte WR *et al.* The effects of implant surface nanoscale features on osteoblast-specific gene expression. *Biomaterials.* 2009; 30: 4053–4062.

Michiardi A, Helary G, Nguyen PCT, Gamble LJ, Anagnostou F, Castner DG *et al.* Bioactive polymer grafting onto titanium alloy surfaces. *Acta Biomater.* 2010; 6: 667–675.

Nauth A, Ristevski B, Begue T, Schemitsch EH. Periprosthetic distal femur fractures: Current concepts. *J Orthop Trau.* 2011; 25: S82–S85.

Negroiu G, Piticescu RM, Chitanu GC, Mihailescu IN, Zdrentu L, Miroiu M. Biocompatibility evaluation of a novel hydroxyapatite-polymer coating for medical implants (*in vitro* tests). *J Mater Sci-Mater Med.* 2008; 19: 1537–1544.

Oh SH, Finones RR, Daraio C, Chen LH, Jin SH. Growth of nano-scale hydroxyapatite using chemically treated titanium oxide nanotubes. *Biomaterials.* 2005; 26: 4938–4943.

de Oliveira MA, Asahi DA, Silveira CAE, Lima L, Glick M, Gallottini M. The effects of zoledronic acid and dexamethasone on osseointegration of endosseous implants: Histological and histomorphometrical evaluation in rats. *Clin Oral Impl Res.* 2015; 26: e17–e21.

Ong JL, Harris LA, Lucas LC, Lacefield WR, Rigney D. X-ray photoelectron-spectroscopy characterization of ion-beam sputter-deposited calcium-phosphate coatings. *J Amer Cer Soc.* 1991; 74: 2301–2304.

Overgaard S, Lind M, Josephsen K, Maunsbach AB, Bunger C, Soballe K. Resorption of hydroxyapatite and fluorapatite ceramic coatings on weight-bearing implants: A quantitative and morphological study in dogs. *J Biomed Mater Res.* 1998; 39: 141–152.

Parcharoen Y, Termsuksawad P, Sirivisoot S. Improved bonding strength of hydroxyapatite on titanium dioxide nanotube arrays following alkaline pretreatment for orthopedic implants. *J Nanomater.* 2016; 2016: 1–13.

Park J, Bauer S, von der Mark K, Schmuki P. Nanosize and vitality: TiO2 nanotube diameter directs cell fate. *Nano Lett.* 2007; 7: 1686–1691.

Petering HG. Pharmacology and toxicology of heavy metals: Silver. *Pharm Therap, Part A: Chem, Tox.* 1976; 1: 127–130.

Pham VH, Jang TS, Jung HD, Kim HE, Koh YH. Creation of nanoporous tantalum (Ta)-incorporated titanium (Ti) surface onto Ti implants by sputtering of Ta in Ar under extremely high negative substrate biases. *J Mater Chem.* 2012; 22: 24798–24804.

Rabe M, Verdes D, Seeger S. Understanding protein adsorption phenomena at solid surfaces. *Adv Coll Inter Sci.* 2011; 162: 87–106.

Rapacz-Kmita A, Slosarczyk A, Paszkiewicz Z. HAp-ZrO2 composite coatings prepared by plasma spraying for biomedical applications. *Ceram Inte.* 2005; 31: 567–571.

Rautray TR, Narayanan R, Kim KH. Ion implantation of titanium based biomaterials. *Prog Mater Sci.* 2011; 56: 1137–1177.

Ren KF, Ji J, Shen JC. Construction and enzymatic degradation of multilayered poly-L-lysine/DNA films. *Biomater.* 2006; 27: 1152–1159.

Ruther C, Ewald H, Mittelmeier W, Fritsche A, Bader R, Kluess D. A novel sensor concept for optimization of loosening diagnostics in total hip replacement. *J Biomech Eng-Trans Asme.* 2011; 133: 5.

Sawada SI, Iwasaki Y, Nakabayashi N, Ishihara K. Stress response of adherent cells on a polymer blend surface composed of a segmented polyurethane and MPC copolymers. *J Biomed Mater Res Part A.* 2006; 79A: 476–484.

Schreurs WJA, Rosenberg H. Effect of silver ions on transport and retention of phosphate by Escherichia coli. *J Bacteriol.* 1982; 152: 7–13.

Scotchford CA, Gilmore CP, Cooper E, Leggett GJ, Downes S. Protein adsorption and human osteoblast-like cell attachment and growth on alkylthiol on gold self-assembled monolayers. *J Biomed Mater Res.* 2002; 59: 84–99.

Sieber IV, Schmuki P. Porous tantalum oxide prepared by electrochemical anodic oxidation. *J Electrochem Soc.* 2005; 152: C639–C644.

Simka W, Sadkowski A, Warczak M, Iwaniak A, Dercz G, Michalska J *et al.* Characterization of passive films formed on titanium during anodic oxidation. *Electroch Acta.* 2011; 56: 8962–8968.

Singh RK, Qian F, Damodaran R, Moudgil S. Laser deposition of hydroxyapatite coatings. *Mater Manufact Proc.* 1996; 11: 481–490.
Song W, Chen H. Protein adsorption on materials surfaces with nano-topography. *Chin Sci Bull.* 2007; 52: 3169–3173.
Tawfick S, De Volder M, Copic D, Park SJ, Oliver CR, Polsen ES *et al.* Engineering of micro- and nanostructured surfaces with anisotropic geometries and properties. *Adv Mater.* 2012; 24: 1628–1674.
Teng SH, Shi JJ, Peng BX, Chen LJ. The effect of alginate addition on the structure and morphology of hydroxyapatite/gelatin nanocomposites. *Comp Sci Tech.* 2006; 66: 1532–1538.
Tsuchiya H, Macak JM, Muller L, Kunze J, Muller F, Greil P *et al.* Hydroxyapatite growth on anodic TiO2 nanotubes. *J Biomed Mater Res Part A.* 2006; 77: 534–541.
Tsuchiya H, Macak JM, Taveira L, Schmuki P. Fabrication and characterization of smooth high aspect ratio zirconia nanotubes. *Chem Phys Lett.* 2005; 410: 188–191.
van den Beucken J, Walboomers XF, Boerman OC, Vos MRJ, Sommerdijk N, Hayakawa T *et al.* Functionalization of multilayered DNA-coatings with bone morphogenetic protein 2. *J Control Release.* 2006; 113: 63–72.
Vetrone F, Variola F, de Oliveira PT, Zalzal SF, Yi JH, Sam J *et al.* Nanoscale oxidative patterning of metallic surfaces to modulate cell activity and fate. *Nano Lett.* 2009; 9: 659–665.
Wada M, Jinno H, Ueda M, Ikeda T, Kitajima M, Konno T *et al.* Efficacy of an MPC-BMA co-polymer as a nanotransporter for paclitaxel. *Antican Res.* 2007; 27: 1431–1435.
Wahl DA, Czernuszka JT. Collagen-hydroxyapatite composites for hard tissue repair. *Eur Cell Mater.* 2006; 11: 43–56.
Wang J, Li J, Qian S, Guo G, Wang G, Wang Q *et al.* Antibacterial surface design of titanium-based biomaterials for enhanced bacteria-killing and cell-assisting functions against periprosthetic joint infection. *ACS Appl Mater Interfaces.* 2016; 8(17): 11162–11178.
Wang LN, Adams A, Luo JL. Enhancement of the capability of hydroxyapatite formation on Zr with anodic ZrO2 nanotubular arrays via an effective dipping pretreatment. *J Biomed Mater Res Part B-App Biomater.* 2011; 99B: 291–301.
Wang LN, Luo JL. Formation of hydroxyapatite coating on anodic titanium dioxide nanotubes via an efficient dipping treatment. *Metall Mater Trans A-Phys Metall Mater Sci.* 2011; 42A: 3255–3264.
Wang SA, Lacefield WR, Lemons JE. Interfacial shear strength and histology of plasma sprayed and sintered hydroxyapatite implants *in vivo. Biomaterials.* 1996; 17: 1965–1970.
Wolke JGC, Vandijk K, Schaeken HG, Degroot K, Jansen JA. Study of the surface characteristics of magnetron-sputter calcium-phosphate coatings. *J Biomed Mater Rese.* 1994; 28: 1477–1484.
Wood KC, Boedicker JQ, Lynn DM, Hammond PT. Tunable drug release from hydrolytically degradable layer-by-layer thin films. *Langmuir.* 2005; 21: 1603–1609.
Xu GQ, Shen XK, Hu Y, Ma PP, Cai KY. Fabrication of tantalum oxide layers onto titanium substrates for improved corrosion resistance and cytocompatibility. *Surf Coat Tech.* 2015; 272: 58–65.
Yamashita K, Yagi T, Umegaki T. Bonelike coatings onto ceramics by reactive magnetron sputtering. *J Amer Ceram Soc.* 1996; 79: 3313–3316.
Yammine P, Pavon-Djavid G, Helary G, Migonney V. Surface modification of silicone intraocular implants to inhibit cell proliferation. *Biomacromol.* 2005; 6: 2630–2637.
Yan WQ, Nakamura T, Kawanabe K, Nishigochi S, Oka M, Kokubo T. Apatite layer-coated titanium for use as bone bonding implants. *Biomaterials.* 1997; 18: 1185–1190.
Yokoyama A, Sato Y, Nodasaka Y, Yamamoto S, Kawasaki T, Shindoh M *et al.* Biological behavior of hat-stacked carbon nanofibers in the subcutaneous tissue in rats. *Nano Lett.* 2005; 5: 157–161.
Zhang J, Senger B, Vautier D, Picart C, Schaaf P, Voegel JC *et al.* Natural polyelectrolyte films based on layer-by layer deposition of collagen and hyaluronic acid. *Biomaterials.* 2005; 26: 3353–3361.
Zhang Y, Bai Y, Yan B. Functionalized carbon nanotubes for potential medicinal applications. *Drug Discov Today.* 2010; 15: 428–435.
Zheng XB, Huang MH, Ding CX. Bond strength of plasma-sprayed hydroxyapatite/Ti composite coatings. *Biomaterials.* 2000; 21: 841–849.

# 9 Nanocrystals of Poorly Water-Soluble Drugs: Production Technologies, Characterization, and Functionalization

*Ana Rita Fernandes, Joana R. Campos, Ana Claudia Santos, Ripandeep Kaur, Shikha Lohan, Maria L. Garcia, and Eliana B. Souto**

## CONTENTS



## 9.1 INTRODUCTION

Nanotechnology is one of the most popular, promising and pragmatic approaches in modern drug delivery and therapeutics. Pharmaceutically, nanocrystals are nanomaterials produced using down-sizing of particles from microscale to nanoscale with an aim to alter their physicochemical characteristics. These nanocrystals are solid in nature and surface-stabilized with a mean diameter below 1000 nm (Merisko-Liversidge *et al.*, 2003).

Currently, one of the most challenging tasks confronting pharmaceutical scientists is poor and inconsistent bioavailability of poorly soluble drugs, and the most extensively employed approach is their solubility enhancement. Numerous strategies have been employed to improve upon the solubility characteristics of such drug candidates. However, all such methods exhibit limited utility and generally face untoward resistance from one or more of other desirable physicochemical attributes of these problematic drugs. Use of surfactants or cosolvents in this context can lead to

* Corresponding author.

increased adverse side effects associated with such delivery systems, especially when the intended route of administration is parenteral (Willems *et al.*, 2001).

One of the approaches to enhance their aqueous solubility is to develop nanocrystals of these drugs. When the particle size of a crystal decreases to 100 nm, the most vital changes result as a consequence of the enhancement of its surface area, which further contributes to the improved drug solubility. Additionally, nanocrystals offer other stellar advantages too, such as simpler structure *vis-à-vis* many other nanostructured systems used for the purpose, simpler method of preparation, and suitability for a variety of administration routes. Micronization of powders of poorly water-soluble drugs tends to increase dissolution velocity; though, a mere enhancement over this process does not always fetch improvement in the bioavailability of these drugs. Further reduction in the size of particles, therefore, from the micro-range to the nano-range has been proposed by numerous researchers to overcome this hiccup. Elan Nanosystems (San Francisco, CA, USA) has reported nanocrystals to be the best available option to improve oral bioavailability, compared to microcrystals (Muller *et al.*, 2011). Nanosuspensions consist of the dispersions of drug nanocrystals in a liquid medium like water or some other non-aqueous solvent. The most often used method to stabilize dispersed particles in a medium is the use of surfactants or polymeric stabilizers (Junghanns and Muller, 2008). Simultaneously, nanonization increases the surface area and saturation solubility *vis-à-vis* micronization (Sattar *et al.*, 2017). In the cases of particles with size above 1 μm, their solubility is constant, depending only on the temperature and the solvent in which the particles are exposed. In some conditions, the saturation solubility is dependent on the particle size, especially when the size is less than 1–2 μm. As a consequence of the improvement in the saturation solubility, there is diminution in the dissolution rate. This increase in the saturation solubility results in formation of a supersaturated solution, which subsequently increases the concentration gradient between the gut lumen and the blood. This phenomenon promotes drug absorption, as the concentration gradient due to supersaturated solution accelerates diffusion and, ultimately, drug permeation (Bansal *et al.*, 2012). Thus, nanocrystals account for an earlier onset of action owing to a faster dissolution rate and absorption. As compared to the conventional dosage forms, the bioavailability of nanocrystals gets significantly enhanced, as the particles are made up of >90% drug. This property is responsible for achieving high therapeutic concentrations at the site of action to exhibit improved pharmacological response. In addition, their high loading capabilities make them highly efficient in transporting drugs (Muller *et al.*, 2011).

Liversidge *et al.* (1992) patented the first nanocrystal formulation and Rapamune® by Wyeth was the first such pharmaceutical product accepted in 2000 by the US Food and Drug Administration (FDA). Its fast acceptance and introduction into the market was attributed to its simple structure without the usage of too many excipients. At the time of its introduction into the market, Rapamune® was formulated as an oral tablet. After only a few years, nanocrystals were studied for their use through various other routes, such as parenteral, ocular, dermal, buccal, and pulmonary (Merisko-Liversidge *et al.*, 2003; Bhol and Schechter, 2005; Rao *et al.*, 2011).

### 9.1.1 Pharmaceutical Formulations

Owing to the capacity of nanocrystals to improve solubility, the bioavailability of poorly water-soluble drugs can be substantially enhanced. Meanwhile, these nanocrystals offer the opportunity to the formulation chemist to formulate oral solid forms, like tablets and capsules, which are far more patient-friendly. Nanocrystals of poorly water-soluble drugs feature high permeability through dermal application, thus making it convenient to incorporate these in cosmetic products (Shegokar *et al.*, 2010). Since the particle size is smaller than the diameter of blood capillaries, these may also be used for intravenous injection, as long as the polydispersity index remains below 0.24. At the same time, if nanocrystals are used for intravenous injections, it is necessary to monitor the absence of particles with a diameter larger than 5 μm, as these may block the blood capillaries (5–6 μm), possibly leading to embolism. Such systems provide nearly 100% bioavailability and simultaneously prevent the use of cosolvents or surfactants, which are often harmful, as well as expensive excipients

(Bansal *et al.*, 2012). Nanocrystals are usually stabilized by electrostatic and/or steric stabilization employing surfactants such as lecithin alone or in combination with sodium cholate or non-ionic surfactants, such as Tween 80, Poloxamer 188, and polyvinylpyrrolidone. These surfactants are accepted for intravenous injection. Use of binary or ternary mixtures of electrostatic and steric surfactants has been found to be effective for long-term stability.

Several methods have been documented in literature to enhance the solubility of poorly water soluble drugs. However, selection of these methods is based on the chemical properties of the drug (i.e., its solubility in organic medium, conformation, and/or molecular size). Though it is possible to formulate systems using surfactants or cosolvents, these options can lead to adverse side effects and toxicity (Jacobs *et al.*, 2000; Rabinow, 2004). In the case of water soluble drugs, it is not possible to formulate nanocrystals, unless a non-aqueous dispersion medium is intended.

Nanocrystals offer a practicable biopharmaceutical solution for problems in the delivery of poorly water-soluble drugs, such as (i) drugs with low bioavailability in the oral administration, (ii) drugs with low dermal bioavailability owing to reduced penetration in the skin, (iii) drugs which require a large injection volume for intravenous administration, and (iv) drugs which need several administrations to achieve the required effect, such as traditional drug solutions formulations. These problems make the treatment suboptimal and less patient-compliant. Verily, nanocrystals have the ability to resolve most of these issues due to their inherent characteristics, including (i) enhanced dissolution rate; (ii) enhanced solubility and, consequently, increased saturation velocity; and (iii) increased adhesion to surfaces/cell membranes. Gao *et al.* (2012) have shown that these three combined effects improved the *in vivo* behavior of some poorly water-soluble drugs when compared to micron-sized drug powders.

The increased saturation solubility, $C_s$, the increased dissolution velocity, dc/dt, and the adhesive properties of ultrafine particles are the most important characteristics of drug nanocrystals. As shown in the Kelvin equation (Equation 9.1), the dissolution pressure of a compound increases due to the decreased particle size, which is attributed to the enhanced curvature of the particle surface:

$$\ln \frac{p}{p_0} = \frac{2\gamma V_m}{rRT} \tag{9.1}$$

where, p is the vapor pressure, $p_0$ the saturated vapor pressure, $\gamma$ is the surface tension, $V_m$ is the molar volume of the liquid, r is the particle radius, R is the universal gas constant, and T is the absolute temperature.

Surrounding nanosized particles, higher saturation solubility occurs in comparison to particles sized above 1–2 μm. This effect is described by the Ostwald–Freundlich equation. Reduction of particle size in a formulation is considered as a nonspecific method, independent of the particles being in the solid state or other physicochemical properties; thus, this can be effectively applied for poorly water-soluble drugs (Moschwitzer and Muller, 2006a). The Noyes–Whitney equation explains that the surface-to-volume ratio tends to grow significantly when particles are broken down to the micrometer range, which results in enhancement of dissolution velocity, subsequently resulting in enhancement of drug absorption from the gastrointestinal tract (Gao *et al.*, 2008). For a faster dissolution rate, it is necessary to increase the surface area of the particles, as shown in the Noyes–Whitney equation (Equation 9.2) (Noyes and Whitney, 1897).

$$\frac{dX}{dt} = \frac{DA}{h_D}(C_s - C_t) \tag{9.2}$$

where, dX/dt is the dissolution velocity, D is the diffusion coefficient, A is the surface area, $h_D$ is the diffusional distance, $C_S$ *is* the saturation solubility, and $C_t$ is the concentration around the particles at time t (Dressman *et al.*, 1998).

As the diameter of the particle decreases, the particles are able to release the drug as a function of their mass. This relationship is described by Equation 9.3:

$$\frac{A}{V} = \frac{3}{r} \tag{9.3}$$

where, A is the surface area, V is the volume, and r is the radius of the particle.

Using this mechanism, it can be clearly explained that, if the size of a particle is reduced from 30 μm to 300 nm, the dissolution rate gets increased by 100-fold, assuring substantial augmentation in bioavailability. It is possible to dissolve a higher amount of drug during the gastrointestinal tract time-limited transit, enhancing the concentration gradient that, consequently, improves the permeability and, thus, the bioavailability of drugs (Peltonen *et al.*, 2014).

Ostwald-Freundlich theory sustains the hypothesis of increase in solubility. Initially framed to explain the dispersibility of liquid droplets in gases, this theory has been further extrapolated to describe the increase in solubility for very small solid particles in liquids too. According to the equation (Equation 9.4), especially in the smaller particles (below nanometer range), the saturation concentration on the surface of these particles is higher when compared to the parameter on the surface of the largest particle (Ostwald, 1900).

$$C_{NP} = C_b \exp\left(\frac{2V_m \gamma}{RT_r}\right) \tag{9.4}$$

where, $C_{NP}$ is the solubility of nanoparticles with a radius r, $C_b$ is the solubility of bulk material, $V_m$ is the molar volume of the drug, $\gamma$ is the interfacial tension between the solvent and the drug material, R is the gas constant, and T is the absolute temperature (Jinno *et al.*, 2006).

Since dc/dt depends on the difference $c_s$–$c_x$ (where $c_x$ is the concentration in the suspending medium), the increased saturation solubility, $c_s$, leads to the increased dissolution velocity. In the same way, since the diffusional distance h is decreased (h is the distance over which the concentration gradient $c_s$–$c_x$ is formed, described by the Prandtl equation), the intrinsic dissolution rate is increased. The increased dissolution pressure and velocity, due to the increase of $c_s$ and decrease of h is described in Figure 9.1.

Based on these properties, the biggest advantage of using nanocrystals is the enhanced surface area per unit mass of the particles, accounting for various desired advantages. In general, submicron-meter particles exhibit increased adhesiveness in comparison to larger particles. As a consequence, an increase in the contact area is reported to occur with improved bioavailability. The best drug candidates for the production of nanocrystals are those depicting high permeability but low solubility.

### 9.1.2 Biopharmaceutics Classification System (BCS)

BCS was developed by Amidon *et al.* (1995) and is based on fundamental variables which control the oral drug absorption in terms of extent and rate. Considering these parameters, this system divides the drugs in four modules, according to their characteristics, such as solubility and permeability. The main objective is to anticipate the performance of the drugs, namely, the *in vivo* pharmacokinetics of numerous compounds based on the permeability and solubility. If the drug in study is administered orally, their efficacy is consequently dependent on the extent of absorption and the oral bioavailability. In turn, oral absorption is dependent on the solubilization of the drug in the intestinal tract and on its permeability. Concentration, temperature, surface area, affinity to a membrane transporter, time, and viscosity are the factors that affect the amount of drug that penetrate through the membrane (Smetanova *et al.*, 2011).

As per BCS, the compounds are considered highly soluble if the highest dose strength present in the market is soluble in 250 mL in a medium at 37°C at a pH between 1 and 6.8. On the other hand, to be considered highly permeable, the drug needs to have an extent of absorption higher or equal to 85% of the administered dose (Cardot *et al.*, 2016; Wu *et al.*, 2016).

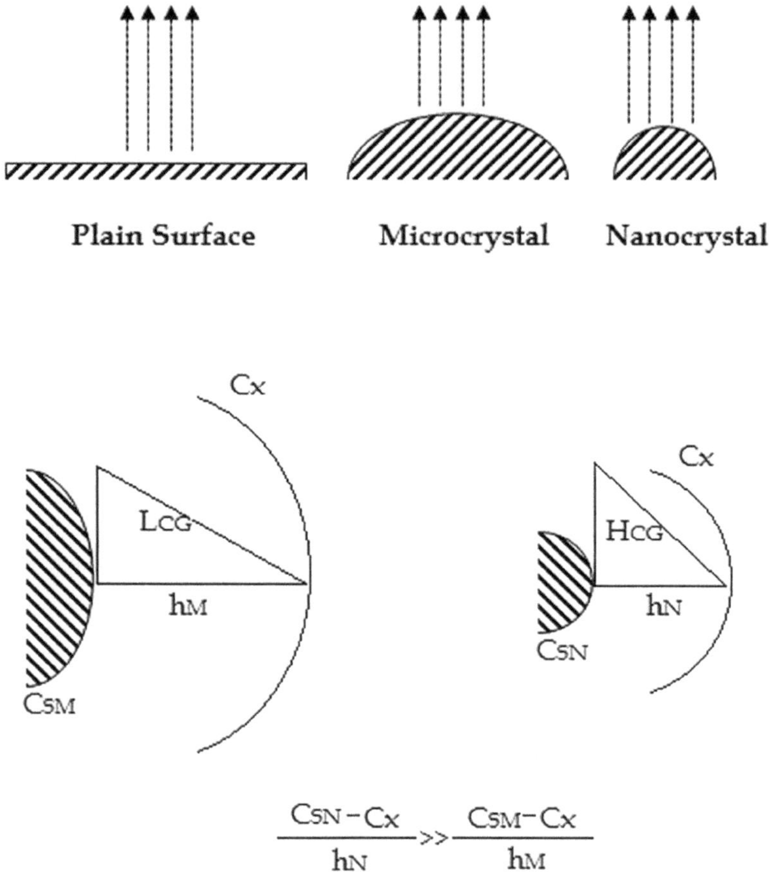


**FIGURE 9.1** Schematic representation of the increase in the saturation solubility $C_S$ and dissolution velocity for drug nanocrystals, suspended in liquid medium ($C_x$, concentration in the surrounding medium; $L_{CG}$, lower concentration gradient; $H_{CG}$, higher concentration gradient; $C_{SM}$, surface saturated solution layer onto microcrystal; $C_{SN}$, surface saturated solution layer onto nanocrystal; $h_M$, high of saturation layer surrounding microcrystal; $h_N$, high of saturation layer surrounding nanocrystal; $C_{SN} > C_{SM}$; $h_N < h_M$).

The compounds belonging to Class I show both high solubility and high permeability (e.g., diazepam, propranolol). In Class II, drugs are highly permeable but poorly water-soluble (e.g., ibuprofen, carbamazepine). Class III drugs display physicochemical characteristics opposite to those of Class II, that is, poor permeability and high solubility (e.g., atenolol, metformin). Class IV compounds possess low permeability and low solubility (e.g., sulfasalazine, bifonazole) (Smetanova *et al.*, 2011).

When the bioavailability of poorly water-soluble drugs, if administered orally, is limited by their dissolution velocity, their low solubility can be increased by different strategies. Improvement of saturation solubility promotes the enhancement of drug penetration through the skin once the saturation solubility creates a higher concentration gradient between the formulations constituted by nanocrystals and the skin when it is applied (Amidon *et al.*, 1995).

## 9.2 NANOCRYSTAL TECHNOLOGY

The standard methodologies to reduce the particle size of poorly water-soluble drugs are bottom-up and top-down techniques; more recently, rational blends of both the approaches are also being used, that is, use of a bottom-up process combined with a top-down stage (Salazar *et al.*, 2012). Figure 9.2

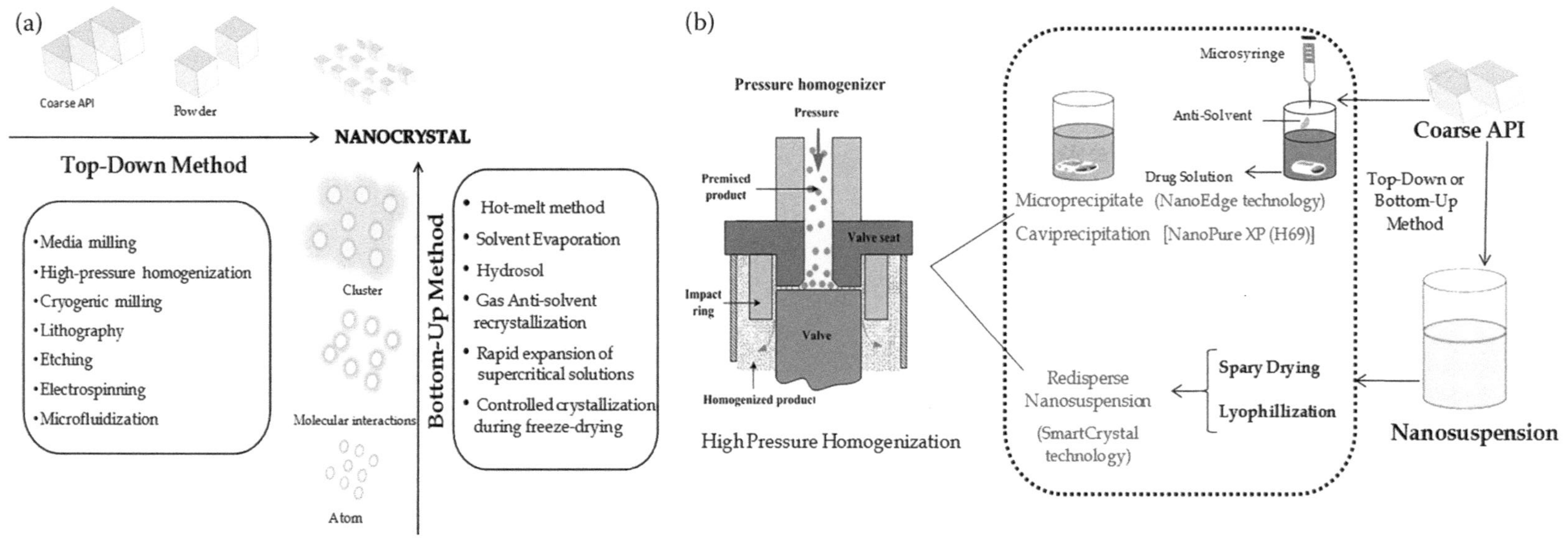


**FIGURE 9.2** Various techniques used for developing drug nanocrystals. (a) Top-down and bottom-up technique (b) Combined/second-generation technique.

illustrates the various techniques used for developing drug nanocrystals. The starting point is the bulk material using top-down methods, followed by a reduction in the particle size, by milling or high-pressure homogenization. For instance, particles can be produced with pre-dissolved molecules using different techniques such as precipitation and bottom-up techniques. Pre-milling, combined with high-pressure homogenization, represents another possibility of a different technology that can be used. In the industrial purposes, all the products are usually prepared using a top-down method (Sharma *et al.*, 2011; Liu *et al.*, 2012). Table 9.1 enlists various drug nanocrystals products in clinical trials and commercially available.

### 9.2.1 Bottom-Up Technique (Precipitation Method)

In the precipitation method, also known as "via *humida paratum*," the drug and the stabilizer are dissolved in a solvent and the obtained solution is spilled onto a non-solvent, which induces a fast decrease in the drug solubility, triggering drug precipitation and, as a result, formation of dispersed drug nanocrystals. The precipitation method is very simple, necessitating low-cost equipment (Junghanns and Muller, 2008; Bansal *et al.*, 2012).

However, this technique presents some disadvantages in the development of new drugs, especially when these drugs are not soluble in both aqueous and organic media. Another problem of this method is that the solvent requires being miscible in one non-solvent, at least. It is difficult to dissolve this type of compound in any solvent, especially due to the miscibility requirement with the anti-solvent. Also, the solvent residues need to be removed, and this issue can pose further problems, namely increased production costs, particularly when bigger solvent volumes are necessary. This occurs if the drugs used in the production have low solubility in water and also in organic solvents (Muller *et al.*, 2011). The control over the particle size is often difficult to achieve, but it is necessary to make sure that crystals do not grow (i.e., stay within the nano-range). To preserve the submicron size range, recommended techniques include spray drying and lyophilization (Speiser, 1998; Müller and Moschwitzer, 2006). Alternatives to preserve the size of nanocrystals consist of processing precipitation under a high-energy zone (e.g., in a microfluidizer) or the use of polymeric growth inhibitors, which can contribute to delay in the particle growth. To prepare nanocrystals, several stabilizers can be used, such as sodium dodecyl sulphate (SDS), polyvinyl alcohol (PVA), Tween® 80, and Poloxamer® 188 (Mauludin *et al.*, 2005). Another problem is high difficulty of many poorly water-soluble compounds to get solubilized in aqueous and/or organic media. The precipitation method is also used to obtain amorphous particles, which has the distinct advantage in the case of solubility; however, this solution presents some issues owing to drug stability. Thus, Soliqs/Abbott developed another precipitation method to enhance the dissolution rate and solubility, namely Nanomorphs (Shackleford *et al.*, 2003). Based on these facts, bottom-up technologies are not currently used to produce commercial formulations.

There are several parameters that should be optimized to obtain uniform nanocrystals: (i) stirring rate, (ii) drug content, (iii) volume ratio of anti-solvent to solvent, and (iv) temperature. An increase in the stirring rate favors the particle size enhancement of nanocrystals due to intensification of the micromixing between the two parts. Higher content of drug increases the possibility of particle aggregation; thus, it is necessary to use moderate/low drug content for this method. Reduction of particle size is favored when using lower temperatures (Gao *et al.*, 2008).

To produce nanocrystals using the precipitation method, some prerequisites should be satisfied (Gao *et al.*, 2008): (i) the solvent needs to be miscible with a non-solvent, (ii) the drug must be soluble at least in one solvent, and (iii) the solvents should be eliminated to maximum possible extent. The main advantages of the bottom-up technique (i.e., precipitation technique) include the production of small particles, good control of the desired size (i.e., monodispersity), and the production of finely dispersed drug nanosuspensions. Main limitations are related to the risk of the solvent residue process being time-consuming during optimization, the difficulty to scale-up, and the need for stabilizers.

**TABLE 9.1**
**List of Nanocrystal Drug Products in the Market and Under Clinical Trials**

| Trade Name | Drug | Indication | Status | Applied Technology | Company | Route |
|---|---|---|---|---|---|---|
| Rapamune® | Sirolimus | Immunosuppressant | Marketed | Elan NanoCrystals® | Elan Corp. | Peroral |
| Emend® | Aprepitant | Anti-emetic | Marketed | Elan NanoCrystals® | Merck | Peroral |
| Avinza® | Morphine sulfate | Moderate to severe pain | Marketed | Elan NanoCrystals® | Elan Corp., Ligand Pharmaceuticals | Peroral |
| TriCor® | Fenofibrate | Antihyperlipidemic | Marketed | Elan NanoCrystals® | Abbott | Peroral |
| Ostim® | Hydroxyapatite | Periodontal Intrabony Defects | Marketed | – | Heraeus Kulzer | Periodontal bone grafting material |
| Triglide® | Fenofibrate | Antihyperlipidemic | Marketed | SkyePharma IDD®-P | SkyePharma First Horizon Pharmaceuticals | Peroral |
| Focalin® XR | Dexmethylphenidate hydrochloride | Antipsychotic | Marketed | Elan NanoCrystals® | Novartis | Peroral |
| Theodur® | Theophylline | Bronchial dilation | Marketed | Elan NanoCrystals® | Mitsubishi Tanabe Pharma | Peroral |
| Megace ES® | Megestrol acetate | Appetizer | Marketed | Elan NanoCrystals® | PAR Pharmaceutical | Peroral |
| Invega® Sustenna™ | Paliperidone palmitate | Anti-Schizophrenic | Marketed | Elan NanoCrystals® | Janssen Pharmaceuticals, Inc., Johnson & Johnson | Intramuscular |
| Gris-PEG® | Griseofulvin | Antifungal | Marketed | Bottom-Up | Novartis | Peroral |
| Cesamet® | Nabilone, synthetic cannabinoid | Antiemetic | Marketed | Bottom-Up | Valeant Pharmaceuticals | Peroral |
| SILCRYST™ | NPI 32101 | Antibacterial, Antimicrobial, Atopic dermatitis | Marketed | Magnetron sputtering | Nucryst Pharmaceuticals | Topical |
| ACTICOAT Absorbent (with SILCRYST™ Nanocrystals) | NPI 32101 | Antimicrobial Barrier Wound Dressing | Marketed | – | Smith & Nephew plc | Topical |
| Paxceed™ | Paclitaxel | Metalloproteinase inhibitor, Rheumatoid Arthritis | Clinical Phase III | Not Available | Angiotech | Intravenous |
| Zolip® | Fenofibrate/Simvastatin | Antihyperlipidemic | Clinical Phase III | Nanocrystal® | Abbott | Peroral |
| Budesonide | Budenoside | Anti-asthmatic | Clinical Phase III | Nanocrystal® | Sheffield Pharmaceuticals | Pulmonary |
| Theralux™ | Thymectacin | Anticancer | Clinical Phase II | Nanocrystal® | Celmed Biosciences | Intravenous |
| Semapimod® | Guanylhydrazone | TNF-α inhibitor | Clinical Phase II | Self-developed | Cytokine Pharmasciences | Peroral |
| Panzem® NCD | 2-methoxy estradiol | Ovarian cancer | Clinical Phase II | Nanocrystal® | EntreMed | Peroral |
| – | Insulin | Diabetes | Clinical Phase I | – | BioSante Pharmaceuticals | Peroral |
| Nanotax® | Paclitaxel | Breast cancer | Clinical Phase I | Supercritical fluid technology | Crititech | Intraperitoneal |

*Source:* Nagarwal R C *et al. Curr Drug Deliv*. 2011; 8: 1–9; Möschwitzer J P. *Int J Pharm*. 2013; 453: 142–156; Chen M *et al. AAPS J*. 2017; 19(3): 642–651.

Ostwald ripening describes the crystal growth by differences in saturation solubility of differently sized particles. In order to avoid it, the content of microcrystals needs to be as low as possible, since the homogeneous size of the suspended particles avoids differences in saturation solubility, therefore delaying Ostwald ripening.

In the bottom-up techniques, for example, the precipitation technique-nanocrystals are usually built from bulk compounds. Using top-down methods, such as the homogenization method and pearl milling, it is possible to fragment the nanocrystals step-by-step from the coarse powder (Gao *et al.*, 2008).

### 9.2.2 Top-Down Methods

Currently, top-down techniques, such as pearl milling and high-pressure homogenization, are widely employed. In both of these methods, it is easy to achieve particle size around 100 nm; however, particles with smaller sizes are more difficult to obtain. In comparison to the bottom-up techniques, nanocrystals are often more polydispersed, but on/by increasing the processing time or the number of cycles of homogenization, it is possible to increase the homogeneity of the product (Van Eerdenbrugh *et al.*, 2008).

Pearl/ball milling was developed by Merisko-Liversidge *et al.* (2003); the drug is placed in combination with the milling media, with water as dispersion medium, usually, and a suitable stabilizer. To obtain a reduction in the particle size, it is necessary to increase the movement of milling media, which generates high shear forces of impact. The pearls or balls are generally composed of ceramic, glass, stainless steel, or highly cross-linked polystyrene resin-coated beads. In the milling process, there are two basic principles involved. In one of the cases, the milling material is stirred using an agitator, or alternatively all the containers can be moved using a combination of complex movements. In the latter, it is difficult to process large batches, therefore mills using agitators for large batches are usually preferred (Buchmann *et al.*, 1996). Depending on several factors, such as drug hardness, surfactant contents, viscosity, temperature, and particle size of milling medium, it is possible to determine the milling time. This period can vary between 30 minutes and several hours. Some advantages of this method are low cost, simple technology, and ability for large scale production. Pearl milling methods have, however, some disadvantages, namely, (i) erosion from the milling material which can lead to product contamination; (ii) adherence of the product to the inside surface of the mill and the surface of the milling pearls; (iii) long milling times, e.g., in case of hard drugs; (iv) potential growth of germs in the water phase, e.g., for long-time milling processes; and (v) time and costs associated with the separation procedure of the milling material from the drug nanoparticle suspension, especially when producing parenteral sterile products. In the case of erosion of glass beads, it is possible to minimize this problem using covered beads with a highly cross-linked polystyrene resin (Merisko-Liversidge *et al.*, 2003). In general, several parameters should be explored in order to obtain the "optimal formulation" (Gao *et al.*, 2008). These parameters consist of controlling the drug amount, number of milling pearls, milling speed, milling time, and temperature.

The first four products present in the market employing nanocrystals were prepared using pearl mill methods by Elan nanosystems: Rapamune®, Emend®, Tricor®, and Megace ES®. The high-pressure homogenization (HPH) technique has been widely used in the production of nanoemulsions and nanosuspensions. A distinct advantage of this method in the production of nanocrystals compared to others is its ease of scale-up. Using the HPH technique, there are three technologies that stand out:

a. The microfluidizer technology (IDD-P™ technology), which is based on the jet-stream principle: Under high pressures (≥1700 bar), two streams of liquid with high velocity (≥1000 m/sec) collide frontally. Due to the high shear force, particle collision, and cavitation, the particle size is reduced. It is possible to obtain the same using jet-stream homogenizers, Microfluidizer®. The collision chamber can be Y-type or Z-type in shape. Phospholipids or surfactants are essential to stabilize particles in the chosen particle size (Tunick *et al.*, 2002).

b. The production with piston gap homogenization in water (Dissocubes® technology) under high pressure: The powdered drug is dispersed in an aqueous surfactant solution, which is forced by a piston through a tiny homogenization gap. The gap breadth is adjusted according to the viscosity of the suspension and the applied pressure. The gap width generally shows the size range between 5 and 20 μm. The static pressure in the gap decreases the vapor pressure of water if worked at room temperature. To obtain gas bubbles, water should start boiling into the gap at room temperature. These bubbles allow pressure waves, which promote the crystal disintegration, a process normally known as "cavitation" (Muller *et al.*, 2001). Reduction of particle size occurs owing to high shear forces and other phenomena such as turbulent flow and shock waves (Sahoo *et al.*, 2011). The biggest disadvantage of this method is the use of water as the dispersion medium, thus promoting the hydrolysis of hydro-sensitive drugs and making the process of drying difficult. Sometimes, in the case of thermolabile compounds or drugs with a low melting point, the process to remove water requires the use of other techniques, such as lyophilization. High cost is another disadvantage of this method (Müller *et al.*, 2003).
c. The piston gap homogenization, in water mixtures or in non-aqueous media (Nanopure® technology), takes advantage of using the non-aqueous medium, or alternative phases with lower water content as the dispersion medium (Keck and Muller, 2006). Oils, water-glycerol mixtures, polyethylene glycols, and hydro-alcoholic mixtures are some of the different media used for homogenization, since these dispersion media show low vapor pressure. In this method, cavitation does not occur because the static pressure in the homogenization gap never falls under the vapor pressure of the liquid, and the liquid does not boil. However, despite without cavitation, there is a decrease in the particle size owing to the nano-range since the forces responsible, in this case, for reduction of the particle size are particle collision and shear forces into the gap (Hao *et al.*, 2012).

### 9.2.3 Combined Techniques

These approaches combine one first step and then high-energy homogenization, e.g., a pre-precipitated sample can once again be homogenized by HPH (cavi-precipitation technique) (Kakran *et al.*, 2012b), or a pre-milled material can be homogenized (Hao *et al.*, 2012). Combination techniques show several advantages, e.g., capacity of production time shortening and production of very small nanocrystals ($\leq$100 nm) that can otherwise be difficult to obtain using only one method. Comparing milling, HPH, and cavi-precipitation as techniques for the production of nanocrystals, it is possible to conclude that is possible to achieve successful nanocrystallizations using any of these techniques. However, the smallest particles size was achieved using the milling process, though the lowest polydispersity index was achieved with HPH (Kakran *et al.*, 2012a). These two referred methods were shown to promote more stable nanocrystals. In the case of cavi-precipitation, stability problems were caused by solvent residues of the precipitation step, resulting from their competition with the surfactant for hydration. These problems caused partial surfactant dehydration and decreased the stability of the nanocrystals (Kakran *et al.*, 2012b).

a. The NANOEDGE® Technology combines the method of microprecipitation and homogenization, i.e., precipitation followed by an annealing process employing high energy as high shear forces and thermal energy (Kipp *et al.*, 2003). When produced using only a precipitation method, nanoparticles have the tendency to grow and be partially amorphous. In addition, amorphous particles may recrystallize and can lead to decreased bioavailability of the drug. These problems can be solved using combination technologies due to their capacity to prevent the crystal growth and also to reduce the uncertainty in formation of either crystalline or amorphous state nanocrystals, once the annealing phenomenon allows the conversion of all precipitated elements into the crystalline state.

b. The SmartCrystal technology enables the use of different combination processes, depending upon the physical characteristics of the compounds. There are multiple combination variants of this technology, such as (i) spray drying and HPH, (ii) precipitation and HPH, (iii) lyophilization and HPH, and (iv) pre-milling followed by HPH (Bansal *et al.*, 2012). Nanocrystal production using SmartCrystal® technology is faster and economically more feasible than using HPH alone. Another advantage is the creation of lower particle size, which is an indicator of better physical stability (Muller *et al.*, 2011). The salient advantages and disadvantages of the different methods for production of nanocrystals, as described above, are summarized in Table 9.2.

A dry dosage form is usually preferred for oral administration due to convenience, necessity to achieve controlled drug delivery, prevention of drug degradation, better drug targeting, enhancement of physical stability, and attainment of non-aggregated suspension in the gastrointestinal tract after oral administration.

There are various techniques used to transform nanosuspensions into solid forms, including spray drying, freeze drying, pelletization, and granulation. Some particularities of each of these approaches are discussed next.

Spray drying is used for drug nanosuspension production using HPH, allowing the production of water-soluble matrix material. In the following phase, the aqueous nanosuspension can be spray-dried under suitable conditions. Drug nanocrystals are still fixed inside the matrix, which is one of the advantages of this method, and their physical contact is prevented, allowing minimization of long-term physical instability (Bansal *et al.*, 2012).

Freeze drying uses a mixture of the compound, solvent, and mannitol. This mixture is cooled quickly, resulting in the separation of the drug in a nanocrystal form covered by a matrix of mannitol. The stability is improved due to the presence of this matrix. However, this method poses some

**TABLE 9.2**
**Advantages and Disadvantages of Different Methods for Production of Nanocrystals**

| Technology | Advantages | Disadvantages |
|---|---|---|
| Precipitation | Finely dispersed drug<br>Good control of desired particle size<br>Low-energy requirement | No universal applicability; restricted for drugs with specific properties<br>Presence of organic solvent residues<br>Necessity of stabilization<br>Risk of particle growth |
| Milling | First four marketed products<br>Low-energy requirement | Presence of impurities due to erosion of milling material<br>Can be slow process taking several days<br>Waste of drug due to adherence to milling pearls<br>Difficulty to transpose to large-scale production due to size of milling chamber<br>Necessity of stabilizers |
| High pressure homogenization (HPH) | Universal applicability<br>Large-scale production possibility<br>Fast process, possibly requiring few minutes<br>Continuity of process<br>Possibility of water-free production | High-energy requirement<br>Training and expertise requirement |

disadvantages, such as being a very complex and expensive process, and the obtained product is highly sensitive, being not appropriate for industrial production (de Waard *et al.*, 2010).

Pelletization results in a multi-particulate dosage form, e.g., coated-pellet system. These dosage forms have some advantages when compared to single unit dosage forms, as they present faster and a more probable gastric discharging and also a more uniform drug distribution in the gastrointestinal tract (Moschwitzer and Muller, 2006b).

Beyond all these techniques, there are many other alternative methods, though industrially less relevant, e.g., solution-enhanced dispersion using the supercritical fluids (Hu *et al.*, 2004), spray freezing into a liquid (Rogers *et al.*, 2001), rapid expansion of supercritical solution (Li and Hsieh, 2008), rapid expansion of supercritical solution into an aqueous solution (Bansal *et al.*, 2012), and evaporative precipitation into an aqueous solution process (Chen *et al.*, 2002).

## 9.3 FUNCTIONALIZATION OF NANOCRYSTAL SURFACE BY LAYER-BY-LAYER SELF-ASSEMBLY

Nanocrystals of poorly water-soluble drugs can be used as substrates for the nanotechnology sonication-assisted layer-by-layer (LbL) assembly. This promising technique enables the conversion of solid materials into stable aqueous nanoparticles, by the use of simultaneous ultrasonication and LbL polyelectrolyte-based coatings upon the surface of drug nanocores. LbL coatings are performed by the alternated adsorption of polyelectrolytes (i.e., polycations and polyanions) to the system under powerful ultrasonication, setting up highly controlled, thin polyelectrolyte shells upon drug nanocrystals surface (Santos *et al.*, 2015). This enables creation of highly functional, thin nanolayered drug-delivery systems with a finely tuned composition (Ariga *et al.*, 2014).

The LbL technique was first introduced in the 1990s and has been demonstrated to be an extremely versatile and universal technique for obtaining an extensive range of thin-film systems (Decher and Hong, 1991; Decher, 1997). LbL was later applied for the microencapsulation of drugs (Donath *et al.*, 1998; Sukhorukov *et al.*, 1998; Caruso, 2001; Lvov *et al.*, 2001; Qiu *et al.*, 2001; Antipov and Sukhorukov, 2004). These works focused on obtaining nanoengineered polyelectrolyte LbL shells upon preformed microtemplates with particle size higher than 2 mm (Ai *et al.*, 2003; Shenoy and Sukhorukov, 2004; Yuan *et al.*, 2005; De Geest *et al.*, 2009; Palankar *et al.*, 2009; Bédard *et al.*, 2010). However, the necessity for new biopharmaceutical characteristics imposed the necessity to reduce the particle size to the nanoscale. Thus, LbL application was channeled for the production of nanoparticles with particle sizes of 100–200 nm, covered with notably thin LbL shells (using polyelectrolyte bilayers with 10–15 nm of thickness). Application of this technique is particularly relevant to increase the solubility of poorly water-soluble drugs. In those cases, the particle size of poorly water-soluble drug-based nanoparticles is appropriate for intravenous administration, whereas, in parallel, the application of the LbL coating at the surface confers a high surface charge (zeta potential around $\pm$40 mV) (Lvov *et al.*, 2011; Shutava *et al.*, 2012).

Before the assembly of LbL coatings, the production of initial drug nanocrystals may be performed using sonication through a top-down or a bottom-up approach, the experimental manufacturing techniques being able to promote a great degree of material dispersion (Keck *et al.*, 2008; Tsuzuki, 2009; Chen *et al.*, 2011), as referred previously. In the top-down approach, aqueous suspensions of micron-sized drugs, without any additive, are submitted to sonication to decrease the drug particle size into the nanoscale. Continuous sonication prevents agglomeration and the application of the LbL technology promotes stabilization of the nanosuspension. Besides, the bottom-up approach relies on the drug nucleation in the co-solvent after the addition of a drug organic solution. Comparatively, the top-down approach has been shown to be more effective in obtaining high amounts of ca. 200 nm nanocapsules. The bottom-up approach promotes a higher degree of disintegration, being more efficient in producing lower-sized capsules with ca. 100 nm, however with inferior process yield (Pattekari *et al.*, 2011; Santos *et al.*, 2015).

Obtaining LbL-coated nanoparticles precisely relies, thus, on the mighty sonication of powders of poorly water-soluble materials, along with polyelectrolytes or surfactants, respectively, by using top-down or bottom-up approaches, which are deposited on newly designed nanosurfaces (Pattekari *et al.*, 2011). After adsorption of the first coating layer of the nanoshell, a colloidal dispersion coated with one highly charged polyelectrolyte layer is formed, providing a high surface zeta-potential magnitude. Adsorption of the second polyelectrolyte joins the chains of the first polyelectrolyte layer together, avoiding their detachment from the surface of the nanocores, whilst reversing the nanoparticle zeta potential (according to the nature of the polyelectrolyte). These highly surface-charged multilayer polyelectrolyte shells prevent re-aggregation when the sonication ends, conferring a colloidal stability for several months (Santos *et al.*, 2015). Each deposited layer ranges in thickness from 1 to 2 nm to a mere hundred nanometers, in accordance to the nature of the adsorbed material and its relative charge density. Usually, during polyelectrolyte assembly, considerable interpenetration of polyelectrolytes occurs over and under the site of deposition, and inter-diffusion may also occur under certain LbL-film construction conditions (Hammond, 2015).

The LbL self-assembly technique has been successful by making use of either synthetic polyelectrolytes, including poly(allylamine hydrochloride) (PAH) (Lvov *et al.*, 2011; Pattekari *et al.*, 2011; Santos *et al.*, 2015), polystyrenesulfonate (PSS) (Lvov *et al.*, 2011; Pattekari *et al.*, 2011; Parekh *et al.*, 2014; Santos *et al.*, 2015), and poly(diallyldimethylammonium chloride) (PDDA) (Lvov *et al.*, 2011), or biopolymers such as bovine serum albumin (BSA) (Pattekari *et al.*, 2011; Shutava *et al.*, 2012; Parekh *et al.*, 2014), protamine sulfate (Lvov *et al.*, 2011; Pattekari *et al.*, 2011), chitosan (Lvov *et al.*, 2011; Pattekari *et al.*, 2011), alginate (Lvov *et al.*, 2011; Pattekari *et al.*, 2011), polyethylenimine (Parekh *et al.*, 2014; Shutava *et al.*, 2012), poly-L-lysine (Shutava *et al.*, 2012), and heparin (Parekh *et al.*, 2014). The experimental conditions of the LbL assembly have to be cautiously optimized, and this is difficult to achieve for a majority of the natural polyelectrolytes. Lo-molecular-weight polyelectrolytes are essential to promote particle size control at the nanoscale, so they are recommended to be used on these systems (Li *et al.*, 1994; Chodanowski and Stoll, 2001; Bantchev *et al.*, 2009). Additionally, the application of PEGylated polyelectrolytes allows for higher stability of LbL nanoparticles in buffer and salt solutions and cell binding resistance (Zheng *et al.*, 2010; Lvov *et al.*, 2011; Pattekari *et al.*, 2011).

During the LbL process, before polyelectrolyte coatings, adsorption is necessary for the removal of polyelectrolyte excesses, which are not hitherto adsorbed on the surface of the LbL shell. For this, centrifugation and filtration, which are used often with 1–5 mm particles in the step-wise polyelectrolyte deposition manner in LbL microencapsulation, are effectively used in handling of nanocapsules. Centrifugation leads to large process losses and also triggers aggregation phenomena. Thus, application of a washless LbL assembly technique to produce a significant number of polyelectrolyte bilayers in the coating shell is necessary for the development of efficient LbL nanocarriers (Bantchev *et al.*, 2009; Parekh *et al.*, 2014; Santos *et al.*, 2015). The washless approach relies on the use of determined and sufficient polyelectrolytes amounts, ensuring their complete adsorption on the nanoparticle/water interface. This avoids the necessity of excess polyelectrolytes removal prior to the addition of the next oppositely charged polyelectrolyte (Ogawa *et al.*, 2004). In practice, the experimental procedure of washless LbL assembly relies on continuously monitoring the zeta potential of the particle surface as the polyelectrolyte is progressively added, in order to assess the quantity of polyelectrolyte necessary for the nanoparticle recharging. The particle size has to be additionally and simultaneously controlled, since the charge reversal has to be accompanied by insignificant particle-size enhancement, which might compromise the stability of the nanosuspension. Immediately after the determination of the first polyelectrolyte amount to be added, the polyelectrolyte is switched to the oppositely charged one, and its stepwise addition to the surface is started until a new surface charge saturation is reached, this time, a reverse zeta-potential value. Such a process is repeated till attainment of the required number of adsorbed polyelectrolyte bilayers. There are some published works that clearly exemplify the recharging process associated with the washless approach (Bantchev *et al.*, 2009; Santos *et al.*, 2015). This washless approach enables the

LbL process scale-up by facilitating the whole production process. In addition, we are facing a low time-consuming and economical procedure due to the use of lower amounts of materials, particularly relevant for modified polyelectrolytes (e.g., PEGylated) that are usually expensive or only available in limited quantities. Both centrifugation and washless approaches enable attainment of equivalent thickness of the polyelectrolyte shell engineered upon the surface of the nanocores (Bantchev *et al.*, 2009). However, the particle size was shown to be lower using the washless approach. Regarding the case of a washless LbL process, coating of paclitaxel nanoparticles with poly-L-lysine and heparin-based bilayers, a significant particle size increase was verified when a traditional centrifugation-based LbL technique was performed and a three-fold excess of polyelectrolytes was applied (Shutava *et al.*, 2012; Parekh *et al.*, 2014).

This technique is capable of originating thin LbL films directly from aqueous medium without the necessity of other chemical modifications, and also promotes direct encapsulation of growth factors, antibodies, DNA, and additional biologic drugs through the establishment of electrostatic or hydrogen interactions. These films may be constructed in order to allow drug release by the inclusion of hydrolytically (Vazquez *et al.*, 2002; Wood *et al.*, 2005; Wood *et al.*, 2006), enzymatically degradable polymers (Benkirane-Jessel *et al.*, 2004; Picart *et al.*, 2005), or by the application of polyelectrolyte pairings, which become unstable at physiologically relevant pH values and ionic strengths (Hong *et al.*, 2011). The versatility and the promising properties of LbL thin-film coatings have led to relevant advances in drug delivery for biomaterials and for biomedical purposes (Hammond, 2015).

Modification of the nanocrystal surfaces using polyelectrolyte LbL shells enables the control of the nanoparticle cell uptake rate, the increase of the nanoparticle colloidal stability, and the control of loading/drug release features, according to the composition and the thickness of the multilayered shell. Moreover, those LbL shells can also provide a template for alteration with tumor-targeting agents (Jessel *et al.*, 2006; DeMuth *et al.*, 2010; Saurer *et al.*, 2013; Santos *et al.*, 2015). In respect of the release of single drugs, application of this technique promotes obtaining stable and regular coatings, capable of releasing in long or short release profiles, or with a conjunction of release profiles (e.g., an initial burst release followed by a sustained release profile). In the case of the release of two or more drugs, the LbL technology may be applied to include different drugs into a uniform thin film, which separately controls the release of each drug. This manipulation of drug release profiles may conduct more synergistic drug-release patterns (Hammond, 2015).

Nanoparticles with high drug content (80%–90%) and concentration up to 5 mg/mL were obtained using LbL through alternating deposition of polyelectrolytes bearing opposite charges (Agarwal *et al.*, 2008; Zheng *et al.*, 2010; Lvov *et al.*, 2011; Pattekari *et al.*, 2011; Santos *et al.*, 2015). For poorly water-soluble drugs, (e.g., curcumin, resveratrol, tamoxifen, paclitaxel), the primary task is to obtain nanoparticles with high concentrations (4–5 mg/mL) that are stable in biological buffers.

These particles can display different properties based on the nature of the polyelectrolytes and the architecture of the charged polyelectrolyte shell or the number of coating cycles (Santos *et al.*, 2015). Additional coatings allow for construction of sophisticated capsule-wall architectures with highly developed characteristics. The core can be used as the template for LbL assembly, and the alternating coating layers may be used to include a second therapeutic agent, e.g., an inhibitor capable of blocking the genetic pathways pertinent to a specific tumor, or siRNA. Extra coating layers may serve to introduce constituents capable of imparting special/unique properties, for example, regulating the cell uptake. The outermost layers of the LbL shell can be designed to provide stealth capabilities, in order to prevent being taken up by immune system. Each of the layer constituents can be separately altered without causing a relevant impact straightaway. These outermost layers of the shell are an effective key of these formulations, as they have the capacity to determine cell-constituent interactions responsible for the fate of the nanoparticles, and thus, the therapeutic effect (Hammond, 2015). LbL-nanocrystal self-assembly technology offers, therefore, scale-up facilities and promising avenues for achieving effective targeted delivery systems tailored to the actual and specific biomedical needs. Figure 9.3 highlights the surface functionalization of nanocrystals for improved site-specific targeting and plausible controlled-release characteristics.

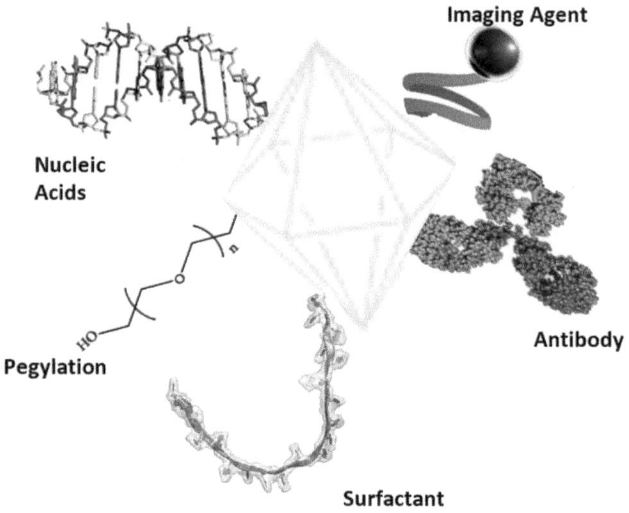


**FIGURE 9.3** Surface functionalizations on nanocrystals for improved site-specific targeting and plausible controlled-release characteristics.

## 9.4 ROUTES OF ADMINISTRATION FOR NANOCRYSTALS

Drug nanocrystals can be formulated in various traditional dosage forms, including tablets, capsules, gels, creams, and solutions, among others, implying that a novel drug-delivery system can be combined with dosage forms well-recognized by the patient.

For oral applications, nanocrystals can be suspended in aqueous media, or be jellified, if necessary, by adding a viscosity enhancer (e.g., bioadhesive polymer) to increase the stickiness to the mouth mucosa. Nanosuspensions for oral administration are one of the most important application area of nanocrystals, enabling effective therapeutic blood concentrations by overcoming the absorption concerns found in the gastrointestinal tract using particle-size reduction (Mauludin *et al.*, 2009). The oral route, due to its numerous advantages (e.g., safety, convenience, compliance), is the most attractive and, thus, the first choice for drug delivery. Most nanosuspension-based references available on the market are actually for oral administration. Drug nanocrystals can be incorporated in tablets, pellets, and in hard or soft gelatin capsules. Nanosuspensions may be applied as granulation media for the obtaining tablets, or as wetting liquid for the extrusion mass for the production of pellets. Nanocrystals can also be spray-dried or freeze-dried for the production of powders for further filling of hard capsules. If a non-aqueous medium is required (e.g., for highly water-soluble drugs), the described Nanopure® production technology can be used for filling soft capsules using low-molecular-weight polyethylene glycols or medium-chain triglycerides as the suspending media.

After administration, the drug undergoes absorption in the gut and enters into the blood circulation in order to be further distributed to various tissues. Solubility in the digestive fluids and the passage throughout the gastrointestinal tract depend on the performance of the compound. The poor drug solubility and the dissolution rate of poorly water-soluble drugs, belonging to the BCS class II, limit the absorption *in vivo* and do not allow for attainment of optimal therapeutic concentrations. Moreover, BCS class II drugs frequently show enhanced or accelerated absorption when co-administered with food (Jinno *et al.*, 2006). There are several problematic issues associated with poorly water-soluble drugs, such as low bioavailability, necessity of high oral doses, alteration in the bioavailability arising from fasted states, and delayed onset of action. A large drug concentration gradient between the gastrointestinal tract and blood vessels is achieved when the drug is administered by a nanosuspension. This occurs owing to the enhanced saturation solubility and the dissolution velocity in the digestive fluids, increasing absorption and thus a high bioavailability, as previously referred. Variations in bioavailability of poorly water-soluble drugs resulting from the fed/fasted condition can be minimized when these are developed as a uniform nanosuspension (Hecq *et al.*, 2006). This occurs because the dissolution rate of nanocrystals is considerably increased owing to the enhanced solubility and immense particle surface. Absorption can show permeability-restricted progress, both in fasted and fed conditions. Another advantage of nanosuspensions is quick onset of action achieved for drugs that exhibit slow dissolution rates. If particle aggregation occurs in the gastrointestinal tract, the characteristics of nanosuspensions and their associated actions may be compromised, thus reducing the oral bioavailability (Muller *et al.*, 2006). Therefore, use of a stabilizer is important to prevent aggregation of the compounds subjected to *in vitro* and *in vivo* conditions.

The parenteral administration route supplies a rapid onset of action, particularly for the intravenous route, and enables the use of a reduced dosage of the drug. Intravenous rapid injection is required to address certain treatment objectives, including immediate action, targeting effect, bypassing first-pass metabolism, and for drugs which do not absorb from the gastrointestinal tract or are degraded there. Albeit drugs can achieve 100% of bioavailability using intravenous administrations only, yet the pharmacokinetic profile and the tissue distribution of nanocrystal-based suspensions of drugs which are soluble in blood are quite comparable to those from drug solutions (Junghanns and Muller, 2008). The intravenous route is an exigent route, for which many preconditions should be met: the products should be sterile, and the constituents should not trigger any biological problems like toxic or allergic reactions. The particle size is also a key factor to determine whether or not a formulation can be administrated by a parenteral route. Particles with particle sizes larger than 5 μm should be monitored severely for intravenous injection, due to the fact that the smallest size of blood capillaries is around 5 μm, and a high level of particles larger than 5 μm can promote embolism and capillary obstruction (Muller and Keck, 2004a).

Formulation of drug nanocrystals in topical/dermal dosage forms (such as creams, hydrogels) can enhance skin bioavailability of a drug due to increased saturation solubility of the drug. Transdermal patches may also be produced containing drug nanocrystals. The retention time of applied drug nanocrystals in the eye may be enhanced by increased contact area with improved drug bioavailability. A depot effect may be created by change of a water-soluble compound in a poorly water-soluble counterpart being formulated into the ultimate dosage form. Preparations, including ointments and suspensions, have been investigated for ocular administration of poorly water-soluble drugs. These formulations have some advantages, such as a prolonged time in the locale of action and a relative higher drug dosage. Limited intrinsic solubility of the drug in lachrymal fluids is responsible for their actual performance. Low intrinsic drug solubility in those fluids conducts low drug concentration for the local action, being difficult to obtain an effective performance (Gao *et al.*, 2008). Intrinsic adhesive characteristics of nanocrystals decrease the precorneal loss of drug produced through the outflow of lachrymal fluids and allow for a sustained drug release. Actually, it is possible to incorporate nanocrystals into a proper mucoadhesive base or ocular inserts to attain sustained drug release for a specific period of time. All those options guarantee high bioavailability and efficient performance. These topical formulations promote drug penetration and maintain therapeutic levels with rational frequency of administration and have the capacity, meanwhile, to reduce undesirable side effects. In order to prove their safety for ocular delivery, nanosuspensions are characterized for particle size, zeta potential, drug loading, *in vitro* drug release, and ocular tolerability (Araujo *et al.*, 2009).

Nanosuspensions can also be nebulized using mechanical or ultrasonic nebulizers for lung/pulmonary delivery. These nanosuspensions offer a strong potential for the pulmonary delivery of poorly water-soluble drugs. Compounds have been effectively transformed into aerosols under the form of microparticles in order to be nebulized for pulmonary administration. However, microparticles have some disadvantages too, such as the unwanted sediments of the microparticles in the pharynx and/or mouth. Moreover, together with the drug clearance by cilia motion, those phenomena are the key drivers triggering the loss of the drug in the lungs, constituting additional disadvantages of microparticles. All these demerits result in lower drug bioavailability. The possible solution for these problems can be the substitution of microparticles by nanosuspensions for pulmonary delivery. Enhanced saturation solubility and dissolution velocity promoted by nanosuspensions can quickly create a higher concentration of the drug in the lungs and, consequently, lead to larger local drug concentrations at the absorption locale. Another benefit of the nanosuspensions is the natural inclination of the nanoparticles to adhere to mucosal surfaces at the absorption locale during a prolonged time period. Compared with microparticles, the longer residence time on the mucosal surface decreases drug losses owing to the clearance by cilia motion, and also contributes to a higher absorption magnitude due to an enhanced dissolution period (Jacobs and Muller, 2002). The increased adhesiveness and prolonged pulmonary residence time are the fundamental advantages of nanocrystals (Peltonen *et al.*, 2014).

Targeted delivery is another possibility for nanosuspensions, due to their surface characteristics and *in vivo* performance that can be simply modified by the choice of adequate stabilizers. The concept of "differential protein absorption" has also been raised by scientists (Muller and Keck, 2004b). Drugs have the tendency to adsorb on the blood proteins, when administered by intravenous injection. Protein adsorption can also govern the *in vivo* destiny of compounds (Muller and Keck, 2004a). Depending upon the physicochemical surface characteristics of the drugs, the qualitative and quantitative composition of the protein's adsorption pattern is altered (Goppert and Muller, 2005b). With the change of the stabilizer, it is easy to change the *in vivo* performance, and drug targeting can be achieved by controlling surface modification of nanoformulation.

## 9.5 PHYSICOCHEMICAL CHARACTERIZATION OF NANOCRYSTALS

Characterization of nanocrystals is essential to have reproducibility regarding the formulation process and the formulation stability. Particle size, shape, surface charge, and crystal morphology are essentials in the formulation performance and, thus, in the drug efficacy and *in vivo* safety patterns (Hirsjarvi *et al.*, 2008).

Particle size distribution of the nanosuspensions is a vital aspect of the characterization process, as it is responsible for other important characteristics such as saturation solubility, dissolution velocity, biological performance, and physical stability. There are several techniques available to evaluate the particle size, including transmission electron microscopy (TEM), scanning electron microscopy (SEM), environmental scanning electron microscopy (ESEM), sedimentation, atomic force microscopy (AFM), and capillary hydrodynamic fractionation (Donini *et al.*, 2002; Shahgaldian *et al.*, 2003; Abdelwahed *et al.*, 2006). It is verily possible to assess the morphology and shape of the particles simultaneously using electron microscopic techniques. Generally, the shape and the morphology of the nanocrystals can be assessed by TEM and/or SEM, and also using AFM, depending on the particle size, even though the latter is evidently a more exigent technique. A wet sample of appropriate concentration is required for the TEM examination. Nanosuspensions are required to be processed into a dried powder. SEM examination is crucial to follow modifications of the particle size prior to and after the process of the water removal. The form of drug crystals depends on their inherent crystalline structure (Gao *et al.*, 2008; Peltonen *et al.*, 2014). Zeta-potential measurements rely on the electrophoretic mobility determinations of particles (Fan *et al.*, 2006), and this analysis enables the estimation regarding the storage stability of colloidal dispersions (Li *et al.*, 2006). The particle-aggregation phenomenon is less likely to happen if particles bear a zeta potential capable of supplying enough electric repulsion, or enough steric barrier for a sufficient steric repulsion. Jacobs *et al.*, specified a zeta potential of a minimum −30 mV for electrostatic and −20 mV for sterically stabilized particulate systems necessary to attain physically stable nanocrystal suspensions (Jacobs *et al.*, 2000). The zeta potential of a nanosuspension is fundamental, because it offers information regarding the physical stability of the nanosuspension. The zeta-potential value of a nanosuspension is dependent on the stabilizer and also on the drug (Patravale *et al.*, 2004). To assess the crystalline structure of the drug nanocrystals differential scanning calorimetry (DSC) and X-ray diffraction (XRD) techniques may be used (Hecq *et al.*, 2006).

**TABLE 9.3**
**Various Characterization Techniques Employed for Nanocrystals**

| Type | Techniques | Purpose |
|---|---|---|
| Solid state characterizations | i. PXRD<br>ii. Thermal analysis<br>*Differential scanning calorimetry*<br>*Hot stage microscopy*<br>*Differentialthermal analysis*<br>iii. Vibrational spectroscopy<br>*Infrared spectroscopy*<br>*Raman spectroscopy* | To determine and examine the solid-state form of nanocrystals.<br>This parameter is vital and has high degree of influence on apparent solubility and thereby the dissolution rate. |
| Chemical interactions | FTIR | To analyze any characteristic interaction with inactive ingredients or any other processing parameter that change or shift the IR peak of the drug. |
| | Surface plasmon resonance analysis (SPRA) | SPRA is useful for studying the interaction of nanocrystal surface with stabilizer.<br>Helpful in the screening of suitable excipients and efficient stabilizer for the preparation of nanocrystals. |
| In-line characterization of crystallization process | Raman spectroscopy | To determine during which stage the crystallization of solute occur. Effect of freezing parameters on particle size and morphology of nanocrystal. |
| Particle size and size distribution | Dynamic light Scattering<br>Optical microscopy<br>Low angle static lightscattering<br>Coulter counter analysis<br>Scanning probe microscopy<br>Scanning tunneling microscopy | To correlate the various properties of nanocrystal such as physical stability, saturation solubility, and dissolution rate. |
| Surface charge | Zeta potential | Gives an idea about physical and storage stability of nanocrystals. |
| Particle morphology | Transmission electron microscope<br>Scanning electron microscope<br>Atomic Force Microscopy | Particularly important in case of pulmonary delivery of dry powder nanocrystal. |
| Cellular interaction | Confocal Raman microscopy | Label-free and non-destructive tool for imaging nanocrystal interaction with cells and tissues. |
| | Coherent anti-stokes Raman scattering microscopy | Important technique to determine the *in vivo* fate of the nanocrystals. |

*Source:* Chogale M M *et al. Pharmaceutics*. 2016; 8(26): 1–18; Srivalli K M R and Mishra B. *Saudi Pharm J*. 2016; 24(4): 386–404; Liu R *et al. R Soc Open Sci*. 2017; 4: 1–9.

Assessment of saturation solubility and dissolution velocity is essential, because not only can these parameters aid to determine the benefits in comparison to the microparticle formulation, but they may aid to predict the *in vivo* behavior as well. It is important to note in this regard that, prior to the investigation of the dissolution behavior, nanosuspensions should be moved into a dried power (Van Eerdenbrugh *et al.*, 2010). Diverse temperatures can be used in the experiments to determine the saturation solubility of a dehydrated powder in a distinctly artificial medium (Gao *et al.*, 2008). Regarding determination of dissolution velocity, the methods prescribed in the Pharmacopoeias may be used. Light-scattering has shown to be quite valuable for accurate investigation of nanocrystal solubility (Anhalt *et al.*, 2012).

Chemical composition of the particle surfaces and the surface analysis of nanocrystals can be evaluated through X-ray photoelectron spectroscopy. Knowledge of the surface characteristics of nanosuspensions is essential, particularly if the desired route is intravenous administration (Soppimath *et al.*, 2001). DSC is used to understand the thermal behavior of crystalline and amorphous materials. This technique has the capacity to detect the transition temperature, the melting point, polymorphic changes and furnish precise quantitative information of the energy, which is adsorbed or released during the experiment (Hyvonen *et al.*, 2005). The infrared spectra are normally used to identify particular functionalities of chemicals and to assess the interactions between distinct compounds (Moon *et al.*, 2009). The *in vivo* behavior of nanocrystals after injection depends on their surface properties, which include surface hydrophobicity and interaction of nanocrystals with plasma proteins (Goppert and Muller, 2005a). Table 9.3 summarizes various characterization techniques of nanocrystals.

## 9.6 CONCLUSIONS

Recently, nanocrystalization has gained widespread attention, primarily to address the bioavailability issues of a growing number of poorly water-soluble drugs. Ever-increasing numbers of nanocrystal products in the pipeline, hundreds of scientific publications, scores of intellectual property claims, and shortening hiatuses between product development research and product launch testify to the enormous industrial significance that these nanocrystals hold as novel drug-carrier systems. Majorly being constituted of a drug and a thin layer of stabilizer, such nanocrystals have shown documented promise not merely to promote drug bioavailability, but for sustained drug release and tissue/organ targeting too. This can be accomplished using surface-functionalization of nanocrystals, especially using layer-by-layer self-assembly. Besides oral and intravenous routes, such nanoconstructs have also demonstrated high potential in delivery of drugs through pulmonary and topical routes. Production techniques, especially HPH, have invariably been employed in large-scale production of nanocrystals.

Besides the pharmaceutical industry, the momentous advantages of nanocrystal technology can also be rationally harvested by the cosmetic and nutraceutical industries. Nevertheless, there are arduous challenges and the ardent need of the hour is the exploration and development of relatively more-advanced, cost-effective, and novel promising manufacturing technologies for producing nanocrystal-based drug formulations with high drug pay-loads, controlled crystal size in the nanometric range, enhanced stability in bio-based buffers at high concentrations, and above all, improved scalability.

## ACKNOWLEDGMENTS

The authors would like to acknowledge the financial support received through the projects M-ERA-NET/0004/2015 and UID/QUI/50006/2013, from the Portuguese Science and Technology Foundation, Ministry of Science and Education (FCT/MEC) through national funds, and co-financed by FEDER, under the Partnership Agreement PT2020. FCT is also acknowledged for the individual fellowship granted to ACS (SFRH/BD/109261/2015).

## REFERENCES

Abdelwahed W, Degobert G, Fessi H. A pilot study of freeze drying of poly(epsilon-caprolactone) nanocapsules stabilized by poly(vinyl alcohol): Formulation and process optimization. *Int J Pharm.* 2006; 309(1–2): 178–188.

Agarwal A, Lvov Y, Sawant R, Torchilin V. Stable nanocolloids of poorly soluble drugs with high drug content prepared using the combination of sonication and layer-by-layer technology. *J Control Release.* 2008; 128(3): 255–260.

Ai H, Jones S A, de Villiers M M, Lvov Y M. Nano-encapsulation of furosemide microcrystals for controlled drug release. *J Control Release.* 2003; 86(1): 59–68.

Amidon G L, Lennernas H, Shah V P, Crison J R. A theoretical basis for a biopharmaceutic drug classification: The correlation of *in vitro* drug product dissolution and *in vivo* bioavailability. *Pharm Res.* 1995; 12(3): 413–420.

Anhalt K, Geissler S, Harms M, Weigandt M, Fricker G. Development of a new method to assess nanocrystal dissolution based on light scattering. *Pharm Res.* 2012; 29(10): 2887–2901.

Antipov A A, Sukhorukov G B. Polyelectrolyte multilayer capsules as vehicles with tunable permeability. *Adv Colloid Interface Sci.* 2004; 111(1–2): 49–61.

Araujo J, Gonzalez E, Egea M A, Garcia M L, Souto E B. Nanomedicines for ocular NSAIDs: Safety on drug delivery. *Nanomedicine.* 2009; 5(4): 394–401.

Ariga K, Yamauchi Y, Rydzek G, Ji Q M, Yonamine Y, Wu K C W *et al.* Layer-by-layer nanoarchitectonics: Invention, innovation, and evolution. *Chem Lett.* 2014; 43(1): 36–68.

Bansal S, Bansal M, Kumria R. Nanocrystals: Current strategies and trends. *J Pharm Biomed Sci.* 2012; 3(1): 406–419.

Bantchev G, Lu Z, Lvov Y. Layer-by-layer nanoshell assembly on colloids through simplified washless process. *J Nanosci Nanotechnol.* 2009; 9(1): 396–403.

Bédard M F, De Geest B G, Skirtach A G, Möhwald H, Sukhorukov G B. Polymeric microcapsules with light responsive properties for encapsulation and release. *Adv Colloid Interface Sci.* 2010; 158(1–2): 2–14.

Benkirane-Jessel N, Schwinté P, Falvey P, Darcy R, Haïkel Y, Schaaf P *et al.* Build-up of polypeptide multilayer coatings with anti-inflammatory properties based on the embedding of piroxicam–cyclodextrin complexes. *Adv Funct Mater.* 2004; 14(2): 174–182.

Bhol K C, Schechter P J. Topical nanocrystalline silver cream suppresses inflammatory cytokines and induces apoptosis of inflammatory cells in a murine model of allergic contact dermatitis. *Br J Dermatol.* 2005; 152(6): 1235–42.

Buchmann S, Fischli W, Thiel F P, Alex R. Aqueous microsuspension, an alternative intravenous formulation for animal studies. In: *42$^{nd}$ annual congress of the International Association for Pharmaceutical Technology (APV).* Mainz. 1996.

Cardot J M, Garcia Arieta A, Paixao P, Tasevska I, Davit B. Implementing the biopharmaceutics classification system in drug development: Reconciling similarities, differences, and shared challenges in the EMA and US-FDA-recommended approaches. *AAPS J.* 2016; 18(4): 1039–1046.

Caruso F. Nanoengineering of particle surfaces. *Adv Mater.* 2001; 13(1): 11–22.

Chen H, Khemtong C, Yang X, Chang X, Gao J. Nanonization strategies for poorly water-soluble drugs. *Drug Discov Today.* 2011; 16(7–8): 354–360.

Chen M, John M, Lee S L, Tyner K M. Development considerations for nanocrystal drug products. *AAPS J.* 2017; 19(3): 642–651.

Chen X, Young T J, Sarkari M, Williams R O 3rd, Johnston K P. Preparation of cyclosporine A nanoparticles by evaporative precipitation into aqueous solution. *Int J Pharm.* 2002; 242(1–2): 3–14.

Chodanowski P, Stoll S. Polyelectrolyte adsorption on charged particles: Size effects. *J Chem Phys.* 2001; 115: 4951.

Chogale M M, Ghodake V N, Patravale V B. Performance parameters and characterizations of nanocrystals: A brief review. *Pharmaceutics.* 2016; 8(26): 1–18.

De Geest B G, Sukhorukov G B, Möhwald H. The pros and cons of polyelectrolyte capsules in drug delivery. *Expert Opin Drug Deliv.* 2009; 6(6): 613–624.

de Waard H, De Beer T, Hinrichs W L, Vervaet C, Remon J P, Frijlink H W. Controlled crystallization of the lipophilic drug fenofibrate during freeze-drying: Elucidation of the mechanism by in-line Raman spectroscopy. *AAPS J.* 2010; 12(4): 569–575.

Decher G. Fuzzy nanoassemblies: Toward layered polymeric multicomposites. *Science.* 1997; 277(5330): 1232–1237.

Decher G, Hong J D. Buildup of ultrathin multilayer films by a self-assembly process II. Consecutive adsorption of anionic and cationic bipolar amphiphiles and polyelectrolytes on charged surfaces. *Ber Bunsenges Phys Chem.* 1991; 95(11): 1430–1434.

DeMuth P C, Su X, Samuel R E, Hammond P T, Irvine D J. Nano-layered microneedles for transcutaneous delivery of polymer nanoparticles and plasmid DNA. *Adv Mater.* 2010; 22(43): 4851–4856.

Donath E, Sukhorukov G B, Caruso F, Davis S A, Möhwald H. Novel hollow polymer shells by colloid-templated assembly of polyelectrolytes. *Angew Chem Int Ed.* 1998; 37(16): 2201–2205.

Donini C, Robinson D N, Colombo P, Giordano F, Peppas N A. Preparation of poly(methacrylic acid-g-poly(ethylene glycol)) nanospheres from methacrylic monomers for pharmaceutical applications. *Int J Pharm.* 2002; 245(1–2): 83–91.

Dressman J B, Amidon G L, Reppas C, Shah V P. Dissolution testing as a prognostic tool for oral drug absorption: Immediate release dosage forms. *Pharm Res.* 1998; 15(1): 11–22.

Fan Y F, Wang Y N, Fan Y G, Ma J B. Preparation of insulin nanoparticles and their encapsulation with biodegradable polyelectrolytes via the layer-by-layer adsorption. *Int J Pharm.* 2006; 324(2): 158–167.

Gao L, Liu G, Ma J, Wang X, Zhou L, Li X. Drug nanocrystals: *In vivo* performances. *J Control Release.* 2012; 160(3): 418–430.

Gao L, Zhang D, Chen M. Drug nanocrystals for the formulation of poorly soluble drugs and its application as a potential drug delivery system. *J Nanopart Res.* 2008; 10: 845–862.

Goppert T M, Muller R H. Adsorption kinetics of plasma proteins on solid lipid nanoparticles for drug targeting. *Int J Pharm.* 2005a; 302(1–2): 172–186.

Goppert T M, Muller R H. Polysorbate-stabilized solid lipid nanoparticles as colloidal carriers for intravenous targeting of drugs to the brain: Comparison of plasma protein adsorption patterns. *J Drug Target.* 2005b; 13(3): 179–187.

Hammond P T. Layer-by-layer approaches to staging medicine from surfaces. *AICHE J.* 2015; 61(4): 1106–1117.

Hao L, Wang X, Zhang D, Xu Q, Song S, Wang F *et al.* Studies on the preparation, characterization and pharmacokinetics of amoitone B nanocrystals. *Int J Pharm.* 2012; 433(1–2): 157–164.

Hecq J, Deleers M, Fanara D, Vranckx H, Boulanger P, Le Lamer S *et al.* Preparation and *in vitro/in vivo* evaluation of nano-sized crystals for dissolution rate enhancement of ucb-35440-3, a highly dosed poorly water-soluble weak base. *Eur J Pharm Biopharm.* 2006; 64(3): 360–368.

Hirsjarvi S, Peltonen L, Hirvonen J. Surface pressure measurements in particle interaction and stability studies of poly(lactic acid) nanoparticles. *Int J Pharm.* 2008; 348(1–2): 153–160.

Hong J, Kim B-S, Char K, Hammond P T. Inherent charge-shifting polyelectrolyte multilayer blends: A facile route for tunable protein release from surfaces. *Biomacromolecules.* 2011; 12(8): 2975–2981.

Hu J, Johnston K P, Williams R O 3rd. Nanoparticle engineering processes for enhancing the dissolution rates of poorly water soluble drugs. *Drug Dev Ind Pharm.* 2004; 30(3): 233–245.

Hyvonen S, Peltonen L, Karjalainen M, Hirvonen J. Effect of nanoprecipitation on the physicochemical properties of low molecular weight poly(L-lactic acid) nanoparticles loaded with salbutamol sulphate and beclomethasone dipropionate. *Int J Pharm.* 2005; 295(1–2): 269–281.

Jacobs C, Kayser O, Muller R H. Nanosuspensions as a new approach for the formulation for the poorly soluble drug tarazepide. *Int J Pharm.* 2000; 196(2): 161–164.

Jacobs C, Muller R H. Production and characterization of a budesonide nanosuspension for pulmonary administration. *Pharm Res.* 2002; 19(2): 189–194.

Jessel N, Oulad-Abdelghani M, Meyer F, Lavalle P, Haikel Y, Schaaf P *et al.* Multiple and time-scheduled *in situ* DNA delivery mediated by beta-cyclodextrin embedded in a polyelectrolyte multilayer. *Proc Natl Acad Sci U S A.* 2006; 103(23): 8618–8621.

Jinno J, Kamada N, Miyake M, Yamada K, Mukai T, Odomi M *et al.* Effect of particle size reduction on dissolution and oral absorption of a poorly water-soluble drug, cilostazol, in beagle dogs. *J Control Release.* 2006; 111(1–2): 56–64.

Junghanns J U, Muller R H. Nanocrystal technology, drug delivery and clinical applications. *Int J Nanomedicine.* 2008; 3(3): 295–309.

Kakran M, Shegokar R, Sahoo N G, Gohla S, Li L, Muller R H. Long-term stability of quercetin nanocrystals prepared by different methods. *J Pharm Pharmacol.* 2012a; 64(10): 1394–1402.

Kakran M, Shegokar R, Sahoo N G, Shaal L A, Li L, Muller R H. Fabrication of quercetin nanocrystals: Comparison of different methods. *Eur J Pharm Biopharm.* 2012b; 80(1): 113–121.

Keck C M, Kobierski S, Mauludin R, Müller R H. Second generation of drug nanocrystals for delivery of poorly soluble drugs: Smart crystals technology. *Dosis.* 2008; 2(24): 124–128.

Keck C M, Muller R H. Drug nanocrystals of poorly soluble drugs produced by high pressure homogenisation. *Eur J Pharm Biopharm.* 2006; 62(1): 3–16.

Kipp J E, Wong J C T, Doty, M J, Rebbeck S B, Werling J, Sriram R. Microprecipitation method for preparing submicron suspensions. USA: 2003. ISBN/ISSN: US Patent 6607784.

Li F, Hsieh Y. Supercritical fluid chromatography-mass spectrometry for chemical analysis. *J Sep Sci.* 2008; 31(8): 1231–1237.

Li Y, Dong L, Jia A, Chang X, Xue H. Preparation and characterization of solid lipid nanoparticles loaded traditional Chinese medicine. *Int J Biol Macromol.* 2006; 38(3–5): 296–299.

Li Y, Xia J, Dubin P L. Complex formation between polyelectrolyte and oppositely charged mixed micelles: Static and dynamic light scattering study of the effect of polyelectrolyte molecular weight and concentration. *Macromolecules.* 1994; 27(24): 7049–7055.

Liu R, Chang Y N, Xing G, Li M, Zhao Y. Study on orally delivered paclitaxel nanocrystals: Modification, characterization and activity in the gastrointestinal tract. *R Soc Open Sci.* 2017; 4: 1–9.

Liversidge G G, Cundy K C, Bishop J F, Czekai D A. Surface modified drug nanoparticles, US Patent 5145684; 1992.

Lvov Y, Antipov A A, Mamedov A, Möhwald H, Sukhorukov G B. Urease encapsulation in nanoorganized microshells. *Nano Lett.* 2001; 1(3): 125–128.

Lvov Y M, Pattekari P, Zhang X, Torchilin V. Converting poorly soluble materials into stable aqueous nanocolloids. *Langmuir.* 2011; 27(3): 1212–1217.

Mauludin R, Möschwitzer J, Müller R H. Comparison of ibuprofen drug nanocrystals produced by high pressure homogenization (HPH) versus ball milling. *AAPS Annual Meeting.* Nashville, T2217. 2005.

Mauludin R, Muller R H, Keck C M. Development of an oral rutin nanocrystal formulation. *Int J Pharm.* 2009; 370(1–2): 202–209.

Merisko-Liversidge E, Liversidge G G, Cooper E R. Nanosizing: A formulation approach for poorly-water-soluble compounds. *Eur J Pharm Sci.* 2003; 18(2): 113–120.

Moon H R, Urban J J, Milliron D J. Size-controlled synthesis and optical properties of monodisperse colloidal magnesium oxide nanocrystals. *Angew Chem Int Ed Engl.* 2009; 48(34): 6278–6281.

Möschwitzer J P. Drug nanocrystals in the commercial pharmaceutical development process. *Int J Pharm.* 2013; 453: 142–156.

Moschwitzer J P, Muller R H. New method for the effective production of ultrafine drug nanocrystals. *J Nanosci Nanotechnol.* 2006a; 6(9–10): 3145–3153.

Moschwitzer J P, Muller R H. Spray coated pellets as carrier system for mucoadhesive drug nanocrystals. *Eur J Pharm Biopharm.* 2006b; 62(3): 282–287.

Muller R H, Gohla S, Keck C M. State of the art of nanocrystals—special features, production, nanotoxicology aspects and intracellular delivery. *Eur J Pharm Biopharm.* 2011; 78(1): 1–9.

Muller R H, Jacobs C, Kayser O. Nanosuspensions as particulate drug formulations in therapy. Rationale for development and what we can expect for the future. *Adv Drug Deliv Rev.* 2001; 47(1): 3–19.

Müller R H, Jacobs C, Kayser O. *DissoCubes–a novel formulation for poorly soluble and poorly bioavailable drugs*, Marcel Dekker; 2003.

Muller R H, Keck C M. Challenges and solutions for the delivery of biotech drugs—a review of drug nanocrystal technology and lipid nanoparticles. *J Biotechnol.* 2004a; 113(1–3): 151–170.

Muller R H, Keck C M. Drug delivery to the brain—realization by novel drug carriers. *J Nanosci Nanotechnol.* 2004b; 4(5): 471–483.

Muller R H, Runge S, Ravelli V, Mehnert W, Thunemann A F, Souto E B. Oral bioavailability of cyclosporine: Solid lipid nanoparticles (SLN) versus drug nanocrystals. *Int J Pharm.* 2006; 317(1): 82–89.

Nagarwal R C, Kumar R, Dhanawat M, Das N, Pandit J K. Nanocrystal technology in the delivery of poorly soluble drugs: An overview. *Curr Drug Deliv.* 2011; 8: 1–9.

Noyes A A, Whitney W R. The rate of solution of solid substances in their own solutions. *J Am Chem Soc.* 1897; 19(12): 930–934.

Ogawa S, Decker E A, McClements D J. Production and characterization of O/W emulsions containing droplets stabilized by lecithin–chitosan–pectin mutilayered membranes. *J Agric Food Chem.* 2004; 52(11): 3595–3600.

Ostwald W. *Phys Z. Chem.* 1900; (34): 495.

Palankar R, Skirtach A G, Kreft O, Bédard M, Garstka M, Gould K *et al.* Controlled intracellular release of peptides from microcapsules enhances antigen presentation on MHC class I molecules. *Small.* 2009; 5(19): 2168–2176.

Parekh G, Pattekari P, Joshi C, Shutava T, DeCoster M, Levchenko T *et al.* Layer-by-layer nanoencapsulation of camptothecin with improved activity. *Int J Pharm.* 2014; 465(1–2): 218–227.

Patravale V B, Date A A, Kulkarni R M. Nanosuspensions: A promising drug delivery strategy. *J Pharm Pharmacol.* 2004; 56(7): 827–840.

Pattekari P, Zheng Z, Zhang X, Levchenko T, Torchilin V, Lvov Y. Top-down and bottom-up approaches in production of aqueous nanocolloids of low solubility drug paclitaxel. *Phys Chem Chem Phys.* 2011; 13(19): 9014–9019.

Peltonen L, Hirvonen J, Laaksonen T. Drug Nanocrystals and Nanosuspensions in Medicine. In: *Handbook of Nanomedical Research, Materials for Nanomedicine.* V. Torchilin; 2014; 3: 169–197.

Picart C, Schneider A, Etienne O, Mutterer J, Schaaf P, Egles C *et al.* Controlled degradability of polysaccharide multilayer films *in vitro* and *in vivo. Adv Funct Mater.* 2005; 15(11): 1771–1780.

Qiu X, Leporatti S, Donath E, Möhwald H. Studies on the drug release properties of polysaccharide multilayers encapsulated ibuprofen microparticles. *Langmuir.* 2001; 17(17): 5375–5380.

Rabinow B E. Nanosuspensions in drug delivery. *Nat Rev Drug Discov*. 2004; 3(9): 785–796.

Rao S S, Camilleri M, Hasler W L, Maurer A H, Parkman H P, Saad R, *et al*. Evaluation of gastrointestinal transit in clinical practice: Position paper of the American and European Neurogastroenterology and Motility Societies. *Neurogastroenterol Motil*. 2011; 23(1): 8–23. ISSN: 1365-2982.

Rogers T L, Johnston K P, Williams R O 3rd, Solution-based particle formation of pharmaceutical powders by supercritical or compressed fluid $CO_2$ and cryogenic spray-freezing technologies. *Drug Dev Ind Pharm*. 2001; 27(10): 1003–1015.

Sahoo N G, Kakran M, Shaal L A, Li L, Muller R H, Pal M *et al*. Preparation and characterization of quercetin nanocrystals. *J Pharm Sci*. 2011; 100(6): 2379–2390.

Salazar J, Ghanem A, Muller R H, Moschwitzer J P. Nanocrystals: Comparison of the size reduction effectiveness of a novel combinative method with conventional top-down approaches. *Eur J Pharm Biopharm*. 2012; 81(1): 82–90.

Santos A C, Pattekari P, Jesus S, Veiga F, Lvov Y, Ribeiro A J. Sonication-assisted layer-by-layer assembly for low solubility drug nanoformulation. *ACS Appl Mater Interfaces*. 2015; 7(22): 11972–11983.

Saurer E M, Jewell C M, Roenneburg D A, Bechler S L, Torrealba J R, Hacker T A *et al*. Polyelectrolyte multilayers promote stent-mediated delivery of DNA to vascular tissue. *Biomacromolecules*. 2013; 14(5): 1696–1704.

Sattar A, Chen D, Jiang L, Pan Y, Tao Y, Huang L *et al*. Preparation, characterization and pharmacokinetics of cyadox nanosuspension. *Sci Rep*. 2017; May 23;7(1):2289.

Shackleford D M, Faassen W A, Houwing N, Lass H, Edwards G A, Porter C J *et al*. Contribution of lymphatically transported testosterone undecanoate to the systemic exposure of testosterone after oral administration of two andriol formulations in conscious lymph duct-cannulated dogs. *J Pharmacol Exp Ther*. 2003; 306(3): 925–933.

Shahgaldian P, Gualbert J, Aissa K, Coleman A W. A study of the freeze-drying conditions of calixarene based solid lipid nanoparticles. *Eur J Pharm Biopharm*. 2003; 55(2): 181–184.

Shegokar R, Al Shaal L, Müller R H. Local Anaesthetic nanocrystals as prolonged release formulation USA, *First Combined Annual Meeting of International Pharmaceutical Federation*, PSWC and AAPS New Orleans Louisiana. 2010.

Shenoy D B, Sukhorukov G B. Engineered microcrystals for direct surface modification with layer-by-layer technique for optimized dissolution. *Eur J Pharm Biopharm*. 2004; 58(3): 521–527.

Shutava T G, Pattekari P P, Arapov K A, Torchilin V P, Lvov Y M. Architectural layer-by-layer assembly of drug nanocapsules with PEGylated polyelectrolytes. *Soft Matter*. 2012; 8(36): 9418–9427.

Smetanova L, Stetinova V, Svoboda Z, Kvetina J. Caco-2 cells, biopharmaceutics classification system (BCS) and biowaiver. *Acta Medica (Hradec Kralove)*. 2011; 54(1): 3–8.

Soppimath K S, Kulkarni A R, Aminabhavi T M. Chemically modified polyacrylamide-g-guar gum-based crosslinked anionic microgels as pH-sensitive drug delivery systems: Preparation and characterization. *J Control Release*. 2001; 75(3): 331–345.

Srivalli K M R, Mishra B. Drug nanocrystals: A way toward scale-up. *Saudi Pharm J*. 2016; 24(4): 386–404.

Sukhorukov G B, Donath E, Davis S, Lichtenfeld H, Caruso F, Popov V I *et al*. Stepwise polyelectrolyte assembly on particle surfaces: A novel approach to colloid design. *Polymer Adv Tech* 1998; 9(10–11): 759–767.

Tsuzuki T. Commercial scale production of inorganic nanoparticles. *Int J Nanotechnol Nanomed*. 2009; 6(5–6): 567–578.

Tunick M H, Van Hekken D L, Cooke P H, Malin E L. Transmission electron microscopy of mozzarella cheeses made from microfluidized milk. *J Agric Food Chem*. 2002; 50(1): 99–103.

Van Eerdenbrugh B, Van den Mooter G, Augustijns P. Top-down production of drug nanocrystals: Nanosuspension stabilization, miniaturization and transformation into solid products. *Int J Pharm*. 2008; 364(1): 64–75.

Van Eerdenbrugh B, Vermant J, Martens J A, Froyen L, Humbeeck J V, Van den Mooter G *et al*. Solubility increases associated with crystalline drug nanoparticles: Methodologies and significance. *Mol Pharm*. 2010; 7(5): 1858–1870.

Vazquez E, Dewitt D M, Hammond P T, Lynn D M. Construction of hydrolytically-degradable thin films via layer-by-layer deposition of degradable polyelectrolytes. *J Am Chem Soc*. 2002; 124(47): 13992–13993.

Willems L, van der Geest R, de Beule K. Itraconazole oral solution and intravenous formulations: A review of pharmacokinetics and pharmacodynamics. *J Clin Pharm Ther*. 2001; 26(3): 159–169.

Wood K C, Boedicker J Q, Lynn D M, Hammond P T. Tunable drug release from hydrolytically degradable layer-by-layer thin films. *Langmuir*. 2005; 21(4): 1603–1609.

Wood K C, Chuang H F, Batten R D, Lynn D M, Hammond P T. Controlling interlayer diffusion to achieve sustained, multiagent delivery from layer-by-layer thin films. *Proc Natl Acad Sci U S A*. 2006; 103(27): 10207–10212.

Wu C, Liu Y, He Z, Sun J. Insight into the development of dissolution media for BCS class II drugs: A review from quality control and prediction of *In Vivo* performance perspectives. *Curr Drug Deliv*. 2016; 13(7):1004–1020.

Yuan J, Zhou S, You B, Wu L. Organic pigment particles coated with colloidal nano-silica particles via layer-by-layer assembly. *Chem Mater*. 2005; 17(14): 3587–3594.

Zheng Z, Zhang X, Carbo D, Clark C, Nathan C-A, Lvov Y. Sonication-assisted synthesis of polyelectrolyte-coated curcumin nanoparticles. *Langmuir*. 2010; 26(11): 7679–7681.

# 10 Nanofibers for Delivery of Drugs and Bioactives

*Amit Kumar Goyal, Saurav Bhandari, Ranjot Kaur, Charan Singh, and Goutam Rath**

## CONTENTS



## 10.1 INTRODUCTION

In the recent era, nanofibers are one of the extensively used drug-delivery carriers for therapeutic management of various diseases. Continuously increasing applications on nanofibers and their utility for delivering various drugs, proteins and DNA prove their importance and accessibility as drug carriers. The physical properties of nanofibers are vital factors for delivering drug(s) effectively to the appropriate site in the body (Doshi and Reneker, 1995). There are some specific requisite properties of nanofibers required for delivery of drugs, as portrayed in Figure 10.1.

Previously, nanofibers were prepared by electrospinning methods, as reported by Formhals (1934). Electrospinning did not receive much initial attention during early stages, but it has become an attractive choice lately for effective fabrication of nanofibers because of the possibility of structural modification of its architecture and delivery of a broad range of drugs. The word, 'electrospinning' is made up of two words, 'electrostatic' and 'spinning', implying an electric, force-driven spinning phenomenon. A number of publications being churned out in the field of nanofibers demonstrate that

* Corresponding author.

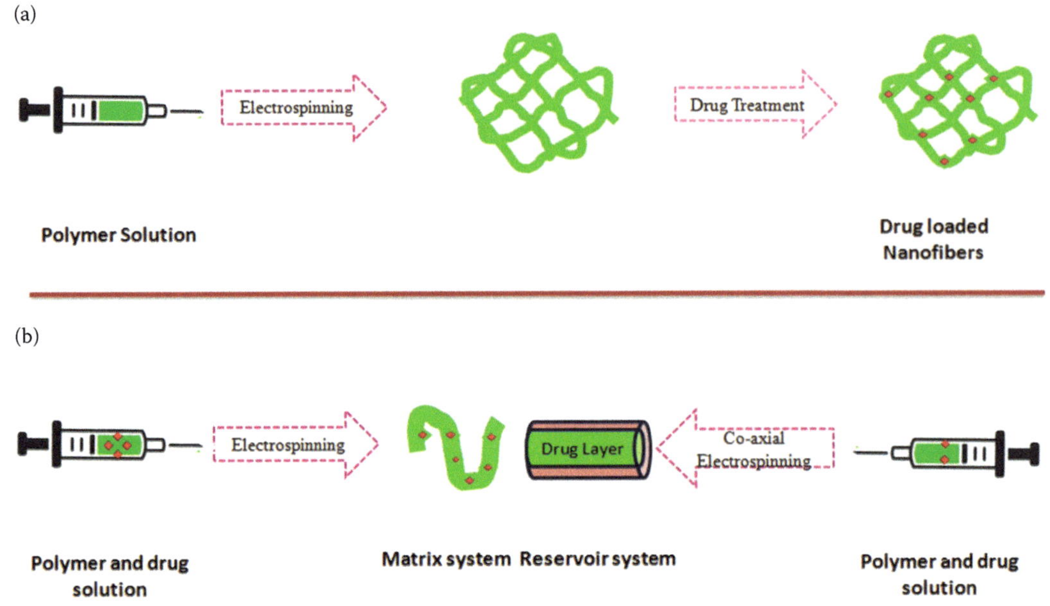


**FIGURE 10.1** Different kinds of nanofibers required for effective drug delivery.

electrospinning has been a popularly used method in recent times (Huang *et al.*, 2003). Figure 10.2 depicts the process of electrospinning of polymeric nanofibers.

Fundamentally, there are three components involved in electrospinning: (i) voltage supplier (1–50 kV), (ii) injector, and (iii) collector. In the process of electrospinning, a high voltage potential is needed to produce an intensely electronegative charge for an injected polymer solution and is connected to a syringe. A voltage gradient is applied between the needle tip to the collector, ranging between 1–50 kV. The needle tip has pendent drop at its tip due to physical properties of polymeric solutions. Applying an electric field makes an electronegative drop molecule, which tends to acquire a tremendous amount of energy. On raising the applied field, the surface of the drop gets stretched and forms a cone-like shape, known as a 'Taylor cone'. This may be due to mutual charge repulsion of the drop molecules and attractive force between the oppositely charged electrodes. Herein, if the applied field is further elevated to a decisive value, such that the mutual repulsive forces exceed the surface tension of the drop, then the polymeric fluids may be dropped out from needle, which may lead to the formation of uneven and non-uniform nanofibers. The solution jet, so formed, is dried and solidified prior to reaching the collector (Shin *et al.*, 2001; Deitzel *et al.*, 2001).

There are various physicochemical parameters, which can be varied over a wide range to produce nanofibers possessing the desired properties, depending upon the drug to be delivered, or the site at which it needs to be delivered, besides other factors. Hence, they eventually make a tangible

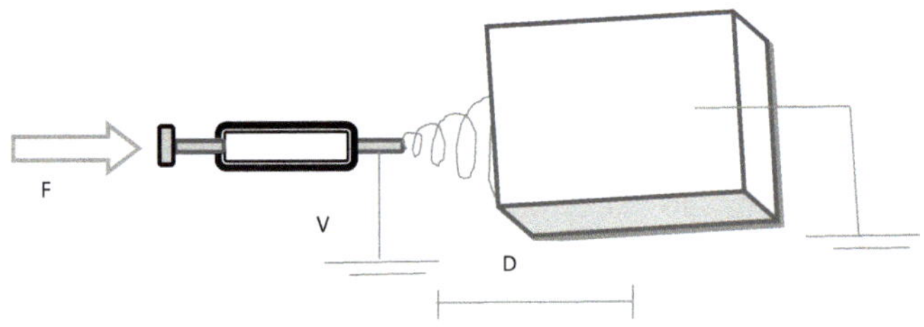


**FIGURE 10.2** Typical electrospinning setup: F, flow rate; D, distance between plate and needle; V, applied voltage.

impact on the drug-delivery process per se. From here onwards, these parameters will be discussed individually, along with their role in making drug delivery more futuristic.

## 10.2 PARAMETERS ON NANOFIBER MORPHOLOGY

### 10.2.1 Applied Voltage

Electrospinning needs a high voltage gradient across the electrode to yield nanofibers (Garg and Bowlin, 2011). The diameter of nanofibers may be regulated by changing the applied voltage with respect to the polymer/solvent systems. Meechaisue *et al.* (2006) studied the effect of applied voltage (10–25 kV/10 cm) on morphology of nanofibers. They observed that nanofibers were formed at 10 kV/10 cm, while nanofibers with beaded morphology were formed on increasing the voltage to 10–15 kV, and number of beads decreased after increasing the voltage from 20 to 25 kV/10 cm. Deitzel *et al.* (2001) observed the effects of voltage gradients on the shape of the Taylor cone on the polymeric solution. Studies suggested that a drop of polymer solution was formed at lower applied voltages, while a Taylor cone was produced at the needle on gradually increasing the applied voltage. This was associated with the bead defects of nanofibers of suboptimal values of applied value. Sohrabi *et al.* (2013) studied the role of voltage gradient with respect to tip collector distance and morphology of the nanofibers. The study suggested that regulating the voltage from 1–15 kV and the distance from 1–15 cm are suitable for the formation of nanofibers of the desired size and shape.

### 10.2.2 Flow Rate

Flow rate of the spinning solution is one of the major critical parameters that controls the physical properties of nanofibers, especially nanofiber diameter and pore size. Incomplete drying leads to the development of a fiber defect, known as ribbon-like fiber or flattened fiber. Megelski *et al.* (2002) observed the effects of flow rate on morphology of fibers. Taylor (1969) observed that it is not possible to maintain a Taylor cone until the flow rate of the solution is properly maintained. Lee *et al.* (2012) studied the role of collector speed and flow rate on nanofiber morphology. The results suggested that slower flow rate with higher collector speed produce uniform nanofibers with suitable architecture.

### 10.2.3 Capillary-Collector Distance

Tip collector distance is important for complete evaporation of solvent from polymeric solution and mainly influences nanofiber size by a magnitude of 1–2 orders. Tip collector distance mainly depends on the solvent properties, polymer nature, applied voltages and viscosity. This may lead to changes in the moisture content and morphology of resultant nanofibers. Jaeger *et al.* (1998) noted that the diameter of polyethylene oxide (PEO)/water solution fibers decreased from 19 to 9 mm after travelling a distance of 1 and 3.5 cm, respectively. Doshi and Reneker (1995) also found that the fiber morphology is inversely proportional to the distance from the Taylor cone. Megelski *et al.* (2002) observed the role of capillary collector distance for formation of beads during the fabrication of polystyrene fibers due to insufficient drying of fibers. In another study, Buchko *et al.* (1999) found that polymeric protein produces ribbon-like fibers at smaller distances, while circular fibers were obtained at increased distances. Yuan *et al.* (2004) reported that polymeric nanofibers were smaller for closer distances between the tip and collector.

### 10.2.4 Polymer Concentration

Polymer concentration is also a vital factor that regulates the morphology of nanofibers. Formulation of nanofibers is completely based on polymeric concentration; when polymer concentration is low, it will not form a proper Taylor cone and will break up before reaching the collector. Conversely, when concentration is high, it is again quite difficult to form a smooth Taylor cone to produce nanofibers (Demir *et al.*, 2002). Koski *et al.* (2004) derived a relationship for minimum polymer concentration and viscosity needed to form electrospun nanofibers: polyvinyl alcohol (PVA) ($[n]c > 5$) and PEO ($[n]c > 10$), where $[n]$ is the intrinsic viscosity and $c$ ias the concentration. McKee *et al.* (2004) reported the correlation of entanglement concentration with spinning of linear and branched polymers. The results showed that higher entanglement concentration produces uniform and bead-free nanofibers. Megelski *et al.* (2002) investigated that increasing polymeric concentration leads to increased fiber diameter, and pore size became narrower. Doshi and Reneker (1995) also reported that lower viscosity of PEO broke up into droplets during electrospinning, while a higher viscosity solution was unable to spin. Studies unequivocally concluded that optimal polymeric concentration is required to fabricate the nanofibers. Deitzel *et al.* (2001) reported a relationship between the fiber diameter of PEO/water solution and PEO concentration. Ranjbar-Mohammadi *et al.* (2013) studied the effect of polymer concentration (i.e., tragacanth gum) and PVA on polymeric nanofiber characteristics. The outcome indicated that the concentration of the polymer affects the antibacterial properties against Gram-positive bacteria. Figure 10.3 shows the interrelation of process and formulation parameters on nanofiber morphology.

### 10.2.5 Solvent Volatility

Volatility of the solvent may be responsible for morphology and overall integrity of nanofiber architecture. High volatility of the solvent is required to complete drying of the polymer during electrospinning, as volatile solvent gets separated from the polymeric solution to form solid nanofibrous residue at the collector. Megelski *et al.* (2002) examined the effect of solvent volatility on fiber morphology using dimethylformamide and tetrahydrofuran (THF). High density of pores with 20%–40% increase in the surface area was demonstrated using 100% THF as the solvent, which is more volatile. Smooth fibers were formed from fibers electrospun by 100% dimethylformamide (DMF), which, being less volatile, completely loses the porous topology.

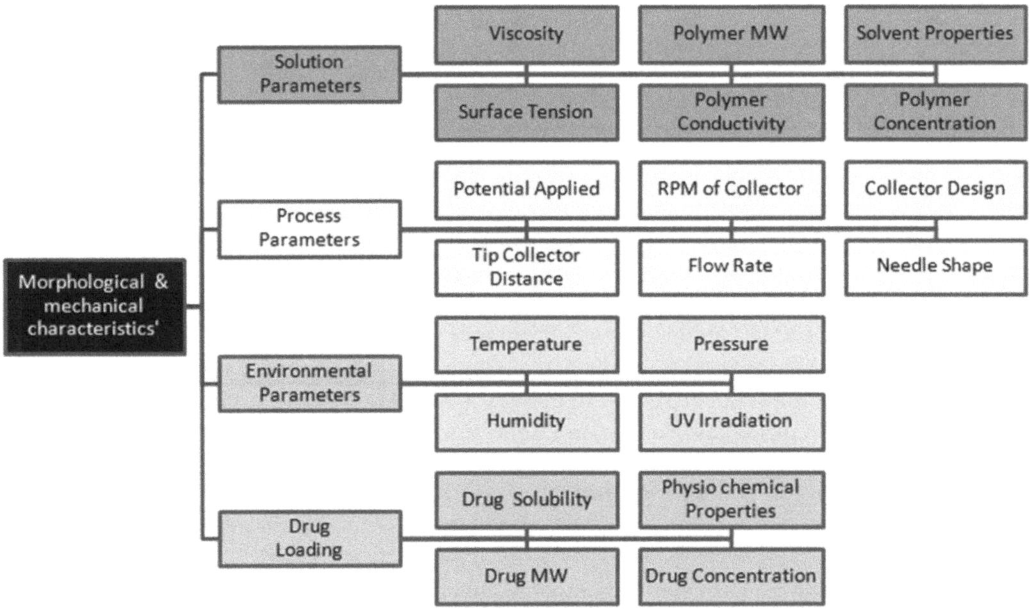


**FIGURE 10.3** Morphological and mechanical characteristics of nanofibers.

### 10.2.6 Solution Conductivity

Solution conductivity is another critical parameter for formulation of the Taylor cone and stretching of polymeric nanofibers. Baumgarten (1971) found that radius of the injector is inversely proportional to the cube root of the solution conductivity. Hayati *et al.* (1987) reported that more conducting solutions produced unstable fibers with wide-diameter distribution in a strong electric field, while low-conducting solutions produced more stable fibers. Zhang *et al.* (2005) reported the role of ions in polymeric solution for formation of diameters of nanofibers. The findings showed that mean fiber diameter decreased significantly by enhancing the concentrations of sodium chloride (NaCl) to increase electrical conductivity at the jet. Jiang *et al.* (2004) also reported the role of Bovine serum albumin (BSA) on electroconductivity for reducing the size of fibers from micrometers to nanometers, which may be attributed to net charge on the polymer jets. Zuo *et al.* (2005) reported that smoother fibers with fewer beads were formed on addition of alcohol to the solvent. Zong *et al.* (2002) investigated the effect of the addition of salt, especially ionic radii, on development of nanofibers. He *et al.* (2011) studied effects of molecular weight, solution viscosity and electrical conductivity on formation of nanofibers. The findings showed that fibers have beads-in-string structures and molecular weight increases after increasing the electrical conductivity.

### 10.2.7 Surface Tension

The surface tension of the polymeric solution may also affect the morphology of the nanofibers. Zuo *et al.* (2005) demonstrated the role of variable surface tension with the same conductivity in the development of nanofibers and beading. The addition of cosolvent of polymeric solution also changes the surface tension and leads ultimately to reduction in fiber diameter.

### 10.2.8 Dipole Moment and Dielectric Constant

Dipole moment and dielectric constant may also indirectly affect the properties of the solvent and solution. Jarusuwannapoom *et al.* (2005) found that polystyrene fibers were successfully formed only when the solvent with the greater value of dipole moment was used. Liu and Kumar (2005) worked on the Polymethyl methacrylate (PMMA) fibers and observed that the number of fibers produced per unit time was correlated to the magnitudes of dipole moment and dielectric constant. Wannatong *et al.* (2004) methodically examined the effect of these parameters of solvents on electrospun polymeric fibers.

### 10.2.9 Ambient Parameters

Ambient parameters have also shown their impact on fiber morphology. Mit-Uppatham *et al.* (2004) worked on spinning of fibers of polyamide-6 in the range of 25°C–60°C and observed that the diameter of fiber got decreased with increasing temperatures. Apart from temperature, high humidity has also been found to result in the formation of small circular pores on the surface of fibers (Casper *et al.*, 2003).

### 10.2.10 Needle-Tip Design and Placement

Design of the needle tip and its displacement pattern also change the morphology of nanofibers. Many types of needle tips have been developed for changing patterns of nanofibers. Li and Xia (2004) prepared hollow fibers using two immiscible liquids of a polymer. Theron *et al.* (2005) used multiple tips to improve the production capacity of nanofibers. Kidoaki *et al.* (2005) used two tips and one collector that can move transversely to mix different polymeric fibers, and this transverse motion resulted in more uniform distribution of each polymer.

### 10.2.11 Collector Composition and Geometry

Commonly, drum and plate collectors are used to collect the nanofibers. The major disadvantage of the above collectors is their unsuitability in separation of the fibers from conductive material. Accordingly, the need of different collectors (such as methanol and water collectors) will improve the collection efficiency of nanofibers. (Kim and Park, 2006 produced poly-L-lactic acid (PLLA) and polylactic glycolic acid (PLGA) nanofibers on metal collectors, water reservoir collectors and methanol collectors. Smooth fibers were obtained from metal collectors, while hydrophobic polymeric fibers shrunk on the water surface and swelled on methanol collectors. Liu *et al.* (2012) used aluminium foil, copper mesh, water and paper as collectors. The collector conductivity mainly influences the packing density of fibers. Charge dissipation was more for greater conductive solutions. When the charge was not dissipated, the similarly charged fibers tend to repel each other to form a more porous structure. Fibers with loose packing were formed when collected on porous collectors such as copper mesh and paper, as compared to those collected on aluminium foil and water. Kenawy el *et al.* (2003) spun directly onto a human hand of poly(ethylene-co-vinyl alcohol). Fridrikh *et al.* (2003) used two parallel plates during the spinning of fibers to produce a uniform electric field. Conductive frame collectors yielded more aligned nanofibers than those collected on non-conductive collectors. Suspended and long fibers (10 cm) were obtained between the rings when one of the rings was rotated, which leads to the production of multifilament yarn. Deitzel *et al.* (2001) used multiple field methods to produce smaller size nanofibers free from beads. Chew *et al.* (2005) used a rotating cylindrical drum to collect fibers. Better alignment was obtained using a copper wire-based drum. Sundaray *et al.* (2004) used a thin steel pin-based collector to obtain fibers greater than 10 cm in length. Theron *et al.* (2005) used a tapered and grounded wheel-like bobbin to collect PEO nanofibers of diameters ranging from 100-300 nm. A summary of all the parameters is presented in Table 10.1, along with their effects on the fiber morphology.

## 10.3 POLYMER SELECTION CRITERIA

Different types of polymeric material have been investigated in healthcare fields. The better the polymer is selected for manufacturing the nanofibers, the better will be the formulation and the more effective will be drug delivery. By proper selection, the range of drugs to be delivered and the range of sites of delivery can be broadened (Grabnar and Kristl, 2011). Selection of the polymer involves consideration of the vital properties, such as (i) biocompatibility, (ii) biodegradability, (iii) nontoxicity, (iv) hydrophilicity, (v) proper mechanical strength and (vi) mucoadhesion.

**TABLE 10.1**
**Plausible Effects of Electrospinning Parameters on Fiber Morphology**

| Physicochemical Parameter | Effect on Nanofiber Morphology |
|---|---|
| Concentration | Fiber diameter and size |
| Surface tension | Fiber strength |
| Conductivity | Fiber strength and diameter |
| Volatility of solvent | Size and shape |
| Dielectric constant | Fiber diameter and size |
| Flow rate | Diameter and density |
| Applied voltage | Diameter and shape |
| Tip collector distance | Size and strength |
| Geometry of collector | Types of fibers |
| Humidity | Fiber integrity |
| Temperature | Size and shape |

**TABLE 10.2**
**Various Vital Characteristics of Commonly Used Natural Polymers**

| Polymer | Biodegradation | Mechanical Properties | Physicochemical Properties |
|---|---|---|---|
| Collagen | Enzymatic | Weak | Hydrophilic |
| Gelatin | Enzymatic | Weak | Hydrophilic |
| Elastin | Enzymatic | Weak | Hydrophilic |
| Chitosan | Enzymatic | Weak | Hydrophilic |
| Silk fibroin | Enzymatic | Strong | Hydrophilic |
| Alginate | Enzymatic | Weak | Hydrophilic |

In general, the process of polymer selection for electrospinning involves limitless considerations, provided that the polymer allows the preparation of a solution with the desirable characteristics needed for effective drug delivery (Rošic *et al.*, 2011). Polymers can be divided into two categories, synthetic or natural.

Most commonly used synthetic polymers are PVA, PEO, polylactic acid (PLA), PLGA and polycaprolactone (PCL), while chitosan, alginate, collagen, gelatine, hyaluronic acid and silk are popular natural polymers (Kristl *et al.*, 1993; Sill and von Recum, 2008). The key biological, mechanical and physicochemical properties of the commonly used polymers are presented in Tables 10.2 and 10.3. Synthetic materials are strong, cheap, reliable and often easily electrospinnable, with their physicochemical characteristics controllable over a broad range through the production process. On the contrary, they lack cell-recognition sites and have poor affinity for cell attachment. Natural polymers are preferred, as these are similar to macromolecular substances present in the human body (De Vrieze *et al.*, 2009). Therefore, the biological environment recognizes and favourably interacts with the natural polymers. As discussed above, as the natural polymers are not easy to electrospin, but these can be blended with synthetic polymers to result in better properties for easier fabrication of nanofibers (Rošic *et al.*, 2012). These polymer blends, so formed, can retain the biological properties of the natural polymer and the durability of synthetic polymer, along with the mechanical properties (Gunn and Zhang, 2010). Using this approach, chitosan and alginate, with addition of PEO or PVA, have been fabricated into nanofibers. But the study was mainly concentrated on chitosan, due to its excellent properties such as biocompatibility, safety, hydrophilicity and tremendous potential as an effective drug delivery system (DDS). Results of above chitosan/PEO solutions have showed significant changes in the fiber morphology when the concentration of chitosan in the blend was reduced. Solvent selection affects many factors of electrospinning, including surface tension and conductivity that eventually affect fiber morphology

**TABLE 10.3**
**Properties of Commonly Used Synthetic Polymers**

| Polymer | Biodegradation | Mechanical Properties | Physicochemical Properties |
|---|---|---|---|
| PGA | Hydrolytic | Strong | Hydrophilic |
| PLA | Hydrolytic | Strong | Hydrophilic |
| PLGA | Hydrolytic | Strong | Hydrophilic/hydrophobic |
| PCL | Hydrolytic | Strong | Hydrophobic |
| PVA | Enzymatic | Strong | Hydrophilic |
| PEO | Enzymatic | Weak | Hydrophilic |

*Abbreviations:* PGA: Polyglycolicacid; PLA: Polylactic acid; PLGA: Poly(lactic-co-glycolic) acid; PCL: Polycaprolactone; PVA: Polyvinyl alcohol; PEO: Polyethylene oxide.

(Luo *et al.*, 2010). Most commonly used solvents in electrospinning are THF, DMF, chloroform, acetone, ethanol, acetic acid and distilled water (Kristl *et al.*, 1993). Distilled water is the most favourable solvent for nanofibers, as there are no safety concerns. However, if the polymer is not water-soluble, the addition of small amounts of cosolvents may be necessary (Gunn and Zhang, 2010). Cosolvents are not always added to increase polymer solubility, but volatile solvents can also be added to enhance the evaporation of solution to facilitate the complete drying of fiber between capillary collector distances. Fiber porosity is known to increase with an increase in solvent volatility (Ramakrishna *et al.*, 2005). Addition of volatile solvent is found to be beneficial in the case in which the polymer is completely soluble in the solvent mixture. Salts and surfactants can also be added to natural and synthetic polymers to get bead-free fiber morphology (Ojha *et al.*, 2008; Arumugam *et al.*, 2009). In a study, the effect of the addition of Tween 80 and NaCl on fiber morphology was investigated, as these are widely employed in the pharmaceutical and cosmetic products. The authors observed that surface tension of the solution decreased, fiber diameter increased and process efficacy improved on the addition of Tween 80. On the other hand, by addition of salt, the conductivity of the polymer solution increased, leading to increased fiber diameter. In general, it is clear that addition of surfactant improves fiber morphology to a greater extent as compared to the addition of salts (Deitzel *et al.*, 2001). With the advent of electrospinning technology and polymer science, currently it is possible to design nanofibers with versatile morphology, such as hollow, coaxial, woven, non-woven and composite nanofibers. In addition, with the surface morphology, it is possible to govern the release of drug from the nanofiber shell, which will allow fabrication of reservoir as well as matrix drug-delivery devices, offering controlled or sustained drug delivery. Many natural and synthetic polymers have been used as excipients in the development of nanofiber-based DDS, as these tend to provide more flexible options to tailor the properties of a DDS. Several natural and synthetic polymers can be utilized for the development of nanofibers, as portrayed in Tables 10.2 and 10.3.

## 10.4 PHARMACEUTICAL APPLICATIONS OF NANOFIBERS

Recently, nanofibers have been extensively used in the healthcare and pharmaceutical sectors. Further, nanofibers have documented commercial applications as textile materials, filter systems, protective and absorption materials, catalytic materials, diagnostic sensors and energy power cells. Some of the advantages and applications of nanofibers have been depicted in Figure 10.4.

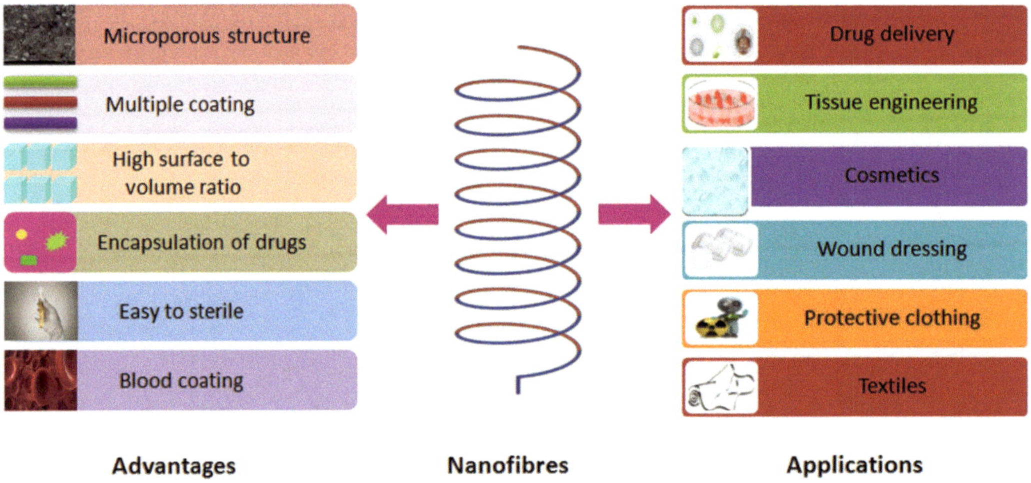


**FIGURE 10.4** Key advantages and applications of nanofiber scaffolds in various fields.

### 10.4.1 Drug-Delivery Applications

Nanofibers provide a larger surface area, nanometric size and architecture, high porosity and the ability to be prepared from all types of polymers for encapsulating hydrophilic as well as lipophilic drugs. On the basis of selection of a polymer in the preparation of nanofibers, it can be used for controlled or instantaneous release of the encapsulated drugs, and may be a biodegradable or non-biodegradable implantable system. It can be used for delivery of both the types of drugs (lipophlic and hydrophilic), macromolecules, proteins, peptides, enzymes and nucleic-acid materials. Hassounah *et al.* (2014) studied the activity of antitubercular drugs embedded in biodegradable electrospun PVA, PEO and PCL nanofibers against *Mycobacterium avium*. The developed antitubercular nanofibrous systems demonstrated significantly higher antibacterial efficiency in immune-compromised patients. Recently, Li *et al.* (2014) used electrospun nanofiber meshes for topical combined delivery of green-tea polyphenols with steroidal dexamethasone for the treatment of keloids. Studies have suggested the developed nanofibers were compatible with skin and significantly improved efficacy in the treatment of keloids. Similarly Karczewski *et al.* (2018) synthesize clindamycin-modified triple antibiotic (metronidazole, ciprofloxacin, and clindamycin [CLIN]) polymer (polydioxanone [PDS]) nanofibers. All CLIN-containing nanofibers and aliquots demonstrated pronounced antimicrobial activity against all bacteria. Overall, a bird's eye view of various drug-delivery applications of nanofibers is summarized as Table 10.4.

### 10.4.2 Tissue-Engineering Applications

Nanofibers are widely used in the field of tissue engineering and provide distinct improvement in performance in cell attachment, proliferation and penetration. They are used to repair, replace, maintain or enhance the function of a particular tissue or organ. Srinivasa Reddy *et al.* (2013) developed composite nanofibrous scaffolds using PCL and bisphenol, an ethoxylated dimethacrylate that demonstrated improved cell adhesion, proliferation and normal functioning of cardiac cells for heart tissue engineering. Kuppan *et al.* (2014) fabricated poly(3-hydroxybutyrate-co-3-hydroxyvalerate) (PHBV)-based nanofibrous scaffolds to support the functional esophageal epithelial cells towards engineering the esophagus. In this study, human esophageal epithelial cells (HEEpiC) cultured on PHBV and PHBV-gelatin nanofibers showed better cell proliferation in a PHBV nanofibrous scaffold than the PHBV-gelatin scaffold after 7 days of culture and could be explored further as extracellular matrix (ECM) substitutes for regeneration of esophageal tissue.

### 10.4.3 Wound-Healing Applications

Electrospun nanofiber mat is a good wound dressing candidate due to its highly porous structure, good interconnectivity and very high specific surface area. Nanofibers are used not only for wound healing, but also to exude extra body fluids from the wound area, decontaminate the exogenous microorganism, improve the appearance and sometimes accelerate the healing process. Cejkova *et al.* (2013) investigated that rabbit bone marrow-derived mesenchymal stem cell (MSC)-loaded nanofiber scaffolds effectively decreased alkali-induced oxidative stress in rabbit cornea. MSCs on nanofiber scaffolds protected the formation of toxic peroxynitrite (detected by NT residues), lowered apoptotic cell death and decreased matrix metalloproteinase and pro-inflammatory cytokine production. This resulted in reduced corneal inflammation as well as neovascularization, and significantly accelerated corneal healing. 2,3-Dihydroxybenzoic acid (DHBA)-containing nanofibers were fabricated for wound dressing which inhibit biofilm formation by *Pseudomonas aeruginosa*. The results indicated that DHBA, electrospun into nanofibers, inhibited cell growth for at least 4 h and could be developed into a wound dressing to treat topical infections caused by *P. aeruginosa*. Some important applications of nanofibers in wound healing are listed in Table 10.5.

**TABLE 10.4**
**Applications of Nanofibers in Drug Delivery**

| Polymer(s) | Drug(s) | Applications/Inferences | Reference(s) |
|---|---|---|---|
| Cellulose acetate | l-tryptophan | Provides a promising transdermal drug delivery | Ghorani *et al.* (2018) |
| PEO, PLA | Doxorubicin | Increased antitumor efficacy | Kuang *et al.* (2018) |
| Cellulose, Polyethylene glycol | Lemongrass oil | Enhanced antibacterial and antioxidant activity | Mishra *et al.* (2018) |
| PEO, CMC | Progesterone | Improved vaginal delivery | Brako *et al.* (2018) |
| PVA | Chlorpheniramine maleate | Exhibited a biphasic release | Jaiturong *et al.* (2018) |
| PDS | Metronidazole, Ciprofloxacin, Clindamycin | Pronounced antimicrobial activity | Ashley *et al.* (2018) |
| Poloxamer | Carvedilol | Prolonged stability and improved dissolution rate | Kajdič *et al.* (2018) |
| Chitosan, PCL, PEO | Ornidazole | Coating material for pulp capping | Tort *et al.* (2017a,b) |
| Gelatin | Ciprofloxacin | Increased antibacterial performance and corrosion resistance | Bakhsheshi-Rad *et al.* (2017) |
| PVP | Artemisinin | *In vitro* inhibition of *P. falciparum* parasites revealed that the developed fibers have the potential to effectively replace traditional antimalarials and showed reduced prostatic cancer cell viability | Bonadies *et al.* (2017) |
| TPU | Naproxen | Mat thickness played a major role in the release rate of naproxen | Akduman *et al.* (2016) |
| Collagen, PVA | Salicylic acid | Electrospun fibers efficiently controlled the release of drugs | Zhang *et al.* (2016) |
| PCL | Vitamin B12 | Better loading and delayed release of encapsulated drugs | Madhaiyan *et al.* (2013) |
| PLGA | siRNA | Better transfection efficiency with significantly higher siRNA activity | Chen *et al.* (2012) |
| Collagen, Chitosan | 5-fluorouracil | Improved treatment and compliance for cancer | Sundar and Sangeetha (2012) |
| Gelatin | Fibroblast growth factor | Improved proliferation of cells | Montero *et al.* (2012) |
| PVP | Silver | Effective against gram-negative and positive microorganisms | Hwang and Jeong, (2011), Nguyen *et al.* (2011) |
| Cellulose, PVA | Model drug | Broad-spectrum antibacterial efficacy | Jia *et al.* (2011) |
| PLGA | Tacrolimus | Effective in recurrent pulmonary venous obstruction | Mutsuga *et al.* (2009) |
| PEA | Nitroxyl radical | Effective delayed release of drug | Li and Chu (2009) |
| Tecophilic polymer | Silver(I)-complex | Broad spectrum antibacterial efficacy | Melaiye *et al.* (2005) |

*Abbreviations:* PCL: Polycaprolactone; CMC: Carboxymethyl cellulose; PDS: Polydioxanone; PEO: Polyethylene oxide; PVP: Polyvinyl pyrrolidone; TPU: Thermoplastic polyurethanes; PVA: Polyvinyl alcohol; PEA: Poly(ester amide); PLGA: Poly (lactide-co-glycolic acid); PLA: (poly(L-lactide).

**TABLE 10.5**
**Applications of Nanofibers in Wound Healing**

| Polymers | Drugs/Cells | Applications/Inferences | References |
|---|---|---|---|
| PCL | Curcumin | Good antibacterial activity and biocompatibility | Reshmi *et al.* (2018) |
| Collagen, Chitosan, Sodium alginate | Doxycycline | Potential to improve healing process | Tort *et al.* (2017a,b) |
| Gum tragacanth, Poly(ε-caprolactone) | Curcumin | Nanofibers show better healing performance | Ranjbar-Mohammadi *et al.* (2016) |
| Hydroxypropyl-beta-cyclodextrin | Sulfisoxazole | Optimized electrospun nanofibers have the potential to be used in wound dressing | Aytac *et al.* (2015) |
| PCL | *Azadirachta indica*, Memecylonedule, Curcumin | Better wound healing recovery | Jin *et al.* (2013), Merrell *et al.* (2009) |
| PEG-PLGA | Salicylic acid | Improved wound healing management | Nguyen *et al.* (2012) |
| Cellulose acetate and polyester urethane | Polyhexamethylene biguanide | Sustained antimicrobial effects with improved healing rate | Liu *et al.* (2012) |
| Chitosan, PVA | Graphene, Lysozyme, Ciprofloxacin HCl | Improved healing rate | Lu *et al.* (2012), Charernsriwilaiwat *et al.* (2012), Jannesari *et al.* (2011) |
| Gelatin | Stem cells | Increased density of cells | Dubsky *et al.* (2012) |
| PLGA | Fusidic acid, Cefazolin | Significantly higher wound-healing capacity | Said *et al.* (2012), Katti *et al.* (2004) |
| PU | Silver | Complete bacterial destruction | Chen and Chiang (2010) |
| Polyglycerol | Calendula officinalis | Better wound-healing recovery | Vargas *et al.* (2010) |

*Abbreviations:* PCL: Polycaprolactone; PEG: Polyethylene glycol; PLGA: Poly (lactide-co-glycolic acid); PVA: Polyvinyl alcohol; PU: Polyurethane.

## 10.5 CONCLUSION AND FUTURE TRENDS

Albeit there are several ways of fabricating nanofibers, electrospinning has caught the eye of the researchers in the recent past owing to its comparatively better results, be it in the terms of the size of the fiber, economy of the process, scale-up feasibility or the modifications possible during the operation. Though still in the nascent stages, nanofibers can offer a plethora of applications. Along with the monoaxial, coaxial and hybrid core–shell nanofibers, these tend to provide better stealth characteristics to micro-/nano-tubes, drug and protein. Coaxial nanofibers are known to provide controlled drug release kinetics for a wide range of pharmaceuticals and bioactives. Virtually any kind of drug can be incorporated within the nanofibers made from biodegradable polymers, which ultimately would allow for controlling the rate of release, based on the degradation rate and pattern it follows inside the body. Apt selection of a polymer based on its compatibility with the respective drug can also allow protection of the active ingredient from the harsh environment inside the body. Future work has been conceived to involve the incorporation of the drug-loaded nanofiber within a capsule (coated/uncoated as per requirement), which may ultimately provide the dual benefits of protection and sustained and targeted release. These nanofibers can also be used as implants inside the body and used as replacements to synthetic implants. Having very low density, these nanofibers have been utilized as floating gastroretentive dosage forms, as well as the mucoadhesive system to adhere to gastrointestinal tract mucosa using a suitable polymer having the desired mucoadhesive strength. The oral delivery of proteins, enzymes and growth factors can also be attempted, along with plausible modifications in the oral route such as the buccal and sublingual. Also, in cases of external applications, these nanofibers can replace the generally used bandages made up of micro-sized non-biodegradable fibers, which often encounter with myriad problems ultimately hampering the recovery process of the wounded tissue. Electrospun drug-loaded nanofibers are often used as the mid-dosage forms. They can be further turned into different kinds of DDS for all types of drug delivery routes, such as transdermal, oral, pulmonary, subcutaneous, intramuscular or intra-cavernosum injection, or for reconstitution into a liquid medium for administrations such as suspensions or solutions. These encompass only a few key instances of the ways in which these nanofibers could be put to use in the near future. With rapid emergence of new technologies within the field of electrospinning, the onus ultimately lies on those currently working in this field to further unravel the potential applications of these nanofibers.

## REFERENCES


Akduman C, Özgüney I, Kumbasar PE. Preparation and characterization of naproxen-loaded electrospun thermoplastic polyurethane nanofibers as a drug delivery system. *Mater Sci Eng C*. 2016;64:383–390.

Arumugam GK, Khan S, Heiden PA. Comparison of the effects of an ionic liquid and other salts on the properties of electrospun fibers, 2 – Poly(vinyl alcohol). *Macromol Mater Eng*. 2009;294(1):45–53.

Aytac Z, Sen HS, Durgun E. Sulfisoxazole/cyclodextrin inclusion complex incorporated in electrospun hydroxypropyl cellulose nanofibers as drug delivery system. *Colloids Surf B*. 2015;128:331–8.

Bakhsheshi-Rad HR, Hadisi Z, Hamzah E, Ismail AF, Aziz M, Kashefian M. Drug delivery and cytocompatibility of ciprofloxacin loaded gelatin nanofibers-coated Mg alloy. *Materials Letters*. 2017 Nov 15;207:179–82.

Baumgarten PK. Electrostatic spinning of acrylic microfibers. *J Colloid Interface Sci*. 1971;36(1):71–9.

Bonadies I, Maglione L, Ambrogi V, Paccez JD, Zerbini LF, e Silva LF, Picanço NS, Tadei WP, Grafova I, Grafov A, Carfagna C. Electrospun core/shell nanofibers as designed devices for efficient Artemisinin delivery. *European Polymer Journal*. 2017 Apr 1;89:211–20.

Brako F, Raimi-Abraham BT, Mahalingam S, Duncan QMC, Edirisinghe M. The development of progesterone-loaded nanofibers using pressurized gyration: A novel approach to vaginal delivery for the prevention of pre-term birth. *Int J Pharm*. 2018;(540):31–39.

Buchko CJ, Chen LC, Shen Y, Martin DC. Processing and microstructural characterization of porous biocompatible protein polymer thin films. *Polymer*. 1999;40(26):7397–407.

Casper CL, Stephens JS, Tassi NG, Chase DB, Rabolt JF. Controlling surface morphology of electrospun polystyrene fibers: effect of humidity and molecular weight in the electrospinning process. *Macromolecules*. 2003;37(2):573–8.

Cejkova J, Trosan P, Cejka C, Lencova A, Zajicova A, Javorkova E *et al*. Suppression of alkali-induced oxidative injury in the cornea by mesenchymal stem cells growing on nanofiber scaffolds and transferred onto the damaged corneal surface. *Exp Eye Res*. 2013;116:312–23.

Charernsriwilaiwat N, Opanasopit P, Rojanarata T, Ngawhirunpat T. Lysozyme-loaded, electrospun chitosan-based nanofiber mats for wound healing. *Int J Pharm*. 2012;427:379–84.

Chen JP, Chiang Y. Bioactive electrospun silver nanoparticles-containing polyurethane nanofibers as wound dressings. *J Nanosci Nanotechnol*. 2010;10:7560–4.

Chen M, Gao S, Dong M, Song J, Yang C, Howard KA *et al*. Chitosan/siRNA nanoparticles encapsulated in PLGA nanofibers for siRNA delivery. *ACS Nano*. 2012;6:4835–44.

Chew SY, Wen J, Yim EKF, Leong KW. Sustained release of proteins from electrospun biodegradable fibers. *Biomacromolecules*. 2005;6:2017–24.

De Vrieze S, Van Camp T, Nelvig A, Hagstrom B, Westbroek P, De Clerck K. The effect of temperature and humidity on electrospinning. *J Mater Sci*. 2009;44(5):1357–62.

Deitzel JM, Kleinmeyer J, Harris D, Beck Tan NC. The effect of processing variables on the morphology of electrospun nanofibers and textiles. *Polymer*. 2001;42(1):261–72.

Demir MM, Yilgor I, Yilgor E, Erman B. Electrospinning of polyurethane fibers. *Polymer*. 2002;43(11):3303–9.

Doshi J, Reneker DH. Electrospinning process and applications of electrospun fibers. *J Electrostat*.1995; 35(2–3):151–60.

Dubsky M, Kubinova S, Sirc J, Voska L, Zajicek R, Zajicova A *et al.* Nanofibers prepared by needleless electrospinning technology as scaffolds for wound healing. *J Mater Sci Mater Med.* 2012;23:931–41.

Formhals. Process and apparatus for preparing artificial threads. US1975504;1934.

Fridrikh SV, Yu JH, Brenner MP, Rutledge GC. Controlling the fiber diameter during electrospinning. *Phys Rev Lett.* 2003;90(14):144502.

Garg K, Bowlin GL. Electrospinning jets and nanofibrous structures. *Biomicrofluidics.* 2011;5(1):13403.

Ghorani B, Goswami P, Blackburn RS, Russell SJ. Enrichment of cellulose acetate nanofibre assemblies for therapeutic delivery of l-tryptophan. *Int J Biol Macromol.* 2018;108:1–8.

Grabnar PA, Kristl J. The manufacturing techniques of drug-loaded polymeric nanoparticles from preformed polymers. *J Microencapsul.* 2011;28(4):323–35.

Gunn J, Zhang M. Poly-blend nanofibers for biomedical applications: Perspectives and challenges. *Trends Biotechnol.* 2010;28(4):189–97.

Hassounah IA, Shehata NA, Kimsawatde GC, Hudson AG, Sriranganathan N, Joseph EG *et al.* Studying the activity of antituberculosis drugs inside electrospun PVA, PEO and PCL nanofibers. *J Biomed Mater Res A.* 2014;102(11):4009–16.

Hayati I, Bailey AI, Tadros TF. Investigations into the mechanisms of electrohydrodynamic spraying of liquids: I. Effect of electric field and the environment on pendant drops and factors affecting the formation of stable jets and atomization. *J Colloid Interface Sci.* 1987;117(1):205–21.

He SW, Li SS, Hu ZM, Yu JR, Chen L, Zhu J. Effects of three parameters on the diameter of electrospun poly(ethylene oxide) nanofibers. *J Nanosci Nanotechnol.* 2011;11(2):1052–9.

Huang ZM, Zhang YZ, Kotaki M, Ramakrishna S. A review on polymer nanofibers by electrospinning and their applications in nanocomposites. *Compos Sci Technol.* 2003;63(15):2223–53.

Hwang S, Jeong S. Electrospun nano composites of poly(vinyl pyrrolidone)/ nano-silver for antibacterial materials. *J Nanosci Nanotechnol.* 2011;11:610–3.

Jaeger R, Bergshoef MM, Batlle CMI, Schönherr H, Julius Vancso G. Electrospinning of ultra-thin polymer fibers. *Macromolecular Symposia.* 1998;127(1):141–50.

Jaiturong P, Sirithunyalug B, Eitsayeam S, Asawahame C, Tipduangta P, Sirithunyalug J. Preparation of glutinous rice starch/polyvinyl alcohol copolymer electrospun fibers for using as a drug delivery carrier. *Asian J Pharm. Sci.* 2018;13:239–47.

Jannesari M, Varshosaz J, Morshed M, Zamani M. Composite poly(vinyl alcohol)/poly(vinyl acetate) electrospun nanofibrous mats as a novel wound dressing matrix for controlled release of drugs. *Int J Nanomedicine.* 2011;6:993–1003.

Jarusuwannapoom T, Hongrojjanawiwat W, Jitjaicham S, Wannatong L, Nithitanakul M, Pattamaprom C. Effect of solvents on electro-spinnability of polystyrene solutions and morphological appearance of resulting electrospun polystyrene fibers. *Eur Polym J.* 2005;41(3):409–21.

Jia B, Zhou J, Zhang L. Electrospun nanofibers mats containing cationic cellulose derivatives and poly (vinyl alcohol) with antibacterial activity. *Carbohydr Res.* 2011;346:1337–41.

Jiang H, Fang D, Hsiao BS, Chu B, Chen W. Optimization and characterization of dextran membranes prepared by electrospinning. *Biomacromolecules.* 2004;5(2):326–33.

Jin G, Prabhakaran MP, Kai D, Annamalai SK, Arunachalam KD, Ramakrishna S. Tissue engineered plant extracts as nanofibrous wound dressing. *Biomaterials.* 2013;34:724–34.

Kajdič S, Vrečera F, Kocbek P. Preparation of poloxamer-based nanofibers for enhanced dissolution of carvedilol. *Eur J Pharm Biopharm.* 2018;117:331–40.

Karczewski A, Feitosa SA, Hamer EI, Pankajakshan D, Gregory RL, Spolnik KJ, Bottino MC. Clindamycin-modified triple antibiotic nanofibers: A stain-free antimicrobial intracanal drug delivery system. *J Endod.* 2018;44(1):155–62.

Katti DS, Robinson KW, Ko FK, Laurencin CT. Bioresorbable nanofiber-based systems for wound healing and drug delivery: Optimization of fabrication parameters. *J Biomed Mater Res B Appl Biomater.* 2004;70:286–96.

Kenawy el R, Layman JM, Watkins JR, Bowlin GL, Matthews JA, Simpson DG, Wnek GE. Electrospinning of poly(ethylene-co-vinyl alcohol) fibers. *Biomaterials.* 2003;24(6):907–13.

Kidoaki S, Kwon IK, Matsuda T. Mesoscopic spatial designs of nano- and microfiber meshes for tissue-engineering matrix and scaffold based on newly devised multilayering and mixing electrospinning techniques. *Biomaterials.* 2005;26(1):37–46.

Kim TG, Park TG. Biomimicking extracellular matrix: cell adhesive RGD peptide modified electrospun poly (D, L-lactic-co-glycolic acid) nanofiber mesh. *Tissue engineering.* 2006 Feb 1;12(2):221–33.

Koski A, Yim K, Shivkumar S. Effect of molecular weight on fibrous PVA produced by electrospinning. *Mater Lett.* 2004;58(3–4):493–7.

Kristl J, Šmid-Korbar J, Štruc E, Schara M, Rupprecht H. Hydrocolloids and gels of chitosan as drug carriers. *Int J Pharm.* 1993;99(1):13–9.

Kuang G, Zhang Z, Dongfang LS, Xiaolan Z, Jing XL, Huang Y. Biphasic drug release from electrospun polyblend nanofibers for optimized local cancer treatment. *Biomater. Sci.* 2018;(6):324–31.

Kuppan P, Sethuraman S, Krishnan UM. Poly(3-hydroxybutyrate-co-3-hydroxyvalerate)-based nanofibrous scaffolds to support functional esophageal epithelial cells towards engineering the esophagus. *J Biomater Sci Polym Ed.* 2014;574–92.

Lee DY, Park JY, Kim BY, Cho NI. Effect of collector speed and flow rate on morphology of Er doped TiO2 nanofibers. *J Nanosci Nanotechnol.* 2012;12(2):1599–603.

Li D, Xia Y. Direct Fabrication of composite and ceramic hollow nanofibers by electrospinning. *Nano Letters.* 2004;4(5):933–8.

Li J, Fu R, Li L, Yang G, Ding S, Zhong Z, Zhou S. Co-delivery of dexamethasone and green tea polyphenols using electrospun ultrafine fibers for effective treatment of keloid. *Pharm Res.* 2014;31(7):1632–43.

Li L, Chu CC. Nitroxyl radical incorporated electrospun biodegradable poly(ester Amide) nanofiber membranes. *J Biomater Sci Polym Ed.* 2009;20:341–61.

Liu J, Kumar S. Microscopic polymer cups by electrospinning. *Polymer.* 2005;46(10):3211–4.

Liu X, Lin T, Gao Y, Xu Z, Huang C, Yao G *et al.* Antimicrobial electrospun nanofibers of cellulose acetate and polyester urethane composite for wound dressing. *J Biomed Mater Res B Appl Biomater.* 2012;100:1556–65.

Lu B, Li T, Zhao H, Li X, Gao C, Zhang S *et al.* Graphene-based composite materials beneficial to wound healing. *Nanoscale.* 2012;4:2978–82.

Luo CJ, Nangrejo M, Edirisinghe M. A novel method of selecting solvents for polymer electrospinning. *Polymer.* 2010;51(7):1654–62.

Madhaiyan K, Sridhar R, Sundarrajan S, Venugopal JR, Ramakrishna S. Vitamin B12 loaded polycaprolactone nanofibers: A novel transdermal route for the water soluble energy supplement delivery. *Int J Pharm.* 2013;444:70–6.

McKee MG, Wilkes GL, Colby RH, Long TE. Correlations of solution rheology with electrospun fiber formation of linear and branched polyesters. *Macromolecules.* 2004;37(5):1760–7.

Meechaisue C, Dubin R, Supaphol P, Hoven VP, Kohn J. Electrospun mat of tyrosine-derived polycarbonate fibers for potential use as tissue scaffolding material. *J Biomater Sci Polym Ed.* 2006;17(9):1039–56.

Megelski S, Stephens JS, Chase DB, Rabolt JF. Micro- and nanostructured surface morphology on electrospun polymer Fibers. *Macromolecules.* 2002;35(22):8456–66.

Melaiye A, Sun Z, Hindi K, Milsted A, Ely D, Reneker DH *et al.* Silver(I)-imidazole cyclophane gem-diol complexes encapsulated by electrospun tecophilic nanofibers: Formation of nanosilver particles and antimicrobial activity. *J Am Chem Soc.* 2005;127:2285–91.

Merrell JG, Mclaughlin SW, Tie L, Laurencin CT, Chen AF, Nair LS. Curcumin-loaded poly(epsilon-caprolactone) nanofibres: Diabetic wound dressing with anti-oxidant and anti-inflammatory properties. *Clin Exp Pharmacol Physiol.* 2009;36:1149–56.

Mishra D, Khare P, Singh DK, Lugman S, Kumar PVA, Yadav A, Das T, Saikia BK. Retention of antibacterial and antioxidant properties of lemongrass oil loaded on cellulose nanofibre-poly ethylene glycol composite. *Ind Crops Prod.* 2018;114:68–80.

Mit-uppatham C, Nithitanakul M, Supaphol P. Ultrafine electrospun polyamide-6 fibers: Effect of solution conditions on morphology and average fiber diameter. *Macromol Chem Phys.* 2004;205(17):2327–38.

Montero RB, Vial X, Nguyen DT, Farhand S, Reardon M, Pham SM *et al.* bFGF-containing electrospun gelatin scaffolds with controlled nano-architectural features for directed angiogenesis. *Acta Biomater.* 2012;8:1778–91.

Mutsuga M, Narita Y, Yamawaki A, Satake M, Kaneko H, Usui A *et al.* Development of novel drug-eluting biodegradable nanofibers for prevention of postoperative pulmonary venous obstruction. *Interact Cardiovasc Thorac Surg.* 2009;8:402–6.

Nguyen TH, Kim YH, Song HY, Lee BT. Nano Ag loaded PVA nano-fibrous mats for skin applications. *J Biomed Mater Res B Appl Biomater.* 2011;96:225–33.

Nguyen TT, Ghosh C, Hwang SG, Chanunpanich N, Park JS. Porous core/sheath composite nanofibers fabricated by coaxial electrospinning as a potential mat for drug release system. *Int J Pharm.* 2012;439:296–306.

Ojha SS, Afshari M, Kotek R, Gorga RE. Morphology of electrospun nylon-6 nanofibers as a function of molecular weight and processing parameters. *J Appl Polym Sci.* 2008;108(1):308–19.

Ramakrishna S, Fujihara K, Teo WE, Lim TC, Ma Z. An introduction to electrospinning and nanofibers. *World Scientific.* 2005;1:91–101.

Ranjbar-Mohammadi M, Bahrami SH, Joghataei MT. Fabrication of novel nanofiber scaffolds from gum tragacanth/poly(vinyl alcohol) for wound dressing application: *In vitro* evaluation and antibacterial properties. *Mater Sci Eng C Mater Biol App.* 2013;33(8):4935–43.

Ranjbar-Mohammadi M, Rabbani S, Bahrami SH, Joghataei MT, Moayer F. Antibacterial performance and *in vivo* diabetic wound healing of curcumin loaded gum tragacanth/poly(ε-caprolactone) electrospun nanofibers. *Mater Sci Eng C Mater Biol Appl.* 2016;69:1183–91.

Reshmi CR, Suja PS, Manaf O, Sanu PP, Sujith A. Nanochitosan enriched poly ε-caprolactone electrospun wound dressing membranes: A fine tuning of physicochemical properties, hemocompatibility and curcumin release profile. *Int J Biol Macromol.* 2018;108:1261–72.

Rošic R, Kocbek P, Baumgartner S, Kristl J. Electro-spun hydroxyethyl cellulose nanofibers: The relationship between structure and process. *J Drug Del Sci Tech.* 2011;21(3):229–36.

Rošic R, Kocbek P, Baumgartner S, Kristl J. Electrospun Chitosan/Peo nanofibers and their relevance in biomedical application. *5th European Conference of the International Federation for Medical and Biological Engineering IFMBE Proceedings.* 2012;37:1296–99.

Said SS, El-halfawy OM, El-gowelli HM, Aloufy AK, Boraei NA, El-khordagui LK. Bioburden-responsive antimicrobial PLGA ultrafine fibers for wound healing. *Eur J Pharm Biopharm.* 2012;80:85–94.

Shin YM, Hohman MM, Brenner MP, Rutledge GC. Electrospinning: A whipping fluid jet generates submicron polymer fibers. *Appl Phys Lett.* 2001;78(8):1149–51.

Sill TJ, von Recum HA. Electrospinning: Applications in drug delivery and tissue engineering. *Biomaterials.* 2008;29(13):1989–2006.

Sohrabi A, Shaibani PM, Thundat T. The effect of applied electric field on the diameter and size distribution of electrospun Nylon6 nanofibers. *Scanning.* 2013;35(3):183–8.

Srinivasa Reddy C, Reddy Venugopal J, Ramakrishna S, Zussman E. Polycaprolactone/oligomer compound scaffolds for cardiac tissue engineering. *J Biomed Mater Res A.* 2013;2014:3713–25.

Sundar SS, Sangeetha D. Fabrication and evaluation of electrospun collagen/poly(N-isopropyl acrylamide)/chitosan mat as blood-contacting biomaterials for drug delivery. *J Mater Sci Mater Med.* 2012;23:1421–30.

Sundaray B, Subramanian V, Natarajan TS. Electrospinning of continuous aligned polymer fibers. *Appl Phys Lett.* 2004;84:1222.

Taylor G. Electrically driven jets. *Proceedings of The Royal Society of London A J Math Phys Sci.* 1969;313(1515):453–75.

Theron SA, Yarin AL, Zussman E, Kroll E. Multiple jets in electrospinning: Experiment and modeling. *Polymer.* 2005;46(9):2889–99.

Tort H, Oktay EA, Tort S, Bayar GR, Topcu FT, Kilic E, Acarturk F. Evaluation of ornidazole-loaded nanofibers as an alternative material for direct pulp capping. *Drug Deliv Sci Technol.* 2017a;41:317–24.

Tort S, Acartürk F, Beşikci A. Evaluation of three-layered doxycycline-collagen loaded nanofiber wound dressing. *Int J Pharm.* 2017b;529(1–2):642–53.

Vargas EA, Do vale baracho NC, De brito J, De queiroz AA. Hyperbranched polyglycerol electrospun nanofibers for wound dressing applications. *Acta Biomater.* 2010;6:1069–78.

Wannatong L, Sirivat A, Supaphol P. Effects of solvents on electrospun polymeric fibers: Preliminary study on polystyrene. *Polym Int.* 2004;53(11):1851–9.

Yuan X, Zhang Y, Dong C, Sheng J. Morphology of ultrafine polysulfone fibers prepared by electrospinning. *Polym Int.* 2004;53(11):1704–10.

Zhang C, Yuan X, Wu L, Han Y, Sheng J. Study on morphology of electrospun poly(vinyl alcohol) mats. *Eur Polym J.* 2005;41(3):423–32.

Zhang X, Tang K, Zheng X. Electrospinning and crosslinking of col/pva nanofiber-microsphere containing salicylic acid for drug delivery. *J Bionic Eng.* 2016;13:143–9.

Zong X, Kim K, Fang D, Ran S, Hsiao BS, Chu B. Structure and process relationship of electrospun bioabsorbable nanofiber membranes. *Polymer.* 2002;43(16):4403–12.

Zuo W, Zhu M, Yang W, Yu H, Chen Y, Zhang Y. Experimental study on relationship between jet instability and formation of beaded fibers during electrospinning. *Polym Eng Sci.* 2005;45(5):704–9.

# 11 Rapid Nano-Tooling in Clinical Dentistry

*Kamaljit Singh Boparai and Rupinder Singh**

## CONTENTS



## 11.1 INTRODUCTION

Nanotechnology is applicable to the matter that can be manipulated at nanoscale levels. Currently, this concept is very widely employed with stellar merits in various fields of science and engineering. In dentistry, biomaterials in nanoscale have included applications such as pathological conditions, diagnosis, orthodontic tooth movement, local anesthesia and periodontics (Govindankutty, 2015). The concept of nanotechnology originated in 1959 by the physicist Richard Feynman, though, in 1974, Dr Nori Taniguchi, a Japanese scientist, coined this term and defined it as "the processing of separation, consolidation, and deformation of materials by one atom or one molecule" (Silva, 2004; Khurshid *et al.*, 2015). It was subsequently reported by Dr Drexler in his book, *Engines of Creation-The Coming Era of Nanotechnology* during late 1980 and by Sumio Lijima in 1991 with the introduction of nanotubes in his book, *Helical Microtubules of Graphitic Carbon* (Koruga *et al.*, 1993). Basically, nanotechnology is concerned with four types of materials: metals, ceramics, polymers and composites. A classification of these materials for manufacturing biomedical devices (dental, orthopedic and bone cement, etc.) is shown in Figure 11.1.

Among these four categories, metallic materials exhibit some limitations, mainly the release of wear debris and metallic ions. However, these are most commonly used owing to their high strength, high toughness and biocompatibility. In recent past, many researchers have investigated different materials and alloys for medical applications. Various alloys systems include stainless steels, titanium (Ti) and cobalt (Co) alloys. Table 11.1 shows different alloys, with respective ASTM standards and their compositions for various biomedical applications.

* Corresponding author.

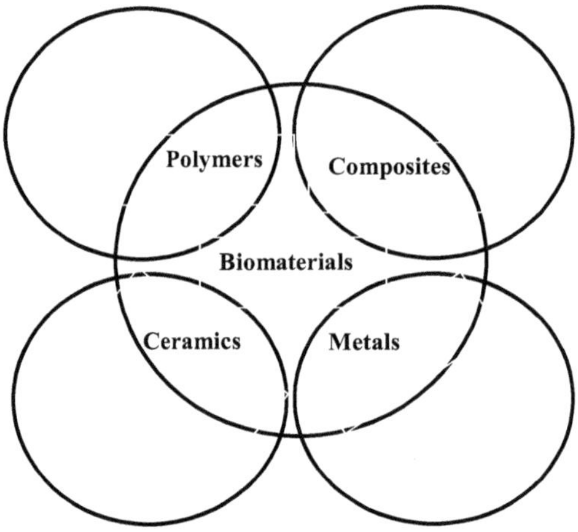


**FIGURE 11.1** Material classification of biomedical devices.

Ceramics generally possess certain outstanding properties such as corrosion resistance, incompressibility, non-absorbable, bioactive and biodegradable (Bhasin *et al.*, 2015). Due to these properties, ceramics demonstrate better biocompatibility than metals, though they suffer from drawbacks such as high brittleness and density, coupled with low fracture strength and poor resilience (Davis and Zanotto, 2017). The commonly used ceramics in biomedical applications are alumina, zirconia, silicone nitrides, carbons, glass, aluminum calcium phosphate, tricalcium phosphate and hydroxyapatite.

Natural polymers have notable biological attributes ideally required in biomaterials, including biocompatibility, biodegradability and low toxicity. Contrary to this, the synthetic polymers possess improved mechanical/physical properties and have better chemical/thermal stability (Asghari *et al.*, 2017). Among polymers, polymethylmethacrylate (PMMA) and ultrahigh-molecular-weight polyethylene are commonly used for bio-implants and bio-devices. Some other types of polymers, which have been used as dental implant materials include polyamide, polyethylene (PE), polyurethane (PU), polypropylene (PP), polydimethylsiloxane, polysulfone (PS) and silicone rubber (Bhasin *et al.*, 2015).

Emergence of nanotechnology fosters the use of nano-filler particles in various dental applications such as tissue engineering and regenerative/restorative dentistry. It also includes prosthodontics, oral implantology periodontics, preventive dentistry, orthodontics and endodontics (Guggenberger *et al.*, 1998; Anusavice and Phillips, 2003; Hatrick *et al.*, 2003; Mitra *et al.*, 2011; Khurshid *et al.*, 2015). Reinforcement of nano-fillers results in significant improvement in mechanical and biological characteristics of composite biomaterials (Wang *et al.*, 2017). Mainly, their purpose is to improve strength and modulus of elasticity, increase wear resistance and micro-hardness, reduce polymerization shrinkage and gain patient satisfaction from the aesthetic appearance. Besides these, they also have a remarkable effect on coefficients of thermal expansion and water absorption (Khurshid *et al.*, 2015). Due to the high surface area per unit volume ratio, nanoparticles possess distinct physiochemical

**TABLE 11.1**
**Vital Biomedical Materials and Their Respective ASTM Standards**

| S. No. | Biomedical Material | ASTM Standard |
|---|---|---|
| 1 | Titanium alloys | ASTM F67-89, F136-84, F620-87, F1108-88, F1295-92, F1713-96, F1813-97 |
| 2 | Stainless steels | ASTM F138-92, F139-96, F621-92, F1314-95, F1586-95 |
| 3 | Co-alloys | ASTM F75-92, F90-96, F562-95, F563-95, F799-96, F1058-91, F1537-94 |

*Source:* Hussein MA *et al.*, *Materials*. 2015; 8: 2749–68.

properties (Gunasekaran *et al.*, 2012). Yee *et al.* (2015) demonstrated the improvement of breaking/adhesive strength and antibacterial efficacy of ophthalmic tissue adhesive (2-octylcyanoacrylate) by doping with silver nanoparticles. Moreno *et al.* (2015) evaluated the potential application of a PMMA/hydroxyapatite nanocomposite in the field of bone tissue regeneration. Padovani *et al.* (2015) reviewed various types of nanocomposite materials such as metallic, polymeric, etc. They also studied their use as nano-cluster fillers, in nanocomposites, bio-mimetic dental materials, medicines and mouthwashes. Significant wear rate reductions were achieved in polytetrafluoroethylene (PTFE) with reinforcement of 5% nano-alumina particles.

Omrani *et al.* (2016) reviewed and demonstrated the improved tribological/wear behavior of natural fiber-reinforced composites and their usability for various tribological applications, which further leads toward the green material world. Vallittu (2015) highlighted the efficacy of fiber-reinforced composites (FRCs), which can be considered as a group of non-metallic dental biomaterials. Yousef *et al.* (2016) recorded an improvement in specific wear rate and mechanical and thermal properties of polyoxymethylene (POM)/carbon nanotube (CNT) nanocomposites due to the inclusion of CNTs. Tai *et al.* (2015) studied various properties of liquid crystalline epoxy nanocomposites such as biocompatibility, micro-hardness and frictional forces of bracket-blocks with different filler contents, and this nanomaterial showed less polymerization shrinkage and rigid bond to tooth structures due to their high efficacy.

In this present work, a biocompatible composite material (BCCM) feed-stock filament has been prepared on a single screw extruder by selecting $Al_2O_3$ as a reinforced material in the matrix of nylon 6. Initially, the suitability of BCCM feed-stock filament for a fused deposition modeling (FDM) system has been verified by a melt flow index (MFI) test. The extent of reinforcement may be limited up to a value so that its MFI should map with the MFI of commercial acrylonitrile butadiene styrene (ABS) material. The MFI of ABS and BCCM were measured as 2.41 and 2.35 g/10 min, respectively. Presently, the commercial FDM setup of 'Stratasys uPrint–SE system, USA' uses support material SR30-XL and model material P430XL. In this study, the objective was to investigate the effect of sliding wear on FDM-printed parts using BCCM and ABS-based feed-stock filaments separately. Further, the wear tracks generated on the abrasive surface were observed by photomicrographs and scanning electron microscope (SEM).

## 11.2 EXPERIMENTATION PROCEDURES

### 11.2.1 Materials

A biocompatible binder material nylon 6, having the specific density of 1.13 g/cm$^3$, was supplied by a supplier in granular form. Its particle size had an average value of 4–5 mm and was reduced to 500–800 μm with a cryogenic grinding process. The reinforced material, alpha phase $Al_2O_3$ (average particle size 20–45 nm) was distributed by Thomas Baker, Mumbai, India. Based upon the work reported by Boparai *et al.* (2016a), the volume % proportions of BCCM material were selected (Table 11.2).

**TABLE 11.2**
**Volume % Proportions of BCCM**

| S. No. | Nylon6 | $Al_2O_3$ | Density (g/cm$^3$) |
|---|---|---|---|
| 1 | 85 | 15 | 1.52 |

*Source:* Courtesy: Unpublished Lab experimental data, Manufacturing Research Lab, Dept. of Production Engineering, GNDEC, Ludhiana, India.

The density of nano composite (NC) was evaluated using the following relationship (Pashaei *et al.*, 2011):

$$\rho_c = \frac{1}{\left\{\left(\frac{Wr}{\rho r}\right) + \left(\frac{Wp}{\rho p}\right)\right\}}$$

where, $\rho_c$ = density of composition, $\rho_p$ = density of nylon 6, $\rho_r$ = density of $Al_2O_3$, $W_p$ = weight proportions of nylon 6 and $W_r$ = weight proportions of $Al_2O_3$.

### 11.2.2 Materials Processing

The selected volumetric proportions of BCCM were prepared in a tumbler mixture (by rotating at 350 rpm for 5 h). In this case, no surfactant(s) or plasticizer(s) were added due to the favorable binding properties of nylon 6. Although the tumbler mixing facilitates uniform composition throughout the mixture, the remaining homogeneity in composition was achieved during the preparation of BCCM with screw extruder (having mixing screw). The various phases for preparation of FDM feed-stock filament are shown in Figure 11.2.

BCCM feedstock filament was fabricated by melt mixing in a single screw extruder having 25 mm of screw diameter (supplied by Binflex Plastic Industry, Ludhiana, India). The extruder barrel had three heaters (capacity 1 KW each) with provision of zone temperature control. The hopper side heater was set at a lower temperature, while the die side heater was set at a higher temperature. Based upon the optimum parametric values achieved (Boparai *et al.*, 2016b), the mean barrel temperature and die temperature were fixed as 170°C and 205°C, respectively. Screw rotation was set at 35 rpm with the pressure of 23 MPa. The equipment was facilitated by a take-up unit and a winder unit. The main function of the take-up unit is to adjust the diameter of filament as per the requirements. The standard ABS filament had diameter of 1.75 ± 0.03 mm and the diameter of BCCM should be controlled within this range. This can be achieved by adjusting the speed of the take-up unit, such that diameter remains within this range. A part of this, the water tank was installed in between the screw extruder and the take-up unit. The filament had to pass through water before moving to the take-up unit. As the water temperature controls the solidification of the filament material, so its temperature depends upon the material to be processed and climate conditions. In this case, the water tank temperature was 25°C. On the other hand, the winder unit provided the facility for automatic winding of filament in spool form so that it became more convenient to load the filament in the FDM system.

### 11.2.3 Sample Preparation

The specimens, as shown in Figure 11.3a,b of 10 mm diameter and 30 mm length, were fabricated on the FDM system. For this comparative study, all the specimens/components were prepared in horizontal orientation mode (0° along the x-axis). For these cylindrically shaped samples, the FDM system was set on solid density mode having raster angle 45°/45°.

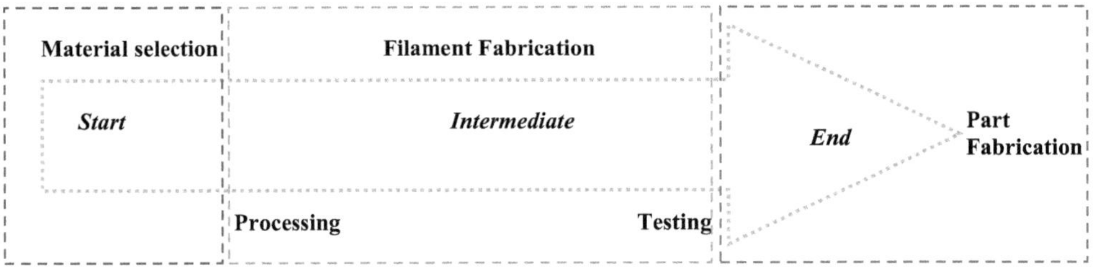


**FIGURE 11.2** Flow chart of feedstock filament development.

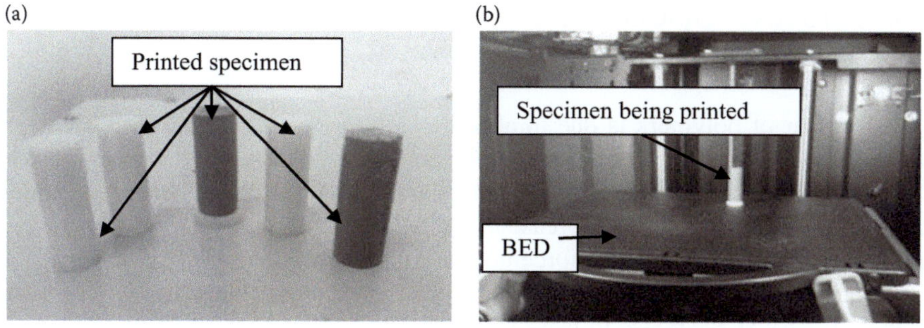


**FIGURE 11.3** Preparation of (a) specimens for wear test; (b) specimens fabricated on FDM system. (Courtesy: Unpublished Lab experimental work, Manufacturing Research Lab, Dept. of Production Engineering, GNDEC, Ludhiana, India.)

### 11.2.4 Wear Testing

The literature suggests that it is most important to study the wear/friction distinctiveness of finally prepared FDM-based biomaterials (Hussein *et al.*, 2015). The various methods used to study the wear manners of metallic biomaterials are block-on-disc test, ball-on-disc test or pin-on-disc test. Turner *et al.* (2015) developed a new tribometer to ascertain the sliding friction between a stationary polymeric sleeve and a rigid metallic rod that moves longitudinally within the sleeve. This setup has a provision to control the normal and pull forces measured at high accuracy.

In this work, the test was performed on the pin on disk setup (DUCOM TL-20, Make: India), as recommended by ASTM (G 99), with dry sliding conditions at room temperature. Many researchers have reported that polymers or polymer composites experience less wear when sliding against steel counter face. This may be attributed to the formation of a thin protective film in between the sliding surfaces (Nagaraju *et al.*, 2011). Emery (SiC) paper (1200 grit size as per ASTM) was pasted on an EN-32 disc (of 65 HRC), which acted as a counter-sliding surface of the pin (Figure 11.4a). Initially, the mating surfaces of specimens were polished manually with emery papers. The tests were carried out at the sliding speed of 0.63 m/s, with loads of 5, 10, 15 and 20N, with the duration of 5 and 10 min in the provided conditions (32 $\pm$ 3°C). Various responses such as friction coefficient, wear ($\mu$m), temperature and velocity were monitored on a real-time basis (Figure 11.4b). The specimen weight loss was also measured. This process was repeated before and after of each wear test. A temperature measuring probe was used to determine the specimen temperature during the test. Micrographs from SEM were taken after each wear test to understand the wear mechanism.

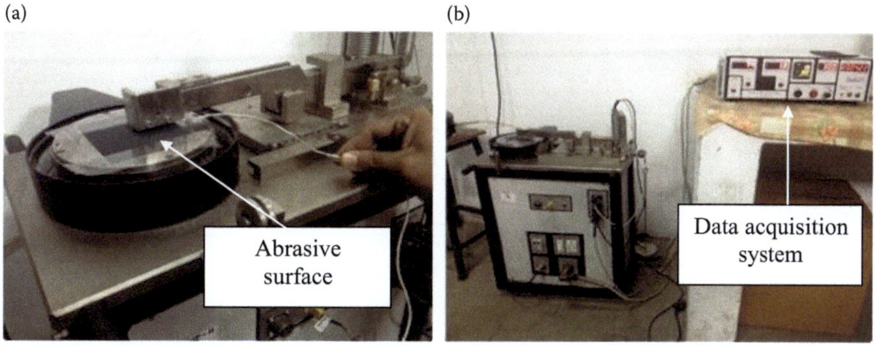


**FIGURE 11.4** Wear testing equipment (a) abrasive surface pasted on steel disc; (b) wear testing with data acquisition system. (Courtesy: Unpublished Lab experimental work, Manufacturing Research Lab, Dept. of Production Engineering, GNDEC, Ludhiana, India.)

## 11.3 RESULTS AND DISCUSSION

### 11.3.1 Volume Loss of Material

Generally, the wear of polymers is due to abrasion/adhesion mechanisms and usually starts with abrasion (Nagaraju *et al.*, 2011). The BCCM specimen showed very small loss of material *vis-à-vis* the ABS specimen by varying the applied load for a 5 min duration, as shown in Figures 11.5 and 11.7. The material loss increases on increase in the duration of test to 10 min, but the trend remains the same (Figures 11.6 and 11.8). Reinforcement of filler materials in polymeric materials increases its wear resistance (Harrass *et al.*, 2010; Kanchanomai *et al.*, 2011; Karsli *et al.*, 2013; Akinci *et al.*, 2014; Zhao *et al.*, 2015). Decrease in wear is due to higher load-bearing capacity of the reinforced material in the polymer matrix (Aigbodion *et al.*, 2012). The wear rate was influenced by the applied load, sliding velocity and operating temperature. It was observed that wear loss was less at lower values of applied load, but as the applied load increased, the reinforced material starts breaking.

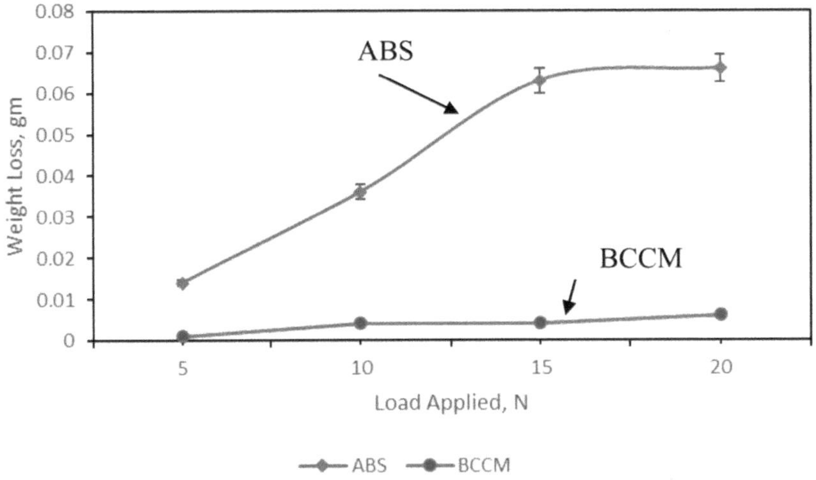


**FIGURE 11.5** Material weight losses with applied load (5 min). (Courtesy: Unpublished Lab experimental work, Manufacturing Research Lab, Dept. of Production Engineering, GNDEC, Ludhiana, India.)

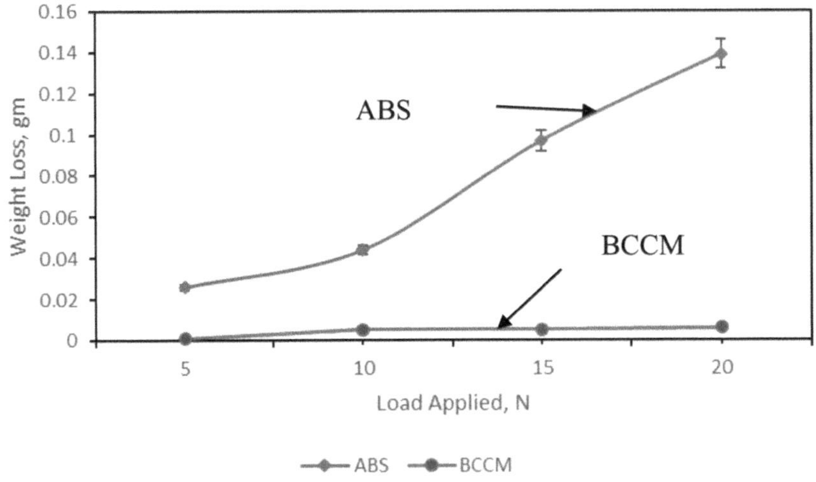


**FIGURE 11.6** Material weight losses with applied load (10 min). (Courtesy: Unpublished Lab experimental work, Manufacturing Research Lab, Dept. of Production Engineering, GNDEC, Ludhiana, India.)

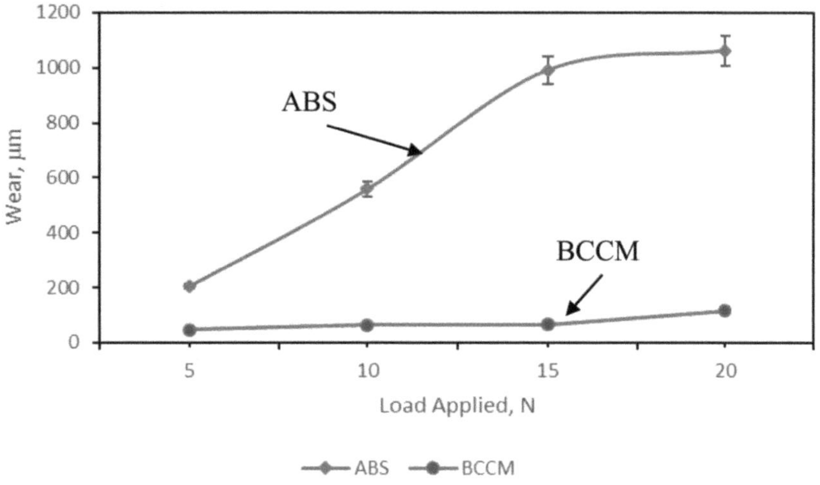


**FIGURE 11.7** Wear with applied load (5 min). (Courtesy: Unpublished Lab experimental work, Manufacturing Research Lab, Dept. of Production Engineering, GNDEC, Ludhiana, India.)

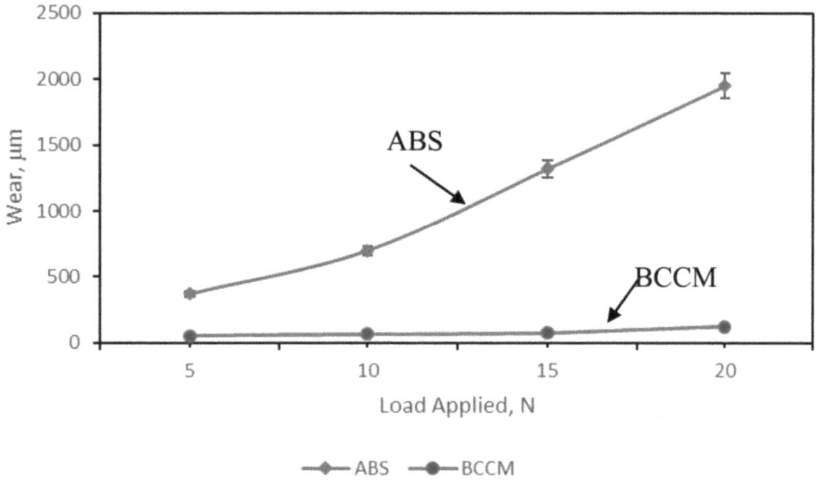


**FIGURE 11.8** Wear with applied load (10 min). (Courtesy: Unpublished Lab experimental work, Manufacturing Research Lab, Dept. of Production Engineering, GNDEC, Ludhiana, India.)

This detachment of filler material from binder material causes extensive loss of the material. Many authors reported that oxidation and adhesion take place at lower loads and delamination at higher loads (Aigbodion *et al.*, 2012). Addition of $Al_2O_3$ increases abrasion and sliding resistance, which prevents seizure due to high thermal stability of $Al_2O_3$. Reinforcement of this material prevents the transition from mild to severe wear at the higher range of applied load.

### 11.3.2 Friction Force

Friction force is the resistance experienced by bodies under sliding conditions. Friction between two polymers may be due to deformation and adhesion (Unal and Mimaroglu, 2012). During the beginning of the wear test, polymeric-based material offers high sliding resistance during sliding on an abrasive surface. As the test continued, it decreases and becomes stable due to the formation of transition layers on the sliding surface. It should be noted that BCCM experiences less friction force than ABS material due to the presence of $Al_2O_3$, as shown in Figures 11.9 and 11.10.

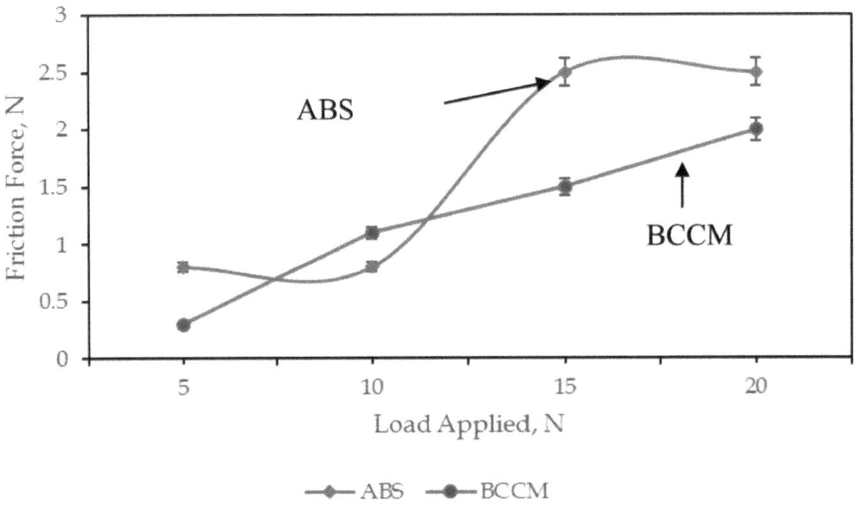


**FIGURE 11.9** Friction force with applied load (5 min). (Courtesy: Unpublished Lab experimental work, Manufacturing Research Lab, Dept. of Production Engineering, GNDEC, Ludhiana, India.)

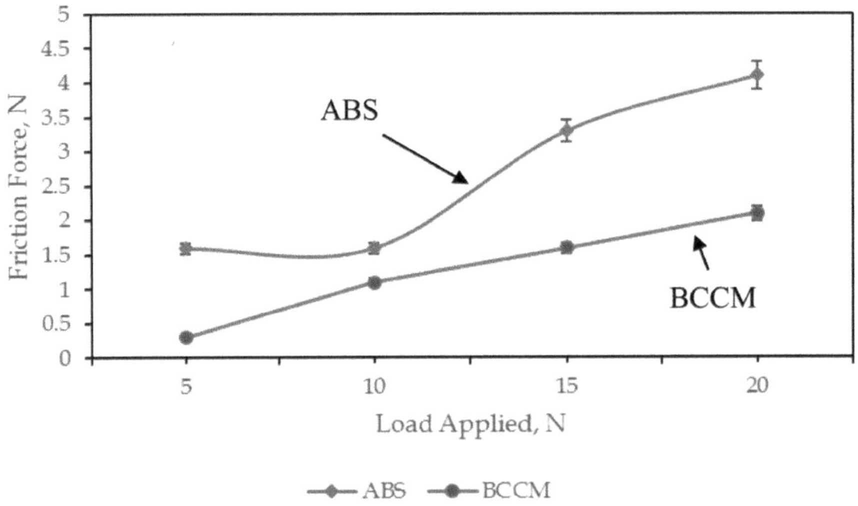


**FIGURE 11.10** Friction force with applied load (10 min). (Courtesy: Unpublished Lab experimental work, Manufacturing Research Lab, Dept. of Production Engineering, GNDEC, Ludhiana, India.)

### 11.3.3 Friction Coefficient

Srinivas *et al.* (2012) and Unal and Mimaroglu (2012) concluded that the coefficient of friction decreases with the filling of graphite in PA6. Analogous results were also observed with PA1212 polymer (Zhang *et al.*, 2011). Inclusion of aramid fiber in PA1010 also decreases the coefficient of friction (Xu *et al.*, 2010). Some researchers have highlighted that, in order to reduce the coefficient of friction between two sliding surfaces, solid lubricants were useful (Unal and Mimaroglu, 2012). In the present case, the coefficient of friction value was higher initially due to high resistance faced by the pin while sliding on the abrasive surface and later, it became stable. Figures 11.11 and 11.12, respectively, show the variation of coefficient of friction with applied load during the trial run of 5/10 min. BCCM has less friction coefficient than ABS material and increases with increase in load.

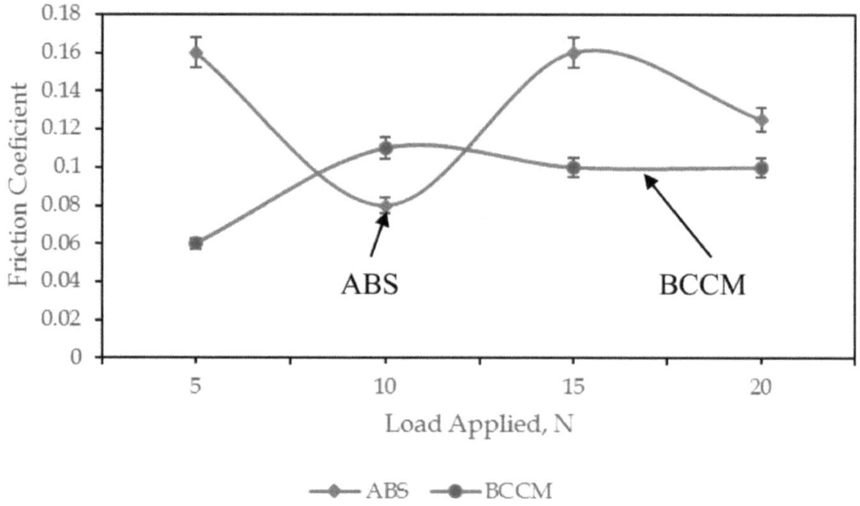


**FIGURE 11.11** Friction coefficient with applied load (5 min). (Courtesy: Unpublished experimental data from our own laboratories, GNDEC, Ludhiana, India.)

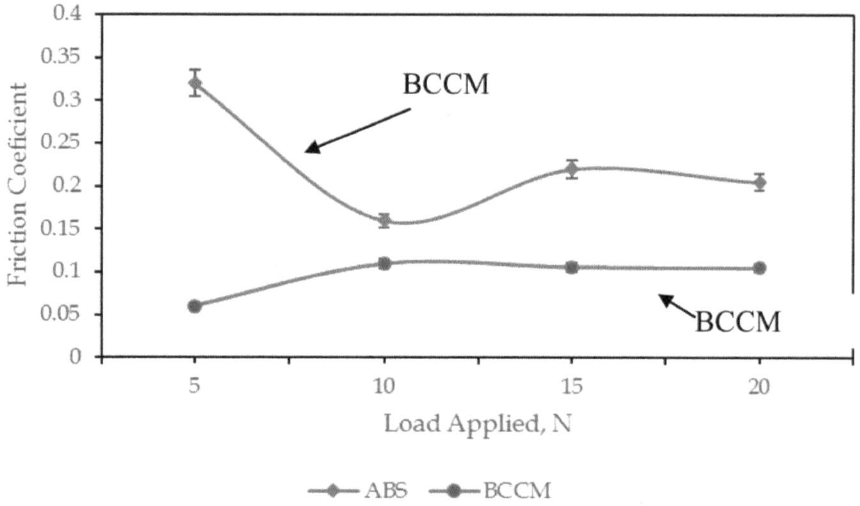


**FIGURE 11.12** Friction coefficient with applied load (10 min). (Courtesy: Unpublished Lab experimental work, Manufacturing Research Lab, Dept. of Production Engineering, GNDEC, Ludhiana, India.)

### 11.3.4 Temperature Change

During the test, it was observed that nano-filler behaves as a thermodynamic sink. These fillers absorb frictional heat. Moreover, heat was continuously dissipated from the contacting surface, which resulted in a small rise in temperature for the ABS material. Zhang *et al.* (2011) highlighted that under dry sliding conditions, friction heat was hard to dissipate, which further aggravates adhesive and fatigue wear. Figures 11.13 and 11.14 show the temperature measurements of specimens during the last test phase. The results highlighted that specimen temperature increases with the increase of applied load and test duration.

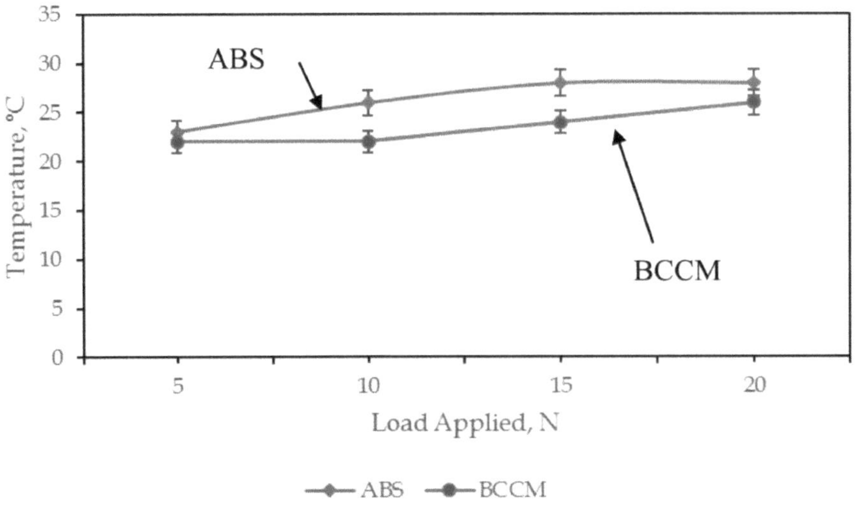


**FIGURE 11.13** Temperature change with applied load (5 min). (Courtesy: Unpublished Lab experimental work, Manufacturing Research Lab, Dept. of Production Engineering, GNDEC, Ludhiana, India.)

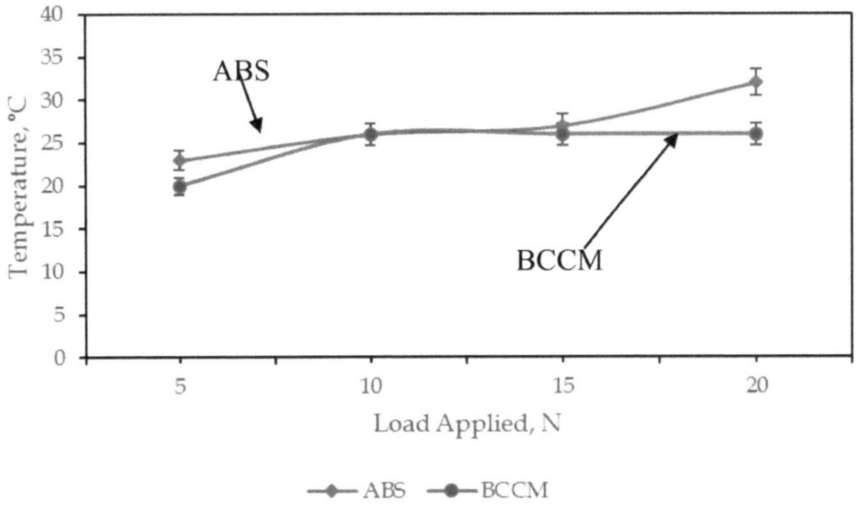


**FIGURE 11.14** Temperature change with applied load (10 min). (Courtesy: Unpublished Lab experimental work, Manufacturing Research Lab, Dept. of Production Engineering, GNDEC, Ludhiana, India.)

### 11.3.5 Surface Morphology

The continuous sliding of the specimens on the abrasive surface caused a wear groove. These wear grooves increase with load and test duration for all types of materials investigated. As mentioned above, wear process starts with abrasion and the material loss in terms of wear debris were subsequently removed from the specimen. In the case of ABS material, the volume loss of material was high. As shown in Figures 11.15 and 11.16, it generally increased with the increase of applied load and test duration. The abrasive wear mechanism continued for ABS material until the completion of a test, due to the sliding rough abrasive surface. Increase of load and run time increased frictional thrust at the interface, which further accelerated wear rate. In BCCM, wear generally started with abrasion. Hard and wear resistant particles of $Al_2O_3$ in nylon 6 matrix act as

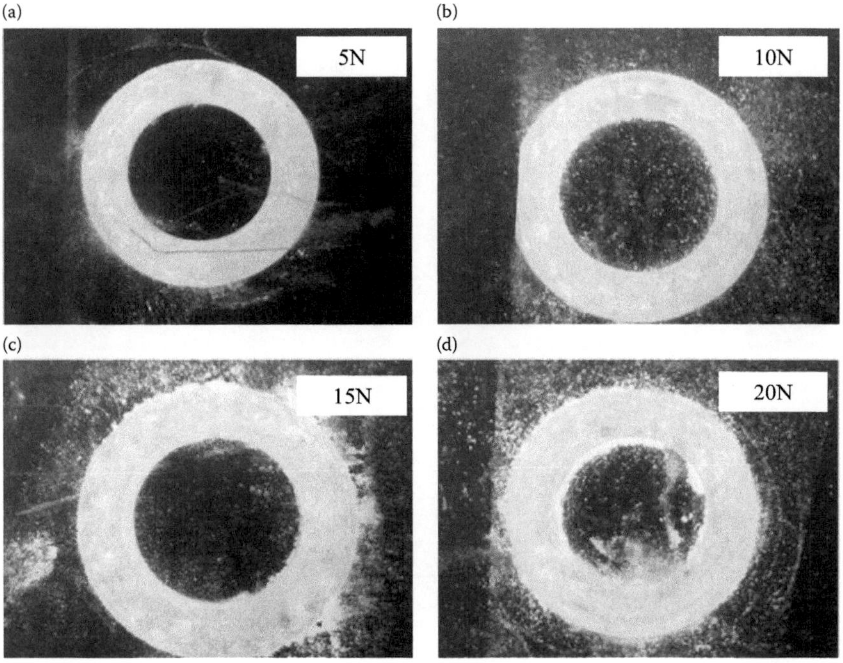


**FIGURE 11.15** Wear track of ABS material (5 min run time) with (a) 5N load; (b) 10N load; (c) 15N load; (d) 20N load. (Courtesy: Unpublished Lab experimental work, Manufacturing Research Lab, Dept. of Production Engineering, GNDEC, Ludhiana, India.)

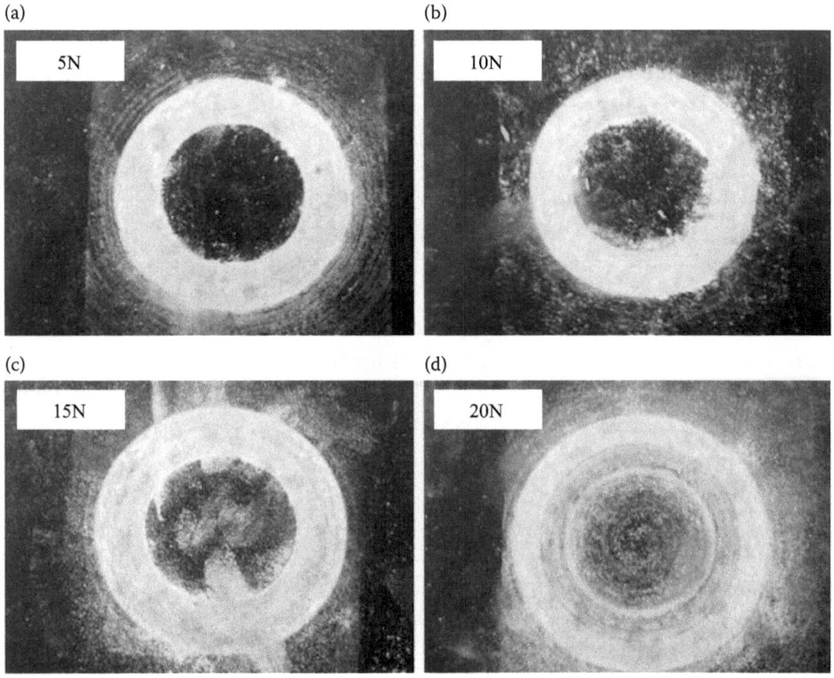


**FIGURE 11.16** Wear track of ABS material (10 min run time) with (a) 5N load; (b) 10N load; (c) 15N load; (d) 20N load. (Courtesy: Unpublished Lab experimental work, Manufacturing Research Lab, Dept. of Production Engineering, GNDEC, Ludhiana, India.)

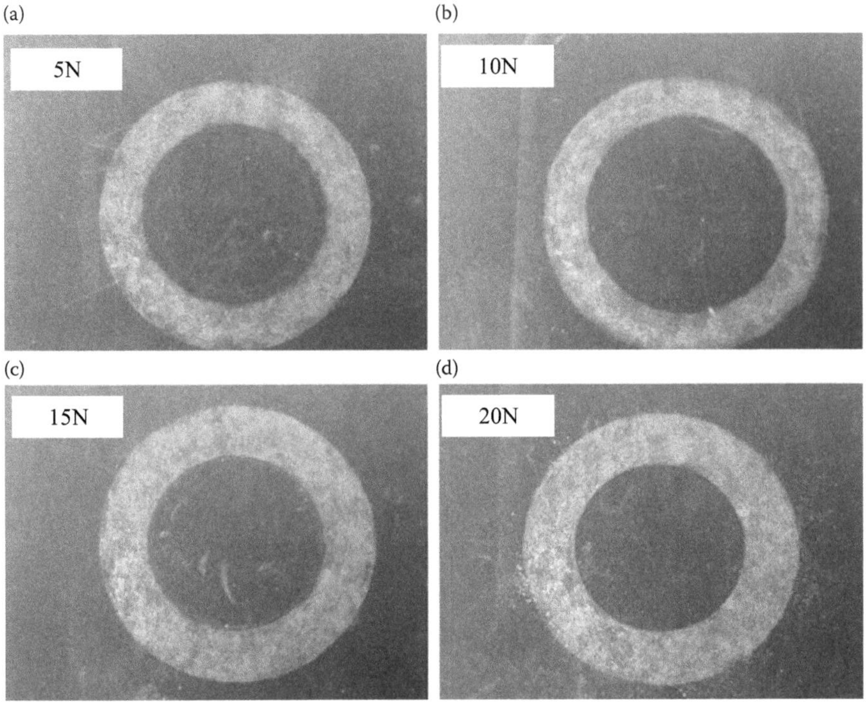


**FIGURE 11.17** Wear track of BCCM (5 min run time) with (a) 5N load; (b) 10N load; (c) 15N load; (d) 20N load. (Courtesy: Unpublished Lab experimental work, Manufacturing Research Lab, Dept. of Production Engineering, GNDEC, Ludhiana, India.)

bearing in three body abrasion mode (Mejia *et al.*, 2010). With the increase of applied load and test duration, heat was generated which promotes adhesive wear and reduces abrasive wear as shown in Figures 11.17 and 11.18.

As observed from Figure 11.19a, the dominant wear mechanisms for BCCM are found to be adhesion and abrasion. The wear grooves and voids due to detachment of filler material from the nylon 6 matrix highlights that wear at the interface of sliding surfaces was due to the transfer of a thin layer of material. As observed from SEM images, the traces of abrasive grooves decreased with the reinforcement of filler material ($Al_2O_3$) in nylon 6 matrix. The other abrasive wear features, such as micro-cutting/grooving and ploughing, were also visible on the worn surface. The surface appearance was relatively clean and shiny without much trace of abrasion. Along with the sign of abrasive wear, the distinguishing feature observed for the specimen was that surface delamination and plastic deformation seemed to be quite less. This could be due to the high load bearing capacity and thermal stability of BCCM *vis-à-vis* ABS material (Figure 11.19b).

## 11.4 CONCLUSIONS

The wear results of ABS and BCCM specimens fabricated on FDM are summarized in Table 11.3. The influence of applied load (5N, 10N, 15N, 20N) and test duration (5 min, 10 min) at constant sliding velocity (0.63 m/s) were examined at room temperature using pin on disk apparatus. The above study may be summarized as follows:

- BCCM has high wear resistance due to the inclusion $Al_2O_3$ in nylon 6 matrix.
- Various wear mechanisms were realized for material loss. In ABS material, the dominant mechanism was abrasion, but in the case of BCCM, wear was due to an adhesive. However,

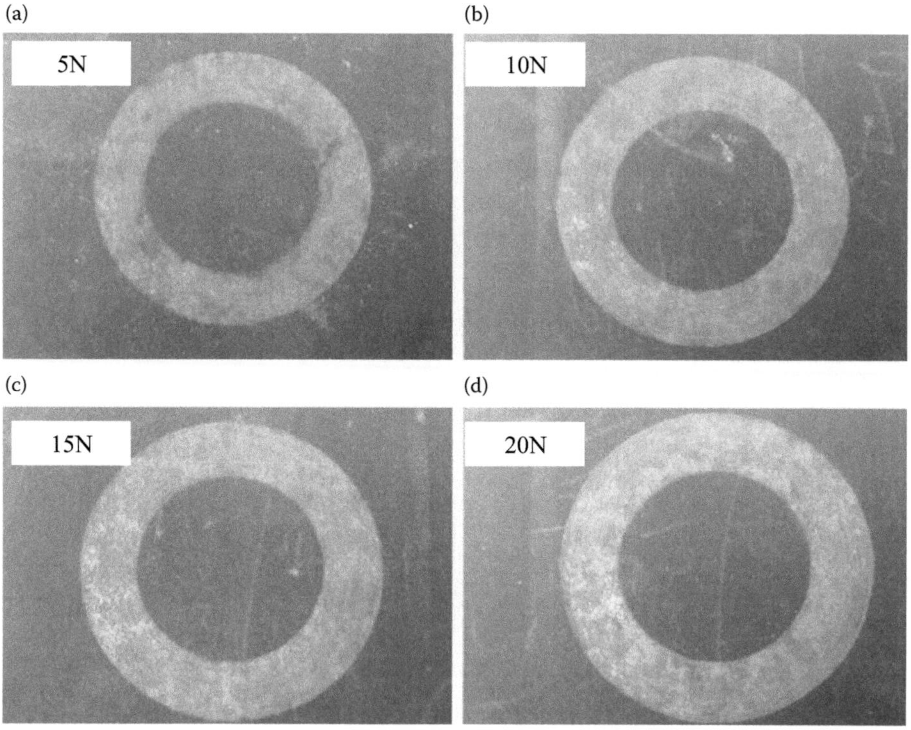


FIGURE 11.18 Wear track of BCCM (10 min run time) with (a) 5N load; (b) 10N load; (c) 15N load; (d) 20N load. (Courtesy: Unpublished Lab experimental work, Manufacturing Research Lab, Dept. of Production Engineering, GNDEC, Ludhiana, India.)

micro-cutting and micro-ploughing were also observed in both cases, when operating at high test conditions.

- BCCM has less friction coefficient (49%), friction force (49%), wear (94%) and material loss (96%) as compared to ABS material, when the applied load is 20N, sliding at a velocity of 0.63 m/s for 10 min duration.
- BCCM shows a little rise in temperature during the test, which shows that the material has high stiffness and stability when operating at high sliding velocity and long duration. As the surface fatigue and adhesive wears are temperature-dependent mechanisms, their influence seems to be relatively less in BCCM than in ABS.

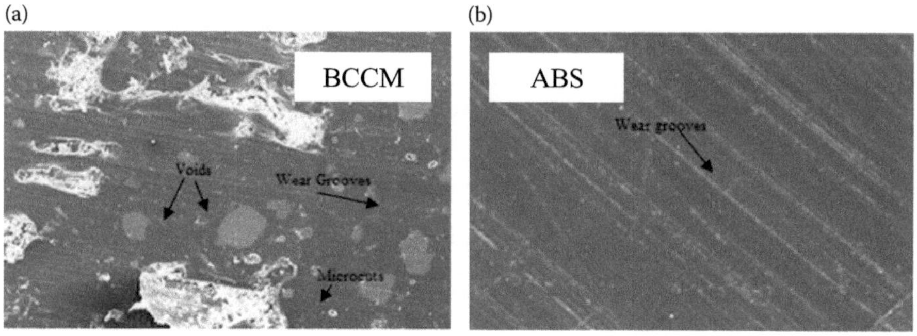


FIGURE 11.19 SEM images (50X) at 20 N load of (a) BCCM; (b) ABS material. (Courtesy: Unpublished Lab experimental work, Manufacturing Research Lab, Dept. of Production Engineering, GNDEC, Ludhiana, India.)

**TABLE 11.3**
**Experimental Log for Wear Test**

| Exp. No | Composition | Load (N) | Run Time (min) | Wear (μm) | FF (N) | FC | Temp.°C | Weight (g) Before | Weight (g) After | Material Removed (g) |
|---|---|---|---|---|---|---|---|---|---|---|
| 1 | ABS | 5 | 5 | 205 | 0.8 | 0.16 | 23 | 1.821 | 1.807 | 0.014 |
| 2 | | 10 | | 559 | 0.8 | 0.08 | 26 | 1.960 | 1.924 | 0.037 |
| 3 | | 15 | | 993 | 2.5 | 0.16 | 28 | 1.924 | 1.861 | 0.063 |
| 4 | | 20 | | 1061 | 2.5 | 0.13 | 28 | 1.861 | 1.795 | 0.066 |
| 5 | | 5 | 10 | 369 | 1.6 | 0.32 | 23 | 2.230 | 2.204 | 0.026 |
| 6 | | 10 | | 699 | 1.6 | 0.16 | 26 | 2.204 | 2.160 | 0.045 |
| 7 | | 15 | | 1320 | 3.3 | 0.22 | 27 | 2.160 | 2.063 | 0.097 |
| 8 | | 20 | | 1950 | 4.1 | 0.21 | 32 | 2.063 | 1.924 | 0.139 |
| 9 | BCCM | 5 | 5 | 47 | 0.3 | 0.06 | 22 | 2.119 | 2.118 | 0.001 |
| 10 | | 10 | | 63 | 1.1 | 0.11 | 22 | 2.118 | 2.114 | 0.004 |
| 11 | | 15 | | 66 | 1.5 | 0.10 | 24 | 2.114 | 2.110 | 0.004 |
| 12 | | 20 | | 116 | 2.0 | 0.10 | 26 | 2.110 | 2.104 | 0.006 |
| 13 | | 5 | 10 | 53 | 0.3 | 0.06 | 20 | 2.100 | 2.099 | 0.001 |
| 14 | | 10 | | 68 | 1.1 | 0.11 | 26 | 2.099 | 2.094 | 0.005 |
| 15 | | 15 | | 75 | 1.6 | 0.11 | 26 | 2.094 | 2.088 | 0.005 |
| 16 | | 20 | | 123 | 2.1 | 0.11 | 26 | 2.088 | 2.082 | 0.006 |

*Source:* Courtesy: Unpublished Lab experimental data, Manufacturing Research Lab, Dept. of Production Engineering, GNDEC, Ludhiana, India.

- The wear track images show that BCCM is highly wear resistant, thermally stable and stiffer than ABS material.
- It should be noted that the wear characteristics of FDM-built parts have high dependence on process input parameters and can be improved with the optimum value of FDM. In this work, the important FDM parameters such as layer thickness (0.254 mm), part built orientation (0°), etc. were kept fixed throughout the study. Future studies may focus on optimization of aforesaid parameters on the sliding wear of test specimens.
- Finally, the efficacy of BCCM suggests its use as rapid nano-tooling for various dental applications.

## ACKNOWLEDGEMENT

The authors are thankful to Manufacturing Research Lab, Production Engineering Department, Guru Nanak Dev Engineering College, Ludhiana, India, DST, Government of India, for providing the requisite facilities to carry out present research work.

## REFERENCES

Aigbodion VS, Hassan SB, Agunsoye JO. Effect of bagasse ash reinforcement on dry sliding wear behaviour of polymer matrix composites. *Mater Des.* 2012; 33: 322–27.

Akinci A, Sen S, Sen U. Friction and wear behavior of zirconium oxide reinforced PMMA composites. *Composites: Part B.* 2014; 56: 42–7.

Anusavice K, Phillips R. Skinners Science of Dental Materials. Saunders: St. Louis. MO. USA; 2003.

Asghari F, Samiei M, Adibkia K, Akbarzadeh A, Davaran S. Biodegradable and biocompatible polymers for tissue engineering application: A review. *Artif Cells Nanomed Biotechnol.* 2017; 45: 185–92.

Bhasin SS, Perwez E, Sachdeva S. Mallick R. Trends in prosthetic biomaterials in implant dentistry. *J Int Clin Dent Res Organ*. 2015; 7: 148–59.

Boparai KS, Singh R, Singh H. Experimental investigations for development of Nylon6-Al-$Al_2O_3$ alternative FDM filament. *Rapid Prototyping J.* 2016a; 22: 217–24.

Boparai KS, Singh R, Singh H. Modeling and optimization of extrusion process parameters for the development of Nylon6-Al-$Al_2O_3$ alternative FDM Filament. *Rapid Prototyping J.* 2016b; 2(4).

Davis MJ, Zanotto ED. Glass-ceramics and realization of the unobtainable: Property combinations that push the envelope. *MRS Bulletin*. 2017; 42: 195–9.

Govindankutty D. Applications of nanotechnology in orthodontics and its future implications: A review. *Int J Appl Dent Sci*. 2015; 1: 166–71.

Guggenberger R, May R, Stefan K. New trends in glass-ionomer chemistry. *Biomaterials*. 1998; 19: 479–83.

Gunasekaran T, Nigusse T, Dhanaraju MD. Silver nano-particles as real topical bullets for wound healing. *J Am Coll Clin Wound Spec*. 2012; 3: 82–96.

Harrass M, Friedrich K, Almajid AA. Tribological behavior of selected engineering polymers under rolling contact. *Tribol Int*. 2010; 43: 635–46.

Hatrick CD, Eakle WS, Bird WF. Dental Materials: Clinical Applications for Dental Assistants and Dental Hygienists. Saunders: St. Louis, MO, USA, 2003.

Hussein MA, Mohammed AS, Al-Aqeeli N. Wear characteristics of metallic biomaterials: A review. *Materials*. 2015; 8: 2749–68.

Kanchanomai C, Noraphaiphipaksa N, Mutoh Y. Wear characteristic of epoxy resin filled with crushed-silica particles. *Composites: Part B*. 2011; 42: 1446–52.

Karsli NG, Yilmaz T, Aytac A, Ozkoc G. Investigation of erosive wear behavior and physical properties of SGF and/or calcite reinforced ABS/PA6 composites. *Composites: Part B*. 2013; 44: 385–93.

Khurshid Z, Zafar M, Qasim S, Shahab S, Naseem M, AbuReqaiba A. Advances in Nanotechnology for Restorative Dentistry. *Materials*. 2015; 8: 717–31.

Koruga D, Hameroff S, Withers J, Loutfy R, Sundareshan M. Fullerene C60: History, Physics, Nanobiology, Nanotechnology. Elsevier: Amsterdam, The Netherlands, 1993.

Mejia OO, Brostow W, Buchman E. Wear resistance and wear mechanisms in polymer + metal composites. *J Nanosci Nanotechnol*. 2010; 10: 1–6.

Mitra SB, Oxman JD, Falsafi A, Ton TT. Fluoride release and recharge behavior of a nano-filled resin-modified glass ionomer compared with that of other fluoride releasing materials. *Am J Dent*. 2011; 24: 372–78.

Moreno KJ, Garcı´a-Miranda JS, Hernández-Navarro C, Ruiz-Guillén F, Aguilera-Camacho LD, Lesso R, Arizmendi-Morquecho A. Preparation and performance evaluation of PMMA/HA nanocomposite as bulk material. *J Compos Mater.* 2015; 49(11): 1345–1353.

Nagaraju B, Ramji K, Prashad VSRK. Studies on tribology properties of ZnO filled polymer nanocomposites. *ARPN JEAS*. 2011; 6(6): 75–82.

Omrani E, Menezes PL, Rohatgi PK. State of the art on tribological behavior of polymer matrix composites reinforced with natural fibers in the green materials world. *ESTIJ*. 2016; 19(2): 717–36.

Padovani GC, Feitosa VP, Sauro S, Tay FR, Durán G, Paula AJ, Durán N. Advances in dental materials through nanotechnology: Facts, perspectives and toxicological aspects. *Trends Biotechnol*. 2015; 33(11): 621–36.

Pashaei S, Siddaramaiah, Avval MM, Syed AA. Thermal degradation kinetics of Nylon6/GF/Crysnano nanoclay nanocomposites by TGA. *Chem Ind & Chem Eng Q*. 2011; 17(2): 141–51.

Silva GA. Introduction to nanotechnology and its applications to medicine. *Surg Neurol*. 2004; 61: 216–20.

Srinivas CL, Sarcar MMM, Suman KNS. Abrasive wear properties of graphite filled PA6 polymer composites. *IJMERR*. 2012; 1(3): 157–62.

Tai Y, Hsu S, Chen R, Su W, Chen M. Liquid crystalline epoxy nanocomposite material for dental application. *J Formos Med Assoc*. 2015; 114(1): 46–51.

Turner TJ, Gorman JM, Sparrow EM. Friction measurement method and modeling of a metallic rod sliding through a flexible polymer tube. *Wear*. 2015; 338–339: 418–21.

Unal H, Mimaroglu A. Friction and wear performance of polyamide 6 and graphite and wax polyamide 6 composites under dry sliding conditions. *Wear*. 2012; 289: 132–37.

Vallittu PK. High-aspect ratio fillers: Fiber-reinforced composites and their anisotropic properties. *Dental Mater*. 2015; 31(1): 1–7.

Wang X, Jiang M, Zhou Z, Gou J, Hui D. 3D printing of polymer matrix composites: A review and prospective. *Composites Part B: Engineering*. 2017; 110: 442–58.

Xu L, Zhu Z, Chen G, Qu C. Effect of loading and sliding velocity on tribological behaviors of aramid fiber reinforced PA1010 composites. *Ind Lub Tribol*. 2010; 62(1): 46–51.

Yee W, Selvaduray G, Hawkins B. Characterization of silver nanoparticle-infused tissue adhesive for ophthalmic use. *J Mech Behav Biomed Mater.* 2015; 55: 67–74.

Yousef S, Visco AM, Galtieri G, Njuguna J. Wear Characterizations of polyoxymethylene (POM) reinforced with carbon nanotubes (POM/CNTs) using the paraffin oil dispersion technique. *JOM.* 2016; 68(1): 288–99.

Zhang S, Wang S, Mao Y. Mechanical and tribological properties of PTFE composites filled with POB. *Adv Mater Res.* 2011; 194–196: 1728–31.

Zhao G, Hussainova I, Antonov M, Wang Q, Wang T, Yung DL. Effect of temperature on sliding and erosive wear of fiber reinforced polyimide hybrids. *Tribol Int.* 2015; 82: 525–33.

# 12 Face Recognition through Nano-Based Neural Networks

*Geetika Singh* and Indu Chhabra*

**CONTENTS**



## 12.1 INTRODUCTION

Biometrics, i.e., identification of individuals through their specific characteristics and traits, has lately been a major area of interest worldwide due to its multiple practical applications in various domains of day-to-day life. It is a key instance of the useful convergence of nanotechnology, biotechnology and information technologies. It employs image-processing-based pattern-recognition technology, wherein

* Corresponding author.

the data or the patterns extracted from an image are identified and classified. Several biometric techniques utilizing the iris, retina, fingerprint, DNA, signature and speech have been extensively deployed and researched over the last decade (Bhardwaj *et al.*, 2014). These techniques have shown several benefits compared with the conventional token and knowledge-based processes like identification cards and passwords and several practical applications as well, though their main limitation is the requirement of an explicit action on behalf of the concerned individual. Alternative biometric techniques, which are efficient, robust and independent of the user's action, are therefore researched and implemented. Face recognition-based technology is emerging as a proficient alternative in this context.

As the name itself suggests, face recognition involves identification of a person through the facial image (Turk and Pentland, 1991). Humans have been known to possess a remarkable identifying ability. Despite large variations in the viewing conditions, ageing or expressions, man is able to recognize all faces uniquely. The number of faces a human being can eventually remember, however, remains a limitation. Automating and inculcating this incredible identifying ability into machines can overcome this memory constraint of human beings, and can provide an excellent biometric identification system (Kanade, 1973).

Being the only identification technique in which the data can be captured at a distance and without the user's knowledge, face recognition technology has drawn considerable interest and attracted researchers from several disciplines (Zhao *et al.*, 2003). It has been found suitable not only for authentication and access control, but has several other practical applications such as facilitating crime investigations, carrying out secure e-commerce transactions, surveillance, finding missing people and human-computer interaction.

Major advances have been made in this field and the current systems have reached a certain level of maturity, but their usage reflects their capability of dealing only with the standard frontal images (Abate *et al.*, 2007; Sharif *et al.*, 2017). Their performance is still limited due to several constraints posed by the real-world scenarios such as pose variations, illumination conditions and changes due to occlusion and expression. Hence, there exists the need for developing invariant algorithms that can tackle these types of variations to some permissible extent and provide good recognition accuracy in real-world scenarios as well.

A face recognition system consists of two important phases of training and testing (Mohanraj *et al.*, 2015). The training phase involves training the recognition system for identification of the enrolled users. The testing phase performs identification of the input test-face image. Each phase comprises three main modules: pre-processing, feature extraction and classification. Input to the system is an unknown test image. The pre-processing module focuses on segmenting the individual face (i.e., input image) and normalizing it for geometrical and photometrical properties such as gray-scale conversion, noise removal and illumination correction. The normalized faces are then fed into the feature extraction module, wherein the important discriminating information (e.g., features) is extracted. The classification module employs a computational algorithm that takes the representation of an object and categorizes it as some known class type. In the training phase, this module learns to match the extracted features with the existing face classes to provide the correct outputs. In the testing phase, the classification module matches the features of the input test image with each face class stored in the database, and classifies the face to the desired class, based on its deviation for different properties from the template classes. The final output from the system is identification of the individual to be recognized (Zhao *et al.*, 2003).

This study develops a face recognition system, comprising a neural network and a facial pattern input unit, wherein the neural network processes the pattern inputs in order to recognize the faces (Figure 12.1). Artificial neural networks (ANNs) are nanotechnology-based computationally intelligent paradigms that supplement the effectiveness of a biometric system (Sivanandam and Deepa, 2013). These play a key role in producing highly accurate results, as well as in the development of future biometric and nanometric applications. They have been used to explore the nonlinear relationship between the input variables and output responses. ANNs always aim to find an adequate configuration of the system automatically through self-learning and, hence, increase the usability of the system.

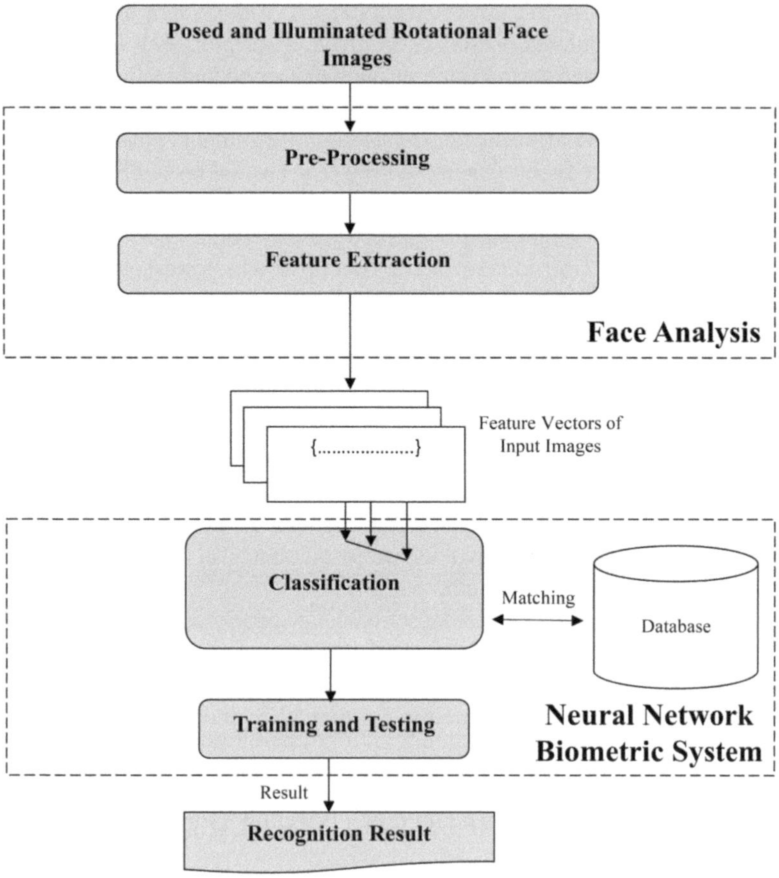


**FIGURE 12.1** Face recognition biometric system.

Feature extraction is also crucial for a face recognition system (Fathi *et al.*, 2016). Hence, a number of approaches have been well-investigated in the literature for facial-feature extraction. These are classified as feature-based approaches, subspace-based methods, spatial frequency techniques and moments-based methods. Feature-based approaches were earlier attempted for facial-feature extraction. These compute the geometrical relationships among the feature points and the relative sizes of the major face components such as width of the eyes or eyebrow thickness to form the feature vector (Kanade, 1973; Manjunath *et al.*, 1992; Brunelli and Poggio, 1993; Lades *et al.*, 1993; Cox *et al.*, 1996). Subspace-based methods are essentially the dimensionality reduction methods that map the face images to a low dimensional space. These include popular methods such as principal component analysis, linear discriminant analysis and independent component analysis (Turk and Pentland, 1991; Belhumeur *et al.*, 1996; Bartlett *et al.*, 2002; Huang and Ma, 2002; Yang *et al.*, 2004; Li and Yuan, 2005; Liu *et al.*, 2005; Martin, 2006; Zhi and Ruan, 2008; Jiang *et al.*, 2017). The spatial-frequency techniques, such as Fourier transforms (FT) and discrete cosine transforms (DCT), extract facial features at some pre-defined frequency (Spies and Ricketts, 2000; Hafed and Levine, 2001; Dabbaghchian *et al.*, 2010). The images are first transformed to the frequency domain, and thereafter, coefficients of low-frequency band are taken as the invariant image features. Moments-based methods are the most widely used face descriptors. These are less susceptible to information redundancy, invariant to the variations in size, pose, scale, illumination, tilt and rotation and are quite insensitive to noise, thus making these quite superior over the existing methods. These consist of the well-established techniques like Zernike moments (ZM), pseudo-Zernike moments (PZMs) and Fourier–Mellin moments (FMM) (Foon *et al.*, 2003;

Haddadnia *et al.*, 2003; Pang *et al.*, 2006; Arnold *et al.*, 2007; Singh *et al.*, 2011a,b, 2012; Rani, 2012). Magnitude of these moments is used as an image descriptor, as it is invariant to rotation and can be made invariant to translation and scale through proper normalization. Amongst the moments-based methods, facial-shape representation with ZM and PZM exhibits efficient recognition ability. The calculated ZM and PZM coefficients are robust to noise and also possess minimum information redundancy and efficient image reconstruction ability. The current research explores the effectiveness of ZMs and PZMs and investigates their efficacy for face recognition.

In the last few years, polar complex exponential transform (PCET), a variation of polar harmonic transforms (PHTs), has been introduced in the literature which possesses invariance properties similar to ZMs, but is computationally much simpler, more robust to noise and highly stable (Yap *et al.*, 2010; Liu *et al.*, 2011). In addition, it also does not suffer from suppression problem, which may cause ZM to focus only on certain parts of an image and neglect the rest. Motivated by these properties of PCET, this chapter also analyses the ability of PCET to represent faces with respect to main parameters of reconstruction ability, rotation, scale invariance and noise robustness.

Efficacy of the ZM, PZM and PCET techniques is evaluated on standard ORL, Yale and FERET face databases for pose, illumination, expression, position, tilt, noise and scale variations. Recognition performance is tested using the traditional nearest neighbour technique and intelligent neural network-based classifiers including back-propagation neural network, support vector machine and adaptive neuro-fuzzy inference system.

The study is organized as follows, with Section 12.2 specifying the method employed for face recognition. Section 12.3 describes the standard databases used for experimentation, while the experimental details and results are presented in Section 12.4. Computational complexity analysis of these techniques is verified and validated in Section 12.5, and conclusions and scope for future work are highlighted in Section 12.6.

## 12.2 IMPLEMENTATION OF THE STUDY UNDERTAKEN

The main objective of the study is to establish the effectiveness of ZMs, PZMs and PCET techniques for face recognition. The scheme for the same is laid out with detailed description, as follows.

### 12.2.1 Pre-Processing

Pre-processing is generally required to make the features to be extracted more robust to changes in noise and illumination. This stage takes the segmented image as its input and processes it according to a normalization strategy followed. In this research, the gamma intensity correction (GIC) method is applied for normalization. This method adjusts the overall brightness of an image and compensates the nonlinear luminance effects. Thus, the effect of varying light conditions is reduced. Figure 12.2 shows the image without normalization as well as the gamma-corrected image.

### 12.2.2 Feature Extraction

Subsequently, the features are drawn from each of the pre-processed images using ZMs, PZMs and PCET separately. The extracted feature sets describe the unique shape characteristics, which is essential for representation of faces. The applied feature extraction methods are discussed.

#### 12.2.2.1 Zernike Moments (ZMs)

ZM features are obtained by transforming an input image on a complex set of orthogonal Zernike polynomials to capture their invariance power. Zernike radial polynomials of order ‘n’ and repetition ‘m’ are computed as:

$$V_{nm}(x,y) = R_{nm}(x,y)e^{im\theta} \tag{12.1}$$

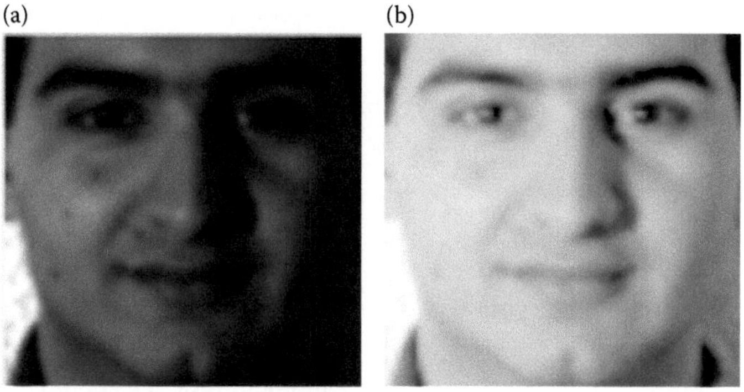


**FIGURE 12.2** (a) Original image and (b) Gamma intensity corrected image. (Reproduced with permission from Computer Vision Laboratory, U.C. San Diego.)

where,

$$i=\sqrt{-1};\ \theta=\tan^{-1}\frac{y}{x} \quad \text{and} \quad R_{nm}(x,y)=\sum_{s=0}^{\frac{n-|m|}{2}}\frac{(-1)^s(n-s)!(x^2+y^2)^{\frac{n-2s}{2}}}{s!\left(\frac{n+|m|}{2}-s\right)!\left(\frac{n-|m|}{2}-s\right)!}$$

such that $n \geq 0$, $|m| \leq n$ and $n - |m| =$ even.

Orthogonality of the above-defined radial kernel allows for maximum separation of data points, thus facilitating differentiation among different classes of images. This makes ZM features suitable for object classification applications like character recognition and face recognition.

After computing the radial kernel, ZMs are calculated for an $N \times N$ digital image through feature calculation of

$$Z_{nm}=\frac{(n+1)}{\pi}\sum_{i=0}^{N-1}\sum_{j=0}^{N-1} f(x_i,y_j)V_{nm}^{*}(x_i,y_j)\Delta x_i \Delta y_j \tag{12.2}$$

where, $V_{nm}^{*}(x,y)$ is the complex conjugate of $V_{nm}(x,y)$;

$$x_i=\frac{2i+1-N}{D};\quad y_j=\frac{2j+1-N}{D}; \tag{12.3}$$

$$x_i^2+y_j^2 \leq 1;\ \Delta x_i=\Delta y_j=\frac{2}{D};\ i=0,1,2\ldots N-1;\ j=0,1,\ldots N-1;$$

$$D=\begin{cases} N, \text{ for the disc inscribed within the image} \\ N\sqrt{2}, \text{ for outer disc containing the complete image}\end{cases}$$

Magnitude coefficients of these moments are hence used as the features, since they are inariant to image rotation, which is a crucial parameter in case the images are aligned. Normalization (Equation 12.3) further makes the Zernike features translation as well as scale as invariant. This whole application of Equations 12.1 to 12.3 makes ZMs invariant to different geometrical transformations.

Images are reconstructed through application of inverse transformation as

$$I = \sum_{n=0}^{n_m} \sum_{m=-n}^{n} Z_{nm} V_{nm}(x_i, y_j) \tag{12.4}$$

where, I is the reconstructed image and $n_m$ is the maximum order of moments.

#### 12.2.2.2 Pseudo-Zernike Moments (PZMs)

These are augmented versions of ZMs having approximately twice the information, which increases their robustness to noise, as they can work at lower moment orders as compared to ZMs. It is well-known that higher moment orders are more prone to noise variations. Technically, they are different from ZMs only in the calculation of their radial kernels. The pseudo-Zernike radial polynomials of order 'n' and repetition 'm' are defined as

$$V_{nm}(x,y) = R_{nm}(x,y)e^{im\theta} \tag{12.5}$$

where

$$i = \sqrt{-1}; \theta = \tan^{-1}\frac{y}{x} \quad \text{and} \quad R_{nm}(x,y) = \sum_{s=0}^{n-|m|} \frac{(-1)^s (2n+1-s)!(x^2+y^2)^{\frac{n-2s}{2}}}{s!(n+|m|+1-s)!(n-|m|-s)!}$$

such that $n \geq 0$ and $|m| \leq n$.

#### 12.2.2.3 Polar Complex Exponential Transform (PCET)

PCET, a category of 2D polar harmonic transforms, is a wave-based transformation, whose kernel definition is harmonic in nature. PCET of order 'n' and repetition 'l' for an $M \times N$ digital image is defined as:

$$M_{nl} = \alpha \sum_{i=0}^{M-1} \sum_{j=0}^{N-1} f(x_i, y_j) H_{nl}^*(x_i, y_j) \Delta x_i \Delta y_j \tag{12.6}$$

such that $|n| + |l| \leq c$, where $|n| = |l| = 0,1,2,...,\infty$ and c is the maximum order used for the computation of PCET coefficients; $H_{nl}(x,y) = e^{i2\pi nr^2} e^{il\theta}$; $i = \sqrt{-1}$; $\theta = \tan^{-1}\frac{y}{x}$; $\alpha = \left\{\frac{1}{\pi}\right\}$, $\forall n$; $H_{nl}^*$ is the complex conjugate of $H_{nl}$, $x_i = \frac{2i-M}{M}$; $y_j = \frac{2j-N}{N}$; $= \frac{2}{M}$; $\Delta y_j = \frac{2}{N}$; $i = 0,1,2...M-1$; $j = 0,1,...N-1$ and $x_i^2 + y_j^2 \leq 1$.

$H_{nl}$ is the orthogonal PCET kernel, which is wave-based and thus accounts for computational simplicity of this transform. In addition, due to its radial nature, PCET features can be computed up to any higher orders without any numerical instability, $\Delta x_i$ and $\Delta y_j$ are the normalization factors and are used to make PCET coefficients translation and scale-invariant.

Images are reconstructed through inverse transformation of

$$I = \sum_{n=-n_m}^{n_m} \sum_{l=-n_m}^{n_m} M_{nl} H_{nl}(x_i, y_j) \tag{12.7}$$

The major advantages of PCET over ZM and PZM include:

- It possesses both the orthogonality and invariance properties of ZM and PZM, but is computationally more durable due to its simple kernel definition.
- PCET coefficients can easily be calculated even at higher orders and thus are free from any numerical instability issues.
- It focuses on all the face regions and thus is capable of capturing even the precise facial details. In contrast, ZM and PZM suffer from suppression problem which may cause them to focus only on certain parts of an image and neglect the rest. Thus, it generates a comprehensive feature set, which greatly facilitates the recognition stage.

### 12.2.3 Classification

For face recognition, classifiers are trained on the extracted feature sets. In the present study, both the traditional as well as intelligent neural network based-classifiers, including Euclidean distance (ED) measure, multi-layer back-propagation neural network (BPNN), adaptive neuro-fuzzy inference system (ANFIS) and support vector machines (SVMs) are investigated. Neural networks are computational systems that permit computers to function in a manner similar to that of the human brain. In general, these are composed of many nonlinear computational elements (equivalent to neurons in the human brain) operating in parallel. The computational elements are connected via variable weights to facilitate communication within the network. The connection weights are automatically adjusted, when the network is made to learn through examples.

#### 12.2.3.1 Euclidean Distance

Euclidean distance (ED) is a well-known method, the oldest and simplest technique of classification. It has also been used in the present work to capture rotation invariance. As moments-based methods are implemented to capture the tilted as well as rotated faces, the classifier employed should also be capable of representing the rotation invariance of the moments-based features. ED is such a classifier which complements with moments-based features and generates superior results.

ED between feature vectors $a = [a_1\ a_2 \ldots a_n]$ and $b = [b_1\ b_2 \ldots b_n]$ of database and query images is calculated as

$$d_{zm}(a,b) = \sqrt{\sum_{i=1}^{n} (a_i - b_i)^2} \tag{12.8}$$

where, n is the size of the drawn feature vector.

#### 12.2.3.2 Support Vector Machines (SVM)

SVMs have been used, as these can easily capture the nonlinearity of faces through their kernel functions. As faces are highly nonlinear data, one important property of the classifier must be the clear separation of nonlinear faces into their respective classes. SVMs inherently possess this capability and have proved to be efficient classifiers especially in the nonlinear context. In addition, as they are based on learning algorithms, these can easily separate the unseen data as well.

Algorithmically, SVMs are based on the principle of finding the equation of a hyperplane that maximally separates all the data points between classes (Lin *et al.*, 2016). SVM kernels are either linear or nonlinear. For the present problem of face recognition, nonlinear kernels, especially radial basis function (RBF) kernel, perform better than linear kernel due to complex distribution of the extracted features, which is nonlinear in nature and therefore, linear kernel cannot represent a face effectively in such cases. Here, libSVM implementation of SVMs is used and the parameters of the kernel are identified through the 5-folds cross-validation technique (Chang and Lin, 2001).

#### 12.2.3.3 Back-Propagation Neural Networks (BPNN)

Neural Networks are the mathematical approximations of the human brain functions, which synthesize algorithms through learning processes. These are nonlinear models, which introduce flexibility in modelling real-world complex problems. These are self-adaptive and self-learning maps, which learn through experience. This nonlinear mapping system trains a neural network on a given set of inputs with known outputs using the supervised learning mechanism, and obtains the good learning and generalization parameters for effective future classification of unrecognized test face images.

#### 12.2.3.4 Adaptive Neuro-Fuzzy Inference System (ANFIS)

ANFIS is a hybrid classifier with the capabilities of neural nets and fuzzy systems both. Thus, it exploits the important properties of self-adaptability and learning capability of the neural networks with the advantage of the fuzzy systems of capturing the prevailing uncertainty and imprecision of actual real face systems. In addition, it also ensures fast convergence to global minima as compared to BPNN and ensures small convergence error to introduce better learning ability for a similar network complexity.

## 12.3 EXPERIMENTAL DATABASES

Performance of the applied and improved techniques is analysed on three well-known and standard face databases: ORL, Yale and FERET, to accommodate the variations in pose, illumination, expression, position and scale of faces in real face biometric systems.

### 12.3.1 AT&T/ORL Database

This database consists of 400 face images for 40 individuals (AT&T/ORL Face Database). There are 10 images in portable gray-map (PGM) format for each person. Images are in gray scale and of size 92 $\times$ 112 pixels. This database contains pose, expression, lighting, occlusion, scale and position variations, with poses ranging from 0° (frontal) to $\pm$20°. Expression variations are presented in a way that either a person is smiling or not smiling and have open/closed eyes. There are also small light variations and occlusion of the eyes with eyeglasses.

### 12.3.2 Yale Database

Yale database has 165 total face images for 15 persons (Yale Face Database). Images per person are 11 and are in gray-scale form. Graphics interchange format (GIF) has been used to store the images, with each image of the size 243 $\times$ 320 pixels. This database presents major variations of light and facial expressions. Therefore, it is a challenging database to check the robustness of face recognition algorithms, especially in case of light changes. To incorporate the light changes, images have been captured by placing the light source on the left, right, front and back of each individual. Images have also been captured at different expressions such as sad, happy and mouth open. There is also occlusion of the eyes with eyeglasses.

### 12.3.3 FERET Database

The FERET database has 14,051 gray-scale images for 1196 individuals (FERET Face Database). Images are of the size 256 $\times$ 384 pixels. Each face has been captured in different variations of facial expression, age, illumination and pose, with poses ranging from 0° (frontal) to $\pm$90°. Usually, experiments are performed on the subsets of this database and it is used specifically to test the

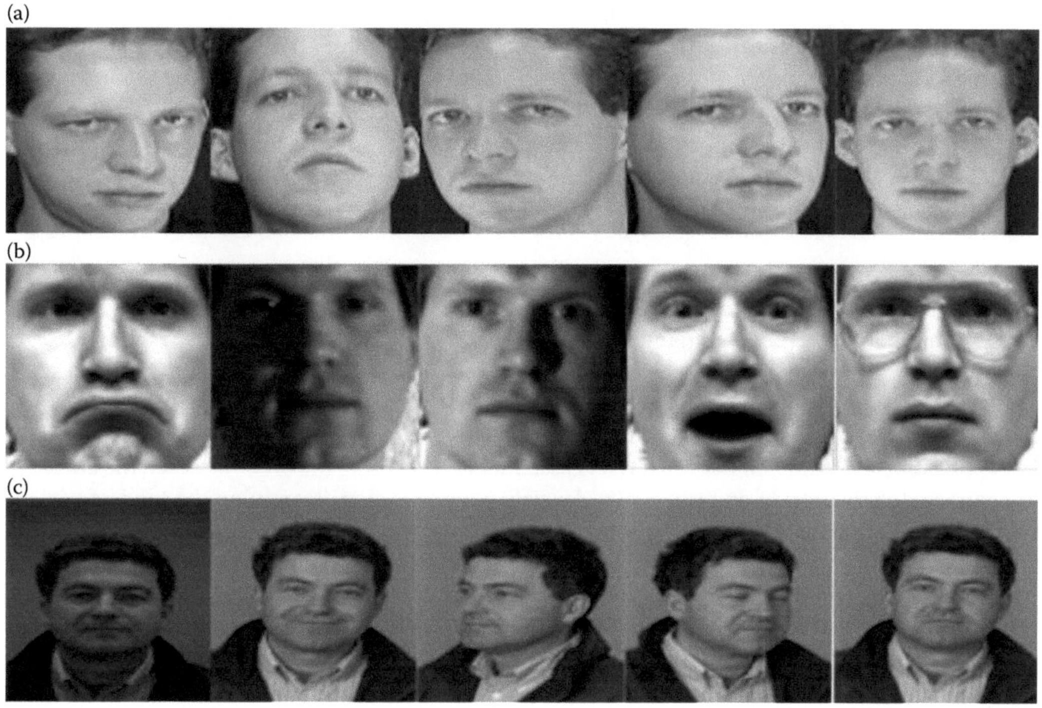

**FIGURE 12.3** Sample face images from (a) ORL, (b) Yale and (c) FERET face databases. ((a) Reproduced with permission from AT&T Laboratories, Cambridge, (b) Reproduced with permission from Computer Vision Laboratory, U.C. San Diego. (c) Reproduced with permission from National Institute of standards and technology, U.S. Department of Commerce.)

pose variations. As the poses are up to ±90°, it throws up a major challenge for checking the pose-invariance properties of face recognition algorithms. It is a benchmark database in the literature, and in the present study, it has been implemented to establish the robustness of the implemented techniques. Figure 12.3 shows sample face images from all three databases.

## 12.4 STANDARDIZED EVALUATION

Robustness of the implemented techniques is evaluated against different types of illumination, facial expression, and scale and pose variations present in face images. The performance is also analysed for images captured in standard environment as well as for noisy images. Techniques have been coded through MATLAB® script under Microsoft Windows environment on a Pentium PC with 2.93 GHz CPU and 4 GB RAM.

### 12.4.1 Experimental Setup for Analysis

As efficiency of the face recognition system is greatly affected by the selection of training and test images, experiments have been performed on differently selected training and test sets to establish the recognition accuracy. The results presented are the means of several possible tests to make the analysis highly representative. In case of the ORL database, out of 10 images of each person, five are used in the training set and the remaining five in the test set. For the FERET database, experiments are performed on a subset of images of 100 randomly selected persons in seven different poses (0°, ±22.5°, ±67.5°, ±90°). For experimentation on each set, three images per person is used for training, and the rest are used for testing. This results in 300 images for training and 400 images for testing. For simplicity, images in both the ORL and FERET databases are normalized to the size of 64 × 64 pixels for analysis. For the Yale database, out of 11 images per person, five are used in the training set and the remaining six in the test set. Images in this database are scaled down to 60 × 80 pixels for experimentation. The recognition rate (in percentage) is measured as

$$\text{Recognition rate} = \frac{N_t - N_r}{N_t} \times 100 \tag{12.9}$$

where,

$N_t$: total number of images in the test set
$N_r$: number of incorrectly recognized images

### 12.4.2 Optimal Order Selection

For application of ZM, PZM and PCET, it is essential to determine the order in which the coefficients are extracted. Selection of an optimum order value depends significantly on the kind of variations present in the image set to be recognized.

#### 12.4.2.1 Selection on Yale Database

Yale database consists of large illumination variations. It is well-established that, in the case of light variations, higher-order moments prove to be more effective, as the lower order coefficients only represent the light changes without including the necessary feature details (Singh *et al.*, 2012). Based on this fact, experimentation has been carried out to select appropriate order for the extraction of ZM, PZM and PCET features. Figure 12.4 shows the investigation results for different values of maximum optimal orders. It is noted that the higher moment order ($n_{max}$) of 12 for PCET performs better; though going beyond this order has lowered the performance. For ZM and PZM, moment order $n_{max} = 14$ and

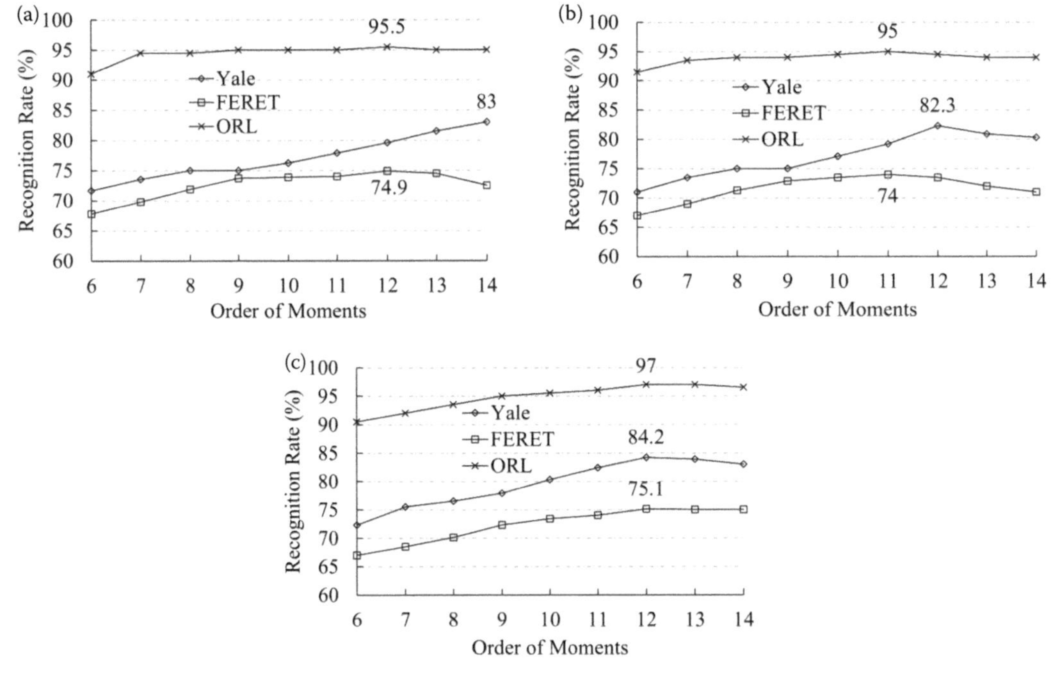


FIGURE 12.4 Performance of (a) ZM, (b) PZM and (c) PCET on ORL, Yale and FERET databases with different groups of orders of moments.

$n_{max} = 12$, respectively, generates high recognition rates. As this database consists of light variations, removal of lower order moments has improved the performance. Hence, moment orders of (3,12), (4,14) and (3,12) are selected for calculating features from PCET, ZM and PZM, respectively.

#### 12.4.2.2 Selection on ORL and FERET Databases

As these databases do not possess any illumination variations, lower order moment coefficients are more discriminative and can perform better over higher order coefficients (Singh *et al.*, 2012). Figure 12.4 shows the experimental results on ORL and FERET databases for different values of maximum orders of PCET, ZM and PZM. It is observed that moment order $n_{max} = 12$ performs better for both ZM and PCET for these databases. With PZM, $n_{max} = 11$ has generated superior results. Removal of lower order moments has degraded the results for these databases as they do not contain any light variations, moment orders of (2,12), (2,12) and (1,11) are selected for PCET, ZM and PZM, respectively, on these databases through experimentation.

### 12.4.3 Recognition Performance

This experiment has compared the performance of the implemented approaches on experimental databases for different classifiers.

#### 12.4.3.1 Baseline Performance Comparison of Implemented Techniques

Table 12.1 presents the mean accuracy obtained by PCET, ZM and PZM for various experimental simulations using ED classifier. The experimental results depict that PCET performs better than ZM and PZM methods due to the fact that PCET can represent even the precise details, in addition to the overall shape features, which is of utmost importance to application of a face recognition system. Therefore, it generates a comprehensive feature set with the complete facial details that facilitate the classification process. Besides, it focuses on all the face regions without neglecting any component which may happen in case of ZMs and PZMs.

#### 12.4.3.2 Performance Comparison of Implemented Techniques with Chosen Classifiers

Table 12.2 shows the results of PCET, ZM and PZM obtained with the undertaken classifiers. On comparing the results of Table 12.2 with those of Table 12.1, it is observed that both ED and BPNN achieve similar recognition accuracy. This is because the moment coefficients are rotation-invariant features and ED is inherently rotation-invariant, which makes it achieve results comparable to that of BPNNs.

SVM classification results have shown significant improvement in performance by 3%–5% over ED and BPNN classifiers both. This is because SVMs always provide a unique solution and thus

**TABLE 12.1**
**Mean Recognition Rate (in %) of Implemented Techniques Using Euclidean Distance (ED) Classifier**

| | Database | | | | | |
|---|---|---|---|---|---|---|
| | ORL | | Yale | | FERET | |
| Method | Nof[a] | ED | Nof | ED | Nof | ED |
| ZM | 47 | 95.0 | 58 | 86.8 | 47 | 73.7 |
| PZM | 77 | 94.5 | 85 | 86.8 | 77 | 71.5 |
| PCET | 81 | 95.5 | 64 | 90.2 | 81 | 75.5 |

[a] Nof: Number of Features.

**TABLE 12.2**
**Mean Recognition Rate (in %) of Implemented Techniques Using ANFIS, BPNN and SVM Classifiers**

| | Database | | | | | | | | |
|---|---|---|---|---|---|---|---|---|---|
| | ORL | | | Yale | | | FERET | | |
| **Method** | **SVM** | **ANFIS** | **BPNN** | **SVM** | **ANFIS** | **BPNN** | **SVM** | **ANFIS** | **BPNN** |
| ZM | 98.0 | 92.5 | 95.5 | 90.0 | 85.0 | 86.8 | ***79.8*** | 71.1 | 73.0 |
| PZM | 97.5 | 92.0 | 95.0 | 89.9 | 85.3 | 87.0 | 76.6 | 70.5 | 71.7 |
| PCET | ***99.0*** | 92.5 | 96.0 | ***94.6*** | 89.0 | 90.2 | 80.1 | 74.0 | 75.3 |

have an edge over BPNNs, which may have multiple solutions associated with local minima. This makes SVMs more robust over different samples. RBF kernel also implicitly contains a nonlinear transformation, which is essential for representing nonlinearity in facial data. Also, SVMs are less prone to over-fitting of training data and more robust to noise. However, improved results are not obtained using ANFIS classifier. This behavior arises since the outputs of this classifier are not whole numbers, and they have to be rounded off to determine class labels which generate unfavorable results.

Figure 12.5 depicts the obtained highest recognition rate through the implemented techniques. It is noted that that PCET provides the highest recognition rates of 80.1%, 94.6% and 99.0% for FERET, Yale and ORL databases, respectively.

### 12.4.4 Image Reconstruction Ability

The image representation ability of orthogonal radial moment-based methods has been tested through their image reconstruction performance. This performance is measured through the mean square reconstruction error (MSE) between the input image I and the reconstructed image $I_r$, which is computed as

$$\mathrm{MSE} = \frac{\sum_{i=0}^{N-1}\sum_{j=0}^{N-1}\{(I(x_i, y_j) - I_r(x_i, y_j)\}^2}{\sum_{i=0}^{N-1}\sum_{j=0}^{N-1}\{(I(x_i, y_j)\}^2} \tag{12.10}$$

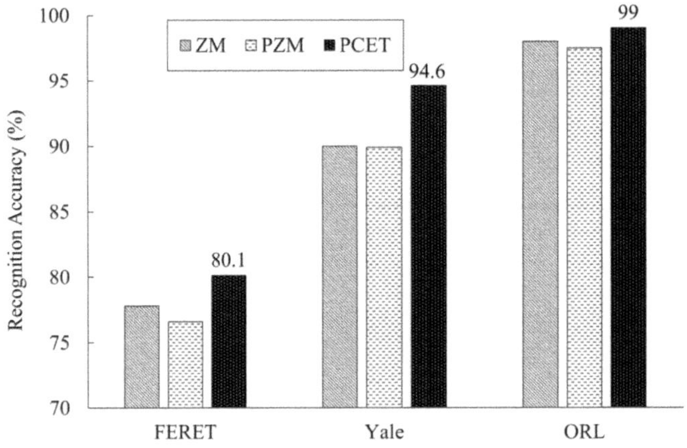


**FIGURE 12.5** Highest recognition rate (in %) achieved by ZM, PZM and PCET on experimental databases.

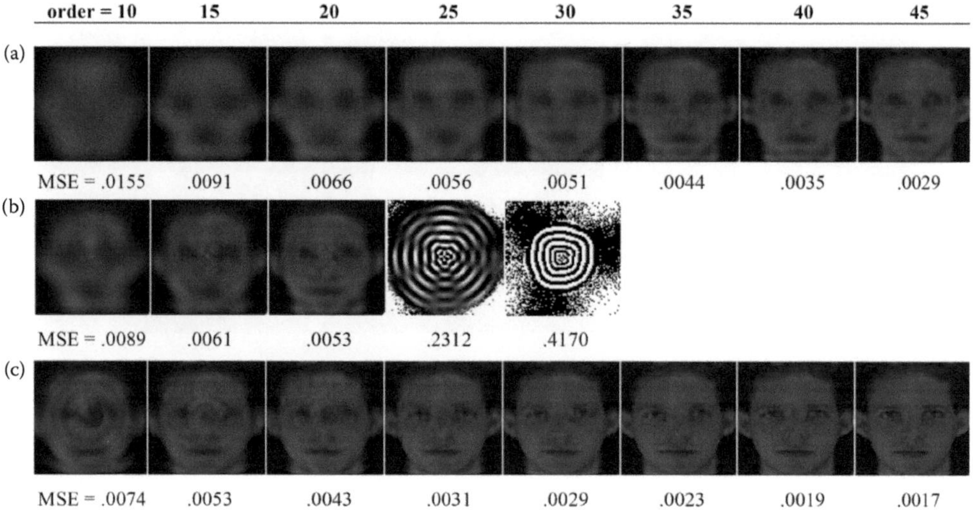


**FIGURE 12.6** Reconstructed images for (a) ZMs, (b) PZMs and (c) PCETs at different orders with their reconstruction errors. (From Singh G, Chhabra, I. Human face recognition through moment descriptors. *IEEE* 2014, 5. With permission.)

In the present experiment, reconstruction capability of the implemented methods is compared at different orders of moments. Figure 12.6 shows the reconstructed face images. It is observed that the output images become quite similar to the original images with increasing orders of moments. Images reconstructed using PCET are closer to the original image with lowest MSE of 0.0017 in contrast to 0.0029 and 0.0053 achieved by ZM and PZM, respectively. It is due to the fact that PCET represents the complete facial characteristics comprehensively without suppression of any facial region like in ZM and PZM. The lowest MSE is achieved by ZMs in the order of 45, though that is achieved by PCET in the order of 30. PZMs achieve the lowest MSE in the order of 20, while it is achieved by PCET in the order of 15. Further, PCET also remains numerically stable in higher orders due to its harmonic and simple kernel definition, while PZMs lose their stability only after order 20, as their kernel definition is polynomial in nature, which gets unstable at higher orders.

### 12.4.5 Noise Robustness

The most important property of moment-based methods is their robustness to noise. In this experiment, the effect of noise on the recognition accuracy of the implemented approaches is analysed. For this, the techniques are tested to two types of noise: Gaussian and salt-and-pepper noise for the experimental databases. To estimate the effect of noise during image acquisition due to illumination and temperature conditions, Gaussian noise of 0 mean, and variance as 0.01 is added in the first set of experiments to the test face images, whereas training is conducted on face images with no added noise. To handle the modalities of data transmission, salt-and-pepper noise of density 0.05 is added to the test images in the second set of experiments. A sample image with added noise is shown in Figure 12.7.

The output results are presented in Table 12.3. As expected, degradation in performance of PZM and PCET methods is less as compared to that of ZM in case of noise variations. This is because both PZM and PCET consider lower orders than that of ZM during feature extraction and it is well-known that higher-order features are more vulnerable to noise variations.

## 12.5 TIME COMPLEXITY ANALYSIS

Table 12.4 shows the comparison of recognition time taken by ZM, PZM and PCET for the experimental databases. For the ORL database, 200 images have been used in training and other

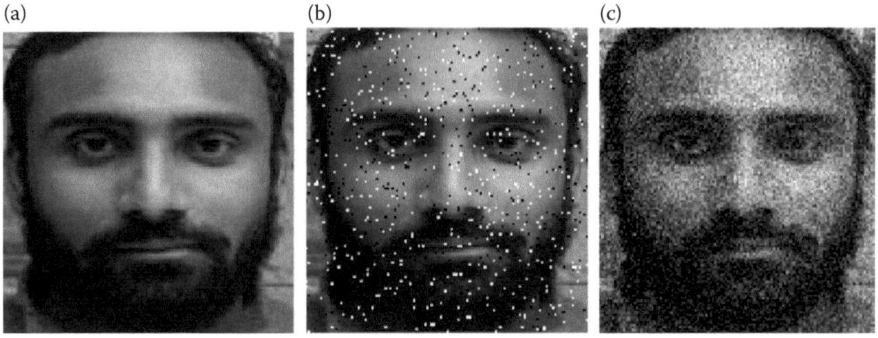


**FIGURE 12.7** (a) Original image, (b) Image with added salt-and-pepper noise and (c) Image with added Gaussian noise.

**TABLE 12.3**
**Highest Recognition Rates of the Proposed and Other Techniques against Noise Variations**

| | Gaussian | | | Salt-and-Pepper | | |
|---|---|---|---|---|---|---|
| **Method** | **ORL** | **Yale** | **FERET** | **ORL** | **Yale** | **FERET** |
| ZM | 94.0 | 85.1 | 71.0 | 92.5 | 84.8 | 70.1 |
| PZM | 94.5 | 86.3 | 70.2 | 94.5 | 85.0 | 70.2 |
| PCET | ***95.0*** | ***89.6*** | ***72.0*** | ***95.0*** | ***88.9*** | ***71.0*** |

**TABLE 12.4**
**Time (in Seconds) Taken by ZM, PZM and PCET to Perform Recognition on Experimental Databases**

| | Time (in sec)[a] | | |
|---|---|---|---|
| **Database** | **ZM** | **PZM** | **PCET** |
| ORL | 59.63 | 71.03 | 18.10 |
| Yale | 20.51 | 35.79 | 8.58 |
| FERET | 66.02 | 79.34 | 28.34 |

[a] Time = Feature vector computation time + recognition time.

200 in testing. For the Yale database, 75 images have been used for training and other 90 images for testing, while 300 images in training and 400 in testing have been used for the FERET database.

As can be observed, PCET exhibits less time, because its radial function is harmonic in nature but not polynomial (as of ZM and PZM), i.e., they are basic waves (Yap *et al.*, 2010). So, the kernel computation time for PCET is more efficient when compared to that of ZM. Thus, PCET requires lower time than the other methods due to its computationally efficient kernel. This indicates that the PCET-based methods are suitable in real-time environments for devices with low computational power.

## 12.6 CONCLUSION

This present study pertains to the application of nanotechnology through neural networks for the development of a secure face recognition system. In a typical face recognition process, feature extraction as well as classification is crucial. The study evaluates the popular moment-based techniques for feature extraction across different scenarios. For classification, the study investigates both the traditional and neural network-based methods including adaptive neuro-fuzzy inference system (ANFIS), back-propagation neural networks (BPNN) and support vector machine (SVM).

Moments-based methods, such as ZM and PZM, have been quite well-analysed in the past for face recognition due to their invariant face representation capabilities. These methods possess important characteristics of good image reconstruction ability and invariance to rotation, translation, scale and noise. Polar complex exponential transform (PCET) is a recently introduced moments-based method that has been proposed in the literature as an alternative to ZM and PZM. The present study proved the effectiveness of PCET for face recognition. Experiments have been performed on ORL, FERET and Yale databases with pose, illumination, expression, noise, scale, position and tilt variations. It is observed that PCET has provided more elaborated face information against ZM and PZM, because of computationally simple kernel definition, thus requiring less time for computation. Also, it does not suffer from the suppression problem which improves its reconstruction ability over ZM and PZM, and also increases its recognition accuracy. It is also more robust to noise, as it requires only lower moment orders for its computation. Owing to these observations and, as investigated in the present study, experimental results clearly depict the efficacy of PCET for face recognition application, as these properties of any feature extraction algorithm are of utmost importance.

Improved results are obtained using SVMs over ED and BPNNs, as SVMs overcome the limitation of being trapped in the local minima and thus always provide with global solutions. High recognition accuracy of 99.0%, 94.6% and 80.1% is achieved with PCET using SVM classifier on ORL, Yale and FERET databases, respectively. The results of the exhaustive experimental studies, in a nutshell, therefore, construe immense promise of proposed nanotechnology-enabled face recognition setup in arriving at successful results in real-time environments despite huge diversity in the prevailing facial variations.

## ACKNOWLEDGMENTS

The authors are grateful to the National Institute of Standards and Technology, AT&T Laboratories and Computer Vision Laboratory, Computer Science and Engineering Department, U.C. San Diego, USA, for providing the FERET, ORL and Yale face databases, respectively. During the course of this study, the first author was funded by INSPIRE-Junior Research Fellowship (IF120810) from the Department of Science and Technology, Govt. of India.

## REFERENCES

Abate AF, Nappi M, Riccio D, Sabatino G. 2D and 3D face recognition: A survey. *Pattern Recognit Lett.* 2007; 28: 1885–906.

Arnold W, Madasu VK, Boles WW, Yarlagadda PK. 2007. A feature based face recognition technique using Zernike moments. In: Mendis, Priyan, Lai, Joseph, Dawson, Ed, Abbas and Hussein, eds. *Proceedings of RNSA Security Technological Conference*, September 28, 2007, Melbourne, Australia, pp. 341–5.

AT&T/ORL Face Database. Available online at: http://www.cl.cam.ac.uk/research/dtg/attarchive/facedatabase.html (Accessed on: 23 July, 2018).

Bartlett MS, Movellan JR, Sejnowski TJ. Face recognition by independent component analysis. *IEEE Trans Neural Netw.* 2002; 13: 1450–64.

Belhumeur PN, Hespanha JP, Kriegman DJ. Eigenfaces vs. fisher faces: Recognition using class specific linear projection. *IEEE Trans Pattern Anal MachIntell.* 1996; 19: 711–20.

Bhardwaj S, Vatsa M, Singh R. Biometric quality: A review of fingerprint, iris, and face. *EURASIP J Image and Video Process.* 2014; 34.

Brunelli R, Poggio T. Face recognition: Features versus templates. *IEEE Transactions Pattern Anal Mach Intell.* 1993; 15: 1042–52.

Chang CC, Lin C-J. 2001. LibSVM: A library for support vector machines. Available online at: https://www.csie.ntu.edu.tw/~cjlin/libsvm/oldfiles/index-1.0.html

Cox J, Ghosn J, Yiaios PN. 1996. Feature based face recognition using mixture-distance. In: *Computer Vision and Pattern Recognition. Proc IEEE Comp Soc Conf Comp Vis Pat Recog*, June 18–20, 1996, San Francisco, CA, pp. 209–16.

Dabbaghchian S, Ghaemmaghami MP, Aghagolzadeh A. Feature extraction using discrete cosine transform and discrimination power analysis with a face recognition technology. *Pattern Recognit.* 2010; 43: 1431–40.

Fathi A, Alirezazadeh P, Abdali-Mohammadi F. A new global-gabor-zernike feature descriptor and its application to face recognition. *J Vis Commun Image Represent.* 2016; 38: 65–72.

FERET Face Database. Available online at: www.nist.gov/itl/iad/ig/colorferet.cfm. (Accessed on: 6 May, 2017)

Foon NH, Pang Y-H, Jin ATB, Ling DNC. 2003. An efficient method for human face recognition using wavelet transform and Zernike moments. *Proc Int Conf Comp Graph Imag Visusal*, July 26–29, 2003, pp. 65–9.

Haddadnia J, Faez K, Ahmadi M. An efficient human face recognition system using pseudo-Zernike moment invariant and radial basis function neural networks. *Int J Pattern Recognit MachIntell.* 2003; 17: 41–62.

Hafed ZM, Levine MD. Face recognition using the discrete cosine transform. *Int J Comput Vis.* 2001; 43: 167–88.

Huang HL, Ma S. 2002. Face recognition using kernel-based fisher discriminant analysis. *Proc Int Conf Autom Face Gesture Recognit*, May 21, 2002, Washington, DC, USA, pp. 197–201.

Jiang TX, Huang TZ, Zhao XL, Ma TH. Patch-based principal component analysis for face recognition. *Comput Intell Neurosci.* 2017. doi. org/10.1155/2017/5317850.

Kanade T. Picture processing system by computer complex recognition. Dissertation 1973, Department of Information Science, Kyoto University.

Lades M, Vorbruggen JC, Buhmann J, Lange J, Malsburg C von der, Wutz RP *et al.* Distortion invariant object recognition in the dynamic link architecture. *IEEE TransComput.* 1993; 42: 300–311.

Li M, Yuan B. 2D-LDA: A statistical linear discriminant analysis for image matrix. *Pattern Recognit Lett.* 2005; 26: 527–32.

Lin W-H, Wang P, Tsai C-F. Face recognition using support vector model classifier for user authentication. *Electron Commer Res Appl.* 2016; 18: 71–82.

Liu M, Jiang X, Kot AC, Yap P-T. 2011. Application of polar harmonic transforms to fingerprint classification. Vol. 1, Emerging Topics in Computer Vision and its Applications (ISBN: 978-981-4340-99-1). In: CH Chen eds., *Computer Vision (4 Volume Series), World Scientific.*

Liu Q, Martiriggiano RT, Leo M, Spaganolo P, Orazio TD, Distante A. 2005. Face recognition by kernel independent component analysis. In: Moonis Ali and Floriana Esposito, eds., *Innovations in Applied Artificial Intelligence. Proc 18th Int Conf Indus Eng App Artific Intel Expert Sys*, June 22–24, 2005, IEA/AIE Bari, Italy, pp. 55–8.

Manjunath BS, Chellappa R, Malsburg C von der. 1992. A feature based approach to face recognition. *Proc IEEE Conf CompVis Pat Recog*, June 15–18, 1992, Champaign, IL, pp. 373–8.

Martin S. 2006. An approximate version of kernel PCA. *Proc 5th Int Conf Mach Learn Appl*, Orlando, FL, pp. 239–44.

Mohanraj V, Vaidehi V, Kumar R, Nakkeeran R. 2015. Face recognition in video using deformable parts model with scale invariant feature transform (DPSIFT). Vol. 396, Advanced Computing and Systems for Security (ISBN:978-81-322-2651-2). In: Rituparna Chaki, Agostino Cortesi, Khalid Saeed and Nabendu Chaki, eds., *Advances in Intelligent Systems and Computing (Series* ISSN: 2194-5357), Springer, India, pp. 69–80.

Pang Y-H, Teoh ABJ, Ngo DCL. A discriminant pseudo-Zernike moments in face recognition. *J Res Pract Inf Technol.* 2006; 38: 97–211.

Rani S. Face recognition using hybrid approach. *Int J Image Grap* 2012; 3: 140–44.

Sharif M, Naz F, Yasmin M, Shahid MA, Rehman A. Face recognition: A survey. *J Eng Sci Technol Rev.* 2017; 10: 166–77.

Singh G, Chhabra I. Human face recognition through moment descriptors. *IEEE.* 2014: 5.

Singh C, Mittal N, Walia E. Face recognition using Zernike and complex Zernike moment features. *Pattern Recognit Image Analy.* 2011a; 21: 71–81.

Singh C, Walia E, Mittal N. Rotation invariant complex Zernike moments features and their applications to human face and character recognition. *IET Comput Vis.* 2011b; 5: 255–66.

Singh C, Walia E, Mittal N. Robust two-stage face recognition approach using global and local features. *Vis Comput.* 2012; 28: 1085–98.

Sivanandam SN, Deepa SN. In: *Principles of Soft Computing*. John Wiley and Sons; 2013.

Spies H, Ricketts I. 2000. Face recognition in Fourier space. *Proc Vis Inter 2000*, pp. 38–44.

Turk M, Pentland A. Eigenfaces for recognition. *J Cognitive Neurosci* 1991; 3: 71–86.

Yale Face Database. Available online at: vision.ucsd.edu/content/yale-face-database. (Accessed on: 6 May, 2017)

Yang J, Zhang D, Frangi AF, Yang JY. Two-dimensional PCA: A new approach to appearance-based face representation and recognition. *IEEE Trans Pattern AnalMachIntell*. 2004; 26: 131–7.

Yap P-T, Jiang X, Kot AC. Two-dimensional polar harmonic transforms for invariant image representation. *IEEE Trans. Pattern Anal MachIntell*. 2010; 32: 1259–70.

Zhao W, Chellappa R, Phillips PJ, Rosenfeld A. Face recognition: A literature survey. *ACM Comput Surv*. 2003; 35: 399–458.

Zhi R, Ruan Q. Two-dimensional direct and weighted linear discriminant analysis for face recognition. *Neurocomputing* 2008; 71: 3607–11.

# 13 Brain Delivery Using Oral Nanoparticles: Promises, Challenges and Future Prospects

*Pramod Kumar, Ruchi Malik, and Kaisar Raza**

## CONTENTS



## 13.1 INTRODUCTION

Nanotechnology is being considered an important tool for drug-delivery scientists and pharmaceutical industries, for the last few decades. A variety of nanotechnology-based products have already been marketed, as shown in Table 13.1 (Pillai, 2014).

* Corresponding author.

**TABLE 13.1**
**A Few Examples of Nanotechnology/Drug Delivery System-Based Products in the Market**

| S. No. | Brand Name (Marketing approval Year) | Active Ingredient | Biomedical Applications | Reference |
|---|---|---|---|---|
| 1. | Rayaldee (2016) | Vitamin D analog (Lipid-based drug delivery system) | Vitamin D deficiency, osteoporosis | Savla *et al.* (2017) |
| 2. | Lipusu (2014) | Paclitaxel (Liposome) | Breast and ovarian cancer | Kumar *et al.* (2016) |
| 3. | Abraxane (2013) | Paclitaxel (Albumin-bound nanoparticles) | Cancers like pancreatic and metastatic cancer | Pillai (2014) |
| 4. | Myocet (2000) | Doxorubicin (Liposome) | Breast cancer | |
| 5. | DepoCyt (1999) | Cytarabine (Liposome) | *Lymphomatous meningitis* | |
| 6. | DaunoXome (1996) | Doxorubicin (Liposome) | HIV-related Kaposi sarcoma | |
| 7. | Doxil (1995) | Doxorubicin (Liposome) | Variety of cancers like metastatic ovarian cancer and AIDS-related Kaposi's sarcoma | |

Numerous nanotechnology-derived drug-delivery vehicles offer promises like better drug delivery, enhanced efficacy, improved pharmacokinetic profile and patient compliance (Farokhzad and Langer, 2009). Analogous to various strata of research, brain delivery is also being explored for the potential benefits. This aspect of drug delivery is regarded as tedious, primarily due to the impervious nature of the so called blood–brain barrier (BBB) to most of the xenobiotics. Other subsidiary concerns, such as inadequate amount of drug at the brain site and the systemic side effects, more frequently associated with conventional brain drug therapy also pave a path for further research in this domain (Kreuter, 2001). Box 13.1 highlights the general concerns for delivery of any drug to the brain (Abbott *et al.*, 2010), whereas Box 13.2 enlists the ideal characteristics of nanocarriers for efficient brain delivery (Olivier, 2005). Under normal physiological conditions, only small lipophilic molecules (400–600 Da) can have access to the brain or cerebrospinal fluid (CSF), either by passive or active transport or a mixture of both, by means of choroid plexus cells or capillary endothelial cells (Lu *et al.*, 2014). In order to develop drug-delivery systems with the capability to penetrate the BBB to exhibit the desired therapeutic effect in the central nervous system (CNS), a proper understanding of various mechanisms involved in the uptake into and efflux from the brain is must. This awareness facilitates better understanding of the complex relationship of transport at the BBB with drug structure and physicochemical properties of the carriers (Misra *et al.*, 2003).

**BOX 13.1 GENERAL CONCERNS IN BRAIN DRUG DELIVERY**

**Issues associated with brain delivery *via* oral route (Abbott *et al.*, 2010)**

Poor absorption
Rapid clearance
Higher molecular weight of molecules
Degradation by various enzymes like esterases
Poor availability to brain
Structure of BBB
Poor BBB permeability

**BOX 13.2 IDEAL CHARACTERISTICS OF NANOCARRIERS FOR BRAIN DELIVERY**

| Ideal properties for brain delivery (Olivier, 2005) |
|---|
| Particle size less than 100 nm |
| Nontoxic, biodegradable and biocompatible |
| Stability of formulation in blood |
| Cost effective and scalable manufacturing process |
| Prolonged blood circulation time |
| Minimum excipient drug interaction |
| Possible alteration of drug release profile amicable for BBB targeting and brain delivery |

## 13.2 BLOOD–BRAIN BARRIER

The brain is one of the most important organs of the human body, and the BBB regulates it as a protective barrier against the entry of foreign toxic substances to the human brain. (Masserini, 2013). Absorption and clearance of xenobiotics can be regarded as the key function of the BBB, quite vital for its proper functioning (Alavijeh *et al.*, 2005). The BBB is documented to be composed of endothelial cells, astroglia, pericytes and perivascular mast cells (Wahl *et al.*, 1988). In the BBB, the extraordinary tightness of junctions is due to brain capillary endothelial cells. These cells primarily regulate two functions, that is, they pose a check on the movement of substances by active transport, and they regulate the protein movement between apical and basolateral membranes of the cells. Proteins, such as occludin, claudins and junction adhesive molecules, are present in these tight junctions. Junction adhesive molecules are responsible for the transport of neutrophils, T-lymphocytes and dendritic cells, and also for immune surveillance as well as inflammatory responses (Abbott *et al.*, 2010). However, the BBB not only maintains the entry and exit, but also regulates protective mechanisms and plays a vital role in metabolism, along with carrier-mediated transport. Glucose, amino acids and ketone bodies enter the brain via specific transporters, whereas receptor-mediated endocytosis plays a significant role in regulation of large molecules like neurotrophins and cytokines. Hydrophilic compounds with mass less than 150 Da and hydrophobic compounds with mass between 400–600 Da can easily cross the BBB, preferably by passive diffusion (Masserini, 2013). A variety of drugs such as morphine, meperidone and methadone can easily cross the BBB, but a majority of antibiotics and anticancer agents are unable to penetrate the BBB (Bellavance *et al.*, 2008). A pictorial depiction of the BBB is shown as Figure 13.1.

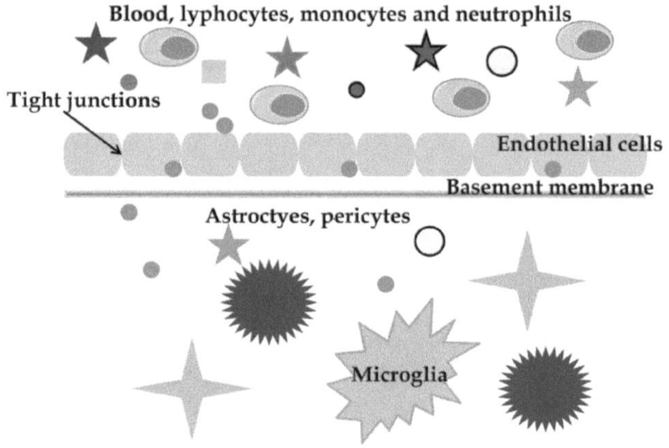


**FIGURE 13.1** Diagrammatic representation of the blood–brain barrier.

## 13.3 DRUG TRANSPORT TO THE BRAIN

Brain delivery of active pharmaceutical agents is generally limited due to lack of specific drug transport systems involving the BBB. Various strategies have been developed to modulate the BBB based on the interactions between nanoparticles and the BBB in the management and diagnosis of neurological disorders. The subsequent sections deal with such noninvasive processes and mechanisms involved thereof.

### 13.3.1 Passive Transport

BBB penetration of small lipid-soluble molecules is generally due to passive diffusion, as illustrated in Figure 13.2a (Liu *et al.*, 2004). Lipophilicity is a physicochemical property, which plays an important role in pharmacokinetics of the active pharmaceutical ingredients (Waterhouse, 2003). Higher polar surface area (more than 80 $\dot{A}^2$), higher molecular mass (more than 450 Da) and higher number of rotational bonds, restrict the entry into the brain (Mikitsh and Chacko, 2014). Various studies speculate that higher plasma protein binding reduces the CNS penetration. For effective penetration of molecules into the CNS, it is not always mandatory that the molecule should follow the Lipinkski's rule of 5 and the afore-discussed physiological parameters. There are numerous CNS effective drugs in the medical practice which do not follow these rules and preset physiological parameters. Positively charged ions can easily cross the BBB due to the interaction with negatively charged glycocalyx (heparin sulphate proteo-glycans), and the phospholipid head group allows entry into the BBB (Abbott *et al.*, 2010). On the other hand, the negatively charged bicarbonate ions possess low passive permeability across the BBB (Bodor and Buchwald, 2003).

### 13.3.2 Cellular Movement

Cellular movement of the vesicles in pathological conditions is due to mononuclear leukocytes, monocytes and macrophages. Brain penetration is possible after brain trauma or ischemia

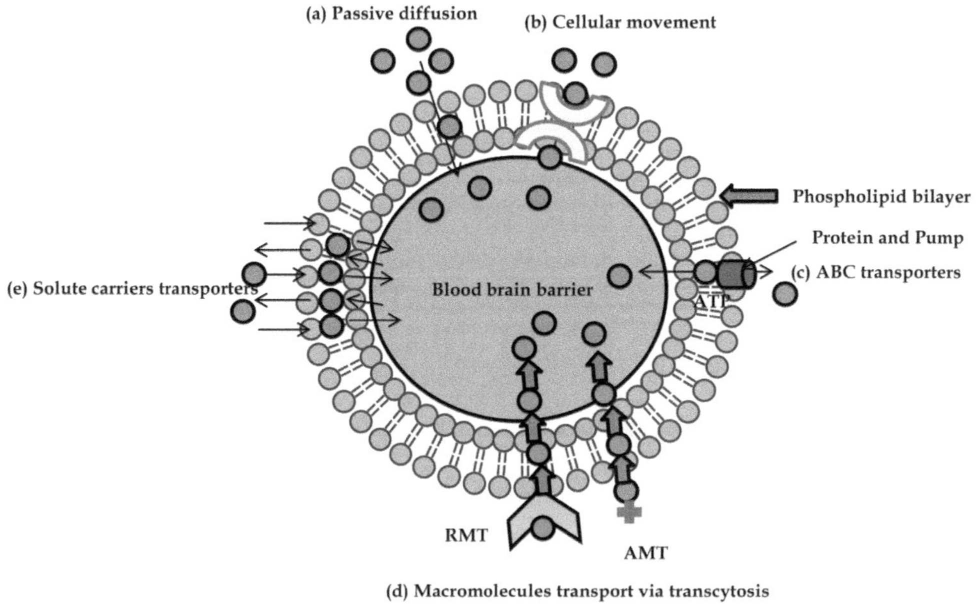


**FIGURE 13.2** Drug transport mechanisms across the blood–brain barrier: (a) Passive diffusion; (b) Cellular movement; (c) ABC transporters; (d) Macromolecules transport via transcytosis; and (e) Solute carriers transporters.

(Davoust *et al.*, 2008). Perivascular macrophages and microglia are being activated during inflammation and other physiological conditions. In normal conditions, mononuclear cells penetrate by the diapedesis process through the cytoplasm of endothelial cells (Engelhardt and Wolburg, 2004). Leucocytes enter into the endothelial cells without disturbing tight junctions by the same mechanism. Whenever a neurological pathological condition occurs, tight junctions between endothelial cells open, allowing entry for the mononuclear cells by both transcellular and paracellular routes, as depicted in Figure 13.2b (Konsman *et al.*, 2007).

### 13.3.3 ABC Transporters

ABC transporters execute their functions by active efflux pump, which consumes ATP and transports the diverse range of molecules inside the brain. The basic role of ABC transporters is to stop the entry of neurotoxic and xenobiotic substances (Cordon-Cardo *et al.*, 1989). Various lipophilic compounds are reported to possess lower CNS entry rates, as these molecules and their metabolites are effluxed from the ATP-binding cassette family (Begley, 2004a,b). Various strategies are being developed for the increment of lipid solubility and thus increase the likelihood of molecules for ABC transporters. ABC transporters verily constitute the superfamily of proteins, having 48 members and seven subfamilies (Dean *et al.*, 2001). P-glycoprotein (P-gp) is reported to efflux the molecules from the BBB inside the brain (Demeule *et al.*, 2002). Various proteins like ABCA2, ABCA1 and ABCG1 have been associated with multi-drug resistance and cholesterol transport, even though the functional significance of these proteins has not been understood (Abbott *et al.*, 2010). This transport mechanism is shown in Figure 13.2c.

### 13.3.4 Macromolecular Transport

Large molecules like proteins and peptides are transported via transcytosis into the BBB, as shown in Figure 13.2d. Endocytotic mechanisms plays a crucial role in the penetration of the BBB by the macromolecules. Large molecules are monitored by the BBB and tight junctions by specific and nonspecific transcytotic mechanisms (Zlokovic *et al.*, 1990). Receptor-mediated transcytosis (RMT) and adsorptive-mediated transcytosis (AMT) are the two mechanisms involved in the transport of the macromolecules. Macromolecule binding to the receptor in the RMT triggers an endocytic event. A receptor and ligand cluster is internalized in the endothelial cells. Separation of ligand and receptor is accomplished during cellular transit or during the exocytotic event. Positively charged macromolecules interact with surface binding site and induce AMT (Sauer *et al.*, 2005).

### 13.3.5 Solute Carriers to the Brain

Tight junctions in the BBB work as a fence and segregate transport proteins and lipid rafts. Most of the polar molecules cannot diffuse across cell membranes and various solute carrier transporters (Zhang *et al.*, 2002). On the contrary, some nutrients such as glucose, amino acids and small peptides are transported via specific solute carrier transporters, as depicted in Figure 13.2e. Orientation of the polar nutrients into or across the endothelial cells and direction of the transport is from blood to brain or brain to blood (Abbott *et al.*, 2010).

## 13.4 STRATEGIES FOR BETTER BRAIN DELIVERY VIA ORAL ROUTE

Various strategies such as chemical modifications and receptor-mediated delivery have been used for brain delivery of nanocarriers for the last few decades. This section deals with various noninvasive techniques, especially oral methods, to target the brain.

### 13.4.1 Chemical Modification of Existing Molecules

Various chemical modifications in the existing active pharmaceutical ingredients have been a practice to improve bioavailability and efficacy and diminish the side effects. As lipophilicity is an important desirable characteristic for the passive diffusion into the BBB, there have been attempts of chemical modification of drugs to enhance the lipophilicity in general and BBB permeability, in particular. On the basis of these facts, a generalization related to octanol-water partition coefficient (i.e., log P) has developed for improved CNS delivery, with the accepted range of log P for this cause being 1.5–2.5 (Lu *et al.*, 2014). Melphalan, a nitrogen mustard alkylating agent, has been chemically linked with phenylalanine (Groothuis *et al.*, 1992). Instances of linking proteins and peptides to antibodies have been reported for BBB receptors in order to enhance the brain permeability (Pardridge *et al.*, 1995). However, the lipidic analogues generally offer low selectivity as well as poor tissue distribution along with numerous concerns. Still, this approach has extensively been used for antiretroviral drugs, but is limited in application against brain tumors (Lu *et al.*, 2014).

### 13.4.2 Pro-Drug Approach

This approach can be regarded as a special class of chemical modification, where an already used drug is changed into pro-forms with several aims, including enhanced permeability, stability and targeting (Testa, 2009). Pro-form of molecules reaches the brain and undergoes enzymatic reactions. Thereafter, the active form of drug binds to the active binding site. Although it changes the fraction of the molecules entering the tissue, it does not change the total body exposure. These approaches have not yet been established for brain tumors (Groothuis, 2000). Pro-drug approaches are certainly known to modulate the drug pharmacokinetics and pharmacodynamics (Testa, 2009). Lalatsa *et al.* (2012b) designed palmitic ester pro-drug of the model drug leucine5-enkephalin. This modified neuropeptide was further coated with chitosan amphiphilic nanoparticles. The developed pro-drug nanoparticles offered superior oral absorption and reduced the plasma degradation, along with the added benefit of improved brain delivery. Dimethyl fumarate, for instance, a widely employed drug for the management of relapsing multiple sclerosis and other neurological disorders, along with psoriasis, can be regarded as a pro-drug, as the functional active moieties at the site of action and plasma are monomethyl fumarate and fumaric acid. The other classical example of this approach is morphine, which has limited access to the brain on its own. However, when latentiated by means of acetyl groups on both the hydroxyl functional groups substantial increment in the BBB penetration has been reported (Bomprezzi, 2015).

### 13.4.3 Receptor-Mediated Delivery

This is a broad area of research, which allows molecules for BBB transport via receptor-mediated delivery for a target. This facilitated transport system depends upon the plasma concentration of the naïve substrate for a particular receptor. A biochemical, i.e., phenylalanine, is available to get access of BBB via large neutral amino acid transport (LNAA); hence the drug, i.e., melphalan, also utilizes the same pathway due to the analogous phenylalanine backbone (Groothuis *et al.*, 1992). However, this transport is competitive in nature and melphalan brain concentrations can be significantly enhanced by lowering the plasma concentrations of the intrinsic amino acid. The known transport systems for protein, peptides, nucleosides, transferrin and insulin do not have high affinity as well as substantial capacity, though these transport systems are being used to increase the brain delivery of pharmaceuticals (Ermisch *et al.*, 1993; Groothuis, 2000).

### 13.4.4 Drug Efflux Inhibition

The entry of the most pharmaceuticals into the brain has been generally believed, for a long time, to take place through passive diffusion. However, numerous studies have challenged this idea by exposing various ATP-dependent efflux pathways, which are responsible for lower brain permeability as well as drug resistance (Schinkel, 1999). P-gp is responsible for efflux of drug in a concentration gradient manner, and various pharmaceuticals have been transported via this system. A variety of anticancer molecules has been established to be transported in the brain via P-gp. This offers an exciting platform for substantial drug delivery to the BBB, as inhibition of the BBB increases both extracellular and intracellular concentrations of drug, even without manipulating the dose (Groothuis, 2000).

## 13.5 NANOPARTICLES AS DRUG DELIVERY TOOLS

With the advancement in the nanotechnology, the domain of medicine, especially drug delivery, has flourished in a significant and synergistic manner. The obvious benefits of high drug loading, drug release control, better drug targeting, enhanced bioavailability, solubility modulations, conducive pharmacokinetic profile and enhanced drug efficacy have attracted researchers from almost every stratum of biological and chemical sciences to refine the domains of drug delivery (Lockman *et al.*, 2004). A plethora of nanotechnology-based drug delivery systems are available in the market for a wide diversity of drugs and diseases (Pillai, 2014). The classical examples of such drug delivery carriers include liposomes, micelles, lipidic nanostructures, niosomes, polymeric nanoconstructs, dendrimers, micro/nanoemulsions, SEDDs/SNEDDs, nanocapsules and inorganic nanoparticles like silica nanoparticles, fullerenes, graphenes and carbon nanotubes (Katare *et al.*, 2010; Kumar *et al.*, 2016). Numerous drug delivery systems have further evolved, and surface tailoring with the desired components fetches the desired drug delivery outcomes (Groothuis, 2000). Figure 13.3 (partially adopted from Raza *et al.*, 2014) shows the pictorial representation of a few vital nanocarriers frequently employed for drug delivery.

## 13.6 NANOPARTICLE CONSIDERATIONS FOR BRAIN DELIVERY

Various micromeritic and other factors are being recognized currently for attaining improved drug delivery to the brain. The subsequent sections highlight the desirable aspects of nanoparticles for effective brain delivery, especially by the oral route.

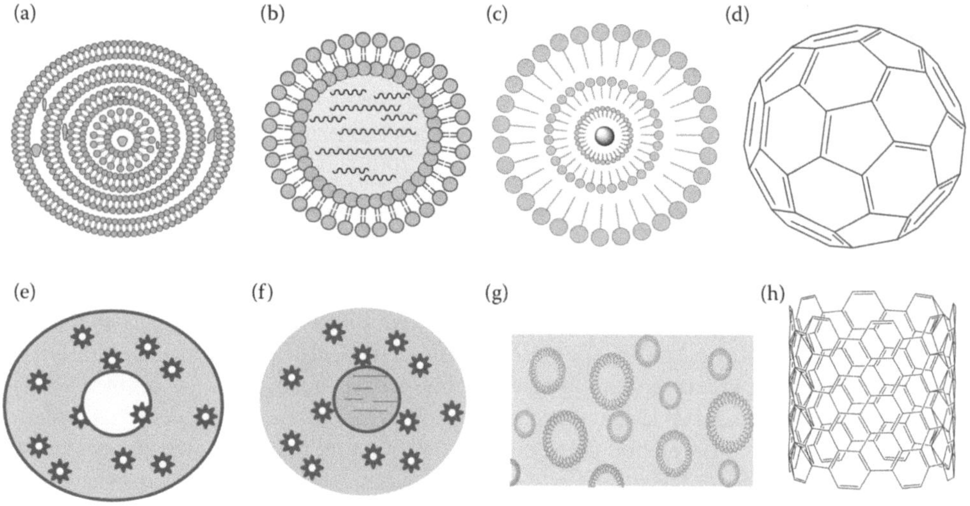


**FIGURE 13.3** Pictorial representation of various nanocarriers: (a) liposomes; (b) ethosomes; (c) niosomes; (d) fullerenes; (e) solid lipid nanoparticles; (f) nanolipodal carriers; (g) micro and nanoemulsions and (h) carbon nanotubes. (From Raza, K *et al.*, *BioMed Res Int.* 2014; 7: Article ID 406731. With permission.)

### 13.6.1 Size

Influence of particle size has been discussed by various scientists. Lower particle size gold nanoparticles (<70 nm) have been reported with better BBB penetration (Gao and Jiang, 2006). In a detailed study, Shilo *et al.* (2015) developed the nanoparticles and performed BBB permeability. They reported that less than 70 nm of the gold nanoparticles can be transported to the brain, and particle size of 20 nm is the optimal size for maximum surface area. It can be inferred through variously reported findings that the lower particle size is not recognized by the reticuloendothelial system (RES) (Prokop and Davidson, 2008). Hence, particles below 100 nm can be easily transported into the brain (Mikitsh and Chacko, 2014).

### 13.6.2 Morphology

It has been observed in various studies that shape and morphology of the nanoparticles affect brain delivery. In one study, it was shown that filamentous nanomicelles were not able to cross the BBB. It was observed that zero-order release in various formulations was shown in spherical shape of nanoparticles. In general, the more symmetric the nanoparticles are, the higher is the propensity for longer circulation in the blood (Christian *et al.*, 2009).

### 13.6.3 Zeta Potential

Zeta potential influences the physical and electrostatic blood–brain barrier entry to brain. In one research study, lower anionic nanoparticles have been shown to result in better brain uptake compared to that from neutral and cationic nanoparticles (Lockman *et al.*, 2004).

### 13.6.4 Surface Modifications

Surface coating, ligand attachment and engineered biopolymers include the various strategies that have now been used for brain delivery. Polysorbate 80, glutathione, albumin, poloxamer, mannose and polyethylene glycol are the most often used coating materials for improved brain delivery. Coating nanoparticles provide protection against phagocytosis (Lahkar and Das, 2013). An honest attempt was made by Dalwadi and Sunderland (2009) for delivery of dalagin (hydrophilic drug) and loperamide (hydrophobic drug). In one study, PEGylated polylactide-co-glycolide (PLGA) nanoparticles were used to transport the cytarabine to the brain (Brigger *et al.*, 2002). Various ligands such as transferrin, lipoprotein, insulin and thiamine can be used for receptor-mediated transport to the brain (Hervé *et al.*, 2008). The hydrophobic coating enables the smaller-sized nanoparticles to bypass the reticuloendothelial system (RES) and mononuclear phagocyte system (MPS) mechanisms and long circulation is assured (Prokop and Davidson, 2008).

## 13.7 ORAL NANOPARTICLES FOR BRAIN DELIVERY

Numerous nanoparticles have been reported for brain delivery of active pharmaceutical ingredients via the oral route, as shown in Table 13.2. Various active pharmaceutical ingredients (APIs) such as dalargin, doxorubicin, estradiol, peptide-leucine-enkephalin, quercetin, leucin-enkephalin, rifampicin, isoniazid, pyrazin aamide and ethambutol have been used for brain delivery using polymers, as indicated in Table 13.2.

Schroeder *et al.* (1998) developed poly (butyl) cyanoacrylate nanoparticles and coated the same with polysorbate 80 to load dalargin. The developed system enhanced bioavailability of the drug by 8–9 fold *vis-à-vis* drug solution. The 288 nm-ized nanoparticles have been reported to cross the BBB and induce analgesia following oral administration.

Mittal *et al.* (2011) developed Tween 80-coated PLGA nanoparticles for brain delivery of estradiol. The authors reported superior brain delivery by surface-coated nanoparticles to that of uncoated

**TABLE 13.2**
**A Bird's Eye View Account on Various Literature Reports on Oral Nanoparticle Systems of Various APIs and Excipients Used for Brain Delivery**

| S. No. | Active Pharmaceutical Ingredient(s) Used | Polymer(s)/Excipients(s) Used | Advantages | Reference |
|---|---|---|---|---|
| 1. | Dimethyl fumarate | Natural fats and vitamins | Enhanced brain bioavailability and substantial reversal of de-myelination of neurons *vis-à-vis* plain drug | Kumar *et al.* (2017a,b, 2018) |
| 2. | Carbamazepine | Poloxamer 188 and PLGA | Manifold enhancement in the anticonvulsive efficacy to that of the free drug with no influence of P-gp inhibition on the efficacy of nanoencapsulated drug | Zybina *et al.* (2018) |
| 3. | Quercetin | Zein, 2-hydroxypropyl-β-cyclodextrin | Enhanced bioavailability | Moreno *et al.* (2017) |
| 4. | 20 (S)-Protopanaxadiol | d-α-tocopheryl polyethylene glycol 1000 succinate | Improved bioavailability | Chen *et al.* (2016) |
| 5. | Quercetin | Poly(n-butylcyanoacrylate) | Enhanced bioavailability | Bagad and Khan (2015) |
| 6. | Peptide-leucine-enkephalin | Quaternary ammonium palmitoyl glycol chitosan | Enhanced permeability | Lalatsa *et al.* (2012a) |
| 7. | Leucin-enkephalin | Chitosan | Enhanced brain uptake levels by 67% and enhanced plasma half-life | Lalatsa *et al.* (2012b) |
| 8. | Estradiol | PLGA (Tween 80 coated PLGA nanoparticle) | Higher brain uptake | Mittal *et al.* (2011) |
| 9. | Rifampicin, isoniazid, pyrazinamide, ethambutol | PLGA | Significant improvement in the pharmacokinetic and relative bioavailability | Pandey and Khuller (2006) |
| 10. | Dalargin | Dextran | Improved brain delivery | Schroeder *et al.* (1998) |

ones. Brain concentrations of the drug by the oral nanoparticles were observed to be comparable to those obtained from the intramuscular route, indicating the potential of coated nanoparticles in brain delivery. The pharmacological outcomes of the coated nanoparticles were also observed to be better.

Lalatsa *et al.* (2012a) developed quaternary ammonium palmitoyl glycol chitosan (GCPQ) nanoparticles for peptide delivery to the brain by the oral route. The formulated nanocomposition not only stabilized the peptide, but also delivered the same in various organs, including the brain. The researchers explored a range of molecular masses of GCPQ polymer and reported that the molecular weight of the polymer should be greater than 6 kDa for better delivery of the peptide to the brain by the oral route. In another study, Lalatsa *et al.* (2012b) employed a pro-drug nanoparticle approach to deliver the palmitic ester of peptide by means of chitosan amphiphilic nanoparticles. The developed nanoparticulate system substantially enhanced brain bioavailability of the pro-drug peptide.

Bagad and Khan (2015) fabricated poly (n-butylcyanoacrylate) quercetin-loaded nanoparticles for oral delivery to brain. The developed nanoparticles significantly enhanced the bioavailability of drug and meanwhile transported the drug to brain tissue in substantially higher amounts. However, the performance of polysorbate nanoparticles was observed to be superior to that of uncoated nanoparticles.

Pandey and Khuller (2006) developed PLGA-based nanoparticles for oral delivery of a combination of anti-tubercular drugs. A single oral dose was able to maintain the plasma drug concentrations up to 5–8 days and brain concentrations up to 9 days in rodents. The developed system was reported to eradicate any detectable levels of *Mycobacterium* in the infected mice during the study period of 10 days.

Kumar *et al.* (2017a,b) developed dimethyl fumarate-loaded solid lipid nanoparticles in conjunction with fat soluble vitamins and reported enhanced brain delivery *vis-à-vis* plain drug. The studies provided evidence of substantial intestinal absorption as well as neuronal uptake of the drug by the developed systems. The designed nanoparticles were able to enhance the oral and brain bioavailability in a daily dose to that of thrice-a-day pure drug.

Ding *et al.* (2017) reported the vinpocetine oral mixed micelles system using PEG-based P123 and reported higher drug loading, improved stability, prolonged *in vitro* release profile, increased bioavailability and enhanced brain targeting.

## 13.8 TOXICITY, BIOCOMPATIBILITY AND COMMERCIAL VIABILITY

Research in pharmaceutical nanotechnology for disease treatment is growing day-by-day, but significant data on the toxicological effects in humans and the environment are still unknown. A lot has been documented for the potential toxicities induced by nanoparticles (Elsaesser and Howard, 2012). Ideally, the excipients used for the preparation of drug delivery carriers should be nontoxic, biocompatible and biodegradable. Nanoparticles composed of biocompatible materials like phospholipids, chitosan, PLGA, biological lipids and poly alkyl cyanoacrylates are generally well-tolerated by the biological system (Blasi *et al.*, 2007). Physiochemical properties of nanomaterials are responsible for the beneficial outcomes as well as for the toxicity of the nanomedicine. Most toxicity has been observed with fine particles *in vivo*, especially by the nanoparticles derived from synthesized polymers. Various bioengineered nanoparticles should also be looked into for toxicity considerations. Due to nanomaterial complexity, various contradictory results in the safety profiles of similar nanoparticles have been pointed out in various studies (Elsaesser and Howard, 2012). Nanoparticles like dendrimers, fullerenes, polymeric nanoparticles, silica and metal-derived nanomaterials are generally tagged as toxic (Sharifi *et al.*, 2012). On the other hand, availability of novel human-use products like NovaC60 cream, derived from so-called "toxic fullerenes" with no reported symptoms of toxicity in humans vouch further research in this domain (Bergeson and Cole, 2012). In general, toxicity of nanomaterials is basically due to positive charge, size and length, surface aggregation and surface coating (Aillon *et al.*, 2009; Elsaesser and Howard, 2012). Meanwhile, various nanotechnology-based formulations, already on the market, as shown in Table 13.1, provide a ray of hope for the availability of more such novel products with an edge over previous systems.

## 13.9 FUTURE PROSPECTS

Nanoparticles possess immense potential in the field of biomedical sciences. For better management of neurological disorders, there is a constant primary need to understand the underlying mechanism and secondarily, exploitation of delivery aspects to cross the BBB is of great significance. A noteworthy improvement in the health outcomes has been reported by use of nanomedicine. Emergence of marketed products like Opaxio® (paclitaxel covalently linked to solid nanoparticles composed of polyglutamate for the management of glioblastoma) and Copaxone® (glatiramer, polypeptide with average MW of 6.4 kDa, composed of four amino acids for the management of multiple sclerosis), after rigorous clinical trials, have established the corner stone for further development in this regard. As far as toxicity of nanoparticles is concerned, there is a need for systematic preclinical study and non-generalization of nanoparticles of diverse origin.

## 13.10 CONCLUSIONS

Neurological and neurodegenerative diseases have been associated with socio-economic problems and quality of life. Albeit treatment therapies for these diseases are emerging, the BBB is the toughest barrier for this. Various approaches have been discussed for the BBB penetration, and influential factors for brain delivery have also been conferred. There is need for development of safer nanoparticles for neurologically active drugs, considering the present understanding of the desired size, charge, surface and other special requirements for effective and improved brain delivery. Strategically designed nanocarriers, with substantial evidence at the preclinical stages, should be carried forward for clinical evaluations via the proper route.

## CONFLICT OF INTEREST

Authors of this manuscript report no conflict of interest for this book chapter.

## ACKNOWLEDGMENT

The fellowship support from University Grant Commission, New Delhi, India, in the form of National Research Fellowship (F./2014-15/NFO-2014-15-OBC-RAJ-8108) to Mr. Pramod Kumar is gratefully acknowledged.

## REFERENCES

Abbott NJ, Patabendige AAK, Dolman DEM, Yusof SR, Begley DJ. Structure and function of the blood-brain barrier. *Neurobiol Dis.* 2010; 37: 13–25.

Aillon KL, Xie Y, El-Gendy N, Berkland CJ, Forrest ML. Effects of nanomaterial physicochemical properties on *in vivo* toxicity. *Adv Drug Deliv Rev.* 2009; 61: 457–466.

Alavijeh MS, Chishty M, Qaiser MZ, Palmer AM. Drug metabolism and pharmacokinetics, the blood-brain barrier, and central nervous system drug discovery. *Neuro Rx.* 2005; 2: 554–571.

Bagad M, Khan ZA. Poly (n-butylcyanoacrylate) nanoparticles for oral delivery of quercetin: Preparation, characterization, and pharmacokinetics and biodistribution studies in wistar rats. *Int J Nanomedicine.* 2015; 10: 3921.

Begley DJ. ABC transporters and the blood-brain barrier. *Curr Pharm Des.* 2004a; 10: 1295–1312.

Begley DJ. Delivery of therapeutic agents to the central nervous system: The problems and the possibilities. *Pharmacol Ther.* 2004b; 104: 29–45.

Bellavance M-A, Blanchette M, Fortin D. Recent advances in blood-brain barrier disruption as a CNS delivery strategy. *AAPS J.* 2008; 10: 166–177.

Bergeson LL, Cole MF. Fullerenes used in skin creams. *Nanotech Bus.* 2012; 9: 114.

Blasi P, Giovagnoli S, Schoubben A, Ricci M, Rossi C. Solid lipid nanoparticles for targeted brain drug delivery. *Adv Drug Deliv Rev.* 2007; 59: 454–477.

Bodor N, Buchwald P. Brain-targeted drug delivery. *Am J Drug Deliv.* 2003; 1: 13–26.

Bomprezzi R. Dimethyl fumarate in the treatment of relapsing–remitting multiple sclerosis: An overview. *Ther Adv Neurol Disord.* 2015; 8: 20–30.

Brigger I, Morizet J, Aubert G, Chacun H, Terrier-Lacombe M-J, Couvreur P *et al.* Poly(ethylene glycol)-coated hexadecylcyanoacrylate nanospheres display a combined effect for brain tumor targeting. *J Pharmacol Exp Ther.* 2002; 303: 928–936.

Chen C, Wang L, Cao F, Miao X, Chen T, Chang Q *et al.* Formulation of 20(S)-protopanaxadiol nanocrystals to improve oral bioavailability and brain delivery. *Int J Pharm.* 2016; 497: 239–247.

Christian DA, Cai S, Garbuzenko OB, Harada T, Zajac AL, Minko T *et al.* Flexible filaments for *in vivo* imaging and delivery: Persistent circulation of filomicelles opens the dosage window for sustained tumor shrinkage. *Mol Pharm.* 2009; 6: 1343–1352.

Cordon-Cardo C, O'Brien JP, Casals D, Rittman-Grauer L, Biedler JL, Melamed MR *et al.* Multidrug-resistance gene (P-glycoprotein) is expressed by endothelial cells at blood-brain barrier sites. *Proc Natl Acad Sci.* 1989; 86: 695–698.

Dalwadi G, Sunderland B. An ion pairing approach to increase the loading of hydrophilic and lipophilic drugs into PEGylated PLGA nanoparticles. *Eur J Pharm Biopharm.* 2009; 71: 231–242.

Davoust N, Vuaillat C, Androdias G, Nataf S. From bone marrow to microglia: Barriers and avenues. *Trends Immunol.* 2008; 29: 227–234.

Dean M, Hamon Y, Chimini G. The human ATP-binding cassette (ABC) transporter superfamily. *J Lipid Res.* 2001; 42: 1007–1017.

Demeule M, Régina A, Jodoin J, Laplante A, Dagenais C, Berthelet F *et al.* Drug transport to the brain: Key roles for the efflux pump P-glycoprotein in the blood–brain barrier. *Vascul Pharmacol.* 2002; 38: 339–348.

Ding J, Sun Y, Li J, Wang H, Mao S. Enhanced blood–brain barrier transport of vinpocetine by oral delivery of mixed micelles in combination with a message guider. *J Drug Target.* 2017; 25:532–540.

Elsaesser A, Howard CV. Toxicology of nanoparticles. *Adv Drug Deliv Rev.* 2012; 64: 129–137.

Engelhardt B, Wolburg H. Mini-review: Transendothelial migration of leukocytes: Through the front door or around the side of the house? *Eur J Immunol.* 2004; 34: 2955–2963.

Ermisch A, Brust P, Kretzschmar R, Rühle HJ. Peptides and blood-brain barrier transport. *Physiol Rev.* 1993; 73: 489–527.

Farokhzad OC, Langer R. Impact of nanotechnology on drug delivery. *ACS Nano.* 2009; 3: 16–20.

Gao K, Jiang X. Influence of particle size on transport of methotrexate across blood brain barrier by polysorbate 80-coated polybutylcyanoacrylate nanoparticles. *Int J Pharm.* 2006; 310: 213–219.

Groothuis DR. The blood-brain and blood-tumor barriers: A review of strategies for increasing drug delivery. *Neuro Oncol.* 2000; 2: 45–59.

Groothuis DR, Lippitz BE, Fekete I, Schlageter KE, Molnar P, Colvin OM *et al.* The effect of an amino acid-lowering diet on the rate of melphalan entry into brain and xenotransplanted glioma. *Cancer Res.* 1992; 52: 5590–5596.

Hervé F, Ghinea N, Scherrmann J-M. CNS delivery *via* adsorptive transcytosis. *AAPS J.* 2008; 10: 455–472.

Katare O, Raza K, Singh B, Dogra S. Novel drug delivery systems in topical treatment of psoriasis: Rigors and vigors. *Indian J Dermatol Venereol Leprol.* 2010; 76: 612.

Konsman JP, Drukarch B, Van Dam A-M. (Peri) vascular production and action of pro-inflammatory cytokines in brain pathology. *Clin Sci.* 2007; 112: 1–25.

Kreuter J. Nanoparticulate systems for brain delivery of drugs. *Adv Drug Deliv Rev.* 2001; 47: 65–81.

Kumar P, Raza K, Kaushik L, Malik R, Arora S, Katare OP. Role of colloidal drug delivery carriers in taxane-mediated chemotherapy: A Review. *Curr Pharm Des.* 2016; 22: 5127–5143.

Kumar P, Sharma G, Kumar R, Malik R, Singh B, Katare OP, Raza K. Enhanced brain delivery of dimethyl fumarate employing tocopherol-acetate-based nanolipidic carriers: Evidence from pharmacokinetic, biodistribution, and cellular uptake studies. *ACS Chem. Neurosci.* 2017a; 08: 860–865.

Kumar P, Sharma G, Kumar R, Malik R, Singh B, Katare OP, Raza K. Vitamin-derived nanolipoidal carriers for brain delivery of dimethyl fumarate: A novel approach with preclinical evidence. *ACS Chem Neurosci.* 2017b; 08: 1390–1396.

Kumar P, Sharma G, Gupta V, Kaur R, Thakur K, Malik R, Kumar A, Kaushal N, Raza K. Preclinical explorative assessment of dimethyl fumarate-based biocompatible nanolipoidal carriers for the management of multiple sclerosis. *ACS Chem Neurosci.* 2018; 09: 1152–1158.

Lahkar S, Das MK. Surface modified polymeric nanoparticles for brain targeted drug delivery. *Curr Trends Biotechnol Pharm.* 2013; 7: 914–931.

Lalatsa A, Garrett NL, Ferrarelli T, Moger J, Schatzlein AG, Uchegbu IF. Delivery of peptides to the blood and brain after oral uptake of quaternary ammonium palmitoyl glycol chitosan nanoparticles. *Mol Pharm.* 2012a; 9: 1764–1774.

Lalatsa A, Lee V, Malkinson JP, Zloh M, Schätzlein AG, Uchegbu IF. A prodrug nanoparticle approach for the oral delivery of a hydrophilic peptide, leucine5-enkephalin, to the brain. *Mol Pharm.* 2012b; 9: 1665–1680.

Liu X, Tu M, Kelly RS, Chen C, Smith BJ. Development of a computational approach to predict blood-brain barrier permeability. *Drug Metab Dispos Biol Fate Chem.* 2004; 32: 132–139.

Lockman PR, Koziara JM, Mumper RJ, Allen DD. Nanoparticle surface charges alter blood–brain barrier integrity and permeability. *J Drug Target.* 2004; 12: 635–641.

Lu CT, Zhao YZ, Wong HL, Cai J, Peng L, Tian XQ. Current approaches to enhance CNS delivery of drugs across the brain barriers. *Int J Nanomedicine.* 2014; 9: 2241–2257.

Masserini M. Nanoparticles for brain drug delivery. *Int Sch Res Not.* 2013; 2013: e238428.

Mikitsh JL, Chacko AM. Pathways for small molecule delivery to the central nervous system across the blood-brain barrier. *Perspect Med Chem.* 2014; 6: 11–24.

Misra A, Ganesh S, Shahiwala A, Shah SP. Drug delivery to the central nervous system: A review. *J Pharm Pharm Sci.* 2003; 6: 252–273.

Mittal G, Carswell H, Brett R, Currie S, Kumar MR. Development and evaluation of polymer nanoparticles for oral delivery of estradiol to rat brain in a model of Alzheimer's pathology. *J Control Rel.* 2011; 150: 220–228.

Moreno LCG, Pureta E, Santiago JES, Magalhaes NSS, Ramirez MJ, Irache JM. Effect of the oral administration of nanoencapsulated quercetin on a mouse model of Alzheimer's disease. *Int J Pharm.* 2017; 517: 50–57.

Olivier J-C. Drug transport to brain with targeted nanoparticles. *NeuroRx.* 2005; 2: 108–119.

Pandey R, Khuller GK. Oral nanoparticle-based antituberculosis drug delivery to the brain in an experimental model. *J Antimicrob Chemother.* 2006; 57: 1146–1152.

Pardridge WM, Kang Y-S, Buciak JL, Yang J. Human insulin receptor monoclonal antibody undergoes high affinity binding to human brain capillaries *in vitro* and rapid transcytosis through the blood brain barrier *in vivo* in the primate. *Pharm Res.* 1995; 12(6): 807–816.

Pillai G. Nanomedicines for cancer therapy: An update of FDA approved and those under various stages of development. *SOJ Pharm Pharm Sci.* 2014; 1: 13.

Prokop A, Davidson JM. Nanovehicular intracellular delivery systems. *J Pharm Sci.* 2008; 97: 3518–3590.

Raza K, Kumar M, Kumar P, Malik R, Sharma G, Kaur M, Katare OP. Topical delivery of aceclofenac: Challenges and promises of novel drug delivery systems. *BioMed Res Int.* 2014; 7: Article ID 406731.

Savla R, Browne J, Plassat V, Wasan KM, Wasan EK. Review and analysis of FDA approved drugs using lipid-based formulations. *Drug Dev Ind Pharm.* 2017; 43: 1743–1758.

Sauer I, Dunay IR, Weisgraber K, Bienert M, Dathe M. An apolipoprotein E-derived peptide mediates uptake of sterically stabilized liposomes into brain capillary endothelial cells. *Biochemistry (Mosc).* 2005; 44: 2021–2029.

Schinkel AH. P-Glycoprotein, a gatekeeper in the blood-brain barrier. *Adv Drug Deliv Rev.* 1999; 36: 179–194.

Schroeder U, Sommerfeld P, Sabel BA. Efficacy of oral dalargin-loaded nanoparticle delivery across the blood–brain barrier. *Peptides.* 1998; 19: 777–780.

Sharifi S, Behzadi S, Laurent S, Forrest ML, Stroeve P, Mahmoudi M. Toxicity of nanomaterials. *Chem Soc Rev.* 2012; 41: 2323–2343.

Shilo M, Sharon A, Baranes K, Motiei M, Lellouche J-PM, Popovtzer R. The effect of nanoparticle size on the probability to cross the blood-brain barrier: An *in-vitro* endothelial cell model. *J Nanobiotechnology.* 2015; 13: 19.

Testa B. Prodrugs: Bridging pharmacodynamic/pharmacokinetic gaps. *Curr Opin Chem Biol.* 2009; 13: 338–344.

Wahl M, Unterberg A, Baethmann A, Schilling L. Mediators of blood-brain barrier dysfunction and formation of vasogenic brain edema. *J Cereb Blood Flow Metab.* 1988; 8: 621–634.

Waterhouse RN. Determination of lipophilicity and its use as a predictor of blood-brain barrier penetration of molecular imaging agents. *Mol Imaging Biol.* 2003; 5: 376–389.

Zhang EY, Knipp GT, Ekins S, Swaan PW. Structural biology and function of solute transporters: Implications for identifying and designing substrates. *Drug Metab Rev.* 2002; 34: 709–750.

Zlokovic BV, Skundric DS, Segal MB, Lipovac MN, Mackic JB, Davson H. A saturable mechanism for transport of immunoglobulin G across the blood-brain barrier of the guinea pig. *Exp Neurol.* 1990; 107: 263–270.

Zybina A, Anshakova A, Malinovskaya J, Melnikov P, Baklaushev V, Chekhonin V, Maksimenko O, Titov S, Balabanyan V, Kreuter J, Gelperina S, Abbasova K. Nanoparticle-based delivery of carbamazepine: A promising approach for the treatment of refractory epilepsy. *Int J Pharm.* 2018; S0378–5173(18)30321–1. doi: 10.1016/j.ijpharm.2018.05.023.

# 14 Microfluidic Technology: A Promising Approach for Nanotoxicity Studies in Physiologically Realistic Environments

*Sanjeev Kumar Mahto* and Suruchi Poddar*

**CONTENTS**



## 14.1 NANOTOXICOLOGY: INTRODUCTION AND CURRENT SCENARIO

In recent years, advancements in design and development of nanomaterials (NMs) have laid the foundation for exploration of commercial (Aitken *et al.*, 2006; Olson and Gurian, 2012), medical (Cavalcanti *et al.*, 2008; Freitas, 2002) and technical opportunities (Roy and Gao, 2009). Owing to the potentiality of NMs to refine material characteristics, they are trending in a range of commercial and healthcare products, such as cosmetics, medicines, paints and fabrics, to name a few (Olson and Gurian, 2012). Notably, several types of NMs are implemented in developing biosensors and clinical diagnostic devices, for example, carbon nanotube (CNT)-based analytical systems (Williams *et al.*, 2002), silicon nanowire (SiNW)-based biosensors (Nair and Alam, 2007) and semiconductor nanomaterials with cadmium selenide as inner core and zinc selenide as outer shell, including functionalized quantum dots (El-Ansary and Al-Daihan, 2009) (QDs). As a result, the growing range of utility of NMs in every corner of our life is leading to an increased demand for mass production of nanomaterials that exhibit unique features and diverse functions. Though the rising needs meet the

* Corresponding author.

supply of nanomaterials, the concern over the acute and chronic effects of NMs on the ecosystem is a growing subject of attention and debate. Recent studies have shed light on the fact that the exposure dosage, frequency and periodicity play a significant role in determining the extent of adverse effects of NMs on human health. (Arora *et al.*, 2012; El-Ansary and Al-Daihan, 2009; Giridharan *et al.*, 2012; Love *et al.*, 2012; Rivera_Gil *et al.*, 2012; Valencia *et al.*, 2012). In addition, it is observed that invasion of NMs at different trophic levels of the food chain leads to their bioaccumulation that, in due course, may result in biomagnification (Judy *et al.*, 2011; Rico *et al.*, 2011).

Unlike their bulk parent material, NMs manifest notably different surface properties due to their nanoscale features (size ranging between 1 and 100 nm). These properties include notable electronic configuration, energy levels and chemical reactivity (Borm *et al.*, 2006; Landsiedel *et al.*, 2010; Liu *et al.*, 2012; Murty *et al.*, 2013). With growing concerns over their widespread use, numerous *in vitro* cytotoxicity studies have indicated that the physicochemical properties of NMs primarily determine their uptake and toxicity in cells (Dowding *et al.*, 2013; Frick *et al.*, 2011; Kroll *et al.*, 2011; Lanone *et al.*, 2009; Liu *et al.*, 2016; Mahto *et al.*, 2010b; Zhu *et al.*, 2013) (e.g., epithelial, endothelial, fibroblasts, macrophages, hepatocytes and neurons). Therefore, establishment of well-characterized data delivering detailed physicochemical features of nanomaterials is critical for nanotoxicity examinations. Furthermore, maintaining similar properties of NMs throughout toxicity investigations remains a crucial aspect to establish a direct and quantitative correlation between NM properties and ensuing effects on cells *in vivo.*

Recently, a number of efforts have explored the potential of microfluidics technology for reestablishing assay procedures for nanotoxicological examinations. (Giridharan *et al.*, 2012; Mahto *et al.*, 2010c; Valencia *et al.*, 2012; Wu *et al.*, 2012; Zheng *et al.*, 2012). With microfluidic devices and other lab-on-chips, microscale systems are able to (i) mimic cell-based *in vitro* models of physiological systems, (ii) synthesize large libraries of nanomaterials in a rather efficient and inexpensive way, (iii) facilitate cytotoxicity testing of NMs in an environment closer to *in vivo* conditions and, finally, (iv) help maintain stable characteristics of NMs during cell exposure (Figure 14.1).

The present chapter attempts to address the contemporary developments of *in vitro* microfluidic platforms for exploration of cytotoxicity of NMs. These efforts have largely arisen in response to obstacles faced with traditional *in vitro* platforms in keeping the physicochemical properties of NMs maintained during exposure. In particular, we emphasize here the potential applications

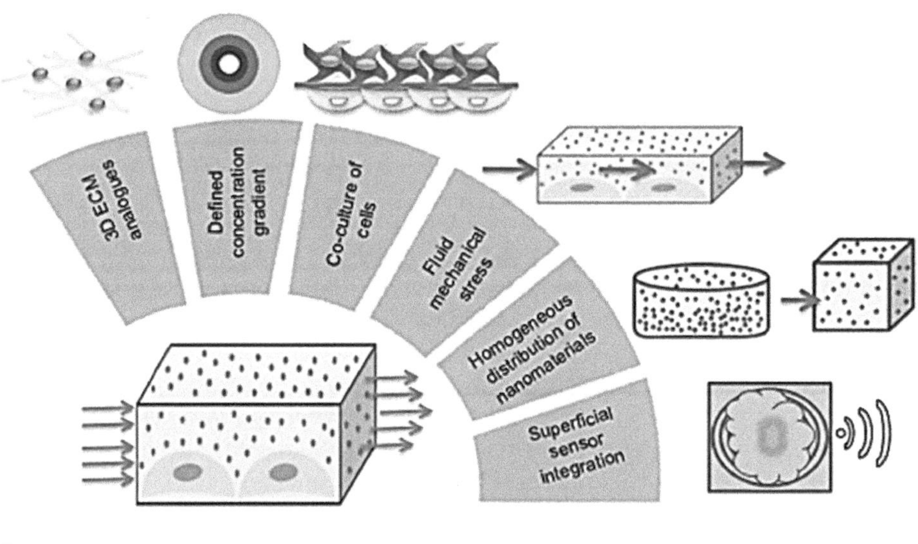


**FIGURE 14.1** Diagram depicting a functional relationship between microfluidic devices and its role in the evaluation of risks posed by NMs. A microfluidic device subjects a cell to *in vivo*-like microarchitecture, posing 3D-scaffold, enabling control over co-culture of cells, creating defined concentration gradients. It can also be used as a facile biosensor tool, when integrated with impedance sensing electrodes. It can be assigned to expose the cell to fluid mechanical stress or shear force resulting in a dynamic, near corporeal *milieu.*

**TABLE 14.1**
**An Overview of Comparison between Conventional *In Vitro* and *In Vivo* Systems, and Microfluidic Platforms**

| | Conventional *In Vitro* and *In Vivo* Test Systems | Microfluidic Systems |
|---|---|---|
| Corporeal importance (e.g., *in vivo* microarchitecture, cell type relevance) | Secondary cell lines, primary cells, mono- and three-dimensional co-cultures are possible while maintaining the complexity of a living organism. A major limitation being the species difference (e.g., rodent-human). | Although not as complex as a living organism, the microarchitecture can be tamed to represent *in vivo* physiology that inclines towards 3D cellular assemblies, human-on-a-chip, and organ-on-a-chip. |
| Standard growth conditions | Established culture conditions for proper growth and proliferation of cells while physiological conditions inside a living organism cannot be determined. | Explicitly defined conditions for consistent and undisturbed culture. |
| Experimental duration | Usually fast (depending on level of automation) for *in vitro* systems, whereas extensive time is consumed while using animals for *in vivo* testing. | As good as standard *in vitro* tests. Systems may be upgraded to *in vivo* conditions. |
| Expense required | *In vitro* tests done in multi-well plates are comparatively affordable considering the extreme financial input required for animal handling and maintenance for *in vivo* cultures. | Can be easily set up with significantly less efforts and processed with average capital investment. |
| Performing multiple experiments | Albeit parallelization is possible in traditional multi-well plate formats by optimization, no real parallelization is possible in animal models. | Parallelization not as good as for standard *in vitro* test formats; nevertheless the versatility and great flexibility paves way for it. |
| Adjustable methodology | The parameters to be controlled however are limited, but implementation in their constituents can be made easily. Sometimes changes are difficult and costly. | Various tasks can be optimized, because of the flexible nature of microfluidics, such as co-cultures, gradient testing, etc. |
| Cells under feigned stress | Particle sedimentation creates a false interpretation of the cell-cell communication because of artificial stress while in *in vivo* animal model systems the physicochemical stress is considered to be a physiologically relevant phenomenon. | The artificial stress created can be controlled and regulated to a physiological level. |
| High-throughput screening technology | Cutting edge technology screening systems for multi-well platforms while the screening tool has far too low throughput for *in vivo* animal model culture. | Although not as well accepted as *in vitro* screening systems, but a microfluidic device provides all the necessary means for high-throughput screening. |
| False positives due to physicochemical properties | While an artifact observed inside an animal model is considered to be of corporeal importance, the formation of artifacts in multi-well plates is because of uneven distribution of NM (sedimentation, buoyancy, aggregation) which further interferes with the colorimetric assays. | The risk of artifacts is minimized by continuous supply of NMs in a laminar flow profile along with label-free sensors for upholding stable suspensions. |

*(Continued)*

**TABLE 14.1 (*Continued*)**
**An Overview of Comparison between Conventional *In Vitro* and *In Vivo* Systems, and Microfluidic Platforms**

| | Conventional *In Vitro* and *In Vivo* Test Systems | Microfluidic Systems |
|---|---|---|
| Power to control dosage | The static culture encourages varying NM behaviors as gravitational settling and buoyancy that leads to false conclusions. Distribution control is lost in an animal model leading to undefined doses in various tissues/cell types. | Dynamic flow in microfluidic devices allows an effortless and continuous application of reagents that lessens the chances of agglomeration of NM. |
| Administration of concentration gradients | Easy administration of concentration gradients is possible in multi-well formats; whereas implementation of concentration gradients in *in vivo* systems seems to be a rigorously cost-limited and time-consuming process. | In addition to concentration gradients, it also allows for automated generation of continuous gradients. |
| Automation | Automated systems are generally in use in commercial and large-scale settings for multi-well plate formats but it is not possible in animal models. | Microfluidic systems provide an apt setting for automation even in low/medium throughput settings (e.g., research/academia). |
| Monitoring of long-term adverse health effects | The design of the multi-well system is well suited for short-term experiments but not for long-term testing. The adverse effects solely depend on the life span of the test animal. | The dynamic flow perfusion system allows for systematic review over extended periods of time. |
| Efficiency required | Standard laboratory techniques with specialized training in animal handling and testing defined by legal regulations. | Expertise required. |
| Interpretation and evaluation | As compared to *in vitro* platforms, *in vivo* systems have a whole set of analysis methods very well known to toxicologists | Conventional assay methods used can be modified for sensor integration. |

of microfluidic devices as alternative platforms for nanotoxicity assays and their advantages over traditional approaches (e.g., culture dishes and multi-well plates) (Table 14.1). Finally, we discuss the prospects of microfluidic technology in recreating biomimetic *in vitro* models of physiological systems for future nanotoxicology studies.

## 14.2 CURRENT CHALLENGES IN NANOTOXICOLOGY: MICROFLUIDICS AS APPEALING CYTOTOXICITY PLATFORMS

### 14.2.1 Size and Shape of Nanomaterials and Their Physicochemical Stability

Nanomaterials fabricated from a similar parent material exhibit a wide range of characteristics such as electronic and optical properties that depend upon their size and shape. Concurrently, the elicited immunological responses within our body differ depending on the size and shape of these same NMs (Guo *et al.*, 2016). For instance, human skin cells subjected to gold nanospheres (GNSs) of variable sizes display no cytotoxicity, whereas gold nanorods (GNRs) induce profound toxicity to the same cell type, albeit the coating material alone was accountable for the cytotoxicity of GNRs (Wang *et al.*, 2008). Comparable facts were reported in an observation akin to the above experiment where cetyltrimethylammonium bromide (CTAB)-encapsulated GNRs were found to show greater extent of cytotoxicity than citrate-stabilized GNSs (Zhang *et al.*, 2012). Such cytotoxicity trends, including unusual cellular responses, were also observed in the diverse nanomaterial (NM) group, revealing that it is not restricted to gold nanomaterials only (Aillon *et al.*, 2009; Zhu *et al.*, 2013). For instance, distinct shapes of titanium oxide ($TiO_2$) nanostructures such as nanospheres and nanobelts (ranging <5 and >15 μm) are responsible for significant differences in their cytotoxic activity (Hamilton *et al.*, 2009). Similar to the cellular effects of silica and asbestos, pulmonary alveolar macrophages elicited inflammatory responses on exposure to long $TiO_2$ nanobelts, indicating its severely toxic nature. Similarly, hydrodynamic size of the QDs is found to be the governing factor in its intracellular uptake (Mahto *et al.*, 2010a).

Given the differential cytotoxic behavior of nanomaterials following size and shape, it is pertinent to choose a suitable method that maintains the physicochemical properties of NMs in culture medium throughout the exposure of cells. Maintenance of such physical and chemical stability of NMs in culture medium is a major challenge for conventional *in vitro* platforms (Teeguarden *et al.*, 2006). In turn, to ascertain cytotoxicity, several modifications are made in the assay protocol as a testing strategy, for example, variations in the procedure of subjecting cells to NMs (Cimpan *et al.*, 2013; Marquis *et al.*, 2009; Soenen and De Cuyper, 2009), replacement of serum-containing medium (Kirchner *et al.*, 2005), or total exclusion of serum in the growth media (Petri-Fink *et al.*, 2008; Zhu *et al.*, 2009), while avoiding the influence of serum proteins present in the medium on the surface-coating materials of NMs that lead to the destabilization of the whole nanostructures. Nevertheless, classic *in vitro* methods are unable to gauge for the concept of kinetics of nanoparticles (NPs) in the culture medium that notably influence the physicochemical stability of NPs and cellular dose. As a matter of fact, in conventional platforms, concentration of NPs is not uniformly stable within the entire volume of a culture medium after a sufficient amount of time. They gradually tend to form sediments and clump, depending on their size, shape, charge and density. Medium around the NPs (Teeguarden *et al.*, 2006) is also responsible for such trends, because of fluidic viscosity and density (Figure 14.2). Thus, the continuous dynamic activity of immersed NPs in the solution prevails in the customary static exposure conditions. One of the major drawbacks, while considering traditional culture platforms is sedimentation of NPs on the cell surface due to their gravitational settling. Therefore, NPs of size >100 nm are large-enough (Teeguarden *et al.*, 2006) to hinder the precise estimation of correlation between cellular dose and toxic responses.

Considering the above limitations posed by traditional platforms for nanotoxicity assays, microfluidics has been brought into practice to create alternative models of *in vivo* systems. For instance, microfluidic technology has the ability to facilitate flow exposure conditions that have been subjugated to reveal the cytotoxicity of CTAB/trioctylphosphine oxide (CTAB/TOPO)-coated QDs suspended in a serum-containing growth medium. This was divulged while abstaining from the sedimentation effects of NPs and changing the dynamics of particles in a culture medium that gives rise to aggregation of NPs over a certain period of time (Mahto *et al.*, 2010c). This flow exposure method is critically important, not only due to the relevance of physiological conditions, but also to avoid any external forces on cells that may result from particles settling. These promising approaches have highlighted, for instance, noticeable variance in the extent of sustainability of BALB 3T3 fibroblast cells in contrast to stagnant conditions (Mahto *et al.*, 2010c); namely, at a concentration of 40 pM of QDs, the percentage of cellular viability was observed to be nearly 30% and 75% in static and flow exposure environments, respectively. These introductory results have highlighted the possibilities of microfluidic platforms as a leading-edge technology for the development of sophisticated models for nanotoxicological examinations.

### 14.2.2 Microfluidics for Monitoring Cellular Dose Exposure

Analyzing the cellular response against the actual amount of nanomaterials that interacts with cells is crucial for establishing their standard cytotoxicity profiles (Cunha-Matos *et al.*, 2016). However, measuring the right cellular dose of nanomaterials is cumbersome, not only due to the lack of appropriate detection tools, but also due to the influence of particle dynamics in the

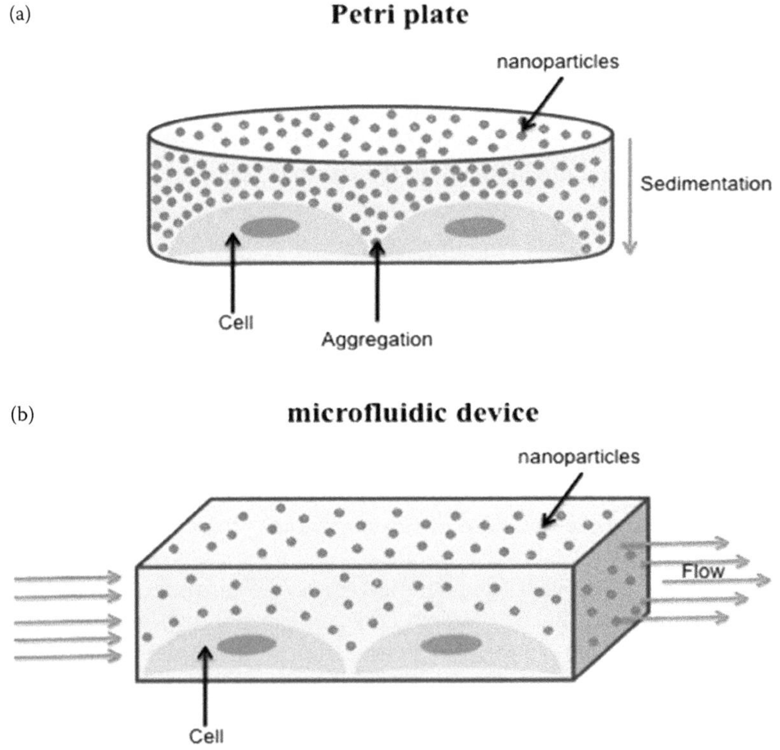


**FIGURE 14.2** Diagrammatic representation of the comparison of sedimentation rates in a conventional (a) cell-culture platform (i.e., petri dishes) and (b) a microfluidic device. The microfluidic device facilitates nanomaterials to a dynamic environment with significant physiological relevance as compared to conventional cell culture platforms that lead to sedimentation and accumulation of the nanomaterials. (Reprinted with permission from Mahto SK, Yoon TH, Rhee SW. A new perspective on *in vitro* assessment method for evaluating quantum dot toxicity by using microfluidics technology. *Biomicrofluidics.* 2010c; 4: 034111–8. doi:10.1063/1.3486610. Copyright 2010, American Institute of Physics.)

solution. Although several analytical methods, including UV-vis spectrofluorometry (Haiss *et al.*, 2007), dynamic light scattering (DLS) (Mahto *et al.*, 2010a), magnetic resonance imaging (Wu, 2012), fluorescence (Mahto *et al.*, 2010a), confocal laser microscopy (Mahto *et al.*, 2010a) and transmission electron microscopy (TEM) (Haiss *et al.*, 2007), are currently used for characterization of nanomaterials, such techniques do not account for the physicochemical instability caused by either particle dynamics or by destabilization of surface ligands coated on the surface of nanomaterials. Moreover, traditional assay platforms enhance the possibility of such instability in culture medium, thus leading eventually to particle settling.

Given the aforementioned advantages, microfluidic devices can help avoid such settling in the main channel populated with cells and thus facilitate the exposure of cells to nanomaterials suspended homogeneously in a culture medium. Alternatively, microfluidic platforms can have the potential to prevent large aggregations of NMs into the cellular region or enable removing such aggregates through continuous flow of medium. In addition, it is anticipated that microfluidic platforms hold great potential in measuring the correct cellular dose of NPs, whether it be in terms of mass per unit surface area of culture plates (i.e., g/cm$^2$), mass per unit volume of culture medium (i.e., g/mL) or the number of NPs per cell. Recently, a multi-compartment microfluidic device, consisting of a series of ten parallel microchannels provided with two inlets and a distinct outlet for each microchannel, was implemented in generating gradient profiles of QDs based on diffusional mixing of two fluidic streams inside the microfluidic channel (Mahto *et al.*, 2010c). Two different fluids, with and without QDs, were introduced into the device through the inlets provided at the same flow rate, and subsequently, fluids were collected from each outlet at fixed intervals of 12 h, and fluorescence spectra of the acquired fluids were then observed. This layout not only unveils the potential of microfluidic platforms in regulating the exposure of cells to NPs under flow exposure environments but also that these platforms confer upon us the chance to control the right cellular dose or concentration of NPs.

Apart from preventing aggregates of nanomaterials for cellular exposure, establishment of an appropriate cell density in gradients is equally essential in helping to estimate the right cellular dose of NMs and their differential cytotoxicity in terms of cellular number. It is known that the difference in cellular densities influences several biological phenomena including the adverse effects of drug molecules and toxins. For example, spatial arrangement of cells in gradients of cellular density shows a noticeable trend in the extent of biochemical signals and extracellular matrix (ECM) produced, which governs the manifestations of cellular behaviors such as growth (Belloni *et al.*, 2001), migration (Liu *et al.*, 2007), cell-cell communication (Rosso *et al.*, 2004) and cytotoxicity (Håkanson *et al.*, 2012). Therefore, the adaptable nature of microfluidics has recently been utilized for designing and establishing cellular density gradients in the device for the assessment of cytotoxicity of QDs (Wu *et al.*, 2013). Such a platform shows enormous scope in the analysis of QD cytotoxicity in a cell density-dependent manner, with better controllability and observability.

### 14.2.3 Measurement of Cytotoxicity Endpoints

Selection of a specific and reliable endpoint for measuring cytotoxicity is critical to precisely estimate lethal values of NMs. However, no reliable cytotoxicity testing strategies are currently available. Furthermore, these methods have their own limitations, depending on the testing models. For example, owing to their intrinsic genetic differences, cytotoxicity profiles of animals and humans might vary from each other, though estimated for the same endpoint (McKim, 2010). Similarly, chronic effects of nanomaterials in animals can remarkably differ from those in humans (Kuempel *et al.*, 2006). In the case of existing conventional *in vitro* models, they do not accurately represent the real scale and microenvironment of physiological systems and, therefore, do not account for the actual toxicity in the human system. Further, during identification of the cytotoxicity endpoints, nanomaterials may directly communicate with the organic dyes, physically or indirectly, by affecting the fluorescence spectra of dyes through their unique optical properties (Alkilany and Murphy, 2010).

As can be well-understood that there are certain inadequacies accompanying the conventional methods, a microfluidic-based lab-on-a-chip platform was designed to develop a novel method for measuring the cytotoxicity of nanomaterials (Richter *et al.*, 2011). This methodology employed the concept of contactless dielectric microsensors, combined with a microfluidic system, for the detection of collagen production of normal human dermal fibroblast cells (NHDF). Briefly, this system exploited the primary function of human dermal fibroblast cells, i.e., collagen production as a parameter for the measurement of cytotoxicity endpoint, indicating the level of cellular stress in the presence of nanomaterials, particularly gold (Au) and silver (Ag) NPs. The platform comprised of a main channel that contains four separate cell-culture chambers, a glass substrate at the bottom, and contactless dielectric microsensors right beneath the monitoring chambers. The study revealed that no significant effects on collagen production were noticed when cells were treated with Au NPs over a 24-h period. On the contrary, cells exposed to Ag NPs (10 μg/mL) for 2 h showed a remarkable decline in the impedance value that typically illustrates the amount of collagen production. Furthermore, cells incubated under the continuous exposure of sub-toxic concentrations of Ag NPs (0.1 μg/mL) through perfusion exhibited significant reduction in collagen production by 40% after 10 h; the basis of such variation was remarked as a characteristic of NP uptake and accumulation. Overall, this study represents substantial advancement in establishing a novel cytotoxicity endpoint using a microfluidic system for exploring risks of nanomaterials.

## 14.3 MICROFLUIDIC PLATFORMS FOR NANOTOXICITY STUDIES

Traditional platforms are unable to mimic the cellular microenvironments of physiological relevance. Therefore, the end result of conventional cytotoxicity assays cannot be expected to work similarly in an *in vivo* environment. Nevertheless, the traditional systems can be used as a standard tool for the basic understanding of the cellular interactions with NMs. Recreation of *in vitro* models that show more fidelity towards biomimicking the cellular niche of anatomical relevance including its microarchitecture, substantial strength, hierarchical arrangement and supply of nutrients through continuous perfusion, which still remains a formidable challenge. Over the past few years, microfluidic devices have provided us with grounds of expectation for evolution of *in vivo* cellular or tissue constructs on a chip that may be used as an advantageous platform for examining the cytotoxicity characteristics of nanomaterials. In the following account, we highlight a growing range of nanotoxicity investigations across various *in vitro* cellular environments.

### 14.3.1 Observation of Changes in Morphology and Behavior of Neurons upon Exposure to Nanomaterials

Evaluation of neuronal cytotoxicity due to nanomaterials has been a competitive assignment, particularly for analyzing the mechanistic role of nanomaterials in inducing the process of axonal degeneration. Neurons manifest compartmentalized microarchitecture and a highly systematic arrangement of its microenvironment in relation to space and time (Ralston, 1990). A compartmentalized microfluidic device has been designed recently for culturing of neurons as well as for the physical isolation and directional sprouting of axons into a fluidically separate environment without the use of growth factors or neurotrophins (Taylor *et al.*, 2003). The device has proved to be an important asset for nanotoxicology, as it eases the administration of physical or chemical agents to axons or somata (i.e., cell bodies) alone (Figure 14.3).

Notably, neuron-like PC12 cells were exposed locally using the compartmentalized microfluidic platforms to examine the effects of QDs on axons or somata alone (Mahto *et al.*, 2010b). Differentiation of PC12 cells and further isolation of axons from their cell bodies were facilitated by the microfluidic platform, thereby establishing somal and axonal compartments. Changes in the cellular morphology and behavior of cells were examined when QDs were introduced into one of the compartments. It was observed that CTAB-coated QDs induced axonal degeneration on account of neuronal cell death as well as of absolute axonal toxicity. In addition, the results indicate oxidative stress and nuclear condensation might be a causal factor for cellular alterations

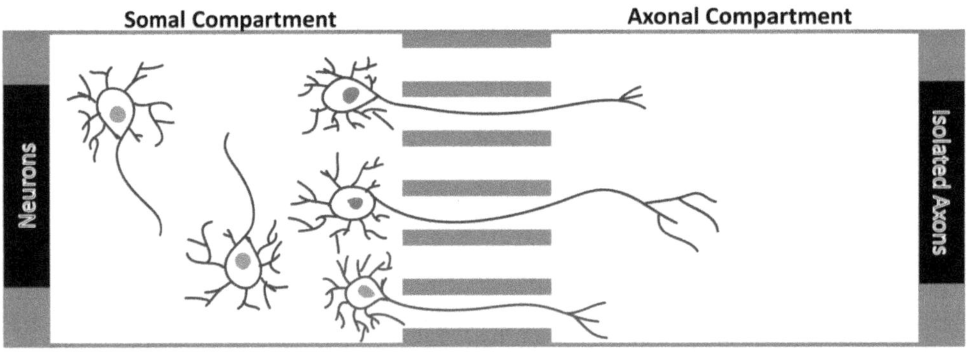


**FIGURE 14.3** Schematic representation of a multi-compartment microfluidic device. The device consists of two compartments for somal and axonal growth, which helps in the toxicity assessment of nanomaterials in a localized area. The two sections are joined by a set of microgrooves that allows only axons to pass through and restricts the cell body.

resulting from QD. This study represents a benchmark for utilization of microfluidic technology for neurotoxicity study of QDs at sub-cellular resolution.

### 14.3.2 Nanotoxicity Examination of Lung Alveoli at the Air-Liquid Interface

Pulmonary alveoli, the distal region of the lungs, are vital for understanding the behavior of NPs at the interface of a thin membranous alveolar-capillary barrier (ACB) (Mahto *et al.*, 2012) that is evidently known to allow the infusion of NMs into the systemic circulation (Choi *et al.*, 2010; Geiser *et al.*, 2005; Mühlfeld *et al.*, 2008). In particular, NPs precipitated at the thin liquid lining layer that coats the interior of alveoli, commonly known as the alveolar air-liquid interface (ALI) (Mahto *et al.*, 2012), account for shape- and size-specific toxicity of NMs to the underlying epithelium (Paur *et al.*, 2011; Xie *et al.*, 2012), thereby leading to destabilization of the epithelial layer and hence, elevation in the penetrability of NPs through the ACB. It is difficult to comprehend NM-induced alveolar damage with prevailing conventional platforms, as they are unable to mimic the cellular microarchitecture at a realistic scale.

Recently, the microfluidic devices have been exploited to develop a multiplexed biochip that mimics structural, functional and mechanical properties of the human ACB (Huh *et al.*, 2010). The microdevice consists of two microchannels placed in close proximity, isolated by a thin, porous and flexible membranous sheet. To recreate the physiological properties of ACB at the micro-scale, human alveolar epithelial cells and human pulmonary microvascular endothelial cells were cultured on opposite sides of the porous membrane. Additionally, the device comprised two larger, lateral microchambers that facilitated the controlled mechanical stretching of microfluidic channels under vacuum. The versatility of the presented "lung-on-a-chip" device further allowed examining the cytotoxicity of NPs, covering at the ALI via monitoring changes in the alveolar epithelium and endothelium. High levels of intercellular adhesion molecules (ICAM-1) expression was observed in the microvascular channel when alveolar epithelial cells were exposed to silica NPs (12 nm) at the ALI, indicating the activation of endothelial cells. Furthermore, breathing movements closer to the physiological range, i.e., at the level of 10% cyclic strains, were applied that resulted in a significant increment in ICAM-1 expression of endothelial cells treated with the silica NPs. For the very first time, it was examined by this microfluidic-based nanotoxicity examination that activation of pro-inflammatory activities of NPs and acute lung inflammation could be one of the resulting effects of physiological breathing movements of the pulmonary alveoli.

### 14.3.3 Vascular Transport of Nanoparticles and Their Toxicity in Tissue Structures

The penetration of capillaries deep within the tissues is fundamental for maintaining the physiological and functional status of organs. Both capillaries and tissues together act as components of a filtering unit, wherein the earlier one aids in the removal of unwanted substances such as ammonia, carbon dioxide and other metabolic waste products from the latter, while feeding the latter with nutrients, oxygen and hormones (Tandon and Kushwaha, 1992). Endothelial cells underlining the capillaries play a key role in governing the exchange of materials across them, and, in particular, diffusion is the process that conducts the relocation of oxygen and nutrients into tissues (Tandon and Kushwaha, 1992). A microfluidic 3D cell-culture device, with an ability to mimic the biochemical processes, was recently fabricated and employed to recreate the liver tissues by encapsulating HepG2 (human hepatocellular liver carcinoma cell line) cells in an agarose matrix, in combination with capillary-like microstructures, to infer the transport of QDs between capillaries and adjacent tissues (Wu *et al.*, 2012). The two-compartmental device was designed in such a way that the main channel resembled blood vessels and was capable of transporting culture medium and QDs. The second compartment served the purpose of cell culture chamber that mimics a model of the liver tissue structures. Two different lengths of stop-flow junctions have also been added to the device. This feature, integrated within the device, enabled the estimation of cytotoxicity of QD in a diffusive length-dependent manner. The chambers namely close and far were located at 0.8 and 2 mm away from the main channel, respectively. The study revealed marked elevation in the number of apoptotic cells, with an increase in the dosage of NPs both in close and far chambers. Besides, cells cultured in the close chambers were found significantly affected due to QD cytotoxicity compared with that of far chambers. Namely, cells displayed that reactive oxygen species (ROS) generation was directly proportional to the QD concentration, while the levels of glutathione (GSH) declined with increased QD concentrations. Overall, this primary study demonstrated that diffusion processes play a pivotal role in the regulation of QD cytotoxicity in tissues.

### 14.3.4 Embedded Impedimetric Assay System in a Microfluidic Device to Assess Cytotoxic Effect of Nanomaterials

Convergence of nanotoxicology and microfluidics, along with integrated impedance biosensors, furnishes us with a noninvasive and continuous technique to monitor dynamic cell-nanoparticle interactions. An electrical impedance sensing (EIS) approach was incorporated, keeping in mind its sensitivity, rapid detection and real-time scrutiny of nano-bio-interface (Shah *et al.*, 2014). A lab-on-a-chip platform has been developed to keep a continuous record of renewal capability of human H441 lung adenocarcinoma cells after exposure to silica nanoparticles under different perfusion conditions. A sandwich of glass-(PDMS) polydimethylsiloxane-glass 3 × 3 cm microdevice (Charwat *et al.*, 2014) was constructed that comprised four cell-loading chambers, each of which was integrated with inter-digitated electrode structures (IDEs) as impedance sensors (Rothbauer *et al.*, 2015). A nanotoxicological model based on the administration of toxic silica nanoparticles (AmSil30) and H441 cells was employed to examine the effects of flow-induced shear force on nanoparticle toxicity. The device was designed to subject the cells to acute (i.e., static flow) and chronic (i.e., varying perfusion) conditions of 4 and 40 μL/min flow rates for a total of 4 hours. The experimental design included treatment of cells with AmSil30 silica nanoparticles at a concentration of 240 μg/mL ($EC_{50}$) for 4 h. The renewal capacity of the H441 cells was observed until a period of 20 hours. An immediate impedance drop was reported in acute and chronic (4 μL/min) conditions, as compared to control (absence of nanoparticles) that registered an increase in impedance, depicting cell growth and proliferation. To analyze the effect of flow rates on nanoparticle uptake and dynamic cell behavior, cells were exposed to a concentration of 30 μg/mL silica nanoparticles (sub-toxic concentration) at 4 and 40 μL/m flow rates. In contrast to the results mentioned above, exposing the cells to AmSil30 nanoparticles at an increased flow rate of 40 μL/m, readily showed an impedance drop because of cell damage and death. This damage to cells is attributed to an increase in nanoparticle uptake at an increased flow rate, thus leading to increased nanotoxicity. Therefore, it was concluded that the approach of administration (i.e., static or dynamic) of nanoparticles plays a significant role in determining its toxicity and its nanotoxicological assays. The scientific procedure described above encourages the fact that impedance-based biosensors can serve to be a useful tool to evaluate cell viability for dynamic nanotoxicological studies.

## 14.4 CONCLUSIONS

The widespread use of NMs in every corner of our lives is posing great risk to human and environmental health (Mahto *et al.*, 2015). As a consequence, the field of nanotoxicology is rapidly growing in an effort to deal with the risks and adverse effects resulting from NMs. In the present chapter, we have highlighted recent developments and alternative solutions using microfluidics technology for nanotoxicity examinations. This novel technology offers numerous varieties of lab-on-a-chip platforms that overcome the experimental limitations of conventional *in vitro* and *in vivo* platforms and helps developing new methodologies for nanotoxicological examinations (Tong *et al.*, 2017). Owing to the unique features exhibited by the microfluidic platforms, they have been exploited for nanotoxicity study in a number of experimental scenarios, as it can aid in avoiding the agglomeration of NMs and thus maintaining the homogeneous suspension of NPs, estimating the right cellular dose of NMs, establishing novel assessment methods and cytotoxicity endpoints, and furthermore mimicking the physiological environments of the cells and tissues more closely (Mahto *et al.*, 2010b,c; Richter *et al.*, 2011; Zheng *et al.*, 2012). Although several studies have addressed the major benefits of the microfluidic-based nanotoxicity evaluation (McCormick *et al.*, 2017), very few have attempted to design and fabricate the microfluidic models of functional tissue structures to realize *in vitro* nanotoxicological studies in physiologically-equivalent tissues (Huh *et al.*, 2010; Wu *et al.*, 2012). In view of advancement of microfluidic platforms specifically for conducting such studies, numerous efforts to recreate multi-organ and human-on-a-chip (Huh *et al.*, 2011; Marx *et al.*, 2012; Meer and Berg, 2012) platforms are now being invested for near-human *in vitro* devices for nanotoxicity examinations. Such improved analytical systems may serve as the benchmark for exploring nanomaterial-induced cellular behavior and responses that are nearly impossible to evaluate with current *in vitro* and *in vivo* models.

## ACKNOWLEDGMENTS

This work was financially supported by a DST-INSPIRE (DST/INSPIRE/04/2013/000836) research grant from the Department of Science and Technology, Government of India.

## REFERENCES

Aillon KL, Xie Y, El-Gendy N, Berkland CJ, Forrest ML. Effects of nanomaterial physicochemical properties on *in vivo* toxicity. *Adv. Drug Deliv. Rev.* 2009; 61: 457–466.

Aitken RJ, Chaudhry MQ, Boxall ABA, Hull M. Manufacture and use of nanomaterials: Current status in the UK and global trends. *Occup Med.* 2006; 56: 300–306.

Alkilany AM, Murphy CJ. Toxicity and cellular uptake of gold nanoparticles: What we have learned so far? *J Nanoparticle Res.* 2010; 12: 2313–2333.

Arora S, Rajwade JM, Paknikar KM. Nanotoxicology and *in vitro* studies: The need of the hour. *Toxicol Appl Pharmacol.* 2012; 258: 151–165.

Belloni AS, Albertin G, Forneris ML, Nussdorfer GG. Proadrenomedullin-derived peptides as autocrine-paracrine regulators of cell growth. *Histol Histopathol.* 2001; 16: 1263–1274.

Borm PJ, Robbins D, Haubold S, Kuhlbusch T, Fissan H, Donaldson K *et al.* The potential risks of nanomaterials: A review carried out for ECETOC. *Part Fibre Toxicol.* 2006; 3: 11.

Cavalcanti A, Shirinzadeh B, Freitas Jr RA, Hogg T. Nanorobot architecture for medical target identification. *Nanotechnology.* 2008; 19: 015103.

Charwat V, Joksch M, Sticker D, Purtscher M, Rothbauer M, Ertl P. Monitoring cellular stress responses using integrated high-frequency impedance spectroscopy and time-resolved ELISA. *Analyst.* 2014; 139: 5271–5282.

Choi HS, Ashitate Y, Lee JH, Kim SH, Matsui A, Insin N *et al.* Rapid translocation of nanoparticles from the lung airspaces to the body. *Nat Biotechnol.* 2010; 28: 1300–1303.

Cimpan MR, Mordal T, Schölermann J, Allouni ZE, Pliquett U, Cimpan E. An impedance-based high-throughput method for evaluating the cytotoxicity of nanoparticles. *J Phys Conf Ser.* 2013; 429: 012026.

Cunha-Matos CA, Millington OR, Wark AW, Zagnoni M. Real-time assessment of nanoparticle-mediated antigen delivery and cell response. *Lab Chip.* 2016; 16: 3374–3381.

Dowding JM, Das S, Kumar A, Dosani T, McCormack R, Gupta A *et al.* Cellular interaction and toxicity depend on physicochemical properties and surface modification of redox-active nanomaterials. *ACS Nano.* 2013; 7: 4855–4868.

El-Ansary A, Al-Daihan S. On the toxicity of therapeutically used nanoparticles: An Overview. *J Toxicol.* 2009; 2009: 1–9.

Freitas RA, 2002. The future of nanofabrication and molecular scale devices in nanomedicine. *Stud Health Technol Inform.* 2002; 80: 45–59.

Frick R, Müller-Edenborn B, Schlicker A, Rothen-Rutishauser B, Raemy DO, Günther D *et al.* Comparison of manganese oxide nanoparticles and manganese sulfate with regard to oxidative stress, uptake and apoptosis in alveolar epithelial cells. *Toxicol Lett.* 2011; 205: 163–172.

Geiser M, Rothen-Rutishauser B, Kapp N, Schürch S, Kreyling W, Schulz H *et al.* Ultrafine particles cross cellular membranes by nonphagocytic mechanisms in lungs and in cultured cells. *Environ Health Perspect.* 2005; 113: 1555–1560.

Giridharan V, Yun Y, Hajdu P, Conforti L, Collins B, Jang Y *et al.* Microfluidic platforms for evaluation of nanobiomaterials: A Review. *J Nanomater.* 2012; 2012: 1–14.

Guo Y, Terazzi E, Seemann R, Fleury JB, Baulin VA. Direct proof of spontaneous translocation of lipid-covered hydrophobic nanoparticles through a phospholipid bilayer. *Science Advances.* 2016; 2: e1600261.

Haiss W, Thanh NTK, Aveyard J, Fernig DG. Determination of size and concentration of Gold nanoparticles from UV – Vis Spectra. *Anal Chem.* 2007; 79: 4215–4221.

Håkanson M, Kobel S, Lutolf MP, Textor M, Cukierman E, Charnley M. Controlled breast cancer microarrays for the deconvolution of cellular multilayering and density effects upon drug responses. *PloS One.* 2012; 7: e40141.

Hamilton RF, Wu N, Porter D, Buford M, Wolfarth M, Holian A. Particle length-dependent titanium dioxide nanomaterials toxicity and bioactivity. *Part Fibre Toxicol.* 2009; 6: 35.

Huh D, Hamilton GA, Ingber DE. From 3D cell culture to organs-on-chips. *Trends Cell Biol.* 2011; 21: 745–754.

Huh D, Matthews BD, Mammoto A, Montoya-Zavala M, Hsin HY, Ingber DE. Reconstituting organ-level lung functions on a chip. *Science.* 2010; 328: 1662–1668.

Judy JD, Unrine JM, Bertsch PM. Evidence for biomagnification of gold nanoparticles within a terrestrial food chain. *Environ Sci Technol.* 2011; 45: 776–781.

Kirchner C, Liedl T, Kudera S, Pellegrino T, Muñoz Javier A, Gaub HE *et al.* Cytotoxicity of colloidal CdSe and CdSe/ZnS nanoparticles. *Nano Lett.* 2005; 5: 331–338.

Kroll A, Dierker C, Rommel C, Hahn D, Wohlleben W, Schulze-Isfort C *et al.* Cytotoxicity screening of 23 engineered nanomaterials using a test matrix of ten cell lines and three different assays. *Part Fibre Toxicol.* 2011; 8: 9.

Kuempel ED, Tran CL, Castranova V, Bailer AJ. Lung dosimetry and risk assessment of nanoparticles: Evaluating and extending current models in rats and humans. *Inhal Toxicol.* 2006; 18: 717–724.

Landsiedel R, Ma-Hock L, Kroll A, Hahn D, Schnekenburger J, Wiench K *et al.* Testing metal-oxide nanomaterials for human safety. *Adv Mater.* 2010; 22: 2601–2627.

Lanone S, Rogerieux F, Geys J, Dupont A, Maillot-Marechal E, Boczkowski J *et al.* Comparative toxicity of 24 manufactured nanoparticles in human alveolar epithelial and macrophage cell lines. *Part Fibre Toxicol.* 2009; 6: 14.

Liu L, Ratner BD, Sage EH, Jiang S. Endothelial cell migration on surface-density gradients of fibronectin, VEGF, or both proteins. *Langmuir ACS J Surf Colloids.* 2007; 23: 11168–11173.

Liu X, Wang D, Li Y. Synthesis and catalytic properties of bimetallic nanomaterials with various architectures. *Nano Today.* 2012; 7: 448–466.

Liu Y, Wang S, Wang Y. Patterned Fibers Embedded Microfluidic Chips Based on PLA and PDMS for Ag Nanoparticle Safety Testing. *Polymers.* 2016; 8: 402–417.

Love SA, Maurer-Jones MA, Thompson JW, Lin Y-S, Haynes CL. Assessing nanoparticle toxicity. *Annu Rev Anal Chem.* 2012; 5: 181–205.

Mahto SK, Charwat V, Ertl P, Rothen-Rutishauser B, Rhee SW, Sznitman J. Microfluidic platforms for advanced risk assessments of nanomaterials. *Nanotoxicology.* 2015; 9: 381–395.

Mahto SK, Park C, Yoon TH, Rhee SW. Assessment of cytocompatibility of surface-modified CdSe/ZnSe quantum dots for BALB/3T3 fibroblast cells. *Toxicol In Vitro.* 2010a; 24: 1070–1077.

Mahto SK, Tenenbaum-Katan J, Sznitman J. Respiratory physiology on a chip. *Scientifica.* 2012; 2012: 1–12.

Mahto SK, Yoon TH, Rhee SW. Cytotoxic effects of surface-modified quantum dots on neuron-like PC12 cells cultured inside microfluidic devices. *BioChip J.* 2010b; 4: 82–88.

Mahto SK, Yoon TH, Rhee SW. A new perspective on *in vitro* assessment method for evaluating quantum dot toxicity by using microfluidics technology. *Biomicrofluidics.* 2010c; 4: 034111–8.

Marquis BJ, Love SA, Braun KL, Haynes CL. Analytical methods to assess nanoparticle toxicity. *Analyst.* 2009; 134: 425–439.

Marx U, Walles H, Hoffmann S, Lindner G, Horland R, Sonntag F *et al.* "Human-on-a-chip" Developments: A translational cutting-edge alternative to systemic safety assessment and efficiency evaluation of substances in laboratory animals and man? *Altern Lab Anim ATLA.* 2012; 40: 235–257.

McCormick SC, Kriel FH, Ivask A, Tong Z, Lombi E, Voelcker NH *et al.* The Use of Microfluidics in Cytotoxicity and Nanotoxicity Experiments. *Micromachines.* 2017; 8: 124–142.

McKim JM. Building a tiered approach to *in vitro* predictive toxicity screening: A focus on assays with *in vivo* relevance. *Comb Chem High Throughput Screen.* 2010; 13: 188–206.

Meer AD van der, Berg A van den. Organs-on-chips: Breaking the *in vitro* impasse. *Integr Biol.* 2012; 4: 461–470.

Mühlfeld C, Rothen-Rutishauser B, Blank F, Vanhecke D, Ochs M, Gehr P. Interactions of nanoparticles with pulmonary structures and cellular responses. *Am J Physiol Lung Cell Mol Physiol.* 2008; 294: L817–829.

Murty BS, Shankar P, Raj B, Rath BB, Murday J. Unique Properties of Nanomaterials. In: *Textbook of Nanoscience and Nanotechnology.* Springer: Berlin Heidelberg; 2013; pp. 29–65.

Nair PR, Alam MA. Design considerations of silicon nanowire biosensors. *IEEE Trans Electron Devices.* 2007; 54: 3400–3408.

Olson MS, Gurian PL. Risk assessment strategies as nanomaterials transition into commercial applications. *J anoparticle Res.* 2012; 14: 1–7.

Paur H-R, Cassee FR, Teeguarden J, Fissan H, Diabate S, Aufderheide M *et al. In-vitro* cell exposure studies for the assessment of nanoparticle toxicity in the lung—A dialog between aerosol science and biology. *J Aerosol Sci.* 2011; 42: 668–692.

Petri-Fink A, Steitz B, Finka A, Salaklang J, Hofmann, H. Effect of cell media on polymer coated superparamagnetic iron oxide nanoparticles (SPIONs): Colloidal stability, cytotoxicity, and cellular uptake studies. *Eur J Pharm Biopharm.* 2008; 68: 129–137.

Ralston HJ. Analysis of neuronal networks: A review of techniques for labeling axonal projections. *J Electron Microsc Tech.* 1990; 15: 322–331.

Richter L, Charwat V, Jungreuthmayer C, Bellutti F, Brueckl H, Ertl P. Monitoring cellular stress responses to nanoparticles using a lab-on-a-chip. *Lab Chip.* 2011; 11: 2551.

Rico CM, Majumdar S, Duarte-Gardea M, Peralta-Videa JR, Gardea-Torresdey JL. Interaction of nanoparticles with edible plants and their possible implications in the food chain. *J Agric Food Chem.* 2011; 59: 3485–3498.

Rivera_Gil P, Clift MJD, Rutishauser BR, Parak WJ. Methods for understanding the interaction between nanoparticles and cells. In: *Nanotoxicity.* Humana Press; Totowa, NJ; 2012; pp. 33–56.

Rosso F, Giordano A, Barbarisi M, Barbarisi A. From cell-ECM interactions to tissue engineering. *J Cell Physiol.* 2004; 199: 174–180.

Rothbauer M, Praisler I, Docter D, Stauber RH, Ertl P. Microfluidic impedimetric cell regeneration assay to monitor the enhanced cytotoxic effect of nanomaterial perfusion. *Biosensors.* 2015; 5: 736–749.

Roy S, Gao Z. Nanostructure-based electrical biosensors. *Nano Today.* 2009; 4: 318–334.

Shah P, Kaushik A, Zhu X, Zhang C, Li C-Z. Chip based single cell analysis for nanotoxicity assessment. *Analyst.* 2014; 139: 2088–2098.

Soenen SJH, De Cuyper M. Assessing cytotoxicity of (iron oxide-based) nanoparticles: An overview of different methods exemplified with cationic magnetoliposomes. *Contrast Media Mol Imaging.* 2009; 4: 207–219.

Tandon PN, Kushwaha K. A study of nutritional transport in capillary-tissue exchange system. *Int J Biomed Comput.* 1992; 30: 1–15.

Taylor AM, Rhee SW, Tu CH, Cribbs DH, Cotman CW, Jeon NL. Microfluidic multicompartment device for neuroscience research. *Langmuir.* 2003; 19: 1551–1556.

Teeguarden JG, Hinderliter PM, Orr G, Thrall BD, Pounds JG. Particokinetics *in vitro*: Dosimetry considerations for *in vitro* nanoparticle toxicity assessments. *Toxicol Sci.* 2006; 95: 300–312.

Tong Z, Ivask A, Guo K, McCormick S, Lombi E, Priest C. Crossed flow microfluidics for high throughput screening of bioactive chemical–cell interactions. *Lab Chip.* 2017; 17: 501–510.

Valencia PM, Farokhzad OC, Karnik R, Langer R. Microfluidic technologies for accelerating the clinical translation of nanoparticles. *Nat Nanotechnol.* 2012; 7: 623–629.

Wang S, Lu W, Tovmachenko O, Rai US, Yu H, Ray PC. Challenge in understanding size and shape dependent toxicity of gold nanomaterials in human skin keratinocytes. *Chem Phys Lett.* 2008; 463: 145–149.

Williams KA, Veenhuizen PTM, de la Torre BG, Eritja R, Dekker C. Nanotechnology: Carbon nanotubes with DNA recognition. *Nature.* 2002; 420: 761.

Wu J, Chen Q, Liu W, Lin J-M. A simple and versatile microfluidic cell density gradient generator for quantum dot cytotoxicity assay. *Lab Chip.* 2013; 13: 1948.

Wu J, Chen Q, Liu W, Zhang Y, Lin J-M. Cytotoxicity of quantum dots assay on a microfluidic 3D-culture device based on modeling diffusion process between blood vessels and tissues. *Lab Chip.* 2012; 12: 3474.

Wu S. Iron Oxide Nanoparticles Imaging Tracking by MR Advanced Techniques: Dual-Contrast Approaches. In: *Smart Nanoparticles Technology.* InTech: Croatia; 2012: 149–172.

Xie Y, Williams NG, Tolic A, Chrisler WB, Teeguarden JG, Maddux BLS *et al.* Aerosolized ZnO nanoparticles induce toxicity in alveolar type II epithelial cells at the Air-Liquid interface. *Toxicol Sci.* 2012; 125: 450–461.

Zhang Y, Xu D, Li W, Yu J, Chen Y. Effect of size, shape, and surface modification on cytotoxicity of gold nanoparticles to human HEp-2 and canine MDCK cells. *J Nanomater.* 2012; 2012: 1–7.
Zheng X, Tian J, Weng L, Wu L, Jin Q, Zhao J *et al.* Cytotoxicity of cadmium-containing quantum dots based on a study using a microfluidic chip. *Nanotechnology.* 2012; 23: 055102.
Zhu M, Nie G, Meng H, Xia T, Nel A, Zhao Y. Physicochemical properties determine nanomaterial cellular uptake, transport, and fate. *Acc Chem Res.* 2013; 46: 622–631.
Zhu Y, Li W, Li Q, Li Y, Zhang X, Huang Q. Effects of serum proteins on intracellular uptake and cytotoxicity of carbon nanoparticles. *Carbon.* 2009; 47: 1351–1358.

# 15 Surface-Engineered Nanomaterials: Environmental and Safety Considerations in Pharmaceutical and Medicinal Products

*Lena Leopold, Chaoyang Zhao, Lisa McConnachie, Rajneet Kaur Khurana, Teenu Sharma, Bhupinder Singh,* and Rodney J. Y. Ho**

## CONTENTS



## 15.1 INTRODUCTION

The concept of making something small in medical and pharmaceutical applications is not new and has been used to make medical devices such as infusion pumps smaller for home use and with respect to medical implants for sustained and long-acting drug dosing for chronic applications. For pharmaceutical drugs, milling and micronizing technologies have been used for decades to increase drug surface area and assist in solubility. Besides, pharmaceutical scientists have used a number of strategies including miniaturization of water-insoluble drugs into smaller nanometer-sized particles (<1000 nm in diameter) to increase surface area for improved water solubility. In some cases, some additive or excipients are added to stabilize the drug particle before size reduction by milling, micro-ionization, or other techniques that may include spray drying of the mixture to generate nano-size drug particles in solid form or in suspension (Raveendran *et al.*, 2017). While some products were referred to as nanocrystals, most dosage forms contained drugs which were amorphous instead of being crystalline (which typically have lower solubility in this physical state).

* Corresponding authors.

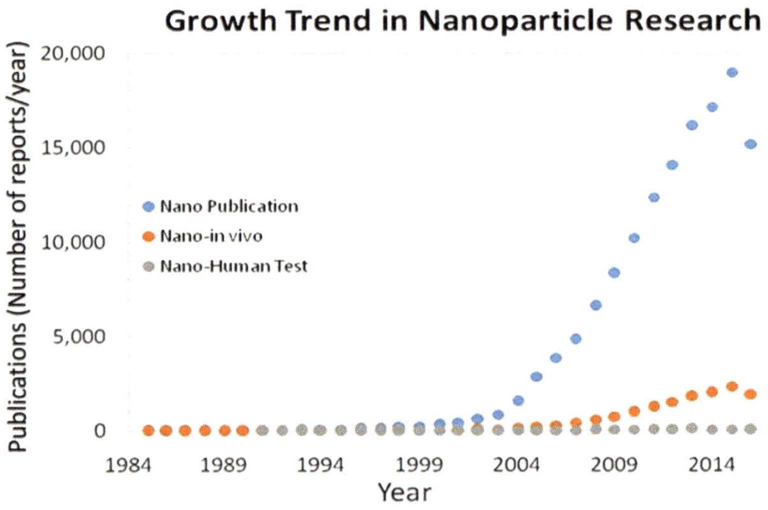


**FIGURE 15.1** Research publications related to nanomaterials and nanoparticles found in PubMed sorted from 1984–2014 and categorized according to their progress translating to animal studies *in vivo*, which typically precedes human testing. A smaller fraction of potential testing *in vivo* are tested in humans and published in human studies involving nanomaterials. These different publications are plotted each year from 1984–2014.

With recent advances in polymer chemistry, material and physical sciences that enable innovation in synthesis of nanomaterials, tremendous growth in nanoparticle research has occurred. As shown in Figure 15.1, an exponential growth in annual publications is witnessed regarding nanoparticle research beginning in 2000, which appeared to have somewhat subsided by 2014. For the past 15 years, the fraction of reported papers involving evaluation *in vivo*, progressing from fabrication and testing in the laboratory to some live animal models, has followed the trend of the nanoparticles publication growth, but lagged somewhat during the period. The fraction of reported papers involving *in vivo* techniques is only about 10% on an average. Even fewer reported papers involve testing of nanoparticle materials on human subjects. Including some of the reviews written, the fraction of reported studies related to human-testing is typically in less than 1%, and for 2016, only 46 reported papers involve human-testing as compared to 15,169 reported papers that cover nanoparticles in general.

In this chapter, we will briefly define what constitutes nanoparticles and nanomaterials, followed by discussion on their proven and potential role in medical, pharmaceutical and consumer applications. We will also discuss the drivers and science behind recent interest in generating new materials for improving product performance. The exponential growth in nanomaterials and related research has also widened the gap between nanomaterial innovations and translation of these innovative materials into pharmaceutical applications. As translation of discovery of new materials and applications of the same to pharmaceutical and medical products requires regulatory, toxicological and safety considerations, as well as consideration of environmental impact, we will address some of these issues and the approaches to evaluate new materials and their applications early in the course of nanomaterials research and development. We will conclude this chapter with some observations regarding the potential outlook of translation of nanomaterials into safe and effective products.

## 15.2 WHAT ARE NANOMATERIALS?

Nanomaterials are small-sized (i.e., less than 1000 nm) materials, with the preferred range for pharmaceutical application as less than 200 nm. The smaller the particle for the solid drug, whether in crystalline or amorphous state, the higher is the surface area exposed per each unit of volume. This high surface-area-per-unit-volume has proved to improve the solubility in water of some of most challenging drug molecules. Increasing surface area per unit enables development of dosage forms of drugs that otherwise exhibit poor solubility and consequently poor bioavailability (Borm *et al.*, 2006). The nanodrug particles may be produced with milling or grinding of drug crystalline or complex with excipients. Alternatively, scientists have also developed nanomaterials as carriers, wherein the drug particles are incorporated, encapsulated or conjugated. Nanomaterials, in their excipients or drug delivery systems, may be derived from natural or synthetic polymers. In some cases, lipids and fatty acid derivatives are also used as inert or functional excipients of drugs to form nanodrug particles. Due to the unique physiochemical characteristics of each nanoparticle's composition and preparation, a method will result in different size, shape, particle and drug stability, crystal structure, surface area, surface energy and surface roughness. Some of these parameters may influence the efficacy in pharmaceutical application with respect to safety of patients, employees and the environment, if significant products and byproducts are discharged into surrounding air, water and land (Gatoo *et al.*, 2014). Additional details on toxicology of nanomaterials and potential environmental impacts are addressed in Sections 15.5 and 15.6, respectively.

## 15.3 NANOMATERIALS USED AS THERAPEUTICS AND IN PHARMACEUTICAL APPLICATIONS

The concept of making drugs small is not new. There are many nanoparticulate pharmaceuticals currently available on the market. Table 15.1 lists some of the available nanoparticle-formulation pharmaceuticals available, along with their objectives of nanoparticle formulation.

There is considerable interest surrounding this approach to drug development because of its potential to increase pharmaceutical performance. Nanoformulation can improve a drug's therapeutic benefit and overall safety profile by improving its solubility, stability and distribution throughout the body (Cornier *et al.*, 2017). Doxil, a liposomal nanoformulation of doxorubicin, and Abraxane, a protein-bound paclitaxel formulation, are examples of how nanoformulation can enhance pharmaceutical efficacy compared to their corresponding non-nanoparticle counterparts (Cornier *et al.*, 2017). Doxorubicin causes severe cardiac myocyte toxicity that is dose-dependent, cumulative and irreversible. The liposomal formulation improves the drug's cardiac safety profile by reducing the off-target effects and

**TABLE 15.1**
**Current Pharmaceutical Products**

| Pharmaceutical Product | Nanoparticle Formulation | Particle Size Diameter (nm) | Objective |
|---|---|---|---|
| Doxil (Doxorubicin) | Liposomes | ∼90–100 | Minimize cardiac drug binding; drug accumulation in tumor |
| DaunoXome (Daunorubicin) | | ∼45 | |
| AmBisome (Amphotericin B) | | ∼100 | Reduce nephrotoxicity and increase potency against fungal infection in spleen |
| Amphotec (Amphotericin B) | Lipidic colloid | ∼115 (Elongated disk) | |
| Abelcet (Amphotericin B) | Flattened liposome vesicles | ∼1600 | |
| Rapamune (Sirolimus) | Nanocrystal | 20–1000 | Improve solubility and bioavailability |
| Emend (Aprepitant) | | | |
| Tricor (Fenofibrate) | | | |
| Copaxone (Glatiramer acetate) | Colloidal peptide suspension | ∼100 (100–1000) | To treat symptoms of multiple sclerosis due to inflammatory response Conner *et al.* (2015) |
| Immunoglobulin | Protein and protein bound drug | 20 | Intrinsic property that provides long acting therapy ≤30 days |
| IgG and monoclonal IgG Gama globulin | | 20 | |
| Abraxane-Albumin-paclitaxel | | 130 | Increase solubility and injection site tolerability |

increases selectivity for tumor association. Doxil improves the toxicity profile of doxorubicin by limiting its off-target effects and improving its tumor selectivity. Encapsulating doxorubicin causes its release to be slower, leading to lower drug plasma concentrations, and thereby reducing off-target binding and its associated toxicity (Tahover *et al.*, 2015). Because the encapsulated doxorubicin is released more slowly, less drug is degraded before it reaches the tumor, thus improving its on-target bioavailability compared to doxorubicin alone (Tahover *et al.*, 2015). In addition, the improved tumor selectivity of Doxil is mediated through the enhanced permeability and retention (EPR) effect. Healthy vasculature endothelium prevents extravasation of liposomes to surrounding tissue, protecting the slow-growing cardiomyocytes. Because tumor vasculature is leakier in nature, the nano-sized liposomes are able to pass through the endothelium, resulting in increased tumor deposition (Tahover *et al.*, 2015).

Another antitumor agent, Abraxane, an albumin-bound paclitaxel formulation, was developed in order to improve the safety and efficacy profile of paclitaxel. Conventional paclitaxel formulations used Cremophor EK, dehydrated ethanol and Tween 80, in order to improve the solubility of the drug for injectable use. However, these solvents caused toxic effects, including severe hypersensitivity reactions, making the drug poorly tolerated by patients (Cornier *et al.*, 2017). Besides toxicity, the solvents potentially limited paclitaxel's permeability through the endothelium, reducing its ability to accumulate in tumors (Joerger, 2016). By binding to albumin, paclitaxel was able to increase its solubility profile and utilize albumin's endogenous pathways to improve its biodistribution and accumulation at the site of the tumor (Joerger, 2016). As a result, Abraxane was able to accumulate better in the tumor due to the exploitation of the aforementioned EPR effect and albumin's ability to translocate to the surrounding tissue through binding endothelium Gp60 receptors (Joerger, 2016).

Novel nanomaterials are being developed by various chemists and bioengineers, which need thorough evaluation before being used in future. A nanomaterial's overall benefit for use as a pharmaceutical product must be weighed against its potential for causing toxicity, and little is known for many of these newer materials about their toxicological impact in the body and the environment. Additional details on the methods and approaches to study nanomaterial safety are discussed below.

## 15.4 RECENT INTERESTS IN GENERATING NEW MATERIALS TO IMPROVE PRODUCT PERFORMANCE

Due to their unique characteristics and potential to improve healthcare and biomedical applications, many nanomaterials are synthesized and reported to produce exciting results in a wide range of disciplines, including chemistry, biophysics, biochemistry, pharmacology and pharmaceutical sciences. Although many nanomaterials were reported in the literature for their potential as drug candidates or nanomedicines, a number of challenges exist, such as formulation complexity, scalability, reproducibility, cost and stability (regarding the particle size and active drug) that may limit some progress in translating these exciting discoveries into viable therapeutic products. Particle size uniformity and reproducibility will be more challenging for injectable formulations, which carry higher standards than oral tablets or capsule dosage forms. In fact, formulation and characterization of nano-drug suspension has propelled evolving areas of regulatory science, which set criteria for the review and regulatory approval of drugs and their generic counter parts. While efficacy or effectiveness in treating specific disease conditions is the intent of developing new drugs, failure of many nanoparticle-formulated drugs and nanomedicine candidates is often driven by their toxicity. Therefore, toxicology is an important area of nanoparticle and nanomaterial research, especially for newly synthesized or modified nanopolymers for which potential impacts on cells and tissues is not yet well understood.

Nanoparticles exist as naturally occurring entities such as proteins and antibodies, or as synthetic nanomaterials derived from sources such as carbon and silica and contain heavy metals. Examples of synthetic nanoparticles include semiconductor quantum dots, carbon nanotubes and fullerenes. The physical structure of nanoparticles can take many forms and configurations such as spheres, cylinders, tubes, fibers and rods, and all of these encompass a wide range of diameters, lengths,

**TABLE 15.2**
**General Classification of Nanomaterials for Toxicology Studies**

| Composition | | | | |
|---|---|---|---|---|
| **Natural** | **Synthetic** | **Structure and Configuration** | **Dimension** | **Charge (Net)** |
| Polypeptides, proteins, antibodies, glycopolymers | Carbon, silica, Zn, Cd, Ag, Au and derivatives | Fullerenes (carbon) | Sphere | Positive |
| Lipids, fatty acids | Poly amino acids and derivatives | Iron oxide | Fiber | Negative |
| | | Zinc oxide | Film | Neutral |
| | | Mesoporous silica | Rod | Polar |
| | | PLGA | Ellipsoid | |
| | | PLA | | |

widths and aspect ratios. Table 15.2 provides a potential framework for general grouping to study nanomaterial toxicology.

A number of strategies have been proposed in order to systematically study the toxicology of nanomaterials. In general, nanomaterials composed of natural substances or their polymers, if previously ingested as food or exposed at varying doses, may be safer or less toxic than nanomaterials that are synthesized *de novo*. However, the toxicology studies generated from the cells with similar nanostructure and physical association of drug molecules may influence composition, which may be the final determinant of safety. In other words, the nanocarrier without specific drug may assume very different physical, chemical, particle size and charge characteristics, such that each individual nanomedicine complex may require consideration in the design of toxicology, formulation and efficacy, as well as in selecting appropriate controls for data interpretation.

## 15.5 SAFETY AND NANOMATERIALS: EVALUATION METHODS AND VALIDATIONS—AN EVOLVING SCIENCE

While the unique physiochemical properties of nanoparticles are responsible for their widespread application, they could also contribute to toxicology profile of nanoparticles. In addition to the toxicology of nanomaterial compositions, there are other physical properties and modifications of nanoparticle structure, and their consequent aggregation state may influence the toxicology profile. Table 15.3 lists some of the factors that influence toxicology.

In general, toxicology of nanoparticles can be physical and chemical in nature. For instance, the particle uptake and cell-surface or protein binding may impair or elicit certain cellular and molecular function(s). The chemical conjugation, oxidation or reduction of nanomaterials may also interfere with electron transport or singlet oxygen or peroxide generation. These processes in theory could be measured in cell-free or cell-based systems *in vitro*. Having a cell-free or cell-based system for screening biocompatibility was postulated to reduce product failure and accelerate discovery of safer biomaterials. These approaches are similar to drug screening strategies, where molecules of similar properties and target affinities are screened *in vitro* for binding or inhibition activity at protein levels and then subsequently screened in cells to identify compounds with limited cell penetration, transport or excessive metabolism and export (e.g., efflux transporter mediated processes). These *in vitro* results are analyzed and integrated before *in vivo* animal testing to understand the overall pharmacokinetics and toxicokinetics of drug candidates. In the analogous manner, *in vivo* toxicity of drug nanoparticles is evaluated in various animal models. Table 15.4 lists various *in vitro* and *in vivo* toxicity studies on nanoparticles.

### 15.5.1 *In Vitro* Studies

Most toxicity studies evaluating nanoparticle toxicity are conducted *in vitro* due to the relative difficulties and limitations of conducting *in vivo* experiments. *In vitro* assays are faster, less

**TABLE 15.3**
**Factors Affecting Toxicology of Nanomaterials**

| Factors | Influence on Toxicology |
|---|---|
| Dose | • Represented as a function of concentration of nanomaterials and the duration of exposure.<br>• Toxicity of nanoparticles is associated with the total surface area of the nanoparticles per dose rather than mass dose alone Buzea *et al.* (2007). |
| Size | • Nanomaterials with particle size $<100$ nm have greater toxic effects compared to larger particles of the same material due to increased<br>✓ Retention time<br>✓ Pulmonary translocation and persistence<br>✓ Alveolar type II cell proliferation<br>✓ Impairment of macrophage function Buzea *et al.* (2007) |
| Surface area | • Due to their smaller size, nanoparticles have a larger surface area and particle number per unit mass compared to larger particles, increasing their reactivity and potentially causing more inflammation Buzea *et al.* (2007). |
| Surface coating | • Surface coatings influence toxicity of nanoparticles due to interactions they make with the biological environment. Surface coatings may make nanoparticles more or less toxic.<br>• Reactive species on the surface of silica particles is associated with its particular cytotoxic effects.<br>• Surface oxygen, ozone, oxygen radicals, and transition metals can increase inflammatory reactions caused by the nanoparticle.<br>• CdSe quantum dots (QDs) can be made nontoxic with the proper surface coating Buzea *et al.* (2007). |
| Functionalization | • Chemical functionalization of nanoparticles can change how they interact with the biochemical environment and influence their overall toxicity.<br>• Nanoparticles covered in polyethylene glycol (PEG) are able to evade phagocytosis of immune cells and thereby have increased residency time Pelaz *et al.* (2015). |
| Concentration | • Compared to lower concentrations, nanoparticles in higher concentrations are more likely to agglomerate. The larger size of the agglomerate aids in the cellular uptake and subsequent removal of the nanoparticle aggregate by phagocytic cells reducing its overall residence time and toxicity Buzea *et al.* (2007).<br>• Most aggregates are larger than 100 nm in size. 100 nm appears to be the size threshold for reduced toxicity of nanoparticles in the body.<br>• Experiments using higher concentrations of nanoparticles show reduced toxicity due to agglomeration compared to studies using lower concentrations of the same nanoparticle Duran *et al.* (2014). |
| Particle chemistry | • Particle chemistry of the nanoparticle can determine its degree of cellular uptake, localization, and the ability to produce reactive oxygen species (ROS), which influences its overall toxicity Buzea *et al.* (2007). For example, the toxicity of QDs is thought to be more closely associated with its surface chemistry rather than its core material. Specifically, –COOH anchored CDSe/ZnSQDs tested in El-4 murine cancer cells appeared to produce less toxicity than those coated with amine Clift *et al.* (2008). |
| Crystalline structure | • Crystallinity and atomic structure of the nanomaterials surface may be the key factors affecting their toxicity.<br>• Differences in crystal structure of $TiO_2$ nanoparticles influenced whether cell death was produced through apoptosis or necrosis Braydich-Stolle *et al.* (2009). |
| Aspect ratio | • Particles with a higher aspect ratio (length/diameter) are more likely to cause toxicity.<br>• Depending on its bio-durability, high aspect nanoparticles can break perpendicularly to or parallel along its longitudinal axis.<br>• Nanoparticles, such as asbestos, that break longitudinally produce particles having the same length but smaller diameter, therefore having an increased aspect ratio and are more difficult to remove by macrophages.<br>• Nanoparticles that breaks perpendicular to their long axes produce shorter pieces of the same diameter that can be easily removed by macrophages. However, these shorter particles still have the potential to produce inflammatory responses such as cytokine release and ROS Buzea *et al.* (2007). |
| Agglomeration | 1. The method of cellular entry, either by passive diffusion or cellular uptake, is largely influenced by agglomeration state.<br>2. Non-phagocytic cell stake up agglomerated nanomaterial more easily than single nanomaterials, potentially increasing cytotoxicity v. |

**TABLE 15.4**
**Summary of Nanomaterial Safety Based on *In Vitro* and *In Vivo* Animal Studies**

| Classification of Nanomaterials | Type of Nanoparticles | Element | Size | Dose/Route | *In Vitro* | *In Vivo* *M = Mice* *R = Rat* | Toxicology | Reference |
|---|---|---|---|---|---|---|---|---|
| Inorganic Nano-metals | Copper (Cu NPs) | Cu | 23.5 nm | 108–1080 mg/kg (acute oral) | – | √(M) | Heavy injuries on kidney, liver and spleen | Chen *et al.* (2006) |
| | Silver (Ag NPs) | Ag | 60 nm | 30–1000 mg/kg (Sub-acute oral for 28 days) | – | √(R) | Dose-dependent effect on alkaline phosphatase and cholesterol. Two-fold more accumulation of NPs in kidneys of female than male | Kim *et al.* (2008) |
| | Iron oxide ($Fe_2O_3$ NPs) | Fe | 22–280 nm | 0.8–20 mg/kg (inhalation) | – | √(R) | Oxidative stress in lungs, inflammation; immune cells increase and pathological changes are there | Zhu *et al.* (2008) |
| | Silicon dioxide ($SiO_2$ NPs) | Si | 50 nm | 25–100 mg/kg (intraperitoneal) | – | √(M) | NPs penetrated blood–brain barrier and distributed in various organs without causing apparent toxicity | Kim *et al.* (2006) |
| | Multi-wall carbon nanotubes (CNTs) | C | 49 ± 13.4 nm | 10–80 μg (aspiration exposure) | – | √(M) | Exposure of CNTs produced significant adverse health outcomes in the lung in addition to pleural penetration | Porter *et al.* (2010) |
| | Single-wall oxidized CNTs | C | 500 nm | 10 ng to 30 μg/mouse | – | √(M) | Early miscarriages and fetal malformations were observed in females exposed to oxidized CNTs. Extensive vascular lesions, increased production of ROS were detected in placentas of malformed fetuses | Pietroiusti *et al.* (2011) |
| | Cadmium telluride Quantum dots (CdTeQDs) | Cd Te | 2.3–5.7 nm | 0.01–100 μg/mL ($1 \times 10^5$ cells/cm$^2$) | √(R) | – | Green QDs were significantly more toxic than red QDs at the highest concentration tested in PC12 cells (100 μg/mL). In N9 cells both types of QDs showed toxic effects | Lovric *et al.* (2005) |

*(Continued)*

**TABLE 15.4 (*Continued*)**
**Summary of Nanomaterial Safety Based on *In Vitro* and *In Vivo* Animal Studies**

| Classification of nanomaterials | Type of Nanoparticles | Element | Size | Dose/Route | *In Vitro* | *In Vivo* M = Mice R = Rat | Toxicology | Reference |
|---|---|---|---|---|---|---|---|---|
| | Graphene QDs (GQD-PEG) | C | 3–5 nm | 10–640 μg/mL ($8\times10^3$ cells per well) (20 mg/kg Intraperitoneal for *in vivo* studies) | √(HeLa and A549 cells) | √(M) | 95% of HeLa cells and A549 cells remained alive after 24 h of incubation with GQD-PEG at concentration of 160 mg/mL. Low cytotoxicity because of PEGylation or the inherent properties of the GQD sample | Chong *et al.* (2014), Guo and Mei (2014) |
| Nano-polymers | Fullerene ($C_{60}$) | $C_{60}$ | – | 100 μg/mL | √ (A549 Lung epithelial cells) | – | No DNA damage; generation of form-amidopyrimidinglyco-silase-sensitive sites indicates oxidation of purines; mutant frequency in cII gene was not affected | Jacobsen *et al.* (2008) |
| | Aqueous suspensions of colloidal $C_{60}$ fullerenes | $C_{60}$ | Aqu/n$C_{60}$: 178 nm; EthOH/n$C_{60}$: 122 nm | Aqu/n$C_{60}$: 0.022 lg/L–110 lg/L; EthOH/n$C_{60}$: 0.42 lg/L–2100 lg/L | √(Human lymphocytes) | – | Genotoxicity observed at: 2.2 lg/L (aqu/n$C_{60}$) and 4.2 lg/L (EthOH/n$C_{60}$) | Dhawan *et al.* (2006) |
| | Dendrimer-platinate | – | – | 95 mg/kg (i.p.) | – | √(M) | Displayed antitumor activity and no adverse weight change in C57 mice bearing B16F10 tumors | Malik *et al.* (1999) |
| | PEGylated dendrimer | – | – | 2.56 g/kg i.p. and 1.28 g/kg i.v. | – | √(M) | No toxicity, mortality or abnormal blood chemistry based on blood urea nitrogen levels or alanine transaminase activity were observed | Chen *et al.* (2004) |
| Nano-composites | PLA-Clay1 nanocomposite | – | – | – | – | √(R) | Different clinical biochemistry, inflammation, and oxidative stress biomarkers were determined showing no apparent damage | Maisanaba *et al.* (2014) |
| | Graphene oxide-silver nanocomposite (GOAg) | Ag | 9.3 ± 2.7 nm | 120.0 μg $mL^{-1}$ | √(M) | – | GOAg nanocomposite was more toxic than pristine GO and pristine AgNP for both macrophages, and it induced remarkable oxidative stress | de Luna *et al.* (2016) |

expensive, have fewer ethical concerns and allow for tighter experimental controls *vis-à-vis in vivo* (Marquis *et al.*, 2009). *In vitro* studies provide initial but significant information about comparative toxicity profiles of varied nanomaterials, with results that can further be tailored to more predictive toxicology.

The most common *in vitro* assessment method of nanoparticle toxicity is culturing cell lines with nanomaterials and then evaluating the subsequent responses. The choice of cell type is critical and screening the nanomaterials with a variety of cell lines is preferable. It is important to consider the distinct advantages between immortalized or primary cell lines. In addition, epithelial, endothelial, dermal, mucosal and alveolar cell types are important to evaluate, as they represent many different potential routes of exposure of nanomaterials. Selected cell lines attempt to simulate the response patterns observed in the *in vivo* system and are intended to yield a reliable and measurable signal. For instance, immune cells, such as macrophages, are appropriately used to study cytokine production or ROS generation in the presence of nanoparticles.

Furthermore, while conducting *in vitro* assays, it is essential to select appropriate endpoints that depict the physiological state or toxicity observed *in vivo* and the appropriate assay to measure that endpoint. For example, *in vitro* assays typically use biomarkers for ROS generation and indicate a toxic inflammatory response caused by nanoparticle exposure. One of the more commonly used probes in evaluating oxidative stress is 2′,7′-dichloro-fluorescein dictate (DCFDA), which is, in its inert state, a non-fluorescent probe, which when metabolized in the cell, gets converted to its active form and reacts with oxygen radicals, producing a fluorescent product allowing for ready evaluation. However, this assay may produce exaggerated results in the presence of apoptosis (Marquis *et al.*, 2009).

### 15.5.2 *In Vivo* Studies

To complement cell-based assays and determine whether a novel nanomaterial may cause toxicity in an intact organism, *in vivo* testing in whole animals is required. Animal models are useful when assessing nanoparticle effects that cannot be evaluated with *in vitro* assays such as toxicity caused by the initial deposition of nanoparticles through various routes of administration and the impact of chronic nanoparticle exposure on various systems (Sharifi *et al.*, 2012). Understanding the pharmacokinetic characteristics of nanoparticles through *in vivo* study allows researchers to determine to what degree particular nanoparticles cause toxicity and in which organs or cell types. Upon entering the body through one of the main routes of exposure, nanoparticles may interact with many biological components, such as cells and proteins, leading eventually to absorption. Following absorption, such nanoparticles distribute to various organs and tissues throughout the body, where they can reside in cells before they are removed from the body through metabolism and other means of elimination.

The choice of animal model is critical when evaluating the *in vivo* effects of nanomaterials. It is well-recognized that there are large differences, even within a species, in responses to nanomaterials, depending on strain (Scoville *et al.*, 2015). Additionally, special consideration must be paid to the route of administration for *in vivo* studies. The main routes of exposure for nanomaterials are dermal, respiratory, gastrointestinal (GI), auditory, ocular and mucosal. The route of exposure can influence the overall biodistribution and elimination of the nanoparticles, and therefore, its overall toxicity. For instance, following oral exposure, nanomaterials distribute to the kidneys, liver, spleen, lung, brain and GI tract, and some can pass through the GI tract for rapid elimination in the feces and urine (Buzea *et al.*, 2007). However, after parenteral administration, nanoparticles are distributed to the colon, lungs, bone marrow, liver, spleen and lymphatics, and are then rapidly cleared from systemic circulation, mainly by liver and spleen macrophages (Liang *et al.*, 2008). Even though there are several different routes of exposure for nanomaterials, most of the current *in vivo* toxicity studies focus on the pulmonary route of exposure in rat and mouse models.

Besides the route of exposure and animal model, physiochemical characteristics and size of nanoparticles can influence their overall residence time and exposure in the body. For example, upon inhalation, nanoparticles smaller than 10 nm preferentially deposit in the nasal cavity and alveolar region of the lung (Oberdorster *et al.*, 2005; Yang *et al.*, 2008). According to *in vivo* studies conducted by Choi *et al.* (2010), nanoparticles smaller than 6 nm tend to translocate to extra-pulmonary sites and, as a consequence, were cleared more quickly in the urine, whereas, particles larger than 34 nm were cleared more quickly via macrophage phagocytosis in the lungs. They also found that positively charged nanoparticles tended to have a longer residence time in the lungs *vis-à-vis* negatively charged particles, thus potentially increasing toxicity (Choi *et al.*, 2010).

The possible consequences of nanoparticle interaction with various biomolecules include generation of ROS, oxidative stress, mitochondrial perturbation, inflammation leading to protein and DNA damage, tissue infiltration along with generation of inflammatory cytokines, loss of enzyme activity and permeability transition. As there are numerous ways by which the nanoparticles tend to distribute and interact with biomolecules in the body, it may be difficult and insufficient to use nanoparticle exposure markers found in the urine or blood to assess overall exposure. Using predictable early-effect biomarkers, originating from the molecular biology of disease, may provide a more biologically relevant context to assess the internal dose of nanoparticles and significance of nanoparticle exposure levels causing such responses (Li and Nel, 2011).

## 15.6 KNOWLEDGE GAP BETWEEN NANOPARTICLES, ENGINEERING INNOVATION AND INTENDED PHARMACEUTICAL APPLICATIONS: REGULATORY AND SAFETY CONSIDERATIONS AS WELL AS ENVIRONMENTAL IMPACT

Despite tremendous efforts directed towards characterizing potential toxicity of engineered nanoparticles, challenges remain in determining the most appropriate screening paradigm in anticipation to complete animal *in vivo* testing. Specifically, nanomaterials may be complex mixtures of constituents, the toxicity testing of which has been quite challenging, historically. Some constituents may have previously generated toxicological data, as with the heavy metal components in semiconductor quantum dots, but others such as silver and gold may act very differently when nano-sized. Nevertheless, *in vitro* cell-based assays are certainly useful as a screening approach to determine the potential for undesirable effects.

However, a challenge of *in vitro* assays, in general, is the difficulty of developing assays that are both selective and sensitive to the toxic effects of interest (Marquis *et al.*, 2009). Even with thoughtful consideration of cell line and assay choice, it can be difficult to extrapolate the findings from the results of *in vitro* work back to the intricate living systems of intact organisms. Results from *in vitro* studies, accordingly, show poor correlation to what is observed *in vivo* (Marquis *et al.*, 2009), and extrapolation of *in vitro* results might lead to mistaken or manipulated conclusions regarding nanoparticle toxicity. One correlation study that compared the toxicity response of nano-sized ZnO particles, in both *in vitro* and *in vivo* models, found conflicting results. During introduction of nanoparticles through inhalation or instillation in rat, exposure to ZnO caused the rats to exhibit a pulmonary inflammatory response similar to that observed in humans. In contrast, ZnO particles did not illicit a cytokine profile from epithelial or macrophage cells *in vitro* that would be predictive of pulmonary inflammation observed in rat models *in vivo*. The authors of the work concluded that the current use of *in vitro* assays to predict the potential of metal-oxide, nanoparticle-related inflammation in the body with biomarkers indicating ROS or free-radical generation, have yet to be successful (Warheit *et al.*, 2009).

There are several potential reasons as to why *in vitro* studies may not translate into *in vivo* results. First, *in vitro* studies typically use larger doses of nanoparticles than the amount typically measured within the body (Marquis *et al.*, 2009). Higher doses may not be relevant or elicit the same toxicity response to what is observed *in vivo*. In one study, the researchers assessed the effects of smaller doses of silver nanoparticles (Ag-NPs) over a longer period, in order to closely mimic *in vivo* nanoparticle exposure conditions. They found that chronic exposure to Ag-NPs did not induce a cytotoxic response, as seen with acute dosing, but rather activated sustained stress and signaling responses, which better correlated to *in vivo* observations (Comfort *et al.*, 2014). Furthermore, it can be difficult to mimic the complex biologic environment inside the body (Marquis *et al.*, 2009). The assaying environment of nanoparticles can influence and potentially alter their physiochemical properties, such as size, surface characteristics, and material integrity resulting in varying toxicity estimates (Maurer-Jones and Haynes, 2012). These ambiguities may lead to different conclusions about the potential of nanoparticles for toxicity, as cellular uptake in the presence of aggregated or protein-bound materials is altered, and degraded nanoproducts may not accurately reflect the toxicity of intact materials. Finally, the multitude of physiochemical forms of nanomaterials lead to inherent complexities, rendering generalizations as difficult and unrealistic. Some nanoparticles can incite the formation of ROS, but others are inert in this regard. Nanomaterials with metals as core constituents may persist in an intact organism, but this would not be observed in the relatively short duration of *in vitro* characterization and not observed at all for metal-free nanomaterials.

Another consideration of nanomaterial development and application is the eventual fate of these materials in the environment. For example, nanopharmaceuticals may be excreted into waste streams, either as unchanged composite drugs or as degraded constituents, and these may accumulate in the environment. Persistence of these nanomaterials is a serious concern due to unknown and potentially damaging ecological impacts.

Despite these concerns and knowledge gaps, there is continued interest in developing nanomaterials for a myriad of applications. This presents a daunting challenge for those agencies responsible for assessing risk and regulating these complex and novel nanomaterials. Moreover, policy makers will ultimately have to rely on these findings to set a prudent regulatory course of action to protect human health and the environment.

## 15.7 FUTURE PROSPECTS

Due to their unique properties, attractive surface and structural characteristics, and other desirable attributes of nanomaterials, research and innovations continue to grow in this area. The growth in nanomaterials has led to a number of commercial applications in electronics, chemical and other industries, leading to smaller and more powerful mobile computing and communication capabilities, efficient chemical production and stronger composites. However, translating new nanomaterial innovations for medical or pharmaceutical application requires additional safety and regulatory considerations. Long-term impact on not only producers, consumers and patients must be taken in to account, but also the metabolized and waste product released into the environment must be tested. While pharmaceutical drug licensing approval requires regulatory review on drug and nanomaterial composites, safety of medical devices used externally for drug delivery is not regulated tightly. It is clear that all the nanomaterials and fabricated nanodrug delivery systems must undergo rigorous toxicology evaluation prior to first-in-human testing. However, toxicology of some of the biomaterials may be altered when combined with drug molecules, due to resulting changes in their properties such as surface area, charge and porosity. These properties may influence how the nanodrug distributes to cells, tissues, and blood. The influence on nanodrug characteristics may also influence drug and carrier dissociation, metabolism and elimination. Modification of the drug disposition profile by nanomaterials, when used appropriately, may improve efficacy, reduce toxicity or both.

Advances in the knowledge in drug disposition, barriers to absorption and the mechanisms of distribution and elimination have led to recent developments of novel drug delivery strategies. Some of these drug delivery strategies include nanodrugs derived from drug nanocrystals, liposomal drugs, lipid-drug complexes, polymeric peptides and other platforms. Advanced pharmaceutical research has enabled improvements in drug safety and efficacy. For instance, some drug formulations can reduce dosing frequency, yet provide better therapeutic response. Ongoing research intended to discover sites of drug action has identified drug targets, some of which could now be mapped for drug target localization during *in vivo* distribution at tissue and cell levels. As the efficiency of identifying drug targets has improved, combinatorial chemistry and high throughput synthesis and drug screening technology has matured, and pharmaceutical formulation and delivery strategies have become the rate-limiting steps in bringing new molecular entities to market. The growing knowledge of these properties, collectively referred to as "biopharmaceutical characteristics," "drug target distribution," and "pharmacokinetics," as these are linked to therapeutic responses, has allowed efficient development of novel drug delivery strategies. These strategies can now be considered in the context of human physiological interactions—referred to as a "systems approach" to drug delivery. Clinical examples exist on a passive or active drug targeting strategy, using nanomaterials that are biocompatible or biodegradable, improves the therapeutic index that leads to dose reduction or minimization of untoward effects at extended doses. The best results are often achieved with strategies that influence drug localization and exposure to target tissues and cells, but avoid exposure to organs linked to toxicity. Biodistribution of all drug targets in cells and tissues within the body is not yet understood fully. However, virus- or bacteria-laden cell distribution, and by extension, their protein target localization in cells and tissues, are now mapped, elucidated and validated for defined organs within the body. Such knowledge has and will continue to influence nanodrug-targeting design for successful development of targeted drug delivery systems to improve the therapeutic index. Together with improvement in development of assays that are predictive of nanomaterial toxicity *in vivo*, and consideration of safety of the material and nanodrug composite design, early in the course of research prior to drug development, is likely to improve the success in capitalizing promising nanomaterials to respective drug delivery and targeting application.

## ACKNOWLEDGMENTS

This work is supported in part by NIH grants UM1 AI120176, AI-077390, AI077390-S1, AI-077390-S2, AI-077390-S3 and 1UL1-RR025014. Also, RJY Ho has been supported by part by Milo Gibaldi endowment.

## REFERENCES

Abdelmonem AM, Pelaz B, Kantner K, Bigall NC, Del Pino P, Parak WJ. Charge and agglomeration dependent *in vitro* uptake and cytotoxicity of zinc oxide nanoparticles. *J Inorg Biochem.* 2015; 153: 334–8.

Borm PJ, Robbins D, Haubold S, Kuhlbusch T, Fissan H, Donaldson K, Schins R *et al.* The potential risks of nanomaterials: A review carried out for ECETOC. *Part Fibre Toxicol.* 2006; 3: 11.

Braydich-Stolle, Laura K, Schaeublin NM, Murdock RC, Jiang J, Biswas P, Schlager JJ, Hussain SM. Crystal structure mediates mode of cell death in $TiO_2$ nanotoxicity. *J. Nanopart. Res.* 2009; 11(6): 1361–74.

Buzea C, Pacheco II, Robbie K. Nanomaterials and nanoparticles: Sources and toxicity. *Biointerphases.* 2007; 2(4): MR17–71.

Chen HT, Neerman MF, Parrish AR, Simanek, EE. Cytotoxicity, hemolysis, and acute *in vivo* toxicity of dendrimers based on melamine, candidate vehicles for drug delivery. *J Am Chem Soc.* 2004; 126(32): 10044–80.

Chen Z, Meng H, Xing G, Chen C, Zhao Y, Jia G, Wang T *et al.* Acute toxicological effects of copper nanoparticles *in vivo. Toxicol Lett.* 2006; 163: 109–20.

Choi HS, Ashitate Y, Lee JH, Kim SH, Matsui A, Insin N, Bawendi MG *et al.* Rapid translocation of nanoparticles from the lung airspaces to the body. *Nat Biotechnol.* 2010; 28(12): 1300–3.

Chong Y, Maa Y, Shen H, Tu X, Zhou X, Xu J, Dai J, Fan S, Zhang Z. The *in vitro* and *in vivo* toxicity of graphene quantum dots. *Biomaterials.* 2014; 35: 5041–8.

Clift MJ, Rothen-Rutishauser B, Brown DM, Duffin R, Donaldson K, Proudfoot L, Guy K, Stone V. The impact of different nanoparticle surface chemistry and size on uptake and toxicity in a murine macrophage cell line. *Toxicol Appl Pharmacol.* 2008; 232(3): 418–27.

Comfort KK, Braydich-Stolle LK, Maure EI. Less is more: Long-term *in vitro* exposure to low levels of silver nanoparticles provides new insights for nanomaterial evaluation. *ACS Nano.* 2014; 8(4): 7939–45.

Conner JB, Bawa RJ, Nicholas JM, Weinstein V. Copaxone in the era of biosimilars and nanosimilars. In: *Handbook of Clinical Nanomedicine: From Bench to Bedside. Pan Stanford Series on Nanomedicine.* Pan Stanford Publishing Pte Ltd; 2015; pp. 783–828.

Cornier J, Kwade A, Owen A, Van de Voorde M. Drug targeting in nanomedicine and nanopharmacy: A systems approach. In: *Pharmaceutical Nanotechnology Innovation and Production*. Weinheim, Germany: Wiley-VCH, Weinheim; 2017; pp. 403–22.

de Luna LA, de Moraes AC, Consonni SR, Pereira CD, Cadore S, Giorgio S, Alves OL. Comparative *in vitro* toxicity of a graphene oxide-silver nanocomposite and the pristine counterparts toward macrophages. *J of Nanobiotech*. 2016; 14(12).

Dhawan A, Taurozzi JS, Pandey AK, Shan W, Miller SM, Hashsham SA, Tarabara V. Stable colloidal dispersions of C60 fullerenes in water: Evidence for genotoxicity. *Environ Sci Technol*. 2006; 40(23): 7394–401.

Duran N, Guterres SS, Alves OL. *Nanotoxicology: Materials, Methodologies, and Assessments. Nanomedicine and Nanotoxicology*. New York: Springer; 2014.

Gatoo MA, Naseem S, Arfat MY, Dar AM, Qasim K, Zubair S. Physicochemical properties of nanomaterials: Implication in associated toxic manifestations. *Biomed Res Int*. 2014; 2014: 1–8.

Guo X, Mei N. Assessment of the toxic potential of grapheme family nanomaterials. *J Food Drug Anal*. 2014; 22: 105–15.

Jacobsen NR, Pojana G, White P, Moller P, Cohn CA, Korsholm KS, Vogel U, Marcomini A, Loft S, Wallin H. Genotoxicity, cytotoxicity, and reactive oxygen species induced by single-walled carbon nanotubes and C(60) fullerenes in the FE1-Mutatrade mark Mouse lung epithelial cells. *Environ Mol Mutagen*. 2008; 49(6): 476–87.

Joerger M. Treatment regimens of classical and newer Taxanes. *Cancer Chemother Pharmacol*. 2016; 77(2): 221–33.

Kim JS, Yoon TJ, Yu KN, Kim BG, Park JJ, Kim HW, Lee KH, Park SB, Lee JK, Cho MH. Toxicity and tissue distribution of magnetic nanoparticles in mice. *Toxicol Sci*. 2006; 89: 338–47.

Kim YS, Kim JS, Cho HS, Rha DS, Kim JM, Park JD, Choi BS, Lim R, Chang HK, Chung YH. Twenty-eight-day oral toxicity, genotoxicity and gender related tissue distribution of silver nanoparticles in Sprague-Dawley rats. *Inhal Toxicol*. 2008; 20: 575–83.

Li N, Nel AE. Feasibility of biomarker studies for engineered nanoparticles: What can be learned from air pollution research. *J Occup Environ Med*. 2011; 53: S74–9.

Liang XJ, Chen C, Zhao Y, Jia L, Wang PC. Biopharmaceutics and therapeutic potential of engineered nanomaterials. *Curr Drug Metab*. 2008; 9(8): 697–709.

Lovric J, Bazzi HS, Cuie Y, Fortin GRA, Winnik FM, Maysinger D. Differences in subcellular distribution and toxicity of green and red emitting CdTe quantum dots. *J Mol Med*. 2005; 83(5): 377–85.

Maisanaba S, Gutiérrez PD, Puerto M, Llana-Ruiz CM, Pichardo S, Moyano R, Blanco A, Jordá-Beneyto M, Jos A. Toxicity *in vivo* toxicity evaluation of the migration extract of an organomodified clay-poly(lactic) acid nanocomposite. *J Toxicol Environ Health A*. 2014; 77(13): 731–46.

Malik N, Evagorou EG, Duncan R. Dendrimer-platinate: A novel approach to cancer chemotherapy. *Anticancer Drugs*. 1999; 10(8): 767–76.

Marquis BJ, Love SA, Braun KL. Analytical methods to assess nanoparticle toxicity. *Analyst*. 2009; 134(3): 425–39.

Maurer-Jones MA, Haynes CL. Toward correlation in *in vivo* and *in vitro* nanotoxicology studies. *J Law Med Ethics*. 2012; 40(4): 795–80.

Oberdorster G, Oberdorster E, Oberdorster J. Nanotoxicology: An emerging discipline evolving from studies of ultrafine particles. *Environ Health Perspect*. 2005; 113(7): 823–39.

Pelaz B, del Pino P, Maffre P, Hartmann R, Gallego M, Rivera-Fernandez S, de la Fuente JM, Nienhaus GU, Parak WJ. Surface functionalization of nanoparticles with polyethylene glycol: Effects on protein adsorption and cellular uptake. *ACS Nano*. 2015; 9(7): 6996–7008.

Pietroiusti A, Massimiani M, Fenoglio I, Colonna M, Valentini F, Palleschi G, Camaioni A *et al*. Low doses of pristine and oxidized single-wall carbon nanotubes affect mammalian embryonic development. *ACS Nano*. 2011; 5(6): 4624–33.

Porter DW, Hubbs AF, Mercer RR, Wu N, Wolfarth MG, Sriram K, Leonard S *et al*. Mouse pulmonary dose- and time course-responses induced by exposure to multi-walled carbon nanotubes. *Toxicology*. 2010; 269(2–3): 136–47.

Raveendran S, Rochani AK, Maekawa T, Kumar DS. Smart carriers and nanohealers: A nanomedical insight on natural polymers. *Materials (Basel)*. 2017; 10(8): pii: E929.

Scoville DK, White CC, Botta D. Susceptibility to quantum dot induced lung inflammation differs widely among the collaborative cross founder mouse strains. *Toxicol Appl Pharmacol*. 2015; 289(2): 240–50.

Sharifi S, Behzadi S, Laurent S, Forrest ML, Stroeve PL, Mahmoudi M. Toxicity of nanomaterials. *Chem Soc Rev*. 2012; 41: 2323–43.

Tahover E, Patil YP, Gabizon A. Emerging delivery systems to reduce doxorubicin cardiotoxicity and improve therapeutic index: Focus on liposomes. *Anticancer Drugs*. 2015; 26: 241–58.
Warheit DB, Sayes CM, Reed KL. Nanoscale and fine zinc oxide particles: Can *in vitro* assays accurately forecast lung hazards following inhalation exposures? *Environ Sci Technol*. 2009; 43: 7939–45.
Yang W, Peters JI, Williams RO 3rd. Inhaled Nanoparticles--a Current Review. *Int J Pharm*. 2008; 356: 239–47.
Zhu MT, Feng WY, Wang B, Wang TC, Gu YQ, Wang M, Wang Y, Ouyang H, Zhao YL, Chai YL. Comparative study of pulmonary responses to nano and submicron sized ferric oxide in rats. *Toxicology*. 2008; 247: 102–11.

# Index

## C



## D



## E

## O



## P

## Q



## R



## S

## T



## U



## V



## W



## Y



## Z